Oxidative-Stress in Human Diseases

Oxidative-Stress in Human Diseases

Editors

Soliman Khatib
Dana Atrahimovich Blatt

Basel • Beijing • Wuhan • Barcelona • Belgrade • Novi Sad • Cluj • Manchester

Editors

Soliman Khatib
MIGAL-Galilee Research
Institute and Tel-Hai College
Kiryat Shmona
Israel

Dana Atrahimovich Blatt
MIGAL-Galilee Research
Institute and Tel-Hai College
Kiryat Shmona
Israel

Editorial Office
MDPI
St. Alban-Anlage 66
4052 Basel, Switzerland

This is a reprint of articles from the Special Issue published online in the open access journal *Antioxidants* (ISSN 2076-3921) (available at: https://www.mdpi.com/journal/antioxidants/special_issues/Oxidative-Stress_Diseases).

For citation purposes, cite each article independently as indicated on the article page online and as indicated below:

Lastname, A.A.; Lastname, B.B. Article Title. *Journal Name* **Year**, *Volume Number*, Page Range.

ISBN 978-3-7258-0729-1 (Hbk)
ISBN 978-3-7258-0730-7 (PDF)
doi.org/10.3390/books978-3-7258-0730-7

Contents

About the Editors

Soliman Khatib

Prof. Soliman Khatib is an organic, analytical, and medicinal chemist who completed his PhD in the faculty of chemistry at the Technicon, Israel, in the field of physical organic chemistry. Now, he is an associate professor in the department of Biotechnology, Tel-Hai college, Israel, and head of the laboratory of Natural Compounds and Analytical Chemistry at MIGAL-Galilee Research Institute, Israel. Khatib's research focuses on the extraction, isolation, and structural elucidation of bioactive compounds from seeds, roots, plant algae and microalgae, and mushrooms for the prevention and treatment of diseases such as atherosclerosis, diabetes, and Alzheimer's disease. Additionally, Khatib analyzes endogenous molecules, oxidative stress levels, volatile organic compounds (VOCs), and targeted and untargeted metabolomics in human and animal fluids, cells, and tissues with the aim of discovering novel biomarkers for the early diagnosis of human diseases such as atherosclerosis and cardiac disease, cancer, and chronic and neurological diseases.

Dana Atrahimovich Blatt

Dr. Dana Atrahimovich Blatt received her Ph.D. in Biochemistry from the Faculty of Medicine, Bar Ilan University, Israel.

She has 11 years of experience within the Academy as a Post-Doctor at the MIGAL-Galilee Research Institute, Israel, and as a Lecturer and Laboratory Coordinator at Tel-Hai College, Israel. Currently, Atrahimovich leads the R&D of Bountica, a food-tech startup company that develops protein-based preservatives to prevent the decay and spoilage of perishable foods and beverages and extend food's shelf-life.

Preface

In the realm of human health, the intricate interplay between oxidative stress and disease has become an area of paramount importance. From the cellular level to systemic manifestations, the repercussions of oxidative stress are profound and multifaceted. This Reprint delves into the intricate mechanisms underlying oxidative stress in human diseases, offering a comprehensive exploration of its role as a critical player in pathophysiology and exploring the potential of natural antioxidants in mitigating such pathological conditions.

We extend our gratitude to the authors of the Articles and Reviews whose expertise and dedication have enriched this Reprint, and we hope that it serves as a valuable resource for researchers, clinicians, and students alike, fostering deeper understanding and innovative approaches in the ongoing battle against human diseases fueled by oxidative stress.

Soliman Khatib and Dana Atrahimovich Blatt

Editors

Review

Flavonoids-Macromolecules Interactions in Human Diseases with Focus on Alzheimer, Atherosclerosis and Cancer

Dana Atrahimovich [1,2], **Dorit Avni** [3] and **Soliman Khatib** [1,2,*]

[1] Lab of Natural Compounds and Analytical Chemistry, MIGAL–Galilee Research Institute, Kiryat Shmona 11016, Israel; Danaa@migal.org.il

[2] Department of biotechnology, Tel-Hai College, Upper Galilee 12210, Israel

[3] Lab of Sphingolipids, Bioactive Metabolites and Immune Modulation, MIGAL—Galilee Research Institute, Kiryat Shmona 11016, Israel; dorita@migal.org.il

* Correspondence: solimankh@migal.org.il; Tel.: +972-4-6953512; Fax: +972-4-6944980

Abstract: Flavonoids, a class of polyphenols, consumed daily in our diet, are associated with a reduced risk for oxidative stress (OS)-related chronic diseases, such as cardiovascular disease, neurodegenerative diseases, cancer, and inflammation. The involvement of flavonoids with OS-related chronic diseases have been traditionally attributed to their antioxidant activity. However, evidence from recent studies indicate that flavonoids' beneficial impact may be assigned to their interaction with cellular macromolecules, rather than exerting a direct antioxidant effect. This review provides an overview of the recent evolving research on interactions between the flavonoids and lipoproteins, proteins, chromatin, DNA, and cell-signaling molecules that are involved in the OS-related chronic diseases; it focuses on the mechanisms by which flavonoids attenuate the development of the aforementioned chronic diseases via direct and indirect effects on gene expression and cellular functions. The current review summarizes data from the literature and from our recent research and then compares specific flavonoids' interactions with their targets, focusing on flavonoid structure–activity relationships. In addition, the various methods of evaluating flavonoid–protein and flavonoid–DNA interactions are presented. Our aim is to shed light on flavonoids action in the body, beyond their well-established, direct antioxidant activity, and to provide insights into the mechanisms by which these small molecules, consumed daily, influence cellular functions.

Keywords: flavonoid; antioxidant; oxidative stress; inflammation; Alzheimer; atherosclerosis; cancer

Citation: Atrahimovich, D.; Avni, D.; Khatib, S. Flavonoids-Macromolecules Interactions in Human Diseases with Focus on Alzheimer, Atherosclerosis and Cancer. *Antioxidants* **2021**, *10*, 423. https://doi.org/10.3390/antiox10030423

Academic Editor: Janusz Gebicki

Received: 28 January 2021
Accepted: 2 March 2021
Published: 10 March 2021

Publisher's Note: MDPI stays neutral with regard to jurisdictional claims in published maps and institutional affiliations.

1. Introduction

Flavonoids are a class of polyphenols in plants that are widely consumed in our diet. They have a general C6–C3–C6 structural backbone, in which the two C6 units (Ring A and Ring B) are of a phenolic nature. Flavonoids can be divided into different subgroups, such as: flavones, flavonols, flavanones, flavanonols, flavan-3-ols, and anthocyanins (Figure 1). Whereas, in most flavonoids, ring B is attached at the C2 position of ring C, in some, such as isoflavones and isoflavans, ring B is connected at the C3 position [1].

Dietary flavonoids are natural products that are widely distributed in the plant kingdom. Many foods and beverages, such as fruits, vegetables, legumes, whole grains, chocolate, spices, tea, and wine, are rich sources of flavonoids [1]. Over decades, researchers and food manufacturers have become increasingly interested in flavonoids, due to their antioxidant properties, their great abundance in our diet, and their suggested role in the prevention of various diseases that are associated with OS, such as cancer, cardiovascular, and neurodegenerative diseases [2–5]. Recent literature provides growing evidence of flavonoids' effects being mediated by mechanisms other than the classical antioxidant activity driven by their chemical property of donating an electron or chelating transition metals [6,7]. Exploring their fundamental modes of action could provide new insights into the mechanisms by which flavonoids influence biological functions.

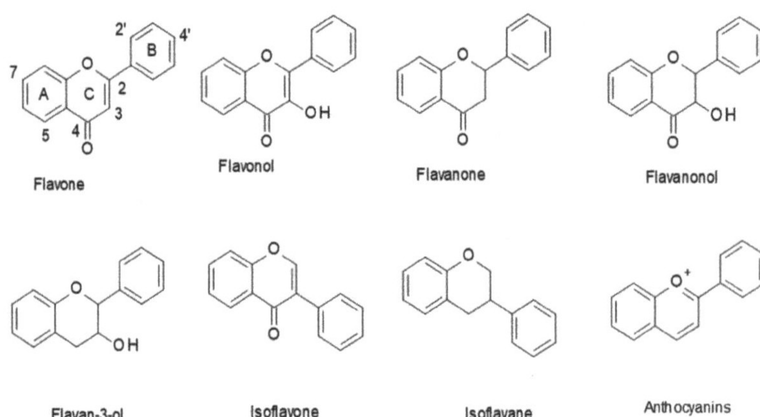

Figure 1. Structures of the main flavonoid subclasses.

2. Biological Activities of Flavonoids

2.1. Flavonoids as Antioxidants

With respect to their antioxidant activity, flavonoids are believed to prevent diseases that are related to OS via the direct scavenging of reactive oxygen species (ROS) through the donation of a hydrogen atom, activation of antioxidant enzymes, metal (such as iron and copper)-chelating activity, and alleviation of oxidative stress that is caused by nitric oxide (NO) [1,8–11]. Antioxidant activity, though, cannot be the sole explanation for the in vivo cellular effects of flavonoids', since antioxidant activity is expressed at flavonoid concentrations that are above 10 μM, but their concentration in the circulation does not exceed 2 μM [12]. Dietary flavonoids are poorly absorbed from the intestine, highly metabolized, or rapidly eliminated. During the course of absorption, flavonoids are conjugated in the small intestine and later in the liver. This process mainly includes methylation, sulfation, and glucuronidation. This is a metabolic detoxification process that is common to many xenobiotics that restricts their potential toxic effects and facilitates their biliary and urinary elimination by increasing their hydrophilicity [13].

Recent studies have suggested that the biological effects of flavonoids may be mediated by different mechanisms that have not yet been fully explored. The present review focuses on flavonoids' mode of action through their interaction with macromolecules, such as lipoproteins, cell and serum proteins, and DNA and RNA (Figure 2).

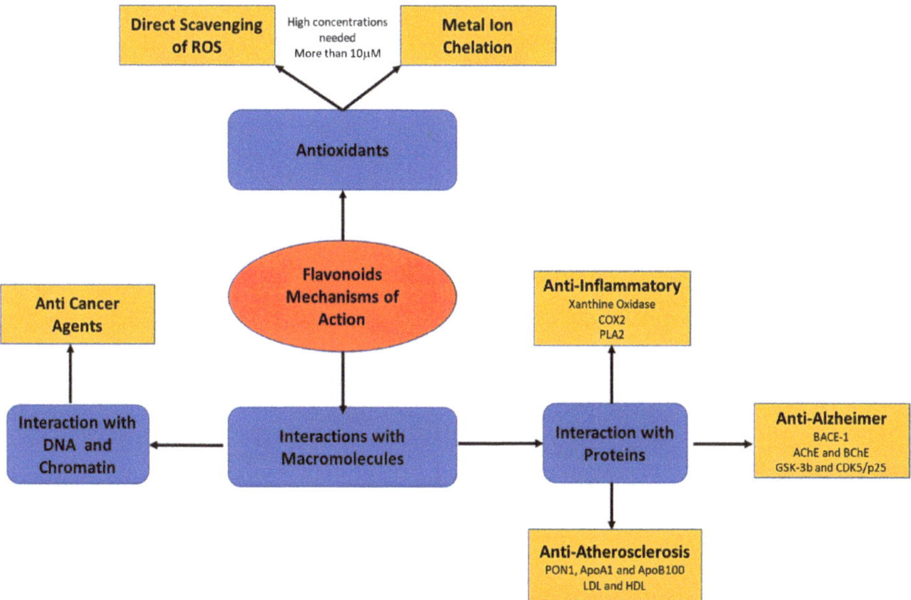

Figure 2. Flavonoids' mode of action through their interaction with macromolecules.

2.2. Flavonoid Interactions with Macromolecules

2.2.1. Flavonoid–Protein Interactions

Molecular interactions of proteins and nucleic acids with low-molecular-weight compounds are an area of fundamental interest [14]. At low concentrations, molecules, such as ions, metabolites, and osmolytes, may affect proteins, such as enzymes, receptors, antibodies, and transcription factors [15]. The effect may be at the structural, functional, or conformational levels [7]. Dietary flavonoids are a good example of small molecules that mediate cellular effects, which are central to intracellular signaling cascades [16]. The effects of flavonoid–enzyme complexes formed by flavonoids' interaction with, for example, hydrolases, oxidases, and kinases, on the enzyme's structure and activity have been widely explored. Investigations have suggested that flavonoids selectively interact with different components of protein kinases and alter their phosphorylation state, thus regulating multiple cell-signaling pathways [17]. Similarly, flavonoids have been found to act as ligands for nuclear receptors, causing their proliferation or activation and modulating energy homeostasis. Apigenin and kaempferol directly suppressed the interaction between estrogen-related receptor γ (ERRγ) and its coactivator peroxisome proliferators-activated receptor γ coactivator-1α (PGC-1α). In contrast, luteolin suppressed PGC-1α activity through promoting the degradation of PGC-1α, leading to suppressed ERRγ activity in HeLa cells [7,18]. Flavonoids, such as glabridin and glabrene, can also interact with, and modulate, the endogenous activities of estrogen receptors in human endothelial and smooth muscle cells, thereby may slow and even prevent cardiovascular diseases and the development of breast and ovarian cancers in post-menopausal women [19]. In addition, the flavonoids' ability to interact with serum albumin and other serum proteins has also been investigated [20,21]. The reversible or irreversible protein–flavonoid interactions depend on pH, temperature, and the protein and flavonoid concentrations [22] Although the biological fate of protein–flavonoid complexes in vivo is still unknown, flavonoids were found to affect various human diseases that were related to OS, such as cancer, and cardiovascular and neurodegenerative diseases [23–25].

Methods for Characterizing Flavonoid–Protein Interactions

Several studies have been performed to characterize the interactions between dietary flavonoids and proteins, mainly serum and food-related proteins, for example, serum albumins and β-casein [26–30]. Flavonoid–protein interactions mainly occur by noncovalent bonding that is derived from hydrophobic, van der Waals, hydrogen bridge-binding, and ionic interactions, which can change protein conformations and enzyme activities [31]. Noncovalent interactions between flavonoids and proteins are weak and reversible. Studies have also provided information on the covalent reactions between flavonoids and proteins. Flavonoids can easily oxidize and covalently react with the amino and thiol side chains of a protein by irreversible binding [32]. Numerous methods, mostly spectroscopic, have been developed to characterize the noncovalent interactions between flavonoids and proteins (Table 1) [33–36].

Table 1. Methods used to characterize flavonoids-macromolecules interactions.

Macromolecule	Method	Type of Interaction	Reference
Flavonoids-Proteins interactions	UV-visible spectroscopy	Covalent complex formation or Protein conformational changes	[37]
	Circular dichroism spectroscopy (CD)	Proteins conformations and secondary structure changes	[38]
	Fourier transform infrared spectroscopy (FTIR)	Proteins conformations and secondary structure changes	[38]
	isothermal titration calorimetry (ITC)	Thermodynamic properties of the binding interaction between flavonoids and proteins	[39]
	Taylor dispersion surface plasmon resonance (SPR)	Binding affinities of non-covalent interactions	[40]
	Tryptophan (Trp)-fluorescence quenching assay	Determines interactions between flavonoids and proteins, quantitative analysis of binding affinities and thermodynamic parameters	[21,41,42]
	docking calculations	Computational models to predict the fit of the evaluated ligand within the protein,	[43]
Flavonoids-DNA interactions	electrochemical and SPR techniques, linear dichroism, absorption, fluorescence and nuclear magnetic resonance spectroscopies	Noncovalent interactions between flavonoids and DNA	[44–46]
	Electrospray ionization mass spectrometry (ESI-MS)	Binding of flavonoids with DNA duplexes	[47]
	Chemical affinity capture coupled with massively parallel DNA sequencing (Chem-seq)	Extracts and sequences DNA regions and captures chromatin regions bound to flavonoids.	[48–50]

UV-visible spectroscopy is used to predict flavonoid–protein interactions and provide information on the nature of those interactions. Protein absorption at 280 nm is related to the aromatic amino acids tryptophan, tyrosine, and phenylalanine, which may be further stimulated upon interaction with flavonoids [37]. *Circular dichroism spectroscopy* is used for quantitative analysis of the conformational changes, α-helix and β-sheet changes, in proteins due to noncovalent interactions with small molecules, such as flavonoids [38]. *Fourier transform infrared spectroscopy* is also used to determine the changes in proteins' secondary structure as a result of flavonoid interactions. This method allows for interpreting the secondary structure from the shape of the amide I band, located around 1650–1660 cm [38].

Thermodynamic properties of the binding interaction between flavonoids and proteins can be studied using *isothermal titration calorimetry*, a method that is based on measuring the heat that evolved during the molecular association [39]. Vichali et al. evaluated the binding interactions between four flavonoids (kaempferol, luteolin, quercetin, and resveratrol) and human serum albumin and glutathione S-transferase Pi isoform 1 using *Taylor dispersion surface plasmon resonance (SPR)*—a highly sensitive, label-free technique to study the noncovalent interactions of biomolecules, especially between proteins, and between proteins and small molecules [40].

Tryptophan (Trp)-fluorescence quenching assay is another sensitive, selective, and widely used method for determining interactions between flavonoids and proteins [21,41,42]. The excitation of proteins at 280–290 nm induces the emission of fluorescence in the range of 340–350 nm due to the presence of Trp. fluorescence quenching in this range can be attributed to flavonoid binding. While using this method, the quenching mechanism—static (complex formation between polyphenol and protein) or dynamic (collision of fluorophore with the quencher)—can be determined using the Stern–Volmer equation and calculating the Stern–Volmer constant and quenching rate constant. For static quenching, the binding constant and number of binding sites in the protein molecule can be calculated, and then thermodynamic properties can be characterized. Finally, *docking calculations* can be used to predict the fit of the evaluated ligand within the protein, where the shape is complementary to the binding site. Computational modeling complements the experimental data on flavonoid–protein binding and it allows for large-scale screening for different protein targets selected from the structures that are available in the Protein Data Bank (PDB) [43].

2.2.2. Flavonoid Interactions with DNA and Chromatin

There is a great deal of evidence in the scientific literature of genome regulation by flavonoids via gene-expression and chromosomal alterations [24,51], although the precise mechanism of action remains unclear [48,52]. Flavonoids, such as quercetin and EGCG, have been shown to penetrate cell membranes and accumulate in the nucleus of human intestinal and hepatic cells [53,54]. The structure of quercetin allows for hydrophobic-nature-type intercalation of its most hydrophobic segment into the interior of the DNA helix [55]. Quercetin intercalates with DNA and RNA duplexes and preferentially binds to triplex and tetraplex DNA in human prostate cancer cells (DU 145) [53]. Although the same number of OH groups, which are mainly involved in the hydrogen-transfer mechanism, are present in kaempferol and luteolin, the latter exhibits slightly higher affinity to DNA. This might be due to the presence of OH at its 3′ position. Structure–activity relationships in flavonoid–DNA interactions have indeed been widely detected. It is proposed that flavonoids' affinity for DNA increases along the same sequence as that exhibited by their biological activity [44]. Upon DNA treatment with EGCG or quercetin, various effects, including DNA damage, in human peripheral lymphocytes, were noted [56,57]. Studies show that EGCG inhibits the activities of various chromatin proteins, such as cAMP-response element-binding protein, DNA polymerase, DNA methyltransferase, and DNA topoisomerase in human lungs and colorectal adenoma cells and in mice liver, lungs, and kidney [6,24]. These reactions are likely affected by EGCG binding to the DNA and RNA, or to the proteins that are attached to nucleic acids in various types of interaction.

While the interactions of flavonoids, such as resveratrol, quercetin, EGCG, and genistein, with DNA are known, the precise location of the flavonoid-binding sites on the DNA, the mode of interaction, and its function in the genome are not fully understood.

Methods for Characterizing Flavonoid–DNA Interactions

Covalent binding of small molecules to DNA was first observed in the early 1980s [58]. After the covalent binding of [^{14}C]quercetin to DNA was determined, it was argued that flavonoids have conflicting biochemical activities (mutagenic effect on the one hand, and anticarcinogenic effect on the other) [44]. In addition to covalent binding, flavonoids can interact with DNA by intercalation, groove-binding, and backbone-binding. Several methods have been used to elucidate the noncovalent interactions between flavonoids and DNA, including electrochemical and SPR techniques, linear dichroism, absorption, fluorescence, and nuclear magnetic resonance spectroscopies [44–46]. The binding of 10 aglycones and flavonoid glycosides with DNA duplexes was investigated while using electrospray ionization mass spectrometry (ESI-MS) [47]. ESI-MS analysis and SPR showed that exactly three molecules of EGCG bind to poly(dT) 18 mer single-stranded DNA oligomers via one hydroxyl group of the trihydroxyphenyl group in EGCG. Upon binding, the EGCG protected double-stranded DNA oligomers from melting to single-stranded DNA [59].

Today, computational simulation and spectroscopy are mainly adopted to explore biophysical information (e.g., interaction mode) on the interactions between flavonoids and DNA [60]. Experiments that were performed in recent years have suggested specific consensus DNA-binding sites for flavonoids. Quercetin, for example, binds to the dodecamer duplex sequence CGCGAATTCGCG, the unbound structure of which was solved many years ago (PDB ID: 1BNA) [61]. Currently, organism's complete genome can be revealed using next-generation sequencing (NGS) technologies, such as Illumina or Sanger massively parallel sequencing machines. Moreover, following specialized protocols, it is possible to extract the DNA in specific regions or with specific functions and then use NGS to obtain the DNA sequence. Chem-seq (chemical affinity capture coupled with massively parallel DNA sequencing) is a new NGS application, which recently used to extract and sequence DNA regions that were bound to small molecules. This method allows for capturing chromatin regions bound to small molecules with no prior information, i.e., with an unbiased, nonspecific marker [49]. The latest studies have already illustrated the ability to isolate known drug–chromatin interactions using Chem-seq [49,50]. Atrahimovich et al. used the Chem-seq technique to characterize the interactions between quercetin and cellular DNA and demonstrated its subsequent effect on downstream transcription [48]. The results show that quercetin binds to monocytes' chromatin and modulates the expression of genes that are involved in the cell cycle and cell development [48]. Using Chem-seq application, flavonoids interactions with DNA and chromatin may be determined to study its significance. This ability could be extremely important to medicine and human health, and beneficial to the design of appropriate dietary interventions and drugs for cancer treatment.

3. Flavonoids Attenuate Human Diseases via Direct Interactions with Proteins, Lipoproteins and DNA

3.1. Flavonoids Interactions with Key Proteins Involved in Inflammation

Inflammation characterizes the protective response of the immune system, involving the production of various proinflammatory cytokines and chemokines, which enhance the production of interferon-γ, proteases, NO, and ROS [62]. Cytokines also induce the expression of cyclooxygenase-2 (COX-2), an enzyme that catalyzes the production of prostaglandins (PGs), which are key mediators of inflammation [63]. Xanthine oxidase (XO) is another critical source of ROS that contributes to inflammation. Inflammatory conditions lead to increased XO levels and, thus, to increased ROS generation and peroxynitrite formation. Peroxynitrite is a powerful reactive nitrogen species (RNS) accompanied by OS, which is produced by the reaction of NO and superoxide radical [64].

Several mechanisms of action have been proposed to explain flavonoids' anti-inflammatory activity in vivo, such as antioxidant activity and the modulation of the production of proinflammatory cytokines and gene expression [11]. Interestingly, flavonoids affect the inflammatory process not only by reducing the expression of cytokines and other related inflammatory markers, but also by interacting with proteins that are related to inflammation. Flavonoids have been shown to modulate the activity of arachidonic acid (AA)-metabolizing enzymes, such as phospholipase A2 (PLA2), COX, and lipoxygenase (LOX), and the NO-producing enzyme nitric oxide synthase (NOS). The inhibition of these enzymes by flavonoids reduces the production of AA, PG, leukotriene, and NO, which are crucial mediators of inflammation. Thus, flavonoid inhibition of these enzymes is definitely one of the important cellular mechanisms of anti-inflammation [65].

Quercetin was the first discovered flavonoid inhibitor of PLA2, from human neutrophils. Quercetin was shown to selectively inhibit group II secretory PLA2 [66]. Likewise, rutin selectively inhibited human PLA2-II from synovial fluid, while it was a weak inhibitor of human PLA2-I from pancreatic juice. When different flavonoids were compared for their ability to inhibit PLA2, small changes in the structure appeared to influence both overall PLA2 inhibition and group II selectivity. The position of the hydroxyl groups was found to be one important aspect of the C-ring-2, 3-double bond. The hydroxyl groups in the 3′ and 4′ positions on the B-ring seemed to be important for the selective inhibition of PLA2-II, whereas the 5-hydroxyl group on the A-ring, the unsaturation, and the 4-oxy on the C-ring seemed to be important for the overall ability of flavonoids to inhibit PLA2 activity [67]; the inhibition of PLA2 was very much dependent on hydroxyl groups position on rings A, B, and C, while hydroxyl groups in positions 5, 6, and 7 on the A-ring were assumed to be necessary for binding to PLA2s. Thus, quercetin, kaempferol, and galangin showed high inhibitory activity on PLA2, while naringin demonstrated a lower inhibitory activity [68].

COX produces PGs and thromboxanes and exists in at least two different isoforms, COX-1 and COX-2. COX-1 is a constitutive enzyme that is present in almost every cell type. While, COX-2 is an inducible enzyme that is highly expressed in the inflammation-related cell types, including macrophages and mast cells [69]. Because it produces PGs, COX-2 is closely associated with acute as well as chronic types of inflammatory disorders. Some flavonoids, such as luteolin, 3′,4′-dihyroxyflavone, galangin and morin, catechin, and epi-catechin, have been found to inhibit the rat renal medulla COX with IC50 of 100–130 μM [70]. In human thrombin-aggregated platelets, certain flavonoids, such as chrysin and apigenin, were revealed to be COX inhibitors with IC50 of 13 and 18 μM, while myricetin and quercetin at 10 μM exerts a strong inhibition of LOX. In particular, the reduction of the C-2, 3-double bond and glycosylation reduced the flavonoids inhibitory activities [71]. In-silico analysis demonstrated that quercetin could partially inhibit the COX-2 enzyme by binding to subunit A, which has peroxidase activity and serves as a ROS source [72].

D'mello et al. modeled the COX-2-inhibitory activity of some flavonoids [73], who used docking studies to determine which flavonoids act as COX-2 inhibitors. It was found that flavonoids have different types of binding patterns, which are similar to synthetic nonsteroidal anti-inflammatory drugs (NSAIDs). Myricetin and luteolin had catechol moiety with 3′, 4′-dihydroxy groups on the B-ring, and baicalein had the catechol-moiety on the A-ring, both being oriented toward the hydrophobic pocket of the enzyme and formed H-bonds with Tyr20, Tyr385, and Ser530. However, myricetin cannot act as a selective COX-2 inhibitor because it was inaccessible toward Arg513, an important residue for the selective inhibition of COX2. The importance of the catechol moiety was further confirmed by its similarity to the existing COX inhibitors [73]. Moreover, the effect of flavonoids on iNOS has been intensively studied. Several flavonoids, including apigenin, luteolin, and quercetin, were found to inhibit NO production in the macrophage cell line (RAW 264.7) that was treated with LPS/cytokine. However, they reduced NO production via downregulation of iNOS induction and not via the inhibition of iNOS by direct interaction [74]. The only exception was echinoisosophoranone, which significantly inhibited iNOS at sufficient concentrations [75]. While they may inhibit endothelial NOS or neuronal NOS, flavonoids are not efficient iNOS inhibitors. Interestingly, flavonoids have been found to be promising potent inhibitors of XO activity in humans. For example, quercetin and its glycosides displayed significant inhibitory activity against XO, thereby affecting ROS formation and the inflammation that is promoted by OS [76,77].

Generally, flavonoids may be mainly involved in the inflammation process via inhibition and regulation of enzymes that modulate pro-inflammatory cytokines or small molecules, such as ROS and RNS.

3.2. Flavonoids Interactions with Key Proteins in Alzheimer's Disease (AD)

AD is a widespread neurodegenerative disease, which is characterized by neurofibrillary tangles, senile plaques, and synaptic loss, eventually resulting in neuronal death [78,79]. AD is a form of dementia, which is characterized by progressive memory loss, a decline in language skills, and other cognitive impairments, and it most commonly affects the elderly [80]. AD's etiology is unclear; however, a variety of factors are considered in the pathophysiology of the disease, such as the formation of amyloid β-protein (Aβ) plaques, low levels of acetylcholine, oxidative stress, and abnormal posttranslational modifications of tau protein [81,82]. The sequential cleavage of amyloid precursor protein forms aggregates of Aβ peptides of 39–43 amino acids, which stick to the neurons as insoluble amyloid plaques. Aβ is generated from the amyloid precursor protein by β-site amyloid precursor protein cleaving enzyme-1 (BACE-1, β-secretase) and γ-secretases [83,84]. Thus, the inhibition of BACE-1 is assumed to play an important role in the prevention of AD [85].

The neurotransmitter acetylcholine plays an important role in the process of learning and memory in the hippocampus. Two enzymes, acetylcholinesterase (AChE) and butyrylcholinesterase (BChE), are involved in the hydrolysis of acetylcholine, lowering its level during the development of AD. Therefore, the inhibition of AChE and BChE is a highly desirable strategy for the treatment of AD [86–88]. The clinically approved drugs tacrine, donepezil, galantamine, and rivastigmine improved short-term memory and cognitive levels via inhibition of AChE. The disadvantages of these drugs and their gradual side effects, such as peripheral side effects, hepatotoxicity, and gastrointestinal tract disorders, have encouraged researchers to develop more effective AChE inhibitors [89–91].

Flavonoids are promising natural products with neuroprotective potential, which either prevent the onset or slow the progression of age-related neurodegenerative diseases. The mechanism by which flavonoids prevent or slow the progression of AD might be via direct interaction with key enzymes that are involved in this disease [81,85,92–95]. Shimmyo et al. examined the potential of flavonols and flavones to inhibit BACE-1. They found that four flavonols: myricetin, quercetin, kaempherol, and morin, and one flavone: apigenin, directly inhibit BACE-1 enzyme activity in a concentration-dependent manner, with IC_{50} values of 2.8, 5.4, 14.7, 21.7, and 38.5 μM, respectively [95]. Studies in aged TASTPM transgenic mice (a model of AD) showed that the oral administration of (-)-epicatechin reduces Aβ pathology through indirect, noncatalytic BACE-1 inhibition and not through modulation of either α- or γ-secretase activity [96]. Epigallocatechin-3-gallate (EGCG) and curcumin were found to reduce Aβ-mediated BACE-1 upregulation in neuronal cultures, which, interestingly, increased the nonamyloidogenic processing of the amyloid precursor protein by enhancing α-secretase cleavage [95]. Pueyo et al. reviewed the literature on natural and synthetic flavonoids with AChE-inhibitory activity. They found 128 such flavonoids: 41 flavones, 21 flavanones, 35 flavonols, 25 isoflavones, and six chalcones. Among them, eight synthetic flavonoids inhibited AChE with $IC_{50} < 100$ nM. Three natural flavonoids, acaciin from *Chrysanthemum indicum* flowers, and desmethylanhydroicaritin and kaempferol from *Sophora flavescens* roots, inhibited AChE, with IC_{50} values of 3.2, 6.7, and 3.3 nM, respectively [97]. Orhan et al. screened various flavonoid derivatives for their inhibition of AChE and BChE. At a concentration of 1 mg/mL, quercetin was the most effective toward AChE, with 76.2% inhibition, and genistein showed the highest inhibition (65.7%) of BChE, followed by luteolin-7-O-rutinoside and silibinin (54.9% and 51.4%, respectively) [98,99]. In another study, *Citrus junos* had a significant inhibitory effect on AChE in vitro and in vivo, and the active compound was identified as naringenin, a major flavanone derivative [100]. Lee et al. examined the inhibitory effect of citrus flavanones on BACE-1, AChE, and BChE. Among all of the examined flavanones, hesperidin demonstrated the best inhibition of BACE-1, AChE, and BChE, with IC_{50} values of 10.02, 22.80, and 48.09 μM, respectively. Kinetic studies revealed that all the flavanones were noncompetitive inhibitors of BACE-1 and cholinesterase [101,102].

Hyperphosphorylation of tau proteins with subsequent accumulation as neurofibrillary tangles is a major contributor to cognitive dysfunctions and one of the earliest AD markers. Several kinases, such as GSK-3b and CDK5/p25, are known to contribute to the phosphorylation of tau proteins and they are implicated in the pathogenesis of AD. Flavonoids that inhibit the activities of several kinases can be used in AD prevention. Therapy with the flavonoid morin has been shown to reduce tau hyperphosphorylation in vitro and in vivo in the hippocampal neurons of transgenic animals (3xTg-AD mice) [103]. Quercetin inhibited the PI3-kinase activity and Cyanidin 3-O-glucoside also provided significant protection against cognitive dysfunctions that are induced by the administration of Aβ in animal models, mediated by the modulation of GSK-3b/tau. [104,105].

Overall, flavonoids may exert their potential neuroprotective actions by interacting with key proteins that are involved in AD. Better understanding the flavonoid–protein interactions in AD could be a promising strategy for developing novel neuroprotective therapies for the prevention and treatment of neurodegenerative diseases.

3.3. Flavonoids Interactions with Key Proteins and Lipoproteins in Atherosclerosis

Atherosclerosis is another disease that flavonoids have been shown to attenuate. The first step in atherosclerosis is the accumulation of low-density lipoprotein (LDL), the main cholesterol carrier, in the arterial wall. High-density lipoproteins (HDL), on the other hand, is a major antiatherogenic factor in the blood, which maintains the whole-body level of cholesterol in a steady state. Over 80 proteins have been identified in the HDL proteome, with apolipoproteins A1 and A2 accounting for approximately 65% and 15% of the protein mass, respectively. The other proteins include a variety of enzymes, such as paraoxonase 1 (PON1). PON1 is responsible for many of HDL's antiatherogenic properties. Correlations between PON1, HDL, and atherosclerosis, both in vivo and in vitro, have been well-established [106,107]. Aside from cholesterol efflux, HDL has other potent biological activities: antioxidative [108], anti-inflammatory [109], antiapoptotic [110], and vasodilatory [111]. These activities do not necessarily depend on HDL quantity, but they probably depend on its quality [112,113].

With respect to cardiovascular health, we have previously shown that the flavonoid glabridin, extracted from licorice root, acts as an excellent antioxidant and demonstrates additive antioxidant and antiatherogenic properties. Glabridin binds to recombinant PON1 (rePON1) and protects its Cys284 from oxidation by the atherosclerotic component linoleic acid hydroperoxide (LA-OOH). This specific capacity of glabridin is unique; the flavonoid catechin does not show any binding affinity to rePON1 [21]. The association between flavonoids structure and their effects on rePON1 activity was further explored. The interactions of 12 representative flavonoids from different chemical subclasses with rePON1 were characterized [42]. In addition, the potential of rePON1–flavonoid complexes to prevent oxidation of LDL, a key process in atherogenesis, was examined. Catechin, which does not bind to rePON1, accelerated LDL oxidation; in contrast, glabridin demonstrated a high binding affinity to rePON1 and enhanced its protective effect against LDL oxidation [42]. Moreover, we have consistently observed interactions of specific flavonoids with the HDL particle or its bound proteins, apolipoprotein A1 and PON1. We have shown that quercetin and punicalagin bind to the HDL particle and increase its anti-inflammatory properties [41], whereas, upon binding to the LDL particle or to its bound apolipiprotein B100, punicalagin induced LDL influx to macrophage J774A.1 cells, which may decrease the circulating LDL levels [114].

Overall, flavonoids, and polyphenols in general, have been found to inhibit symptoms of atherosclerosis and reduce its development via specific flavonoids interactions with cell and serum proteins and lipoproteins.

3.4. Flavonoids as Anticancer Agents via Interaction with DNA and Chromatin

Flavonoids' anticancer activities might be a result of the interaction of these natural compounds with biomolecules (DNA, RNA, and protein). We recognize that dietary flavonoids can bind the DNA specifically or stochastically and change its function [115]. Extensive in-vitro studies suggest that flavonoids effectively decrease cell proliferation, induce apoptosis, and lower the risk of metastasis [24]. The chemo-preventive effects of flavonoids, including luteolin, epigallocatechin gallate, quercetin, apigenin, and chrysin, were shown with a focus on protection against DNA damage that is caused by various carcinogenic factors. Those flavonoids selectively protect normal cells and induce cell-death mechanisms in cancer cells in human lungs and colorectal adenoma cells during chemotherapy or radiotherapy [24]. It was found that flavonoids, namely quercetin, myricetin, kaempferol, apigenin, and luteolin, which are lipid-soluble and weakly acidic, can freely diffuse across the cell membrane and specifically accumulate inside K562 leukaemic cells [116]. Therefore, it is implied that flavonoids are more likely to bind DNA or proteins in the cancer cell nucleus and to specifically interrupt cancer genome regulation. In addition, in-silico results have shown that quercetin, in particular, interacts well with G-quadruplex DNA, which is related to telomerase. Quercetin acts as a therapeutic anticancer agent via the regulation of telomerase activity [117]. By comparing computational and experimental binding profiles, a novel study confirmed that quercetin has the strongest binding affinity to DNA among the studied flavonoids. Furthermore, the study revealed that flavonoids can alter the conformation of DNA and inhibit DNA amplification, they show impressive induction of cell-cycle arrest, and they may promote apoptosis in HepG2, MCF-7, and A549 cancer cells [60].

In order to achieve the effective therapeutic doses used in preclinical studies, importance must be given to improved and targeted drug delivery techniques, so as to achieve maximum efficiency with minimal adverse side effects. Advances in nanotechnology-based drug delivery systems open up better opportunities for increasing solubility, improving bioavailability, and enhancing the targeting capabilities of flavonoids [118]. Nanoparticles based on liposomes, poly-ethylene glycol liposomes, nickel-based, lecithin-based, and nanoribbon are suitable molecular carriers for the delivery of flavonoid drugs to target tissues. It was reported that nanoparticles successfully used to deliver quercetin into solid tumors in in vitro and in vivo models of cancers of the central nervous system, lungs, colon, liver, and breasts [119].

Thus, numerous studies support the potential of flavonoids as natural health products in cancer chemoprevention. However, more studies are needed in order to configure their mechanism of action to improve our understanding of epigenetic processes that may provide a more rational basis for combining specific dietary compounds in a clinical setting [24].

4. Conclusions

Flavonoids are antioxidant molecules that are constantly consumed in our diets; they are found in our cells and serum. Therefore, it is interesting and highly important to determine their biological functions and mechanisms of action. Literature is mainly using classical antioxidant mechanisms to explain those molecules' various cell and serum functions. However, it is more than reasonable that small, hydrophobic molecules at low concentrations will penetrate cytoplasm, organelle, and nucleus membranes and accumulate in specific cells and tissues to bind proteins via different types of interactions and affect signal transduction, gene expression, chromosomal alterations, epigenetic modifications, and so on.

Further investigation at the proteome and genome levels is needed in order to characterize the mode of action by which flavonoids are involved in the reviewed OS related diseases. Mapping the direct interactions of flavonoids with proteins and chromatin genome-wide could provide new insights into the mechanisms by which flavonoids influence cellular functions, and pave the way to understanding, predicting, and controlling flavonoid responses in humans.

Author Contributions: D.A. (Dana Atrahimovich), writing—original draft preparation and editing, D.A. (Dorit Avni) writing the section of 'Flavonoids Interactions with Key Proteins Involved in Inflammation' and editing; S.K. supervision, writing—review and editing. All authors have read and agreed to the published version of the manuscript.

Funding: This research received no external funding.

Conflicts of Interest: The authors declare no conflict of interest.

References

1.　Procházková, D.; Boušová, I.; Wilhelmová, N. Antioxidant and prooxidant properties of flavonoids. *Fitoterapia* **2011**, *82*, 513–523. [CrossRef]

2. Duthie, G.G.; Duthie, S.J.; Kyle, J.A.M. Plant polyphenols in cancer and heart disease: Implications as nutritional antioxidants. *Nutr. Res. Rev.* **2000**, *13*, 79–106. [CrossRef] [PubMed]

3. Ramos, S. Cancer chemoprevention and chemotherapy: Dietary polyphenols and signalling pathways. *Mol. Nutr. Food Res.* **2008**, *52*, 507–526. [CrossRef] [PubMed]

4. Jaeger, B.N.; Parylak, S.L.; Gage, F.H. Mechanisms of dietary flavonoid action in neuronal function and neuroinflammation. *Mol. Aspects Med.* **2018**, *61*, 50–62. [CrossRef] [PubMed]

5. Devi, S.; Kumar, V.; Singh, S.K.; Dubey, A.K.; Kim, J.J. Flavonoids: Potential candidates for the treatment of neurodegenerative disorders. *Biomedicines* **2021**, *9*, 99. [CrossRef]

6. Williams, R.J.; Spencer, J.P.E.; Rice-Evans, C. Flavonoids: Antioxidants or signalling molecules? *Free Radic. Biol. Med.* **2004**, *36*, 838–849. [CrossRef]

7. Virgili, F.; Marino, M. Regulation of cellular signals from nutritional molecules: A specific role for phytochemicals, beyond antioxidant activity. *Free Radic. Biol. Med.* **2008**, *45*, 1205–1216. [CrossRef]

8. Grotewold, E. *The Science of Flavonoids*; Springer: Columbus, OH, USA, 2006; ISBN 9780387288215.

9. Agati, G.; Brunetti, C.; Fini, A.; Gori, A.; Guidi, L.; Landi, M.; Sebastiani, F.; Tattini, M. Are flavonoids effective antioxidants in plants? Twenty years of our investigation. *Antioxidants* **2020**, *9*, 1098. [CrossRef]

10. Liu, Y.; Weng, W.; Gao, R.; Liu, Y.; Monacelli, F. New Insights for Cellular and Molecular Mechanisms of Aging and Aging-Related Diseases: Herbal Medicine as Potential Therapeutic Approach. *Oxid. Med. Cell. Longev.* **2019**, *2019*. [CrossRef]

11. Rolt, A.; Cox, L.S. Structural basis of the anti-ageing effects of polyphenolics: Mitigation of oxidative stress. *BMC Chem.* **2020**, *14*, 1–13. [CrossRef]

12. Manach, C.; Scalbert, A.; Morand, C.; Rémésy, C.; Jiménez, L. Polyphenols: Food sources and bioavailability. *Am. J. Clin. Nutr.* **2004**, *79*, 727–747. [CrossRef]

13. Thilakarathna, S.H.; Vasantha Rupasinghe, H.P. Flavonoid bioavailability and attempts for bioavailability enhancement. *Nutrients* **2013**, *5*, 3367–3387. [CrossRef] [PubMed]

14. Haq, I. Thermodynamics of drug-DNA interactions. *Arch. Biochem. Biophys.* **2002**, *403*, 1–15. [CrossRef]

15. Uversky, V.N. Intrinsically disordered proteins and their environment: Effects of strong denaturants, temperature, pH, counter ions, membranes, binding partners, osmolytes, and macromolecular crowding. *Protein J.* **2009**, *28*, 305–325. [CrossRef]

16. Hou, D.-X.; Kumamoto, T. Flavonoids as protein kinase inhibitors for cancer chemoprevention: Direct binding and molecular modeling. *Antioxid. Redox Signal.* **2010**, *13*, 691–719. [CrossRef]

17. Spencer, J.P.E. Beyond antioxidants: The cellular and molecular interactions of flavonoids and how these underpin their actions on the brain. *Proc. Nutr. Soc.* **2010**, *69*, 244–260. [CrossRef] [PubMed]

18. Huang, Z.; Fang, F.; Wang, J.; Wong, C.W. Structural activity relationship of flavonoids with estrogen-related receptor gamma. *FEBS Lett.* **2010**, *584*, 22–26. [CrossRef] [PubMed]

19. Somjen, D.; Knoll, E.; Vaya, J.; Stern, N.; Tamir, S. Estrogen-like activity of licorice root constituents: Glabridin and glabrene, in vascular tissues in vitro and in vivo. *J. Steroid Biochem. Mol. Biol.* **2004**, *91*, 147–155. [CrossRef]

20. Jin, X.-L.; Wei, X.; Qi, F.-M.; Yu, S.-S.; Zhou, B.; Bai, S. Characterization of hydroxycinnamic acid derivatives binding to bovine serum albumin. *Org. Biomol. Chem.* **2012**, *10*, 3424–3431. [CrossRef]

21. Atrahimovich, D.; Vaya, J.; Tavori, H.; Khatib, S. Glabridin protects paraoxonase 1 from linoleic acid hydroperoxide inhibition via specific interaction: A fluorescence-quenching study. *J. Agric. Food Chem.* **2012**, *60*, 3679–3685. [CrossRef]

22. Luck, G.; Liao, H.; Murray, N.J.; Grimmer, H.R.; Warminski, E.E.; Williamson, M.P.; Lilley, T.H.; Haslam, E. Polyphenols, astringency and proline-rich proteins. *Phytochemistry* **1994**, *37*, 357–371. [CrossRef]

23. Ciumărnean, L.; Milaciu, M.V.; Runcan, O.; Vesa, S.C.; Răchisan, A.L.; Negrean, V.; Perné, M.G.; Donca, V.I.; Alexescu, T.G.; Para, I.; et al. The effects of flavonoids in cardiovascular diseases. *Molecules* **2020**, *25*, 4320. [CrossRef] [PubMed]

24. Cijo, V.; Dellaire, G.; Rupasinghe, H.P.V. ScienceDirect Plant flavonoids in cancer chemoprevention: Role in genome stability. *J. Nutr. Biochem.* **2017**, *45*, 1–14. [CrossRef]

25. Maher, P. The potential of flavonoids for the treatment of neurodegenerative diseases. *Int. J. Mol. Sci.* **2019**, *20*, 3056. [CrossRef] [PubMed]

26. Gecibesler, I.H.; Aydin, M. Plasma protein binding of herbal-flavonoids to human serum albumin and their anti-proliferative activities. *An. Acad. Bras. Cienc.* **2020**, *92*, 1–16. [CrossRef]

27. Lin, C.Z.; Hu, M.; Wu, A.Z.; Zhu, C.C. Investigation on the differences of four flavonoids with similar structure binding to human serum albumin. *J. Pharm. Anal.* **2014**, *4*, 392–398. [CrossRef]

28. Mondal, P.; Bose, A. Spectroscopic overview of quercetin and its Cu(II) complex interaction with serum albumins. *BioImpacts* **2019**, *9*, 115–121. [CrossRef]

29. Geng, R.; Ma, L.; Liu, L.; Xie, Y. Influence of bovine serum albumin-flavonoid interaction on the antioxidant activity of dietary flavonoids: New evidence from electrochemical quantification. *Molecules* **2019**, *24*, 70. [CrossRef] [PubMed]

30. Ma, C.M.; Zhao, X.H. Depicting the non-covalent interaction of whey proteins with galangin or genistein using the multi-spectroscopic techniques and molecular docking. *Foods* **2019**, *8*, 360. [CrossRef]

31. Tang, F.; Xie, Y.; Cao, H.; Yang, H.; Chen, X.; Xiao, J. Fetal bovine serum influences the stability and bioactivity of resveratrol analogues: A polyphenol-protein interaction approach. *Food Chem.* **2017**, *219*, 321–328. [CrossRef]

32. Czubinski, J.; Dwiecki, K. A review of methods used for investigation of protein–phenolic compound interactions. *Int. J. Food Sci. Technol.* **2017**, *52*, 573–585. [CrossRef]
33. Cao, H.; Wu, D.; Wang, H.; Xu, M. Effect of the glycosylation of flavonoids on interaction with protein. *Spectrochim. Acta Part A Mol. Biomol. Spectrosc.* **2009**, *73*, 972–975. [CrossRef]
34. Cao, X.; He, Y.; Kong, Y.; Mei, X.; Huo, Y.; He, Y.; Liu, J. Elucidating the interaction mechanism of eriocitrin with β-casein by multi-spectroscopic and molecular simulation methods. *Food Hydrocoll.* **2019**, *94*, 63–70. [CrossRef]
35. Arroyo-Maya, I.J.; Campos-Terán, J.; Hernández-Arana, A.; McClements, D.J. Characterization of flavonoid-protein interactions using fluorescence spectroscopy: Binding of pelargonidin to dairy proteins. *Food Chem.* **2016**, *213*, 431–439. [CrossRef]
36. Xiao, J.; Kai, G. A review of dietary polyphenol-plasma protein interactions: Characterization, influence on the bioactivity, and structure-affinity relationship. *Crit. Rev. Food Sci. Nutr.* **2012**, *52*, 85–101. [CrossRef]
37. Huang, J.; Liu, Z.; Ma, Q.; He, Z.; Niu, Z.; Zhang, M.; Pan, L.; Qu, X.; Yu, J.; Niu, B. Studies on the Interaction between Three Small Flavonoid Molecules and Bovine Lactoferrin. *Biomed. Res. Int.* **2018**, *2018*. [CrossRef]
38. Hegde, A.H.; Sandhya, B.; Seetharamappa, J. Evaluation of binding and thermodynamic characteristics of interactions between a citrus flavonoid hesperitin with protein and effects of metal ions on binding. *Mol. Biol. Rep.* **2011**, *38*, 4921–4929. [CrossRef]
39. Omidvar, Z.; Asoodeh, A.; Chamani, J. Studies on the antagonistic behavior between cyclophosphamide hydrochloride and aspirin with human serum albumin: Time-resolved fluorescence spectroscopy and isothermal titration calorimetry. *J. Solut. Chem.* **2013**, *42*, 1005–1017. [CrossRef]
40. Vachali, P.P.; Li, B.; Besch, B.M.; Bernstein, P.S. Protein-flavonoid interaction studies by a Taylor dispersion surface plasmon resonance (SPR) technique: A novel method to assess biomolecular interactions. *Biosensors* **2016**, *6*, 6. [CrossRef]
41. Dahli, L.; Atrahimovich, D.; Vaya, J.; Khatib, S. Lyso-DGTS lipid isolated from microalgae enhances PON1 activities in vitro and in vivo, increases PON1 penetration into macrophages and decreases cellular lipid accumulation. *BioFactors* **2018**, *44*, 299–310. [CrossRef]
42. Atrahimovich, D.; Vaya, J.; Khatib, S. The effects and mechanism of flavonoid-rePON1 interactions. structure-activity relationship study. *Bioorg. Med. Chem.* **2013**, *21*, 3348–3355. [CrossRef] [PubMed]
43. Song, S.S.; Sun, C.P.; Zhou, J.J.; Chu, L. Flavonoids as human carboxylesterase 2 inhibitors: Inhibition potentials and molecular docking simulations. *Int. J. Biol. Macromol.* **2019**, *131*, 201–208. [CrossRef] [PubMed]
44. Solimani, R.; Bayon, F.; Domini, I.; Pifferi, P.G.; Todesco, P.E.; Bologna, U.; Frae-cnr, I.; Gobetti, P.; Chimica, D.; Calabria, U.; et al. Flavonoid-DNA Interaction Studied with Flow Linear Dichroism Technique. *J. Agric. Food. Chem.* **1995**, *43*, 876–882. [CrossRef]
45. Bocian, W.; Kawecki, R.; Bednarek, E.; Sitkowski, G.; Ulkowska, A.; Kozerski, L. Interaction of flavonoid topoisomerase I and II inhibitors with DNA oligomers. *New J. Chem.* **2006**, *30*, 467–472. [CrossRef]
46. Thulstrup, P.W.; Thormann, T.; Spanget-Larsen, J.; Bisgaard, H.C. Interaction between ellagic acid and calf thymus DNA studied with flow linear dichroism UV-VIS spectroscopy. *Biochem. Biophys. Res. Commun.* **1999**, *265*, 416–421. [CrossRef]
47. Wang, Z.; Cui, M.; Song, F.; Lu, L.; Liu, Z.; Liu, S. Evaluation of Flavonoids Binding to DNA Duplexes by Electrospray Ionization Mass Spectrometry. *J. Am. Soc. Mass Spectrom.* **2008**, *19*, 914–922. [CrossRef]
48. Atrahimovich, D.; Samson, A.O.; Barsheshet, Y.; Vaya, J.; Khatib, S.; Reuveni, E. Genome-wide localization of the polyphenol quercetin in human monocytes. *BMC Genom.* **2019**, *20*, 1–9. [CrossRef]
49. Anders, L.; Guenther, M.G.; Qi, J.; Fan, Z.P.; Marineau, J.J.; Rahl, P.B.; Lovén, J.; Sigova, A.A.; Smith, W.B.; Lee, T.I.; et al. Genome-wide determination of drug localization. *Nat. Biotechnol.* **2014**, *32*, 92–96. [CrossRef] [PubMed]
50. Rodriguez, R.; Miller, K.M. Unravelling the genomic targets of small molecules using high-throughput sequencing. *Nat. Rev. Genet.* **2014**, *15*, 783–796. [CrossRef]
51. Giuliani, C.; Noguchi, Y.; Harii, N.; Napolitano, G.; Tatone, D.; Bucci, I.; Piantelli, M.; Monaco, F.; Kohn, L.D. The flavonoid Quercetin Regulates Growth and Gene expression in rat FRTL-5 tyroid cells. *Endocrinology* **2008**, *149*, 84–92. [CrossRef] [PubMed]
52. Bell, O.; Tiwari, V.K.; Thomä, N.H.; Schübeler, D. Determinants and dynamics of genome accessibility. *Nat. Publ. Gr.* **2011**, *12*, 554–564. [CrossRef]
53. Hosseinimehr, S.J.; Tolmachev, V.; Stenerlöw, B. ^{125}I-Labeled Quercetin as a Novel DNA-Targeted Radiotracer. *Cancer Biother. Radiopharm.* **2011**, *26*, 469–475. [CrossRef]
54. Walle, T.; Vincent, T.S.; Walle, U.K. Evidence of covalent binding of the dietary flavonoid quercetin to DNA and protein in human intestinal and hepatic cells. *Biochem. Pharmacol.* **2003**, *65*, 1603–1610. [CrossRef]
55. Solimani, R. Quercetin and DNA in solution: Analysis of the dynamics of their interaction with a linear dichroism study. *Int. J. Biol. Macromol.* **1996**, *18*, 287–295. [CrossRef]
56. Bertram, B.; Bollow, U.; Rajaee-Behbahani, N.; Bürkle, A.; Schmezer, P. Induction of poly (ADP-ribosyl) ation and DNA damage in human peripheral lymphocytes after treatment with (−)-epigallocatechin-gallate. *Mutat. Res. Genet. Toxicol. Environ. Mutagen.* **2003**, *534*, 77–84. [CrossRef]
57. Johnson, M.K.; Loo, G. Effects of epigallocatechin gallate and quercetin on oxidative damage to cellular DNA. *Mutat. Res. DNA Repair* **2000**, *459*, 211–218. [CrossRef]
58. Teel, R.W. Ellagic acid binding to DNA as a possible mechanism for its antimutagenic and anticarcinogenic action. *Cancer Lett.* **1986**, *30*, 329–336. [CrossRef]
59. Kuzuhara, T.; Sei, Y.; Yamaguchi, K.; Suganuma, M.; Fujiki, H. DNA and RNA as new binding targets of green tea catechins. *J. Biol. Chem.* **2006**, *281*, 17446–17456. [CrossRef]

60. Chen, X.; He, Z.; Wu, X.; Mao, D.; Feng, C.; Zhang, J.; Chen, G. Comprehensive study of the interaction between Puerariae Radix flavonoids and DNA: From theoretical simulation to structural analysis to functional analysis. *Spectrochim. Acta Part A Mol. Biomol. Spectrosc.* **2020**, *231*, 118109. [CrossRef] [PubMed]

61. Arif, H.; Rehmani, N.; Farhan, M.; Ahmad, A.; Hadi, S.M. Mobilization of copper ions by flavonoids in human peripheral lymphocytes leads to oxidative DNA breakage: A structure activity study. *Int. J. Mol. Sci.* **2015**, *16*, 26754–26769. [CrossRef] [PubMed]

62. Afonina, I.S.; Zhong, Z.; Karin, M.; Beyaert, R. Limiting inflammation—The negative regulation of NF-B and the NLRP3 inflammasome. *Nat. Immunol.* **2017**, *18*, 861–869. [CrossRef]

63. Taniura, S.; Kamitani, H.; Watanabe, T.; Eling, E.T.; Banerjee, T.K. Induction of cyclooxygenase-2 expression by interleukin-1β in human glioma cell line, U87MG. *Neurol. Med. Chir.* **2008**, *48*, 500–505. [CrossRef] [PubMed]

64. Catarino, M.; Alves-Silva, J.; Pereira, O.; Cardoso, S. Antioxidant Capacities of Flavones and Benefits in Oxidative-Stress Related Diseases. *Curr. Top. Med. Chem.* **2014**, *15*, 105–119. [CrossRef]

65. Yoon, J.H.; Baek, S.J. Molecular targets of dietary polyphenols with anti-inflammatory properties. *Yonsei Med. J.* **2005**, *46*, 585–596. [CrossRef]

66. Lindahl, M.; Tagesson, C. Selective inhibition of group II phospholipase A2 by quercetin. *Inflammation* **1993**, *17*, 573–582. [CrossRef] [PubMed]

67. Kumar, S.; Pandey, A.K. Chemistry and biological activities of flavonoids: An overview. *Sci. World J.* **2013**, *2013*, 162750. [CrossRef] [PubMed]

68. Lättig, J.; Böhl, M.; Fischer, P.; Tischer, S.; Tietböhl, C.; Menschikowski, M.; Gutzeit, H.O.; Metz, P.; Pisabarro, M.T. Mechanism of inhibition of human secretory phospholipase A2 by flavonoids: Rationale for lead design. *J. Comput. Aided. Mol. Des.* **2007**, *21*, 473–483. [CrossRef]

69. Needleman, P.; Isakson, P.C. The discovery and function of COX-2. *J. Rheumatol.* **1997**, *49*, 6–8.

70. Baumann, J.; Bruchhausen, F.V.; Wurm, G. Flavonoids and related compounds as inhibitors of arachidonic acid peroxidation. *Prostaglandins* **1980**, *20*, 627–639. [CrossRef]

71. Landolfi, R.; Mower, R.L.; Steiner, M. Modification of platelet function and arachidonic acid metabolism by bioflavonoids. Structure-activity relations. *Biochem. Pharmacol.* **1984**, *33*, 1525–1530. [CrossRef]

72. Raja, S.B.; Rajendiran, V.; Kasinathan, N.K.; Amrithalakshmi, A.P.; Venkatabalasubramanian, S.; Murali, M.R.; Devaraj, H.; Devaraj, S.N. Differential cytotoxic activity of Quercetin on colonic cancer cells depends on ROS generation through COX-2 expression. *Food Chem. Toxicol.* **2017**, *106*, 92–106. [CrossRef]

73. D'mello, P.; Gadhwal, M.K.; Joshi, U.; Shetgiri, P. Modeling of COX-2 inhibotory activity of flavonoids. *Int. J. Pharm. Pharm. Sci.* **2011**, *3*, 33–40.

74. Kim, H.K.; Cheon, B.S.; Kim, Y.H.; Kim, S.Y.; Kim, H.P. Effects of naturally occurring flavonoids on nitric oxide production in the macrophage cell line RAW 264.7 and their structure-activity relationships. *Biochem. Pharmacol.* **1999**, *58*, 759–765. [CrossRef]

75. Cheon, B.S.; Kim, Y.H.; Son, K.S.; Chang, H.W.; Kang, S.S.; Kim, H.P. Effects of prenylated flavonoids and biflavonoids on lipopolysaccharide-induced nitric oxide production from the mouse macrophage cell line RAW 264.7. *Planta Med.* **2000**, *66*, 596–600. [CrossRef]

76. Nile, S.H.; Nile, A.S.; Keum, Y.S.; Sharma, K. Utilization of quercetin and quercetin glycosides from onion (Allium cepa L.) solid waste as an antioxidant, urease and xanthine oxidase inhibitors. *Food Chem.* **2017**, *235*, 119–126. [CrossRef] [PubMed]

77. Middleton, E.; Kandaswami, C. Effects of flavonoids on immune and inflammatory cell functions. *Biochem. Pharmacol.* **1992**, *43*, 1167–1179. [CrossRef]

78. Ovais, M.; Zia, N.; Ahmad, I.; Khalil, A.T.; Raza, A.; Ayaz, M.; Sadiq, A.; Ullah, F.; Shinwari, Z.K. Phyto-Therapeutic and Nanomedicinal Approaches to Cure Alzheimer's Disease: Present Status and Future Opportunities. *Front. Aging Neurosci.* **2018**, *10*. [CrossRef] [PubMed]

79. Ayaz, M.; Junaid, M.; Ullah, F.; Subhan, F.; Sadiq, A.; Ali, G.; Ovais, M.; Shahid, M.; Ahmad, A.; Wadood, A.; et al. Anti-Alzheimer's studies on ß-sitosterol isolated from Polygonum hydropiper L. *Front. Pharmacol.* **2017**, *8*, 1–16. [CrossRef]

80. Burns, A.; Iliffe, S. Alzheimer's disease. *BMJ* **2009**, *338*, 467–471. [CrossRef]

81. Faraji, L.; Nadri, H.; Moradi, A.; Bukhari, S.N.A.; Pakseresht, B.; Moghadam, F.H.; Moghimi, S.; Abdollahi, M.; Khoobi, M.; Foroumadi, A. Aminoalkyl-substituted flavonoids: Synthesis, cholinesterase inhibition, β-amyloid aggregation, and neuroprotective study. *Med. Chem. Res.* **2019**, *28*, 974–983. [CrossRef]

82. Tumiatti, V.; Minarini, A.; Bolognesi, M.L.; Milelli, A.; Rosini, M.; Melchiorre, C. Tacrine Derivatives and Alzheimers Disease. *Curr. Med. Chem.* **2010**, *17*, 1825–1838. [CrossRef] [PubMed]

83. Cole, S.L.; Vassar, R. The Alzheimer's disease β-secretase enzyme, BACE1. *Mol. Neurodegener.* **2007**, *2*, 1–25. [CrossRef] [PubMed]

84. Cai, H.; Wang, Y.; McCarthy, D.; Wen, H.; Borchelt, D.R.; Price, D.L.; Wong, P.C. BACE1 is the major β-secretase for generation of Aβ peptides by neurons. *Nat. Neurosci.* **2001**, *4*, 233–234. [CrossRef]

85. Zhang, D.; Lv, J.T.; Zhang, B.; Sa, R.N.; Ma, B.B.; Zhang, X.M.; Lin, Z.J. Molecular insight into the therapeutic promise of xuebijing injection against coronavirus disease 2019. *World J. Tradit. Chin. Med.* **2020**, *6*, 203–215. [CrossRef]

86. Balducci, C.; Forloni, G. Novel targets in Alzheimer's disease: A special focus on microglia. *Pharmacol. Res.* **2018**, *130*, 402–413. [CrossRef] [PubMed]

87. Chaudhary, A.; Maurya, P.K.; Yadav, B.S.; Singh, S.; Mani, A. Current Therapeutic Targets for Alzheimer's Disease. *J. Biomed.* **2018**, *3*, 74–84. [CrossRef]
88. Jannat, S.; Balupuri, A.; Ali, M.Y.; Hong, S.S.; Choi, C.W.; Choi, Y.H.; Ku, J.M.; Kim, W.J.; Leem, J.Y.; Kim, J.E.; et al. Inhibition of β-site amyloid precursor protein cleaving enzyme 1 and cholinesterases by pterosins via a specific structure–activity relationship with a strong BBB permeability. *Exp. Mol. Med.* **2019**, *51*, 1–18. [CrossRef] [PubMed]
89. Ahmad, S.; Ullah, F.; Ayaz, M.; Sadiq, A.; Imran, M. Antioxidant and anticholinesterase investigations of Rumex hastatus D. Don: Potential effectiveness in oxidative stress and neurological disorders. *Biol. Res.* **2015**, *48*, 1–8. [CrossRef]
90. Ayaz, M.; Junaid, M.; Ullah, F.; Sadiq, A.; Khan, M.A.; Ahmad, W.; Shah, M.R.; Imran, M.; Ahmad, S. Comparative chemical profiling, cholinesterase inhibitions and anti-radicals properties of essential oils from Polygonum hydropiper L: A Preliminary anti-Alzheimer's study. *Lipids Health Dis.* **2015**, *14*, 1–12. [CrossRef]
91. Sharma, K. Cholinesterase inhibitors as Alzheimer's therapeutics (Review). *Mol. Med. Rep.* **2019**, *20*, 1479–1487. [CrossRef]
92. Ayaz, M.; Sadiq, A.; Junaid, M.; Ullah, F.; Ovais, M.; Ullah, I.; Ahmed, J.; Shahid, M. Flavonoids as prospective neuroprotectants and their therapeutic propensity in aging associated neurological disorders. *Front. Aging Neurosci.* **2019**, *11*, 155. [CrossRef] [PubMed]
93. Baptista, F.I.; Henriques, A.G.; Silva, A.M.S.; Wiltfang, J.; Da Cruz, E.; Silva, O.A.B. Flavonoids as therapeutic compounds targeting key proteins involved in Alzheimer's disease. *ACS Chem. Neurosci.* **2014**, *5*, 83–92. [CrossRef] [PubMed]
94. Singh, D.; Hembrom, S. Neuroprotective Effect of Flavonoids: A Systematic Review. *Int. J. Aging Res.* **2018**, *2*, 1–17.
95. Shimmyo, Y.; Kihara, T.; Akaike, A.; Niidome, T.; Sugimoto, H. Flavonols and flavones as BACE-1 inhibitors: Structure-activity relationship in cell-free, cell-based and in silico studies reveal novel pharmacophore features. *Biochim. Biophys. Acta Gen. Subj.* **2008**, *1780*, 819–825. [CrossRef]
96. Cox, C.J.; Choudhry, F.; Peacey, E.; Perkinton, M.S.; Richardson, J.C.; Howlett, D.R.; Lichtenthaler, S.F.; Francis, P.T.; Williams, R.J. Dietary (–)-epicatechin as a potent inhibitor of βγ-secretase amyloid precursor protein processing. *Neurobiol. Aging* **2015**, *36*, 178–187. [CrossRef]
97. Uriarte-Pueyo, I.; Calvo, M.I. Flavonoids as Acetylcholinesterase Inhibitors. *Curr. Med. Chem.* **2011**, *18*, 5289–5302. [CrossRef]
98. Orhan, I.; Kartal, M.; Tosun, F.; Şener, B. Screening of various phenolic acids and flavonoid derivatives for their anticholinesterase potential. *Z. Naturforsch. Sect. C J. Biosci.* **2007**, *62*, 829–832. [CrossRef]
99. Orhan, I.; Şenol, F.S.; Kartal, M.; Dvorská, M.; Žemlička, M.; Šmejkal, K.; Mokrý, P. Cholinesterase inhibitory effects of the extracts and compounds of Maclura pomifera (Rafin.) Schneider. *Food Chem. Toxicol.* **2009**, *47*, 1747–1751. [CrossRef]
100. Heo, H.J.; Kim, M.J.; Lee, J.M.; Choi, S.J.; Cho, H.Y.; Hong, B.; Kim, H.K.; Kim, E.; Shin, D.H. Naringenin from Citrus junos has an inhibitory effect on acetylcholinesterase and a mitigating effect on amnesia. *Dement. Geriatr. Cogn. Disord.* **2004**, *17*, 151–157. [CrossRef]
101. Sun, D.Y.; Cheng, C.; Moschke, K.; Huang, J.; Fang, W.S. Extensive structure modification on luteolin-cinnamic acid conjugates leading to BACE1 inhibitors with optimal pharmacological properties. *Molecules* **2020**, *25*, 102. [CrossRef] [PubMed]
102. Lee, S.; Youn, K.; Lim, G.T.; Lee, J.; Jun, M. In silico docking and in vitro approaches towards BACE1 and cholinesterases inhibitory effect of citrus flavanones. *Molecules* **2018**, *23*, 1509. [CrossRef]
103. Gong, E.J.; Park, H.R.; Kim, M.E.; Piao, S.; Lee, E.; Jo, D.G.; Chung, H.Y.; Ha, N.C.; Mattson, M.P.; Lee, J. Morin attenuates tau hyperphosphorylation by inhibiting GSK3β. *Neurobiol. Dis.* **2011**, *44*, 223–230. [CrossRef]
104. Spencer, J.P.E.; Rice-Evans, C.; Williams, R.J. Modulation of pro-survival Akt/protein kinase B and ERK1/2 signaling cascades by quercetin and its in vivo metabolites underlie their action on neuronal viability. *J. Biol. Chem.* **2003**, *278*, 34783–34793. [CrossRef] [PubMed]
105. Qin, L.; Zhang, J.; Qin, M. Protective effect of cyanidin 3-O-glucoside on beta-amyloid peptide-induced cognitive impairment in rats. *Neurosci. Lett.* **2013**, *534*, 285–288. [CrossRef] [PubMed]
106. Mackness, B.; Quarck, R.; Verreth, W.; Mackness, M.; Holvoet, P. Human paraoxonase-1 overexpression inhibits atherosclerosis in a mouse model of metabolic syndrome. *Arterioscler. Thromb. Vasc. Biol.* **2006**, *26*, 1545–1550. [CrossRef] [PubMed]
107. Gur, M.; Cayli, M.; Ucar, H.; Elbasan, Z.; Sahin, D.Y.; Gozukara, M.Y.; Selek, S.; Koyunsever, N.Y.; Seker, T.; Turkoglu, C.; et al. Paraoxonase (PON1) activity in patients with subclinical thoracic aortic atherosclerosis. *Int. J. Cardiovasc. Imaging* **2014**, *30*, 889–895. [CrossRef]
108. Kontush, A.; Chantepie, S.; Chapman, M.J. Small, dense HDL particles exert potent protection of atherogenic LDL against oxidative stress. *Arterioscler. Thromb. Vasc. Biol.* **2003**, *23*, 1881–1888. [CrossRef]
109. Barter, P.J.; Puranik, R.; Rye, K.A. New Insights Into the Role of HDL as an Anti-inflammatory Agent in the Prevention of Cardiovascular Disease. *Curr. Cardiol. Rep.* **2007**, *9*, 493–498. [CrossRef]
110. Nofer, J.R.; Levkau, B.; Wolinska, I.; Junker, R.; Fobker, M.; von Eckardstein, A.; Seedorf, U.; Assmann, G. Suppression of endothelial cell apoptosis by high density lipoproteins (HDL) and HDL-associated lysosphingolipids. *J. Biol. Chem.* **2001**, *276*, 34480–34485. [CrossRef]
111. Yuhanna, I.S.; Zhu, Y.; Cox, B.E.; Hahner, L.D.; Osborne-Lawrence, S.; Lu, P.; Marcel, Y.L.; Anderson, R.G.; Mendelsohn, M.E.; Hobbs, H.H.; et al. High-density lipoprotein binding to scavenger receptor-BI activates endothelial nitric oxide synthase. *Nat. Med.* **2001**, *7*, 853–857. [CrossRef]
112. Camont, L.; Chapman, M.J.; Kontush, A. Biological activities of HDL subpopulations and their relevance to cardiovascular disease. *Trends Mol. Med.* **2011**, *17*, 594–603. [CrossRef]

113. Krauss, R.M. Lipoprotein subfractions and cardiovascular disease risk. *Curr. Opin. Lipidol.* **2010**, *21*, 305–311. [CrossRef] [PubMed]
114. Atrahimovich, D.; Khatib, S.; Sela, S.; Vaya, J.; Samson, A.O. Punicalagin Induces Serum Low-Density Lipoprotein Influx to Macrophages. *Oxid. Med. Cell. Longev.* **2016**, 9. [CrossRef] [PubMed]
115. Chien, S.; Shi, M.; Lee, Y.; Te, C.; Shih, Y. Galangin, a novel dietary flavonoid, attenuates metastatic feature via PKC/ERK signaling pathway in TPA-treated liver cancer HepG2 cells. *Cancer Cell Int.* **2015**, 1–11. [CrossRef] [PubMed]
116. Das, A.; Majumder, D.; Saha, C. Correlation of binding efficacies of DNA to flavonoids and their induced cellular damage. *J. Photochem. Photobiol. B Biol.* **2017**, *170*, 256–262. [CrossRef] [PubMed]
117. Vinnarasi, S.; Radhika, R.; Vijayakumar, S.; Shankar, R. Structural insights into the anti-cancer activity of quercetin on G-tetrad, mixed G-tetrad, and G-quadruplex DNA using quantum chemical and molecular dynamics simulations. *J. Biomol. Struct. Dyn.* **2020**, *38*, 317–339. [CrossRef]
118. Wang, S.; Zhang, J.; Chen, M.; Wang, Y. Delivering flavonoids into solid tumors using nanotechnologies. *Expert Opin. Drug Deliv.* **2013**, *10*, 1411–1428. [CrossRef]
119. Abotaleb, M.; Samuel, S.M.; Varghese, E.; Varghese, S.; Kubatka, P.; Liskova, A.; Büsselberg, D. Flavonoids in cancer and apoptosis. *Cancers* **2019**, *11*, 28. [CrossRef]

Article

Lyso-DGTS Lipid Derivatives Enhance PON1 Activities and Prevent Oxidation of LDL: A Structure–Activity Relationship Study

Ali Khattib [1,2], Sanaa Musa [1,3], Majdi Halabi [4], Tony Hayek [2] and Soliman Khatib [1,3,*]

[1] Natural Compounds and Analytical Chemistry Laboratory, MIGAL—Galilee Research Institute, P.O. Box 831, Kiryat Shemona 11016, Israel

[2] The Rappaport Family Institute for Research in the Medical Sciences, Rambam Medical Center, Haifa 31096, Israel

[3] Department of Biotechnology, Tel-Hai College, Upper Galilee 12210, Israel

[4] Ziv Medical Center, Safed 13100, Israel

* Correspondence: solimankh@migal.org.il; Tel.: +972-4-6953512; Fax: +972-4-6944980

Abstract: Paraoxonase 1 (PON1) plays a role in regulating reverse cholesterol transport and has antioxidative, anti-inflammatory, antiapoptotic, vasodilative, and antithrombotic activities. Scientists are currently focused on the modulation of PON1 expression using different pharmacological, nutritional, and lifestyle approaches. We previously isolated a novel active compound from *Nannochloropsis* microalgae—lyso-diacylglyceryltrimethylhomoserine (lyso-DGTS)—which increased PON1 activity, HDL-cholesterol efflux, and endothelial nitric oxide release. Here, to explore this important lipid moiety's effect on PON1 activities, we examined the effect of synthesized lipid derivatives and endogenous analogs of lyso-DGTS on PON1 lactonase and arylesterase activities and LDL oxidation using structure–activity relationship (SAR) methods. Six lipids significantly elevated recombinant PON1 (rePON1) lactonase activity in a dose-dependent manner, and four lipids significantly increased rePON1 arylesterase activity. Using tryptophan fluorescence-quenching assay and a molecular docking method, lipid–PON1 interactions were characterized. An inverse correlation was obtained between the lactonase activity of PON1 and the docking energy of the lipid–PON1 complex. Furthermore, five of the lipids increased the LDL oxidation lag time and inhibited its propagation. Our findings suggest a beneficial effect of lyso-DGTS or lyso-DGTS derivatives through increased PON1 activity and prevention of LDL oxidation.

Keywords: lyso-DGTS; paraoxonase 1; docking; structure-activity relationship; antioxidant activities; LDL oxidation

Citation: Khattib, A.; Musa, S.; Halabi, M.; Hayek, T.; Khatib, S. Lyso-DGTS Lipid Derivatives Enhance PON1 Activities and Prevent Oxidation of LDL: A Structure–Activity Relationship Study. *Antioxidants* **2022**, *11*, 2058. https://doi.org/10.3390/antiox11102058

Academic Editor: Stanley Omaye

Received: 26 September 2022
Accepted: 14 October 2022
Published: 19 October 2022

Publisher's Note: MDPI stays neutral with regard to jurisdictional claims in published maps and institutional affiliations.

1. Introduction

Atherosclerosis is the accumulation of fatty acids/lipids and oxidized lipids in the innermost layer, the intima, of the arteries. Advanced atherosclerotic plaques can penetrate the arterial lumen, inhibit blood flow, and lead to tissue ischemia. Atherosclerotic cardiovascular disease (CVD) remains a leading cause of vascular disease worldwide [1].

Epidemiological studies have demonstrated that low-density lipoprotein (LDL) and high-density lipoprotein (HDL) are independent causes of CVD. Clinically, lipoprotein disorders, such as an increase in LDL-cholesterol or decrease in HDL-cholesterol levels, are common characteristics of CVD [2].

Many lines of evidence suggest that the oxidation of LDL plays an important role in the pathogenesis of atherosclerosis. Animal models have shown that the accumulation of oxidized LDL (ox-LDL) in the circulation can be a well-known risk marker of human CVD, caused by atherosclerosis [3]. HDL has a key role in atherosclerosis inhibition due to its antiatherogenic properties, such as reverse cholesterol transport, antioxidant, anti-inflammatory, antiapoptotic, vasodilatory and cytoprotective effects, and improvement of endothelial function [4,5]. HDL loses its beneficial properties in certain pathological situations, such as diabetes, acute infections, and coronary artery disease (CAD), termed "dysfunctional HDL". Oxidative modifications of HDL proteins and lipids have been demonstrated to be one of the main causes of HDL dysfunction [6].

There is considerable evidence that the antioxidant activity of HDL is largely related to its associated antioxidant enzyme paraoxonase 1 (PON1). Epidemiological evidence has shown that low PON1 activity is associated with an increased risk of cardiovascular events [7,8]. PON1 has many activities—hydrolysis of organophosphate triesters, aryl esters, cyclic carbamates, glucuronides, estrogen esters, and thiolactones—while its "natural" substrates are assumed to be lactones [9,10]. The aromatic nature of amino acids in the active site of PON1 could explain why the enzyme prefers lipophilic substrates [11]. All three members of the PON family (PON1, PON2, and PON3) share this property of being lactonases.

PON1 can hydrolyze lipid peroxide in lipoproteins through its lipolactonase activity, thereby decreasing oxidative stress in serum lipoproteins and macrophages and reducing inflammation [12,13]. These important features can markedly affect the development of atherosclerotic plaques and cardiovascular events [14]. Experiments with PON1-knockout mice have indicated that the absence of PON1 leads to an increase in endothelial adhesion molecules and oxidative stress, confirming this enzyme's role in preventing the onset of atherosclerosis [9,15].

A clinical study suggested that the serum antioxidant activity of PON1 (arylesterase activity) is an important factor in protecting from oxidative stress and lipid peroxidation in CAD. Thus, evaluating the effects of PON 1 on CAD patients may be promising for the treatment and prognosis of this disease [16].

PON1 enzymatic activities can be affected by its interaction with elements in its environment. For example, dietary lipids and lipid-peroxidation products can decrease PON1 activity and gene expression; following fish oil supplementation, polyunsaturated fatty acids significantly decreased postprandial serum PON1 activity, associated with an increase in susceptibility of the serum to lipid peroxidation [17]. The effect has been confirmed in humans [18]. Conversely, the consumption of pomegranate juice, which is rich in polyphenols and several antioxidants, resulted in higher PON1 activity [19,20]. Other antioxidants, such as flavonoids, quercetin, and glabridin, also protect PON1 from oxidation and increased its lactonase activity [21,22].

We hypothesize that natural agents with the potential to increase PON1 function can improve its atheroprotective effects and may reduce CVD risk. We previously isolated a promising agent from *Nannochloropsis* sp. microalgae, determined to be lyso-diacylglyceryltrimethylhomoserine (lyso-DGTS), which significantly increased recombinant PON1 (rePON1) and HDL-cholesterol efflux and enhanced endothelial nitric oxide release [23,24]. In our previous studies, lyso-DGTS was found to protect macrophages and LDL from oxidation and potentially improve HDL function by increasing the activity of PON1 [24]. Moreover, we showed that lyso-DGTS lipid interacts with HDL proteins and with the HDL core, improving HDL functions in vitro and in vivo [25].

In the present study, to characterize this important lipid moiety's effect on PON1 activities, we investigated the effect of synthesized lipid derivatives and endogenous analogs of lyso-DGTS on rePON1 lactonase and arylesterase activities using structure–activity relationship (SAR) methods. In addition, the biological importance of these interactions was evaluated by examining the effect of these bioactive lipids on LDL oxidation.

2. Experimental Procedures

2.1. Materials

Lyso-DGTS was isolated from *Nannochloropsis* microalgae in our laboratory. Lyso-DGTS analogs palmitic acid, cis-5,8,11,14,17-eicosapentaenoic acid (EPA), oleic acid, oleoyl chloride, EDC-HCl, DMAP, and dihydrocoumarin were purchased from Sigma-Aldrich.

EPA chloride was purchased from Cayman Chemical. RePON1, generated in *Escherichia coli* via direct evolution, was purchased from the Weizmann Institute of Science (Rehovot, Israel).

2.2. Isolation of Lyso-DGTS

Lyso-DGTS was isolated and purified as described previously [23]. *Nannochlorpsis* sp. was extracted using 70% ethanol-water solvent to obtain extract in 20% yield. Lyso-DGTS was purified using preparative chromatography methods to obtain the pure compound, as determined via HPLC (purity = 97%), in ~1% yield of the extract.

2.3. Synthesis of Lyso-DGTS Derivatives

2.3.1. General Procedure for the Synthesis of (*S*)-1-Carboxy-3-hydroxy-*N,N,N*-trimethylpropan-1-aminium (L-Homoserine Betaine) (Scheme 1)

L-Homoserine (1 g, 8.39 mmol) was dissolved in MeOH (20 mL) and the solution was cooled to 0 °C under a nitrogen atmosphere. $SOCl_2$ (0.7 mL, 10.07 mmol) was added slowly while stirring. The reaction was stirred for 24 h at room temperature and the solvent was evaporated to obtain (*S*)-methyl 2-amino-4-hydroxybutanoate.

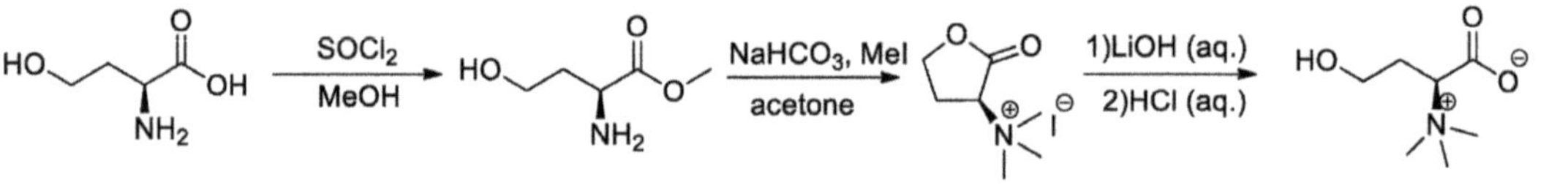

Scheme 1. Synthesis of (*S*)-1-carboxy-3-hydroxy-*N,N,N*-trimethylpropan-1-aminium (L-homoserine betaine).

(*S*)-methyl 2-amino-4-hydroxybutanoate (1.118 g, 8.39 mmol) and $NaHCO_3$ (5 g, 59.52 mmol) were suspended in 70 mL acetone. A solution of methyl iodide (3.13 mL, 50.42 mmol) in 3 mL acetone was slowly added to the suspension. The reaction mixture was stirred overnight at room temperature under a nitrogen atmosphere. The solution was filtered, and the filtrate was evaporated to dryness and used in the next step without further purification.

N,N,N-trimethyl-L-homoserine lactone (8.39 mmol) was dissolved in H_2O (20 mL). LiOH (100 mg, 4.17 mmol) was added and the mixture was stirred for 30 min at room temperature. The reaction mixture was neutralized with 7% aqueous HCl, followed by evaporation. The residue was subjected to chromatography in a silica gel column eluted with (20%) methanol in dichloromethane to give (*S*)-1-carboxy-3-hydroxy-*N,N,N*-trimethylpropan-1-aminium (1.24 g, 95%) (L-homoserine betaine).

2.3.2. General Procedure for the Synthesis of L-Homoserine Betaine Fatty Acid Ester Derivatives (Scheme 2)

(*S*)-1-carboxy-3-hydroxy-*N,N,N*-trimethylpropan-1-aminium (1 equiv.) was dissolved in 2 mL anhydrous N,N-dimethylformamide (DMF) and cooled to 0 °C. N-(3-Dimethyla-minopropyl)-*N′*-ethylcarbodiimide hydrochloride (EDC-HCl) (1.5 equiv.) and DMAP (1.2 equiv.) were added to the cooled solution. The mixture was stirred for 20 min under a nitrogen atmosphere; then, fatty acid derivative (1 equiv.) was added to the solution. The reaction mixture was stirred overnight at room temperature and then extracted with ethyl acetate. The organic layer was dried over sodium sulfate, filtered, and evaporated to dryness. The crude product was purified via chromatography on a silica gel column using dichloromethane: methanol as the eluent.

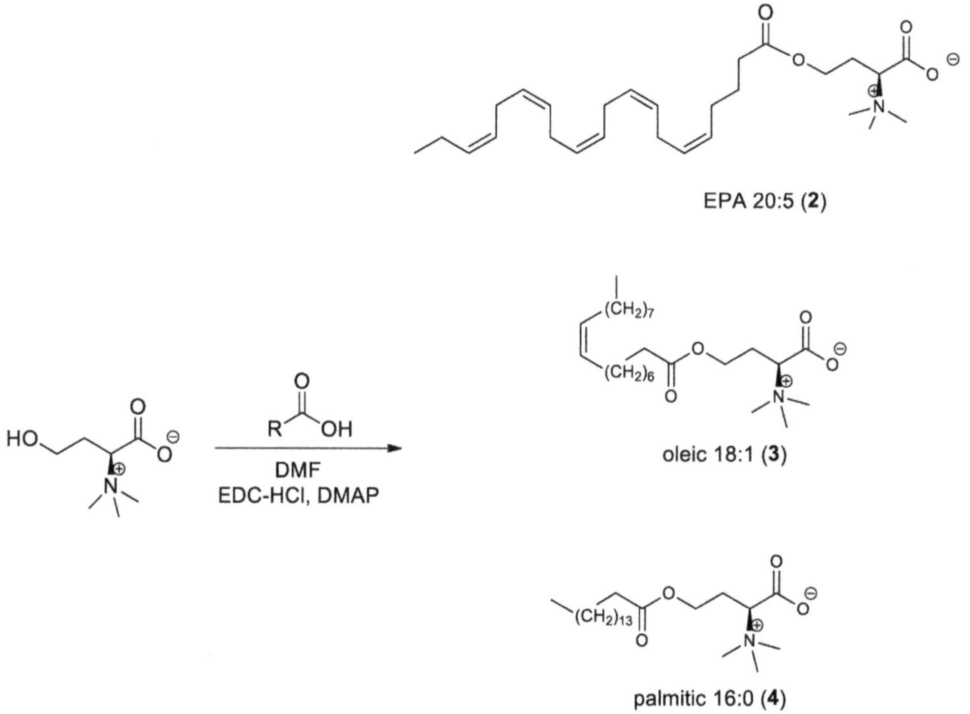

EPA 20:5 (**2**)

oleic 18:1 (**3**)

palmitic 16:0 (**4**)

Scheme 2. Synthesis of L-homoserine betaine fatty acid ester derivatives.

2.3.3. General Procedure for the Synthesis of Amide-Fatty Acid Derivatives

(*S*)-2-((Tert-butoxycarbonyl) amino)-4-oleylamido butanoic acid (C2) and
(*S*)-2-((Tert-butoxycarbonyl)amino)-4-((5*Z*,8*Z*,11*Z*,14*Z*,17*Z*)-eicosa-5,8,11,14,17-pentaenamido)butanoic acid (C3) (Scheme 3)

(*S*)-4-Amino-2-(tert-butoxycarbonylamino) butanoic acid (C1) (100 mg, 0.458 mmol) and *N,N*-diisopropylethylamine (290 mg, 2.243 mmol) were dissolved in 5 mL anhydrous DMF. Oleoyl chloride (135 mg, 0.448 mmol) was dissolved in anhydrous DMF (1 mL) and stirred at room temperature for 10 min under a nitrogen atmosphere. The mixture was added slowly to the solution of **C1** and the reaction mixture was stirred for 4 h at room temperature. DDW (20 mL) and chloroform (40 mL) were added to the mixture, and the aqueous layer was treated with 7% HCl solution until pH 3 was achieved. Then, it was extracted twice with chloroform (40 mL) and the organic layers were collected, dried over anhydrous Na_2SO_4, filtered, and evaporated. The residue was subjected to chromatography in a silica gel column eluted with 25% methanol in dichloromethane to give compound **C2**. Compound **C3** was prepared from eicosapentaenoyl chloride using a similar procedure.

(*S*)-2-Amino-4-oleylamido Butanoic Acid (C4) and
(*S*)-2-Amino-4-((5*Z*,8*Z*,11*Z*,14*Z*,17*Z*)-eicosa-5,8,11,14,17-pentaenamido)butanoic Acid (C5) (Scheme 4)

(*S*)-2-((tert-butoxycarbonyl)amino)-4-((5*Z*,8*Z*,11*Z*,14*Z*,17*Z*)-eicosa-5,8,11,14,17- pentacnamido)butanoic acid (**C3**) was dissolved in DCM (10 mL). Trifluoroacetic acid (TFA; 2 mL, 26.09 mmol) was added and the reaction mixture was stirred at room temperature for 24 h to give **C5**, which was purified in a silica gel column eluted with 20% methanol in dichloromethane. Using a similar procedure, compound **C4** was prepared from **C2**.

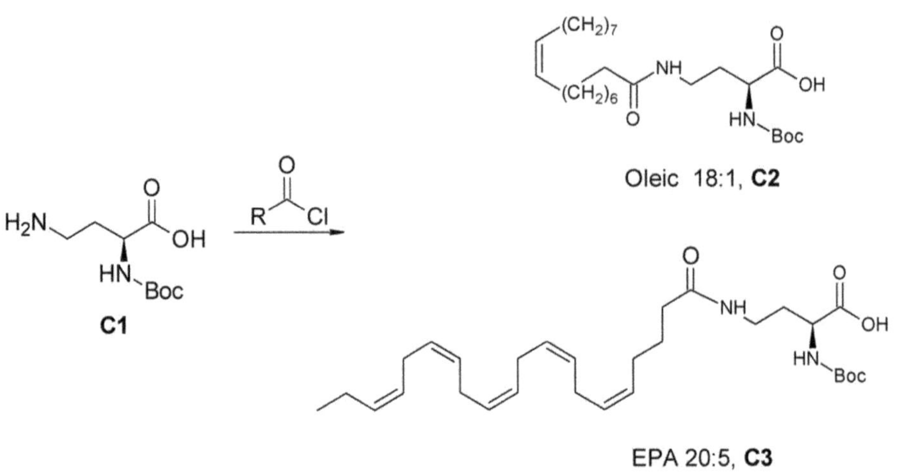

Scheme 3. Synthesis of (*S*)-2-((tert-butoxycarbonyl) amino)-4-oleylamido butanoic acid (**C2**) and (*S*)-2-((tert-butoxycarbonyl)amino)-4-((5Z,8Z,11Z,14Z,17Z)-eicosa-5,8,11,14,17-pentaenamido)butanoic acid (**C3**).

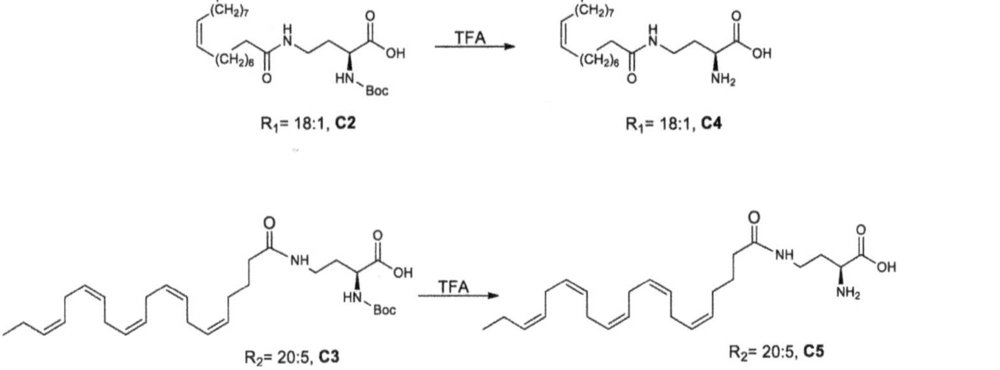

Scheme 4. Synthesis of (*S*)-2-amino-4-oleylamido butanoic acid (**C4**) and (*S*)-2-amino-4-((5Z,8Z,11Z,14Z,17Z)-eicosa-5,8,11,14,17-pentaenamido)butanoic acid (**C5**).

3-(*S*)-4-((5Z,8Z,11Z,14Z,17Z)-Eicosa-5,8,11,14,17-pentaenamido)-1-methoxy-*N,N,N*-trimethyl-1-oxobutan-2-aminium (**5**)) (Scheme 5)

(*S*)-2-amino-4-((5Z,8Z,11Z,14Z,17Z)-eicosa-5,8,11,14,17-pentaenamido)butanoic acid (**C5**) was dissolved in methanol (10 mL), and NaHCO$_3$ (2.19 g, 5.956 mmol) was added to the mixture while stirring. Iodomethane (MeI; 200 μL, 3.2 mmol) was dissolved in methanol (1 mL) and added to the reaction mixture. The reaction mixture was stirred at room temperature for 24 h and then evaporated. The crude product was subjected to chromatography in a silica gel column eluted with 35% methanol in dichloromethane to give compound **5**. Using a similar procedure, compound **6** was prepared from **C4**.

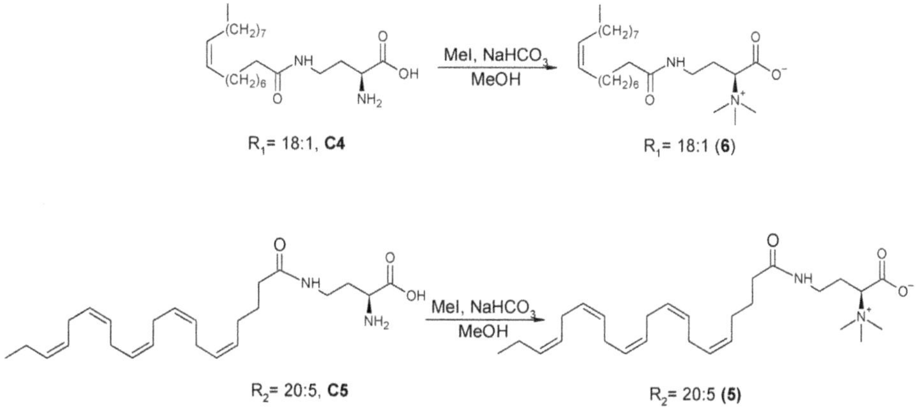

Scheme 5. Synthesis of 3-(S)-4-((5Z,8Z,11Z,14Z,17Z)-eicosa-5,8,11,14,17-pentaenamido)-1-methoxy-N,N,N-trimethyl-1-oxobutan-2-aminium (**5**) and 1-carboxy-N-N-N-trimethyl-3-oleamidopropan-1-aminium (**6**).

2.4. Chromatography Analysis

2.4.1. HPLC Analysis

HPLC analysis was performed with UHPLC connected to a photodiode array detector (Dionex Ultimate 3000), with a reverse-phase column (Phenomenex RP-18, 150 × 4.0 mm, 3 μm). The mobile phase was a mixture of (A) DDW with 0.1% formic acid and (B) acetonitrile with 0.1% formic acid with a flow of 1ml/min and gradient starting with 5% B and increasing in a concentration to 95% B for 30 min and then kept at 95% B for an additional 5 min.

2.4.2. LC–MS Analysis

LC–MS analysis was performed with a Waters ZQ-mass spec system connected to an API single-quadrupole mass detector. ESI capillary voltage was set to 3800 V, capillary temperature to 250 °C, and gas flow to 60 mL/min.

2.5. NMR Analysis

The synthesized compounds were dissolved in deuterated methanol (CD_3OD) or water (D_2O), and ^{1}H-NMR spectra were recorded at room temperature with a Bruker 400 MHz instrument with chemical shifts reported in ppm relative to the residual deuterated solvent.

2.5.1. L-Homoserine Betaine

^{1}H-NMR (400 MHz, CD_3OD), δ: 3.69 (2H, m), 3.52 (1H, m), 3.13 (9H, s), 2.14 (1H, m), 1.99 (1H, m). (ESI+) *m/z*: 162. (NMR purity = 98%).

2.5.2. Homoserine Betaine Eicosapentaenoate (2)

^{1}H-NMR (400 MHz, CD_3OD), δ: 5.45 (10H, m), 4.21 (1H, m), 4.14 (1H, m), 3.73 (1H, m), 3.29 (9H, s), 2.92 (8H, m), 2.08–2.51 (6H, m), 1.78 (2H, t), 1.29 (2H, m), 1.04 (3H, t). (ESI+) *m/z*: 446. HPLC Purity: 95%

2.5.3. Homoserine Betaine Oleate (3)

^{1}H-NMR (400 MHz, CD_3OD), δ: 5.35 (2H, dt), 4.27 (1H, m), 4.13 (1H, m), 3.19 (1H, m), 3.21 (9H, s), 2.35 (2H, t), 2.04 (4H, m), 1.61 (2H, m), 1.30 (m, 22H), 0.90 (3H, t). (ESI+) *m/z*: 426. HPLC purity: 95%.

2.5.4. Homoserine Betaine Palmitate (4)

^{1}H-NMR (CD$_3$OD, 400 MHZ), δ: 4.25 (1H, m), 4.12 (1H, m), 3.67 (1H, m), 3.19 (9H, s), 2.31 (2H, t), 2.22 (2H, m), 2.12 (2H, m) 1.61 (2H, q), 1.31 (25H, s), 0.90 (3H, t). (ESI+) *m/z*: 400. HPLC purity: 96%.

2.5.5. EPA Amide-Lyso (5)

^{1}H-NMR (600 MHz, CD$_3$OD), δ: 1.01 (3H, t, J = 7 Hz), 1.72 (2H, q), 2.02(1H, m), 2.14 (4H, m), 2.22 (1H, m), 2.25 (2H, t, J = 7 Hz), 2.87 (8H, m), 3.25 (10H, s), 3.30 (1H, m), 3.37 (1H, m), 3.62 (1H, dd, J = 7 Hz), 5.40 (10H, m). (ESI+) *m/z*: 445. HPLC purity: >98%.

2.5.6. Oleic Amide-Lyso (6)

^{1}H-NMR (600 MHz, CD$_3$OD), δ: 0.94 (3H, t, J = 7 Hz), 1.36 (21H, m), 1.65 (2H, m), 2.05 (5H, m), 2.23 (2H, t, J = 7 Hz), 2.27 (1H, m), 3.25 (9H, s), 3.30 (1H, m), 3.38 (1H, m), 3.61 (1H, dd, J = 7 Hz), 5.38 (2H, m). (ESI+) *m/z*: 425. HPLC purity: >98%.

2.6. Biological Assays

2.6.1. Effect of the Lipids on rePON1 Lactonase and Arylesterase Activities

The synthesized and purchased lyso-DGTS derivatives and analogs (final concentration 10 and 20 µg/mL) were incubated with rePON1 for 2 and 4 h at 37 °C in Tris buffer pH 8.4 containing 1 mM CaCl$_2$. These concentrations were chosen in accordance with our previous published studies which showed that, in these concentrations, lyso-DGTS has a positive effect on PON1 and HDL activities in vitro and in vivo without inducing a cytotoxic response [23–25].

Lactonase and arylesterase activities were performed using our previous procedure [24]. Briefly, a 10-µL aliquot of each solution was added to a 96-well UV microplate containing 90 µL Tris buffer and 1 mM CaCl$_2$ in each well. Then, 100 µL of 2 mM dihydrocoumarin (for lactonase activity) or phenylacetate (for arylesterase activity) was added.

The substrate hydrolysis rate was measured at 270 nm every 15 s for 10 min using a Tecan Infinite 200 PRO plate reader. Nonenzymatic hydrolysis of the substrate was subtracted from the total rate of hydrolysis. One unit of activity was equal to the hydrolysis of 1 µmol dihydrocoumarin or phenylacetate per minute.

2.6.2. Tryptophan (Trp) Fluorescence-Quenching Measurements

The assay was performed as previously published in our studies [25,26]. Briefly, in 200 µL Tris buffer in a black 96-well plate, 2 µL of bioactive lipid (in DMSO) at various final concentrations (1, 5, 10, 25, 50, 75, and 100 µg/mL) and rePON1 (final concentration 50 µg/mL) were incubated at 37 °C or 25 °C for 4 h; then, fluorescence measurements were performed in a plate reader (Tecan Infinite 200 PRO). The slit widths for both excitation and emission were set to 5 nm. Emission spectra were recorded from 300 to 450 nm with an excitation wavelength of 290 nm. The relative change in fluorescence (Δf) was calculated as $(Ix - Io)/Io$, where Ix and Io are the fluorescence intensities of rePON1 in the presence and absence of the bioactive lipid, respectively.

2.7. Molecular Docking

The crystal structure of PON1 (1V04) was retrieved from the Protein Data Bank (PDB). The enzyme was prepared for docking using the AutoDock Tools program, an accessory program that allows the user to interact with AutoDock 4.2 from a graphical user interface. Water molecules were removed from the PDB file. Polar hydrogen atoms were added and Kollman united atom charges were assigned. Ligand docking was carried out with the AutoDock 4.2 Lamarckian Genetic Algorithm (GA-LS).

2.8. Cu^{2+}-Induced LDL Oxidation

Oxidation of LDL was carried out in a UV ELISA plate at 37 °C and measured using a Tecan Infinite 200 PRO plate reader. To each well 10 µL of rePON1 (0.005 mg/mL in PBS pH 7.4) incubated with the tested compounds (final concentration of 20 µg/mL in PBS pH 7.4 with 0.1% *v/v* DMSO) for 2 h at 37 °C or 2 µL of the tested compounds (final concentration of 20 µg/mL in PBS pH 7.4 with 0.1% DMSO) without rePON1 was added. Then, 6 µL LDL (final concentration of 0.05 mg protein/mL in PBS pH 7.4) was added to the reaction. Finally, 10 µL $CuSO_4$ (10 µM in DDW) was added and PBS was added to a final volume of 100 µL. LDL oxidation was determined via continuous monitoring (every 5 min for 3 h) of conjugated diene formation at 234 nm. Lag time and rate constant (K) of the propagation phase were calculated via nonlinear regression using the Gompertz growth equation. Percent inhibition of LDL oxidation was calculated as: $100 - (K(LDL + sample)/K(LDL) \times 100)$.

3. Results

3.1. Synthesis of Lyso-DGTS Derivatives

Five derivatives of lyso-DGTS were synthesized by binding different fatty acids—EPA, oleic acid, or palmitic acid—directly to the N-methylated homoserine via ester or amide bonds (compounds **2–6**; Figure 1A). In addition, six endogenous lipids which are analogs of lyso-DGTS were purchased (compounds **7–11**; Figure 1B). The effect of the lipids on the hydrolytic activities of the antioxidant enzyme PON1 and their ability to prevent LDL oxidation induced by copper ion were investigated, and SARs were determined to better understand the lipid moieties responsible for enhancing PON1 activities and preventing LDL oxidation.

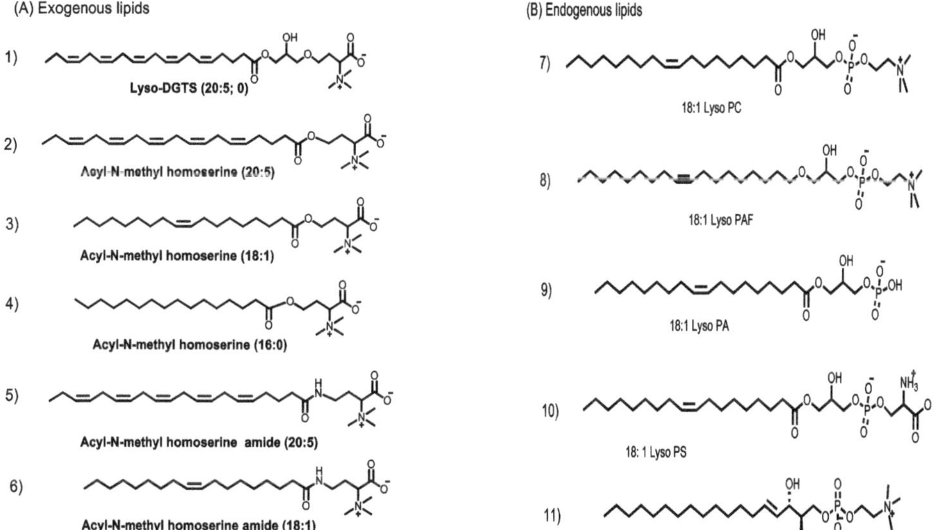

Figure 1. Synthesized derivatives of Lyso-DGTS (**A**) and lyso-DGTS analogs were purchased from (Sigma–Aldrich–Merck) (**B**). PC, phosphatidylcholine; PAF, platelet-activating factor; PA, phosphatidic acid; PS, phosphatidylserine; SM, sphingomyelin.

3.2. Effect of Lyso-DGTS Derivatives and Endogenous Analogs on Lactonase Activity of rePON1

The effect of the synthesized lyso-DGTS derivatives and analogs on rePON1 lactonase activity was examined and compared to that of natural lyso-DGTS isolated from *Nannochloropsis* microalgae. RePON1 lactonase activity was significantly increased, in a dose dependent manner, when incubated with the natural lyso-DGTS for 2 h at 37 °C—from 5 units to 10 units and 12 units after incubation with lyso-DGTS at a concentration of 10 and 20 µg/mL, respectively. Similarly, incubation of rePON1 with the synthesized compounds **4**, **5**, and **6** at concentrations of 10 and 20 µg/mL increased its lactonase activity to 10 and 13 units, respectively. Compounds **2** and **3** did not affect the lactonase activity of rePON1—it was similar to rePON1 without lipids (Figure 2A). Incubation of rePON1 with the free fatty acids EPA, oleic acid, palmitic acid, and L-homoserine betaine also had no effect on its lactonase activity (data not shown).

(A) rePON1 lactonase activity (synthesized derivatives of lyso-DGTS)

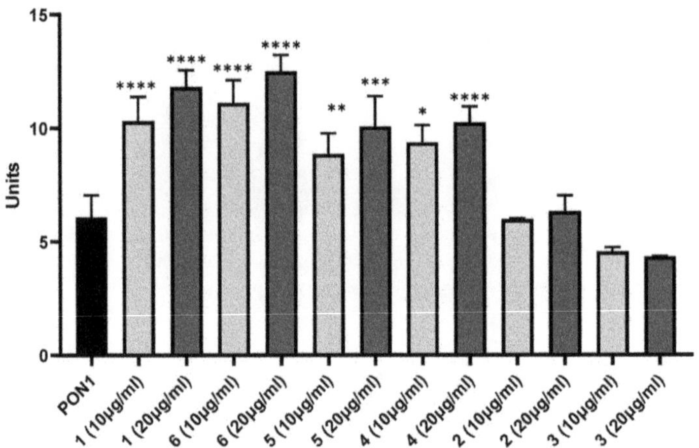

(B) rePON1 lactonase activity (endogenous analogs of lyso-DGTS)

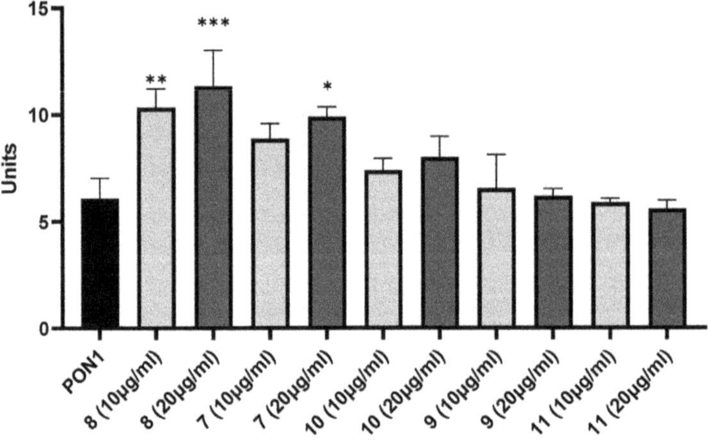

Figure 2. Lactonase activity of rePON1 after incubation with (**A**) synthesized derivatives of lyso-DGTS and (**B**) endogenous analogs of lyso-DGTS. RePON1 (5 μg/mL) was incubated with the lipid at concentrations of 10 and 20 μg/mL for 2 h at 37 °C in Tris buffer pH 8.4. Then, lactonase activity of rePON1 was measured via dihydrocoumarin assay and the results were expressed as units defined by the ability of rePON1 to hydrolyze 1 μM dihydrocoumarin in 1 min. Each experiment was repeated separately at least three times. * $p \leq 0.05$, ** $p \leq 0.01$, *** $p \leq 0.001$, **** $p \leq 0.0001$ relative to the control (PON1; rePON1 without bioactive lipid). Statistical analysis was via one-way ANOVA and Graphpad prism 9 software. See Figure 1 for compounds corresponding to each number.

The effect of the purchased endogenous lipids on rePON1 lactonase activity was also examined. Only two lipids—lyso-PAF (**8**) and lyso-PC (**7**)—increased the lactonase activity of rePON1 in a dose-dependent manner, from 5 units to 9 units and 13 units when the enzyme was incubated with the lipids at a concentration of 10 and 20 μg/mL, respectively. The other endogenous lipids—lyso-PS (**10**), lyso-PA (**9**), and lyso-SM (**11**)—did not affect rePON1 lactonase activity (Figure 2B).

3.3. Effect of Lyso-DGTS Derivatives and Endogenous Analogs on Arylesterase Activity of rePON1

The effect of the synthesized lyso-DGTS derivatives and analogs on rePON1 arylesterase activity was also examined. The arylesterase activity of rePON1 was significantly increased in a dose-dependent manner when incubated with the exogenous compounds **1** and **6** for 2 h at 37 °C, from 4 units without the lipid to 8 units at a concentration of 10 μg/mL and 10 units when incubated with 20 μg/mL. Compounds **2**, **3**, **4**, and **5** had no effect on the arylesterase activity of rePON1 compared to the control activity (Figure 3A).

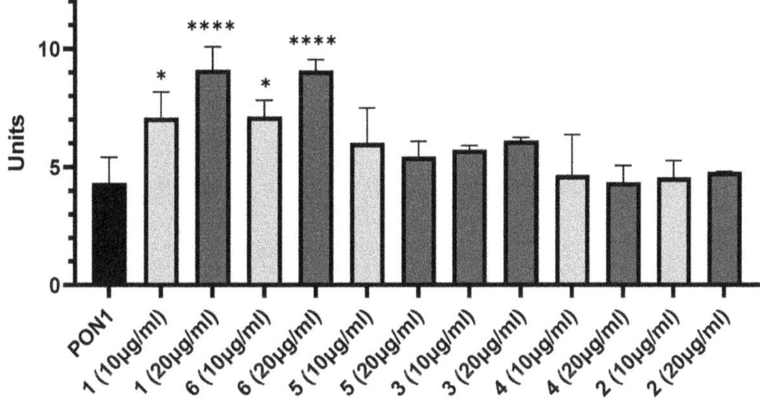

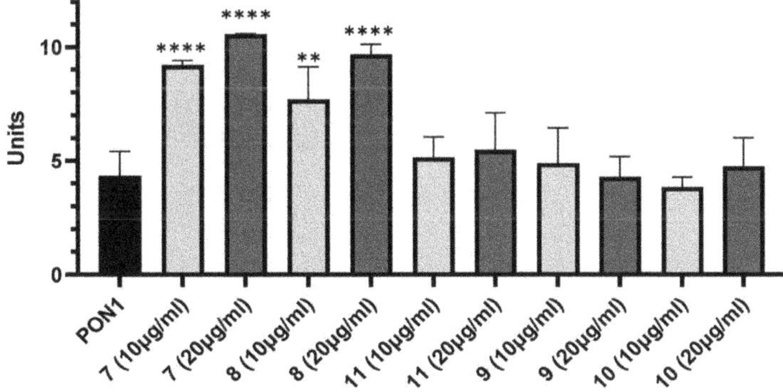

Figure 3. Arylesterase activity of rePON1 after incubation with (**A**) synthesized derivatives of lyso-DGTS and (**B**) endogenous analogs of lyso-DGTS. rePON1 (5 μg/mL) was incubated with the lipid at concentrations of 10 and 20 μg/mL for 2 h at 37 °C in Tris buffer pH 8.4. Then, arylesterase activity of rePON1 was measured via phenylacetate assay and the results were expressed as units defined by the ability of rePON1 to hydrolyze 1 mM phenylacetate in 1 min. Each experiment was repeated separately at least three times. * $p \leq 0.05$, ** $p \leq 0.01$, **** $p \leq 0.0001$ relative to the control (PON1; rePON1 without bioactive lipid). Statistical analysis was via one-way ANOVA and Graphpad prism 9 software. See Figure 1 for compounds corresponding to each number.

Only two purchased endogenous lipids—lyso-PC (**7**) and lyso-PAF (**8**)—increased re-PON1 arylesterase activity in a dose-dependent manner to about 10 units at a concentration of 10 µg/mL and to 12 units at a concentration of 20 µg/mL. The other endogenous lipids—lyso-PS (**10**), lyso-PA (**9**), and lyso-SM (**11**)—did not affect PON1 arylesterase activity (Figure 3B).

3.4. Interaction of Lyso-DGTS Derivatives and Endogenous Analogs with rePON1, Determined via the Trp Fluorescence-Quenching Method

Possible interactions between the bioactive lipids and rePON1 were examined by measuring the effect of the lipids on the fluorescence spectrum of Trp residues in rePON1. As demonstrated in Figure 4, three lipids—lyso-PA (**9**), lyso-PAF (**8**), and compound **6**—enhanced the Trp fluorescence intensity of rePON1 in a concentration-dependent manner up to 20 µM. However, three different lipids—the natural lyso-DGTS (**1**), compound **2**, and compound **3**—quenched the Trp fluorescence of rePON1 in a concentration-dependent manner. The other lipids, including lyso-PS (**10**), did not affect the Trp fluorescence of rePON1 (Figure 4).

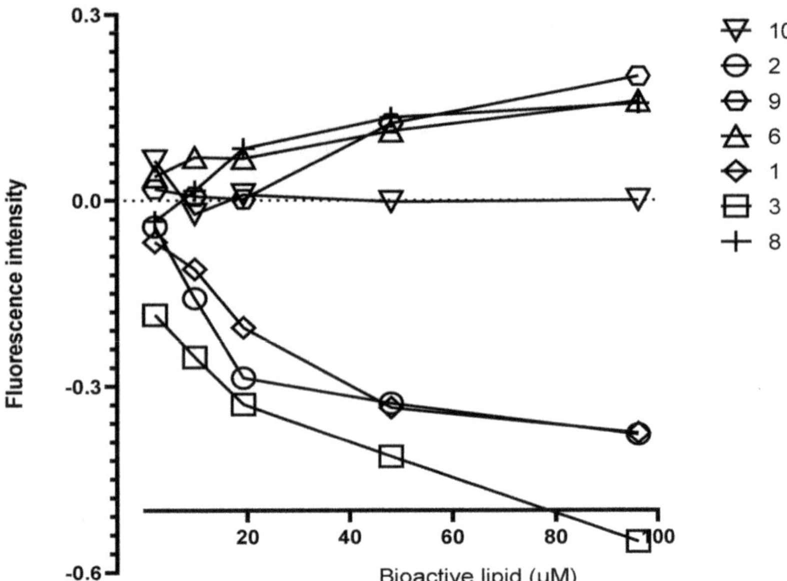

Figure 4. Effect of bioactive lipids on the Trp fluorescence of rePON1. rePON1 (50 µg/mL) was incubated with different bioactive lipids at various concentrations (0, 1, 5, 10, 25, 50, 75, or 100 µg/mL) in Tris buffer (pH 8.4). The fluorescence spectra were measured at an excitation of 290 nm and emission in the 320–340 nm range. Change in fluorescence was calculated as described in Experimental Procedures. Each experiment was repeated separately at least three times. See Figure 1 for compounds corresponding to each number.

3.5. Interaction between rePON1 and Lyso-DGTS Derivatives and Endogenous Analogs Using Molecular Modeling Calculation

PON1's crystal structure was adjusted, and the lipids were constructed and docked on the enzyme's crystallographic structure as described in Experimental Procedures. All of the bioactive lipids entered the same PON1 groove nearest the α-helix (H2), with very similar orientation, as shown in Figure 5. The fatty acid substructure of the lipid was oriented to the hydrophobic part of the groove (blue) and the polar substructure was oriented to the hydrophilic part (red). The inverse correlation ($R^2 = 0.5578$, $p < 0.05$) between the lactonase activity of rePON1 (X-axis) and the theoretical docking energy of the interactions between rePON1 and the lipids obtained from the docking calculations (Y-axis) are shown in Figure 6 (lower docking energy represents stronger enzyme–lipid interaction).

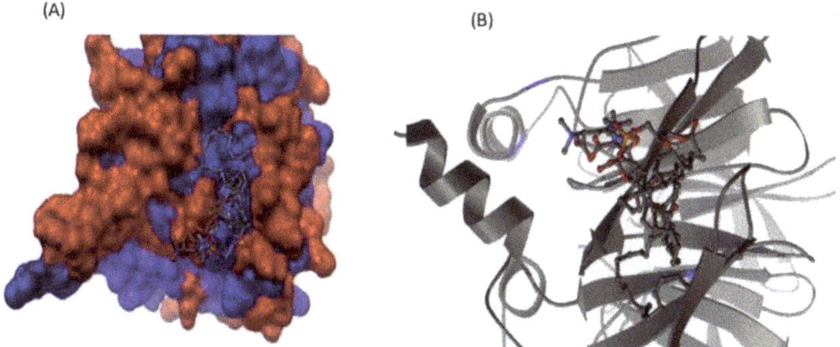

Figure 5. Some of the lipids in the rePON1 groove are orientated nearest α-helix H2. (**A**) The molecular surface of PON1 shows hydrophilic (red) and hydrophobic (blue) zones docked with the lipid. (**B**) Location of the lipid within the rePON1 groove showing interactions with the rePON1 groove using the secondary and tertiary structure of rePON1.

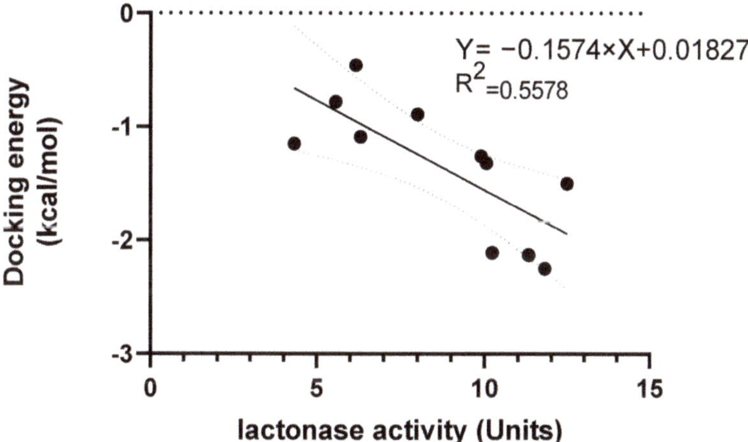

Figure 6. Correlation between the binding energy calculated from the AutoDock Tools program (Y-axis) and the rate at which PON1 catalyzes lactone hydrolysis (X-axis). The anticipated inverse correlation was obtained, with $R^2 = 0.5578$ ($p < 0.05$).

3.6. Effect of rePON1–Lyso-DGTS Derivative and Analog Complexes on Cu^{2+}-Induced LDL Oxidation

The ability of the bioactive lipids, with or without rePON1, to prevent LDL oxidation induced by copper ions was examined. LDL oxidation is a radical reaction characterized by two parameters, lag time and propagation phase, which were calculated using nonlinear regression with the Gompertz growth equation. A representative kinetic analysis of LDL oxidation induced by the copper ions is shown in Figures 7 and 8, and the oxidation parameters are summarized in Tables 1 and 2. The LDL oxidation lag time increased from 24.2 min to 96.8, 99.4, 35.37, 29.37, and 36.25 min after incubation with lyso-DGTS (**1**) and the synthesized compounds **5**, **6**, **2**, and **4**, respectively. In addition, LDL oxidation was inhibited by 75%, 75%, 31.5%, 17.5%, and 33.16% after incubation with lyso-DGTS (**1**) and the synthesized compounds **5**, **6**, **2**, and **4**, respectively (Table 1). Compound **4** and the endogenous compounds **7** to **11** had no protective effect on LDL oxidation induced by copper ions (Figure 7B).

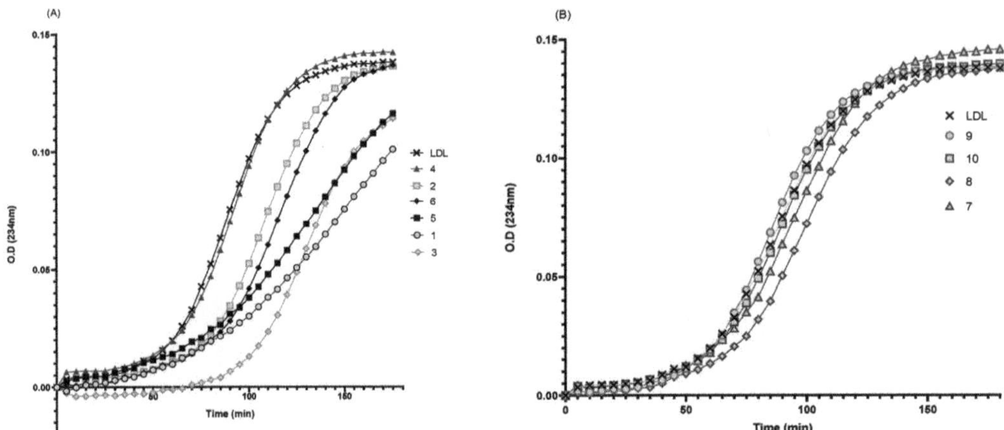

Figure 7. Effect of bioactive lipids on the oxidation of LDL induced by Cu^{2+}. (**A**) The synthesized exogenous lipids 1–6 and (**B**) The purchased endogenous lipids 7–10. Conjugated diene formation during copper-mediated LDL oxidation was monitored for 180 min for changes in absorbance at 234 nm. Each experiment was repeated separately three times. Results are presented as mean ± SD. See Figure 1 for compounds corresponding to each number.

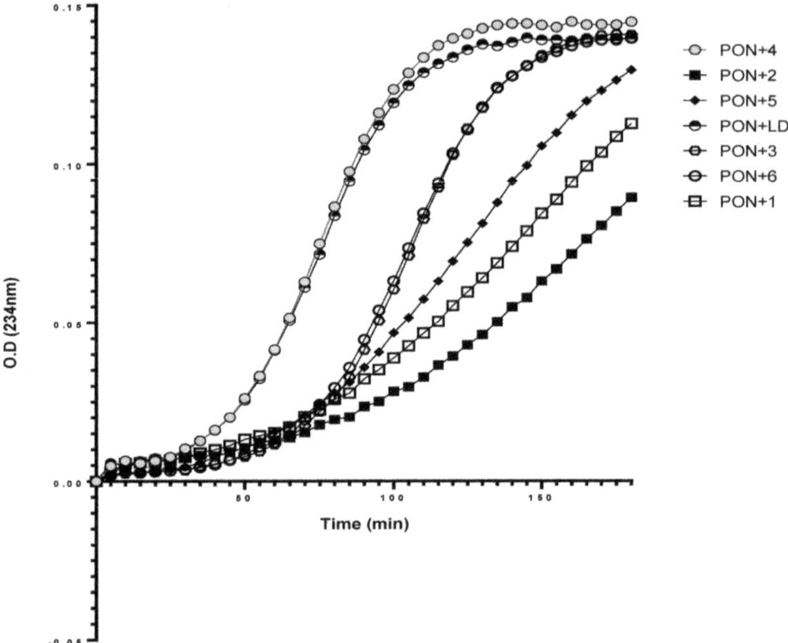

Figure 8. Effect of rePON1 with the exogenous bioactive lipids on the oxidation of LDL induced by Cu^{2+}. Conjugated diene formation during copper-mediated LDL oxidation was monitored for 180 min for changes in absorbance at 234 nm. Each experiment was repeated separately at least three times. Results are presented as mean ± SD. See Figure 1 for compounds corresponding to each number.

Table 1. Effect of the synthesized lipids on lag time, rate constant of the reaction propagation (K), and % inhibition of LDL oxidation. See Figure 1 for compounds corresponding to each number.

Compound	Lag Time	K	% Inhibition
LDL	24.22	0.04128	0
1	96.8	0.0103	75.04
5	99.4	0.01006	75.62
6	35.37	0.02827	31.51
2	29.37	0.03405	17.51
4	25.56	0.03913	5.20
3	80.50	0.02759	33.16

Table 2. Effect of rePON1 in a complex with the synthesized lipids on lag time, rate constant of the oxidation propagation (K), and % inhibition of LDL oxidation. See Figure 1 for compounds corresponding to each number.

Compound	Lag Time	K	% Inhibition
LDL	24.08	0.04152	0
1	118	0.00844	79.67
5	67.87	0.01473	64.52
6	29.65	0.03373	18.76
2	199	0.005013	87.9
4	23.73	0.04215	0
3	90.30	0.03481	16.16

RePON1 was incubated with the exogenous lipids lyso-DGTS (**1**), **2**, **3**, **5**, and **6**, and its protective effect on LDL oxidation was also examined (Figure 8). As shown in Table 2, the complex of rePON1 with lyso-DGTS (**1**) inhibited LDL oxidation by 79%, and compounds **5**, **6**, **2**, and **3** inhibited LDL oxidation by 64%, 31%, 87%, and 16%, respectively. Compound **4** did not inhibit LDL oxidation. In addition, LDL oxidation lag time increased from 24.08 to 118, 67.87, 29.65, 199, and 28.73 min after incubation with rePON1 and lyso-DGTS (**1**) or the synthesized compounds **5**, **6**, **2**, and **4**, respectively.

4. Discussion

PON1 is a calcium-dependent HDL-bound enzyme that catalyzes the hydrolysis of multiple compounds, such as paraoxon (paraoxonase activity), arylesterase (arylesterase activity), and lactones (lactonase activity). Many of the antiatherogenic functions of HDL are attributed to PON1 [27]. PON1 enhances HDL cholesterol-mediated efflux from macrophages, protects LDL from oxidation by lowering lipid peroxide levels, inhibits ox-LDL uptake by macrophages thus inhibiting macrophage foam cell formation, and inhibits macrophage cholesterol biosynthesis [28]. PON1 antioxidant activity is inversely correlated to carotid intima-media thickness. The hydrolytic lactonase, arylesterase, and paraoxonase activities of PON1 are all inactivated under oxidative stress, and epidemiological evidence shows that low serum PON1 activity is associated with an increased risk of CVD [29]. Thus, enhancement of PON1 activity could be a useful approach to attenuating atherosclerosis development.

A previous study showed that phosphatidylcholine (PC) (16:0/18:2) specifically interacts with rePON1 and increases its lactonase and paraoxonase activities [28]. Recently, a betaine lipid derivative composed of EPA (C20:5) fatty acid connected to glyceryltrimethyl-homoserine (C20:5 lyso DGTS lipid) was isolated from an ethanol–water (70:30%) extract of *Nannochloropsis* sp. microalgae and was found to increase rePON1 lactonase and esterase activities [23]. In the present study, five new derivatives of lyso-DGTS were synthesized, and five additional endogenous lyso-lipids were purchased, and their effects on rePON1 lactonase and arylesterase activities were examined. The role of the glycerol moiety in lyso-DGTS activity was examined by connecting the fatty acids EPA, oleic acid, or palmitic acid directly to the N-methylated homoserine via an ester bond (compounds **2**, **3**, and **4**, respectively). In the two other synthesized compounds, EPA and oleic acid were connected directly to the N-trimethylhomoserine via an amide bond (**5** and **6**, respectively).

The natural compound lyso-DGTS (**1**) and the synthesized compound **6** strongly increased lactonase and esterase activities of rePON1, more than the other synthesized compounds. Compounds **4** and **5** increased the lactonase activity of rePON1, but to a lower level than compounds **1** and **6**. In contrast, compounds **2** and **3** did not affect the lactonase or esterase activities of rePON1 (Figures 2 and 3). These results demonstrated that connecting the fatty acids oleic and EPA directly to the N-trimethylhomoserine through an ester bond without the glyceryl moiety attenuates its ability to enhance lactonase and esterase activities of rePON1; however, connecting it through an amide bond preserved these abilities, most probably due to the higher stability of the amide bond compared to its ester analogue. The lyso-DGTS amide derivative with oleic acid enhanced the lactonase and esterase activities of rePON1 more than that with EPA. It is worth noting that incubation of rePON1 with free fatty acids (EPA, oleic acid, or palmitic acid) or N-trimethylhomoserine did not have any effect on its lactonase and esterase activities (data not shown).

Five endogenous active lipids analogous to lyso-DGTS were purchased and their effect on the lactonase and esterase activities of rePON1 were examined. The structures of lyso-PC and lyso-PAF (platelet-activating factor) are similar to that of lyso-DGTS; however, their negative charge is located on the phosphate oxygen instead of the carboxylic acid moiety. Lyso-PC differs from lyso-PAF in that the ester group is replaced by an ether group in the latter. Lyso-PA (phosphatidic acid) and lyso-PS (phosphatidylserine) are negatively charged lipids, whereas lyso-SM (sphingomyelin) is a positively charged lipid under physiological conditions. Therefore, the effect of the lipid charge on rePON1 activities was also examined. Two lipids, lyso-PC and lyso-PAF (compounds **7** and **8**, respectively), enhanced lactonase and esterase activities of rePON1 by the same level, meaning that replacing the ester bond in the glycerol moiety with an ether bond did not affect the ability to enhance rePON1 activities. In contrast, the other lipids (compounds **9**, **10**, and **11**) did not affect rePON1 activities. These results demonstrate that phospholipids with a neutral charge under physiological conditions (compounds **7** and **8**) better enhance rePON1 activities than the negatively or positively charged phospholipids (compounds **9**, **10**, and **11**). Thus, lyso-DGTS lipid derivatives and analogs have specific interactions with PON1 with respect to enhancing its activity, independent of their emulsification properties.

PON1 is known to contain three helical regions; of these, two helices (H1 and H2) can associate with amphiphilic phospholipids. Three out of four Trp residues reside in or close to helical regions H2 and H3 of PON1 [28,30]. Therefore, a Trp fluorescence-quenching assay was carried out to determine possible interactions between rePON1 and the lyso-DGTS derivatives and analogs. The fluorescence-quenching behavior varied widely, from no effect (or even slight enhancement) to strong quenching, up to 100%. As demonstrated in Figure 4, lyso-PA (**9**), lyso-PAF (**8**), and compound **6** enhanced the Trp fluorescence intensity of rePON1 in a concentration-dependent manner, suggesting that some Trp residue-containing regions of rePON1 might be exposed to a relatively low level of these compounds. However, three different lipids, the natural lyso-DGTS (**1**), compound **2**, and compound **3**, quenched the Trp fluorescence of rePON1 in a concentration-dependent manner, suggesting a specific interaction between PON1 and these lipids. Lyso-PS (**10**) did not have any quenching effect. There was no correlation between the tested activities of rePON1 and the change in Trp fluorescence.

It was previously found that a specific PC interacts with PON1 near α-helix H2 [28]. The results obtained in the present study from the molecular docking analysis also showed that the examined lipids interact near the H2 site. There was a significant correlation between the lactonase activity of rePON1 and the binding energy (ΔG) obtained from molecular docking. H1 at the N-terminus of PON1 mediates its anchoring to HDL, and together, H2 and H1 form a hydrophobic patch that binds to the HDL surface. In addition to its role in HDL binding, H2 comprises part of the active site wall of PON1. The interaction site of the lipids with PON1 was near the α-helix H2. It is possible that this binding of the lipids to the allosteric site of PON1, near H2, might change its active-site conformation and enhance its affinity to the substrates, thereby elevating its activity.

Oxidation of LDL lipids and apolipoprotein B100 results in LDL in a proatherogenic process [31]. A clinical study suggested that a high ox-LDL concentration and a short LDL oxidation lag time might be independent risk factors for CAD; serum ox-LDL concentration was significantly elevated in CAD patients and the lag time of LDL oxidation was significantly shorter compared to controls [32]. We examined the ability of the bioactive lipids and their complexes with rePON1 to prevent LDL oxidation induced by copper ions. Two parameters were measured: the ability of lipids and the rePON1–lipid complex to inhibit LDL oxidation—represented by the lag time required for initiation of LDL oxidation and the reduction in lipoprotein-associated peroxides formed during 3 h of incubation—represented by the slope of the linear increase of the curve. As shown in Figure 7B, the endogenous lipids did not have any additional protective effect on Cu^{2+}-induced LDL oxidation. Thus, lipids **7**, **8**, **9** and **10**, with fatty acids based on oleic acid, did not show any effect on antioxidant capacity. On the other hand, when the bioactive lipid was based on EPA (e.g., **1** and **5**), it increased the lag time by up to 95 min, and LDL oxidation was inhibited by about 75%, with or without incubation with rePON1.

EPA exhibited the most favorable molecular structure; the multiplicity of double bonds enhanced its ability to prevent LDL oxidation, as expected, and their absence had no effect on LDL oxidation (such as lipids **4** and **11**). A previous study showed that EPA has the optimal chain length and degree of unsaturation to inhibit oxidation of small dense LDL and membrane cholesterol domains [33].

5. Conclusions

In conclusion, the current study showed that the interaction of lyso-DGTS lipid and some of its derivatives with PON1 increases the latter's activity, improves its lactonase and arylesterase activities, and prevents oxidation of LDL induced by copper ion. A clear understanding of the mechanism underlying the lipids' role as antiatherogenic compounds, via the influence of their binding on the structural and biological activities of proteins (PON1) and lipoproteins, is key to the evaluation of these potent biomolecules' contribution to human health and is a topic of active investigation in our laboratory.

Author Contributions: Conceptualization, S.K.; methodology, A.K., S.M. and S.K.; validation, A.K., S.M., M.H., T.H. and S.K.; formal analysis, A.K.; investigation, A.K., S.M., M.H., T.H. and S.K.; data curation, A.K., S.M., M.H., T.H. and S.K.; writing—original draft preparation, A.K.; writing—review and editing, A.K., S.M., M.H., T.H. and S.K.; visualization, A.K. and S.K.; supervision, T.H. and S.K. All authors have read and agreed to the published version of the manuscript.

Funding: This research received no external funding.

Institutional Review Board Statement: Not applicable.

Informed Consent Statement: Not applicable.

Data Availability Statement: The data presented in this study are available in the article.

Conflicts of Interest: The authors declare no conflict of interest.

References

1. Libby, P.; Buring, J.E.; Badimon, L.; Hansson, G.K.; Deanfield, J.; Bittencourt, M.S.; Tokgozoglu, L.; Lewis, E.F. Atherosclerosis. *Nat. Rev. Dis. Prim.* **2019**, *5*, 56. [CrossRef] [PubMed]
2. Pirillo, A.; Danilo Norata, G.; Luigi Catapano, A. Treating High Density Lipoprotein Cholesterol (HDL C): Quantity Versus Quality. *Curr. Pharm. Des.* **2013**, *19*, 3841–3857. [CrossRef] [PubMed]

3. Steinberg, D.; Parthasarathy, S.; Carew, T.E.; Khoo, J.C.; Witztum, J.L. Modifications of low-density lipoprotein. *N. Engl. J. Med.* **1989**, *320*, 915–924. [PubMed]

4. Hafiane, A.; Genest, J. HDL, Atherosclerosis, and Emerging Therapies. *Hindawi* **2013**, *2013*, 891403. [CrossRef]

5. Rosenson, R.S.; Brewer, H.B.; Ansell, B.; Barter, P.; Chapman, M.J.; Heinecke, J.W.; Kontush, A.; Tall, A.R.; Webb, N.R. Translation of high-density lipoprotein function into clinical practice: Current prospects and future challenges. *Circulation* **2013**, *128*, 1256–1267. [CrossRef]

6. Rosenson, R.S.; Brewer, H.B.; Ansell, B.J.; Barter, P.; Chapman, M.J.; Heinecke, J.W.; Kontush, A.; Tall, A.R.; Webb, N.R. Dysfunctional HDL and atherosclerotic cardiovascular disease. *Nat. Rev. Cardiol.* **2016**, *13*, 48–60. [CrossRef] [PubMed]

7. Gür, M.; Çayli, M.; Uçar, H.; Elbasan, Z.; Şahin, D.Y.; Gözükara, M.Y.; Selek, Ş.; Koyunsever, N.Y.; Şeker, T.; Türkoğlu, C.; et al. Paraoxonase (PON1) activity in patients with subclinical thoracic aortic atherosclerosis. *Int. J. Cardiovasc. Imaging* **2014**, *30*, 889–895. [CrossRef] [PubMed]

8. Gugliucci, A.; Caccavello, R.; Nassar, H.; Ahmad, W.A.; Sinnreich, R.; Kark, J.D. Low protective PON1 lactonase activity in an Arab population with high rates of coronary heart disease and diabetes. *Clin. Chim. Acta* **2015**, *445*, 41–47. [CrossRef]

9. Guns, P.J.; Van Assche, T.; Verreth, W.; Fransen, P.; Mackness, B.; Mackness, M.; Holvoet, P.; Bult, H. Paraoxonase 1 gene transfer lowers vascular oxidative stress and improves vasomotor function in apolipoprotein E-deficient mice with pre-existing atherosclerosis. *Br. J. Pharmacol.* **2008**, *153*, 508–516. [CrossRef]

10. Tavori, H.; Khatib, S.; Aviram, M.; Vaya, J. Bioorganic & Medicinal Chemistry Characterization of the PON1 active site using modeling simulation, in relation to PON1 lactonase activity. *Bioorganic Med. Chem.* **2008**, *16*, 7504–7509. [CrossRef]

11. Bargota, R.S.; Akhtar, M.; Biggadike, K.; Gani, D.; Allemann, R.K. Structure-activity relationship on human serum paraoxonase (PON1) using substrate analogues and inhibitors. *Bioorganic Med. Chem. Lett.* **2003**, *13*, 1623–1626. [CrossRef]

12. Rosenblat, M.; Gaidukov, L.; Khersonsky, O.; Vaya, J.; Oren, R.; Tawfik, D.S.; Aviram, M. The Catalytic Histidine Dyad of High Density Lipoprotein- associated Serum Paraoxonase-1 (PON1) Is Essential for PON1-mediated Inhibition of Low Density Lipoprotein Oxidation and Stimulation of Macrophage Cholesterol Efflux. *J. Biol. Chem.* **2006**, *281*, 7657–7665. [CrossRef] [PubMed]

13. Ben-David, M.; Elias, M.; Filippi, J.; Duñach, E.; Silman, I.; Sussman, J.L.; Tawfik, D.S. Catalytic Versatility and Backups in Enzyme Active Sites: The Case of Serum Paraoxonase 1. *J. Mol. Biol.* **2012**, *418*, 181–196. [CrossRef] [PubMed]

14. Mackness, M.; Mackness, B. Human paraoxonase-1 (PON1): Gene structure and expression, promiscuous activities and multiple physiological roles. *Gene* **2015**, *567*, 12–21. [CrossRef] [PubMed]

15. Ng, D.S.; Chu, T.; Esposito, B.; Hui, P.; Connelly, P.W.; Gross, P.L. Paraoxonase-1 deficiency in mice predisposes to vascular inflammation, oxidative stress, and thrombogenicity in the absence of hyperlipidemia. *Cardiovasc. Pathol.* **2008**, *17*, 226–232. [CrossRef]

16. Shekhanawar, M.; Shekhanawar, S.M.; Krisnaswamy, D.; Indumati, V.; Satishkumar, D.; Vijay, V.; Rajeshwari, T.; Amareshwar, M. The role of "paraoxonase-1 activity" as an antioxidant in coronary artery diseases. *J. Clin. Diagnostic Res.* **2013**, *7*, 1284–1287. [CrossRef]

17. Ferretti, G.; Bacchetti, T. Effect of dietary lipids on paraoxonase-1 activity and gene expression. *Nutr. Metab. Cardiovasc. Dis.* **2012**, *22*, 88–94. [CrossRef]

18. Freese, R.; Alfthan, G.; Jauhiainen, M.; Basu, S.; Erlund, I.; Salminen, I.; Aro, A.; Mutanen, M. High intakes of vegetables, berries, and apples combined with a high intake of linoleic or oleic acid only slightly affect markers of lipid peroxidation and lipoprotein metabolism in healthy subjects. *Am. J. Clin. Nutr.* **2002**, *76*, 950–960. [CrossRef]

19. Fuhrman, B.; Volkova, N.; Aviram, M. Pomegranate juice polyphenols increase recombinant paraoxonase-1 binding to high-density lipoprotein: Studies in vitro and in diabetic patients. *Nutrition* **2010**, *26*, 359–366. [CrossRef]

20. Betanzos-Cabrera, G.; Guerrero-Solano, J.A.; Martínez-Pérez, M.M.; Calderón-Ramos, Z.G.; Belefant-Miller, H.; Cancino-Diaz, J.C. Pomegranate juice increases levels of paraoxonase1 (PON1) expression and enzymatic activity in streptozotocin-induced diabetic mice fed with a high-fat diet. *Food Res. Int.* **2011**, *44*, 1381–1385. [CrossRef]

21. Atrahimovich, D.; Avni, D.; Khatib, S. Flavonoids-macromolecules interactions in human diseases with focus on alzheimer, atherosclerosis and cancer. *Antioxidants* **2021**, *10*, 423. [CrossRef] [PubMed]

22. Atrahimovich, D.; Samson, A.O.; Khattib, A.; Vaya, J.; Khatib, S. Punicalagin Decreases Serum Glucose Levels and Increases PON1 Activity and HDL Anti-Inflammatory Values in Balb/c Mice Fed a High-Fat Diet. *Oxid. Med. Cell. Longev.* **2018**, *2018*, 2673076. [CrossRef] [PubMed]

23. Khatib, S.; Artoul, F.; Paluy, I.; Boluchevsky, L.; Kvitnitsky, E.; Vaya, J. Nannochloropsis sp. ethanol extract prevents macrophage and LDL oxidation and enhances PON1 activity through the principal active compound lyso-diacylglyceryltrimethylhomoserine (lyso-DGTS). *J. Appl. Phycol.* **2018**, *30*, 1679–1689. [CrossRef]

24. Dahli, L.; Atrahimovich, D.; Vaya, J.; Khatib, S. Lyso-DGTS lipid isolated from microalgae enhances PON1 activities in vitro and in vivo, increases PON1 penetration into macrophages and decreases cellular lipid accumulation. *BioFactors* **2018**, *44*, 299–310. [CrossRef] [PubMed]

25. Khattib, A.; Atrahimovich, D.; Dahli, L.; Vaya, J.; Khatib, S. Lyso-diacylglyceryltrimethylhomoserine (lyso-DGTS) isolated from Nannochloropsis microalgae improves high-density lipoprotein (HDL) functions. *BioFactors* **2020**, *46*, 146–157. [CrossRef]

26. Atrahimovich, D.; Khatib, S.; Sela, S.; Vaya, J.; Samson, A.O. Punicalagin Induces Serum Low-Density Lipoprotein Influx to Macrophages. *Oxid. Med. Cell. Longev.* **2016**, *2016*, 7124251. [CrossRef]

27. Aviram, M.; Vaya, J. Paraoxonase 1 activities, regulation, and interactions with atherosclerotic lesion. *Curr. Opin. Lipidol.* **2013**, *24*, 339–344. [CrossRef]
28. Cohen, E.; Aviram, M.; Khatib, S.; Artoul, F.; Rabin, A. Free Radical Biology and Medicine Human carotid plaque phosphatidyl-choline speci fi cally interacts with paraoxonase 1, increases its activity, and enhances its uptake by macrophage at the expense of its binding to HDL. *Free Radic. Biol. Med.* **2014**, *76*, 14–24. [CrossRef]
29. Corsetti, J.P.; Sparks, C.E.; James, R.W.; Bakker, S.J.L.; Dullaart, R.P.F. Low serum paraoxonase-1 activity associates with incident cardiovascular disease risk in subjects with concurrently high levels of high-density lipoprotein cholesterol and c-reactive protein. *J. Clin. Med.* **2019**, *8*, 1357. [CrossRef]
30. Khersonsky, O.; Tawfik, D.S. Structure-reactivity studies of serum paraoxonase PON1 suggest that its native activity is lactonase. *Biochemistry* **2005**, *44*, 6371–6382. [CrossRef]
31. Niki, E. Do free radicals play causal role in atherosclerosis? Low density lipoprotein oxidation and vitamin E revisited. *J. Clin. Biochem. Nutr.* **2011**, *48*, 3–7. [CrossRef] [PubMed]
32. Nayeri, H.; Naderi, G.A.; Asgary, S.; Pakmehr, F.; Sadeghi, M.; Javadi, E. Correlation Between Lag Time of Ldl To in Vitro Oxidation and in Vivo Ox-Ldl in the Patients With Cad. *Atheroscler. Suppl.* **2010**, *11*, 130. [CrossRef]
33. Sherratt, S.C.R.; Juliano, R.A.; Mason, R.P. Eicosapentaenoic acid (EPA) has optimal chain length and degree of unsaturation to inhibit oxidation of small dense LDL and membrane cholesterol domains as compared to related fatty acids in vitro. *Biochim. Biophys. Acta Biomembr.* **2020**, *1862*, 183254. [CrossRef] [PubMed]

Article

Metabolic Alterations Identified in Urine, Plasma and Aortic Smooth Muscle Cells Reflect Cardiovascular Risk in Patients with Programmed Coronary Artery Bypass Grafting

Aranzazu Santiago-Hernandez [1], Paula J. Martinez [1], Marta Agudiez [1], Angeles Heredero [2], Laura Gonzalez-Calero [1], Alma Yuste-Montalvo [3], Vanesa Esteban [3,4,5], Gonzalo Aldamiz-Echevarria [2], Marta Martin-Lorenzo [1,†] and Gloria Alvarez-Llamas [1,6,*,†]

[1] Immunoallergy and Proteomics Laboratory, Department of Immunology, IIS-Fundacion Jimenez Diaz, UAM, 28040 Madrid, Spain; aranzazu_sant@hotmail.com (A.S.-H.); paula_jmg@hotmail.com (P.J.M.); magudiezperez18@gmail.com (M.A.); lauragon.k@gmail.com (L.G.-C.); marta.martin@fjd.es (M.M.-L.)
[2] Department of Cardiac Surgery, Fundacion Jimenez Diaz, UAM, 28040 Madrid, Spain; angeles.heredero@quironsalud.es (A.H.); galdamiz@quironsalud.es (G.A.-E.)
[3] Allergy and Inmunology Department, IIS-Fundacion Jimenez Diaz, UAM, 28040 Madrid, Spain; almamonyu@gmail.com (A.Y.-M.); vesteban@fjd.es (V.E.)
[4] Red de Asma, Reacciones Adversas y Alergicas, Instituto de Salud Carlos III, 28040 Madrid, Spain
[5] Faculty of Medicine and Biomedicine, Alfonso X El Sabio University, 28691 Madrid, Spain
[6] Red de Investigacion Renal (REDINREN), Instituto de Salud Carlos III, 28040 Madrid, Spain
*   Correspondence: galvarez@fjd.es; Tel.: +34-915504800 (ext. 2203)
†   Both senior authors contributed equally to this work.

Citation: Santiago-Hernandez, A.; Martinez, P.J.; Agudiez, M.; Heredero, A.; Gonzalez-Calero, L.; Yuste-Montalvo, A.; Esteban, V.; Aldamiz-Echevarria, G.; Martin-Lorenzo, M.; Alvarez-Llamas, G. Metabolic Alterations Identified in Urine, Plasma and Aortic Smooth Muscle Cells Reflect Cardiovascular Risk in Patients with Programmed Coronary Artery Bypass Grafting. *Antioxidants* 2021, *10*, 1369. https://doi.org/10.3390/antiox10091369

Academic Editors: Soliman Khatib and Dana Atrahimovich Blatt

Received: 3 August 2021
Accepted: 25 August 2021
Published: 27 August 2021

Publisher's Note: MDPI stays neutral with regard to jurisdictional claims in published maps and institutional affiliations.

Abstract: Atherosclerosis is the predominant pathology associated to premature deaths due to cardiovascular disease. However, early intervention based on a personalized diagnosis of cardiovascular risk is very limited. We have previously identified metabolic alterations during atherosclerosis development in a rabbit model and in subjects suffering from an acute coronary syndrome. Here we aim to identify specific metabolic signatures which may set the basis for novel tools aiding cardiovascular risk diagnosis in clinical practice. In a cohort of subjects with programmed coronary artery bypass grafting (CABG), we have performed liquid chromatography and targeted mass spectrometry analysis in urine and plasma. The role of vascular smooth muscle cells from human aorta (HA-VSMCs) was also investigated by analyzing the intra and extracellular metabolites in response to a pro-atherosclerotic stimulus. Statistically significant variation was considered if p value < 0.05 (Mann-Whitney test). Urinary trimethylamine N-oxide (TMAO), arabitol and spermidine showed higher levels in the CVrisk group compared with a control group; while glutamine and pantothenate showed lower levels. The same trend was found for plasma TMAO and glutamine. Plasma choline, acetylcholine and valine were also decreased in CVrisk group, while pyruvate was found increased. In the secretome of HA-VSMCs, TMAO, pantothenate, glycerophosphocholine, glutathion, spermidine and acetylcholine increased after pro-atherosclerotic stimulus, while secreted glutamine decreased. At intracellular level, TMAO, pantothenate and glycerophosphocholine increased with stimulation. Observed metabolic deregulations pointed to an inflammatory response together with a deregulation of oxidative stress counteraction.

Keywords: cardiovascular risk; atherosclerosis; chronic kidney disease; vascular smooth muscle cells; oxidative stress; metabolites; metabolomics; biomarkers

1. Introduction

Cardiovascular disease continues being the leading cause of premature death worldwide despite continuing efforts in primary prevention. The predominant pathology of cardiovascular disease is atherosclerosis, an immunoinflammatory disease of arterial vessels. Formation of the atheroma plaque and progressive arterial obstruction takes place silently and asymptomatically, making it extremely challenging to promptly identify those individuals at high cardiovascular risk to prevent a fatal event. Main risk factors are age, male gender, lipid profile, blood pressure, smoking habits and diabetes. Despite that over 50% of the observed decrease in mortality caused by coronary heart disease can be attributed to changes in those risk factors, further knowledge is needed to better stratify individual risk and improve prevention [1,2]. No early diagnostic markers are available to date that may undoubtedly predict future events on healthy subjects or stratify individual risk. In this clinical scenario, there is an urgent need to find out novel molecular indicators, alone or combined with existing ones, together with a better knowledge of the operating mechanisms in atherosclerosis development.

The metabolome reflects the ultimate response of an organism to a patho(physio)logical condition and provides an integrated profile of biological status and metabolic health. Metabolomics integrates the comprehensive analysis of low molecular weight molecules in biological systems, providing a global picture of individual response to intrinsic and extrinsic exposures to genetic, dietary, lifestyle and gut microbial factors [3]. Accordingly, defining the chemical phenotypes of health or disease using metabolic signatures is gaining attraction in cardiovascular risk stratification [4–7]. In this line, an improved prediction of cardiovascular events based on metabolic features mainly composed by intermediates of fatty acid oxidation has been shown [8]. Metabolic signatures associated with coronary artery disease plaque phenotypes [4] and correlating with severity of coronary stenosis [9] have been recently described. Previous studies from our group have evidenced metabolic patterns associated with increased cardiovascular risk in various clinical contexts. In particular, urinary metabolic alterations were identified in hypertensive subjects with increased cardiovascular risk (i.e., resistant hypertension or albuminuria development) [10,11]. Specific urinary panels were also shown to be altered during atherosclerosis development in a high fat diet animal model and in human subjects in response to an acute coronary syndrome [12]. In a different approach, the study of molecular mechanisms taking place directly in the atherosclerotic aorta revealed novel molecular actors pointing to oxidative stress and arterial remodeling [13]. In a step further in cardiovascular prevention, the aim of this study is to evaluate the potential of the previously identified metabolic signatures in association with cardiovascular risk in a cohort of subjects with programmed coronary arterial bypass grafting (CABG). Besides, a potential role for the vascular smooth muscle cells from human aorta (HA-VSMCs) in observed changes was also investigated.

2. Materials and Methods

2.1. Patient Selection and Samples Collection

A total of 27 patients undergoing CABG at Fundación Jiménez Díaz Hospital (Madrid, Spain) were recruited (CVrisk group). Inclusion criteria were diagnosis of coronary artery disease, and programmed CABG surgery. Exclusion criteria were previous open cardiac surgery, valvulopathy, neoplasia, emergency, hemodynamic instability, treatment with corticoids or immunosuppressants, active infection or in-hospital exitus. As control group (C), a cohort of 24 healthy subjects was recruited from Donation Unit at the same hospital without previous history of hypertension, diabetes or cardiovascular disease. Subjects' clinical characteristics are compiled in Table 1. A spot urine sample was collected from each participant in a sterile container. Blood samples were collected in EDTA tubes. Urine and blood samples were centrifuged and supernatants were collected and stored at −80 °C until analysis. For surgery patients, samples were collected at two timepoints: before surgery (CVrisk group) and at clinical follow-up after 3 months (recovery, R).

2.2. In Vitro Human Cell Cultures

Primary cultures of vascular smooth muscle cells were obtained from a whole aortic arterial segment of patient undergoing cardiac surgery (HA-VSMC) according to previously described [14]. The arterial segment was cleaned and cut into small pieces. They were then incubated with Dulbecco's modified Eagle's medium (DMEM-F12; Lonza Walkersville) containing 0.1% collagenase type 1 (Gibco) overnight at 37 °C. After incubation, the reaction was stopped with 0.5% fetal bovine serum (FBS) in DMEM-F12 and after centrifugation at 1200 rpm for 5 min, the explants were seeded together with the cells in DMEM-F12 medium supplemented with 0.1 mg/mL heparin (Merck Life Science), 30 µg/mL endothelial cell growth factor (ECGF; Merck Life Science), 1% penicillin/streptomycin (P/E; Lonza Walkersville), 1% L-glutamine (Lonza Walkersville), 10% Fungizone (Lonza Walkersville) and 20% FBS. Petri dishes used for seeding were pre-coated with 0.5% sterile gelatin. After 3–5 days of incubation, HA-VSMCs were negatively selected with the human CD31 antibody (BDbiosciences) and secondary antibody-associated magnetic beads (Dynabeads® anti-mouse IgG from the CELLection™ Pan Mouse IgGy kit; Invitrogen). The cell populations were then separated, the endothelial aortic cells were retained on the surface of the tube in contact with the magnet, while the HA-VSMCs remained in suspension without adhering to the magnet. HA-VSMCs were seeded and maintained in DMEM medium (Lonza Walkersville) with 10% FBS.

2.3. Stimulation Assay

HA-VSMCs were seeded and grown in four 60-mm polystyrene plates (Corning, Cultek), indicated for the cultivation of adherent cells until monolayer formation, and were allowed to reach confluence. Once FBS depletion was completed for 18 h at 0.5%, stimulation was performed as follow: DMEM medium in control plate vs DMEM medium with an inflammatory cocktail of Interleukin-1 (10 ng/mL), Interferon-α (10 ng/mL), Angiotensin-II (1 ng/mL) in stimulated plates. After 1 h, 4 h and 24 h of contact the medium was collected (secretome), and the cells were washed five times with cold PBS, lifted from the surface of the plates using a scraper and processed for metabolites extraction in methanol. The experiment was performed four times (technical replicates).

2.4. Metabolites Extraction

For metabolite analysis, protein removal in urine and secretome from HA-VSMCs was performed by precipitation in acetonitrile/0.1% formic acid (1:1) based on previous analyses [10–12]. In the case of plasma samples, protein removal was performed with prechilled methanol (1:1). Metabolites were extracted from HA-VSMCs in 50% methanol [13]. In all cases, following centrifugation, the supernatant was diluted with mobile phase A (0.1% formic acid in Milli-Q water) and filtered through 0.22 µm for subsequent LC-MS/MS analysis.

2.5. Targeted Analysis by Liquid Chromatography and Mass Spectrometry

Targeted mass spectrometry analysis was accomplished by Selected Reaction Monitoring (SRM-LC-MS/MS) methodology as previously published [12,15]. A 6460 Triple Quadrupole QQQ on-line connected to HPLC (1200 Series) controlled by Mass Hunter Acquisition Software (v4.01) (Agilent Technologies) was used. Samples were analyzed in a reversed-phase column (Atlantis T3, 3 µm, 2.1 × 100 mm, Waters) thermostatically controlled at 40 °C. A sample volume of 10 µL was injected and separation took place at 0.4 mL/min in an acetonitrile gradient. Instrumental conditions (collision energy and fragmentor potential) were optimized for each SRM transition (see Supplementary Table S1). Delta electron multiplier voltage was set to 600 V in negative ion mode or 400 V in positive ion mode. Peak areas were calculated from individual signals for intergroup comparison. Urine metabolic content was normalized by creatinine values. A detailed list of the measured metabolites is compiled in Supplementary Table S2.

2.6. Statistical Analysis

The ROUT method was applied to detect outliers based on the FDR, setting Q value to 5%. Mann-Whitney non-parametric test (95% confidence level) was performed with GraphPad Prism 6 software (version 6.01). Individual and combined diagnostic capacity was evaluated by receiving operating characteristics (ROC) curves using Metabonalyst

5.0 platform. The combined multivariate ROC curves were obtained using random forest algorithm as built-in method. The area under the curve (AUC) and the 95% confidence interval were estimated for both individual and multivariate analysis.

3. Results

A cohort of subjects with programmed CABG was recruited as a representative population of high cardiovascular risk (CVrisk group) (Table 1). The control group (age 48 ± 6; 38% male) have no history of diabetes, hypertension or cardiovascular disease. Those metabolites previously identified by our group with significant alteration in urine and aorta from atherosclerotic rabbits [12,13], in urine from hypertensive subjects with higher cardiovascular risk in terms of albuminuria development [10], and in urine from subjects at the onset of an acute coronary syndrome [12] were here analyzed by targeted mass spectrometry to investigate their alteration during human atherosclerosis development. Supplementary Table S2 compiles these altered metabolites previously identified in the referred studies and here analyzed in a new cohort and in four different matrices.

Table 1. Clinical data of CABG patients included in the study.

	CVrisk
Age	68 ± 9
Sex, male (%)	52
Diabetes mellitus (%)	26
Arterial hypertension (%)	89
Smoking habits (%)	26
Estimated glomerular filtration rate (mL/min/1.73 m^2)	66 ± 21

We first analyzed the metabolites abundance in urine according to previous findings, confirming their association with cardiovascular risk for five metabolites: trimethylamine N-oxide (TMAO), arabitol and spermidine showed higher levels in pre CABG patients compared with the control group; while glutamine and pantothenate showed lower levels (Figure 1). When analyzed 3 months after CABG surgery, spermidine levels normalize towards control values, contrary to TMAO and arabitol which showed increased alteration.

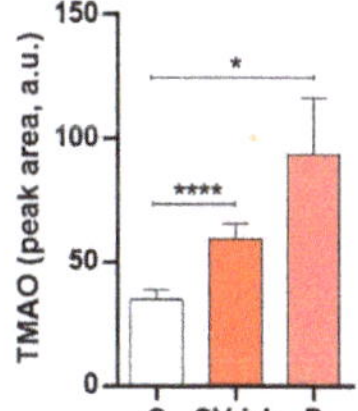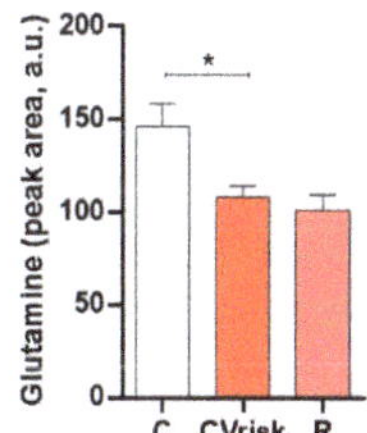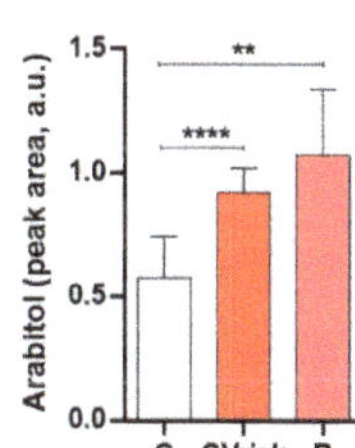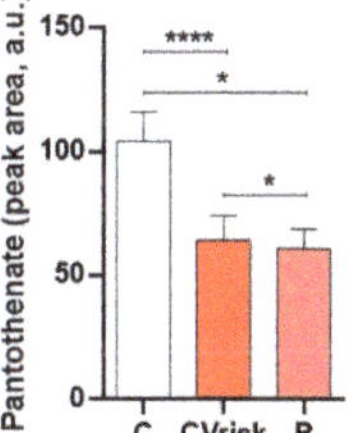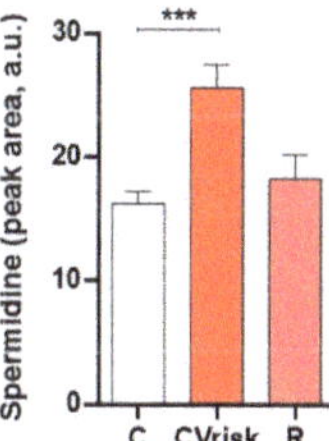

Figure 1. Metabolite abundance in urine in subjects with programmed bypass surgery (CVrisk group) compared to healthy subjects (C group). Data during clinical follow-up Table 0. * p value < 0.05, ** p value < 0.01, *** p value < 0.001, **** p value < 0.0001.

Metabolites were also analyzed in plasma (Figure 2). The same variation trends previously observed in urine were found in plasma for TMAO and glutamine, which normalized to control values following surgery. Choline, acetylcholine and valine were found to be decreased in plasma from CVrisk subjects, while pyruvate was found to be increased. Normalized levels were observed post-surgery also for valine and pyruvate

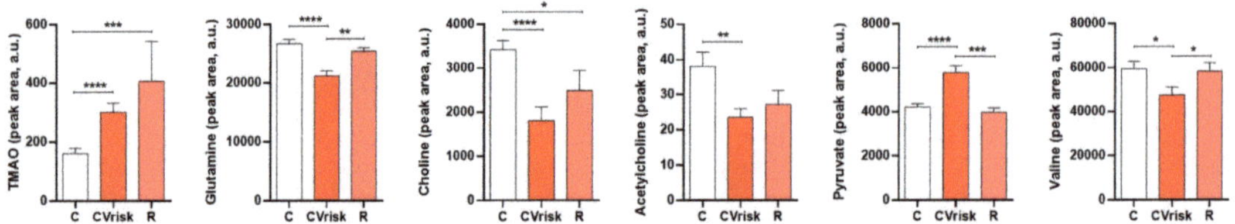

Figure 2. Metabolite abundance in plasma in subjects with programmed bypass surgery (CVrisk group) compared to healthy subjects (C group). Data during clinical follow-up 3 months after Scheme 0. * *p* value < 0.05, ** *p* value < 0.01, *** *p* value < 0.001, **** *p* value < 0.0001.

The metabolites diagnostic potential was evaluated by ROC curves in terms of sensitivity and specificity. Table 2 shows the area under the curve (AUC) obtained for each metabolite found significantly altered in urine or plasma and for combined models.

Table 2. Metabolites potential as biomarkers of cardiovascular risk, evaluated by ROC curves calculation. AUC: area under the curve. CI: confidence interval.

	AUC	**CI (95%)**
Individual metabolites urine		
Arabitol	0.763	(0.671–0.860)
Glutamine	0.528	(0.419–0.639)
Pantothenate	0.686	(0.590–0.800)
Spermidine	0.679	(0.573–0.785)
TMAO	0.720	(0.612–0.816)
Individual metabolites plasma		
Acetylcholine	0.682	(0.575–0.772)
Choline	0.740	(0.635–0.829)
Glutamine	0.748	(0.656–0.836)
Pyruvate	0.723	(0.627–0.811)
TMAO	0.741	(0.641–0.832)
Valine	0.617	(0.488–0.724)
Combined metabolites		
Urine	0.889	(0.778–0.980)
Plasma	0.902	(0.823–0.983)
Glutamine (urine and plasma)	0.743	(0.602–0.855)
TMAO (urine and plasma)	0.779	(0.656–0.909)

Best performing biomarkers of cardiovascular risk were arabitol and TMAO in urine, and glutamine, TMAO and choline in plasma (AUC = 0.720–0.763). All urine metabolites resulted in an AUC = 0.889 when combined. Combined plasma metabolites showed AUC = 0.902. Glutamine and TMAO were found significantly altered in both matrices giving AUC = 0.743 and AUC = 0.779, respectively, if analyzed in both biological fluids.

Finally, the potential contribution of HA-VSMCs to metabolic de-regulation observed in biological fluids was investigated. HA-VSMCs intracellular and extracellular (secretome) metabolic content was analyzed in response to a pro-atherosclerotic stimulus at different time points (1 h, 4 h and 24 h) (Figure 3). Increased secreted values compared to cells without stimulation (control) were found for TMAO, pantothenate, glycerophosphocholine, glutathion, spermidine and acetylcholine. Secreted glutamine was found to be decreased in the pro-atherosclerotic environment. Pantothenate and spermidine showed the earliest response, while extracellular levels of glutathione, glutamine, acetylcholine and glycerophosphocholine significantly varied only after 24 h. At an intracellular level, TMAO, pantothenate and glycerophosphocholine varied with stimulation.

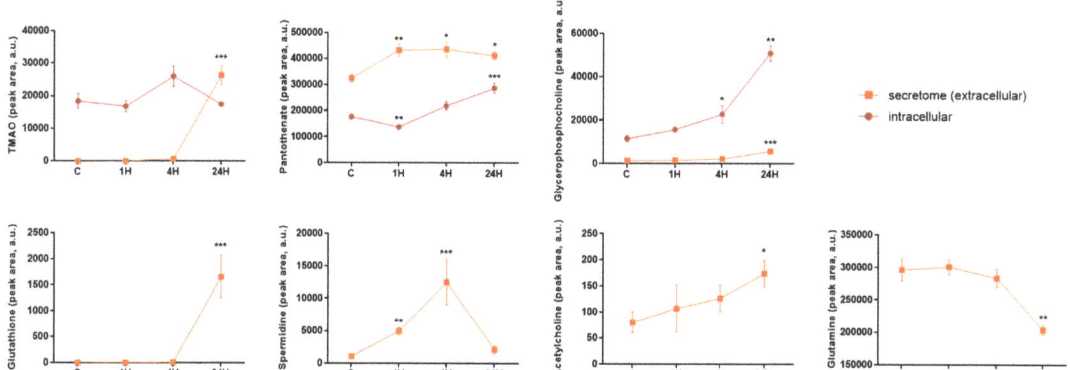

Figure 3. Intracellular and extracellular metabolic content of HA-VSMCs exposed to pro-atherogenic stimulus. Metabolites were extracted from HA-VSMCs and from the secreted media after 1 h, 4 h and 24 h following stimulation. Statistical significance was calculated for every time point versus control group (C). * p value < 0.05, ** p value < 0.01, *** p value < 0.001.

4. Discussion

Metabolomics is an approach that has shown great potential in cardiovascular research, opening new avenues for the study of underlying mechanisms and revealing new therapeutic targets and markers. In this work, we have applied targeted metabolomics to four different matrices, urine, plasma, HA-VSMCs and secretome. That allows us, on the one hand, to better understand the metabolisms associated with the pathology and, on the other hand, to identify the most robust molecules as markers. Our study reveals an imbalance in the defense mechanisms against oxidative stress and a response to atherosclerosis related inflammation. Additionally, it identifies TMAO and glutamine as the most altered molecules in the pathology in different accessible fluids.

4.1. Insufficient Oxidative Stress Counteraction during Atherosclerosis Development

Oxidative stress is one of the main operating mechanisms in atherosclerosis [16], induced by inflammation, mitochondria, autophagy and apoptotic processes taking place in the course of the disease [17]. Elevated levels of reactive oxygen species (ROS) resulting in antioxidative stress acts as a crucial mechanical effect in atherosclerosis progression. In this line, increasing antioxidant capacity may be one of the key points in future atherosclerosis therapy.

In response to the exposure of ROS released in oxidative stress conditions, synthesis of polyamines is induced. Spermidine is a natural polyamine with protective effects in human cardiovascular health [18]. In particular, spermidine is exported from the cell by the polyamine transporter TPO1 in response to oxidative stress [19]. In agreement with an attempt to counteract oxidative stress conditions, we observed here increased spermidine in the VSMCs secretome following inflammatory stimulus, and in urine of subjects with programmed CABG surgery, following normalization to pre-surgery values 3 months after CABG intervention. We previously found spermidine levels increased in urine collected after an acute coronary syndrome [12]. Endothelial nitric oxide synthase (eNOS) has a protective function in the cardiovascular system by generating antioxidant nitric oxide (NO). Arginase competes with eNOS for arginine and converts it in ornithine, a precursor of spermidine. Valine inhibits arginase, and reduced levels here observed for valine may also reflect a deleterious balancing in oxidative stress counteraction and increased production of spermidine.

Glutamine is another metabolite which counteracts oxidative stress and exerts an anti-inflammatory role, showing a negative correlation between its circulating levels and cardiometabolic disease [20]. As a precursor of glutathione, glutamine plays a key role in the prevention of oxidative damage and contributes to the maintenance of the endothelial function of the blood vessels [21]. Glutamine was here found decreased in both biological fluids from cardiovascular risk patients, and also in the secretome of VSMCs pointing to a deleterious protective effect against atherosclerosis development. These data are in consonance with our previously reported diminished levels in urine from middle age (50–70 years) and elderly (>70 years) subjects with cardiovascular risk factors [22].

4.2. Metabolic Response to Inflammation in Atherosclerosis Development

Choline and acetylcholine were also found decreased in plasma of CABG subjects. Choline is a structural component of cell membranes; it participates in the transport and metabolism of cholesterol and helps maintaining circulating homocysteine levels. Increased plasma homocysteine is an independent risk factor for atherosclerosis development which can be caused by decreased methylation of homocysteine to form methionine, being choline the methyl donor [22]. Our data support a role for choline in the increased cardiovascular risk associated with circulating homocysteine. Choline is a precursor of acetylcholine and the involvement of its receptor a7 nicotinic acetylcholine receptor (α7nAChR) in the development of atherosclerosis is an expanding field. In vivo studies revealed both anti- or pro-atherogenic effects of α7nAChR stimulation [23]. In this study, the observed reduced levels of plasma acetylcholine are in consonance with the same trend observed for its precursor choline. On the contrary, VSMCs secreted higher amount of acetylcholine after inflammatory stimulation, in line with inflammation and oxidative stress suppression observed in mice aorta after vascular injury mediated by α7nAChRs activation [24].

Choline is also one of the trimethylamine-containing nutrients and, as such, a TMAO precursor. Plasma TMAO levels depend on liver production, ingestion and renal excretion. Exposure of VSMCs to TMAO induce inflammation and suggest atherosclerosis enhancement by a TMAO-dependent mechanism [25]. In our study, secreted TMAO levels increase in response to a pro-inflammatory stimulus in what can be seen as a contribution to observed higher levels in plasma and urine from cardiovascular risk subjects with moderately preserved renal function. Urinary TMAO was found increased in young (30–50 years), middle age (50–70 years) and elderly (>70 years) population with cardiovascular risk factors in our previous study [22], in agreement with data shown here and pointing to a strong association with CV risk independently of age.

Figure 4 shows a representative image including all metabolites with significantly altered levels in the different matrices investigated. It has to be considered that not all metabolites could be detected in all matrices (urine, plasma, HA-VSCMs and cellular secretome). Discussion was thus focused on confirmed significant differences between CVrisk subjects and stimulated HA-VSMCs in comparison with their respective controls.

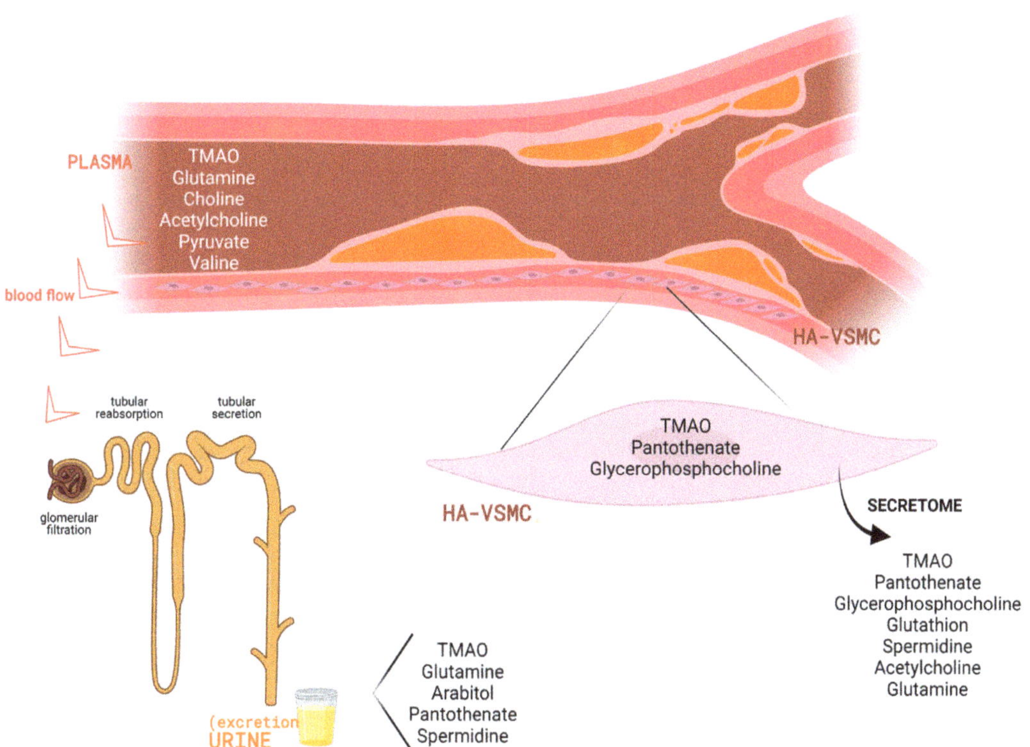

Figure 4. Overview of metabolic alterations identified in association with CVrisk and atherosclerosis development. Specifically, changes were observed in urine and plasma from subjects with programmed CABG surgery, and in the intracellullar and extracellular (secretome) content of human aortic VSMCs (HA-VSMCs) after exposure to a pro-atherosclerotic stimulus. Created with BioRender.com (accessed on 29 July 2021).

4.3. Urine and Plasma Metabolites with Diagnostic Potential in Cardiovascular Risk Evaluation

Metabolites with diagnostic potential should be easily accessible in routine analyses. In this sense, urine and plasma are the most used biological fluids in daily clinical practice. Regarding cardiovascular risk estimation, novel tools based on molecular profiling should ideally respond to subjacent atherosclerosis development despite its silent progression. In this study, we show the stronger potential of multimarker panels compared to metabolites individual performance. This combination may include best performing metabolites according to sensitivity/specificity criteria, or analyses in various biological matrices.

We and others have previously demonstrated urinary metabolites alteration after an acute coronary syndrome (ACS). In that case, two metabolic responses may be overlapping, i.e., responding both to the acute event itself and to the pathophysiological deregulation taking place in the atherosclerotic arterial vessel. Arabitol and spermidine show here the same increased trend than previously observed in ACS. In the case of arabitol, no recovery towards control values was observed in any case (following the acute event, or after 3 months from CABG surgery). However, spermidine reflects recovery following surgery indicating a quicker response to therapeutic intervention.

TMAO and glutamine showed the same trend in urine and plasma from CABG patients. In the aorta from atherosclerotic rabbits, the same trend was observed for glutamine suggesting a direct reflection in fluids of diminished content in the atherosclerotic vessel. However, TMAO showed reduced aortic levels pointing to a plausible diet effect on its urine/plasma levels. Plasma levels of choline, acetylcholine, pyruvate and valine are also in agreement with aortic levels during atherosclerosis progression.

The analysis of combined and individual features and their predictor potential showed, as expected, that metabolic panels present a better performance than individual molecules in risk prediction. Particularly, in this study, the combination of five urine metabolites or six plasma metabolites led to marker panels with an AUC of 0.9 while individual features average AUC was 0.7. Previous analyses performed in plasma in large cohorts have showed the value of metabolomics for biomarker discovery. Phenyalanine, MUFA relative to total fatty acids (MUFA%), Omega-6 and DHA combined are described to improve prediction of cardiovascular risk in comparison with established risk factors [26]. More recently, a study performed in more than 10,500 individuals showed that five phosphocholines have a comparable discrimination of risk of coronary heart disease with classic estimators [27]. Regarding urine and CV risk estimation, large studies are missed. By analysing the same set of metabolites in plasma, urine and HA-VSMCs here, we cannot only estimate the predicting value of the features but also the different performance of the measurable matrix. Our results demonstrate the value of urine marker panels, in accordance to our previous studies, and showed a similar performance to those performed in plasma. Interestingly, in this particular case, there is not an improvement when combining plasma and urine measurements. Pending studies in large and multiple cohorts, these metabolites being quantifiable in accessible fluids may aid in assessing individual CV risk during asymptomatic stages in atherosclerosis development, facilitating earlier diagnosis and patients' risk stratification.

5. Conclusions

Here we demonstrate how molecular deregulation directly observed in the atherosclerotic aortic tissue can be detected in an accessible biological fluid, setting the basis for novel diagnostic tools complementing existing algorithms in cardiovascular risk prediction or individual stratification. Previous metabolic alterations shown in response to an acute coronary syndrome can also take place during atherosclerosis development. We show here that metabolite deregulations identified during atherosclerosis development point to an inflammatory response together with a defective or insufficient action of mechanisms counteracting oxidative stress.

Supplementary Materials: The following are available online at https://www.mdpi.com/article/10.3390/antiox10091369/s1, Table S1: Experimental conditions for metabolites analysis by SRM, Table S2: Metabolites previously identified by our group in association with cardiorenal risk uEA: early atherosclerosis, detected in urine [12]. aEA: early atherosclerosis, detected in aorta [13]. ACS: acute coronary syndrome at the onset [12]. ALB: subjects with moderately increased albuminuria [10].

Author Contributions: Conceptualization, G.A.-L. and G.A.-E.; methodology, A.S.-H., P.J.M., M.A., A.H., M.M.-L. and G.A.-L.; investigation, A.S.-H., P.J.M., L.G.-C., A.Y.-M. and V.E.; writing—original draft preparation, G.A.-L.; writing—review and editing, all authors; supervision, M.M.-L. and G.A.-L.; project administration, G.A.-L.; funding acquisition, G.A.-L. and G.A.-E. All authors have read and agreed to the published version of the manuscript.

Funding: This work was supported by the Instituto de Salud Carlos III, co-supported by FEDER grants (PI16/01334, PI18/00348, PI20/01103, IF08/3667-1, CPII15/00027, PT13/0001/0013 and PRB3 [IPT17/0019-ISCIII-SGEFI/ERDF]), REDinREN (RD16/0009), ARADyAL (RD/0006/0013), Fundación SENEFRO/SEN, CAM (PEJD-2019-PRE/BMD-16992, 2018-T2/BMD-11561, CM_P2018/BAAA-4574), Fundación Conchita Rábago de Jiménez Díaz, Getinge Group Spain SL, SEAIC (19_A08) and Alfonso X el Sabio University Foundations.

Institutional Review Board Statement: The study was conducted according to the recommendations of the Declaration of Helsinki and was approved by the local ethics committee (PIC51-2013, PIC143-2016).

Informed Consent Statement: In all cases informed consent was requested from subjects.

Data Availability Statement: Data are contained within the article or in Supplementary Material.

Acknowledgments: Authors acknowledge personnel from Donation Unit in Fundación Jiménez Díaz Hospital for the collection of control samples.

References

1. Spring, B.; Moller, A.C.; Colangelo, L.A.; Siddique, J.; Roehrig, M.; Daviglus, M.L.; Polak, J.F.; Reis, J.P.; Sidney, S.; Liu, K. Healthy lifestyle change and subclinical atherosclerosis in young adults: Coronary Artery Risk Development in Young Adults (CARDIA) study. *Circulation* **2014**, *130*, 10–17. [CrossRef]
2. Zethelius, B.; Berglund, L.; Sundström, J.; Ingelsson, E.; Basu, S.; Larsson, A.; Venge, P.; Arnlöv, J. Use of multiple biomarkers to improve the prediction of death from cardiovascular causes. *N. Engl. J. Med.* **2008**, *358*, 2107–2116. [CrossRef]
3. Iliou, A.; Mikros, E.; Karaman, I.; Elliott, F.; Griffin, J.L.; Tzoulaki, I.; Elliott, P. Metabolic phenotyping and cardiovascular disease: An overview of evidence from epidemiological settings. *Heart Br. Card. Soc.* **2021**, *107*, 1123–1129.
4. Deidda, M.; Noto, A.; Cadeddu Dessalvi, C.; Andreini, D.; Andreotti, F.; Ferrannini, E.; Latini, R.; Maggioni, A.P.; Magnoni, M.; Maseri, A.; et al. Metabolomic correlates of coronary atherosclerosis, cardiovascular risk, both or neither. Results of the 2 × 2 phenotypic CAPIRE study. *Int. J. Cardiol.* **2021**, *336*, 14–21.
5. Yap, I.K.S.; Brown, I.J.; Chan, Q.; Wijeyesekera, A.; Garcia-Perez, I.; Bictash, M.; Loo, R.L.; Chadeau-Hyam, M.; Ebbels, T.; De Iorio, M.; et al. Metabolome-wide association study identifies multiple biomarkers that discriminate north and south Chinese populations at differing risks of cardiovascular disease: INTERMAP study. *J. Proteome Res.* **2010**, *9*, 6647–6654. [CrossRef] [PubMed]
6. Ruiz-Canela, M.; Hruby, A.; Clish, C.B.; Liang, L.; Martínez-González, M.A.; Hu, F.B. Comprehensive Metabolomic Profiling and Incident Cardiovascular Disease: A Systematic Review. *J. Am. Heart Assoc.* **2017**, *6*, e005705. [CrossRef] [PubMed]
7. Newgard, C.B. Metabolomics and Metabolic Diseases: Where Do We Stand? *Cell Metab.* **2017**, *25*, 43–56. [CrossRef]
8. Rizza, S.; Copetti, M.; Rossi, C.; Cianfarani, M.A.; Zucchelli, M.; Luzi, A.; Pecchioli, C.; Porzio, O.; Di Cola, G.; Urbani, A.; et al. Metabolomics signature improves the prediction of cardiovascular events in elderly subjects. *Atherosclerosis* **2014**, *232*, 260–264. [CrossRef]
9. Huang, M.; Zhao, H.; Gao, S.; Liu, Y.; Liu, Y.; Zhang, T.; Cai, X.; Li, Z.; Li, L.; Li, Y.; et al. Identification of coronary heart disease biomarkers with different severities of coronary stenosis in human urine using non-targeted metabolomics based on UPLC-Q-TOF/MS. *Clin. Chim. Acta Int. J. Clin. Chem.* **2019**, *497*, 95–103. [CrossRef]
10. Gonzalez-Calero, L.; Martin Lorenzo, M.; Martínez, P.J.; Baldan-Martin, M.; Ruiz-Hurtado, G.; Segura, J.; de la Cuesta, F.; Barderas, M.G.; Ruilope, L.M.; Vivanco, F.; et al. Hypertensive patients exhibit an altered metabolism. A specific metabolite signature in urine is able to predict albuminuria progression. *Transl. Res. J. Lab. Clin. Med.* **2016**, *178*, 25–37.e7.
11. Martin-Lorenzo, M.; Martinez, P.J.; Baldan-Martin, M.; Ruiz-Hurtado, G.; Prado, J.C.; Segura, J.; de la Cuesta, F.; Barderas, M.G.; Vivanco, F.; Ruilope, L.M.; et al. Citric Acid Metabolism in Resistant Hypertension: Underlying Mechanisms and Metabolic Prediction of Treatment Response. *Hypertens. Dallas Tex* **2017**, *70*, 1049–1056. [CrossRef] [PubMed]
12. Martin-Lorenzo, M.; Zubiri, I.; Maroto, A.S.; Gonzalez-Calero, L.; Posada-Ayala, M.; de la Cuesta, F.; Mourino-Alvarez, L.; Lopez-Almodovar, L.F.; Calvo-Bonacho, E.; Ruilope, L.M.; et al. KLK1 and ZG16B proteins and arginine-proline metabolism identified as novel targets to monitor atherosclerosis, acute coronary syndrome and recovery. *Metab. Off. J. Metab. Soc.* **2015**, *11*, 1056–1067. [CrossRef]
13. Martin-Lorenzo, M.; Gonzalez-Calero, L.; Maroto, A.S.; Martinez, P.J.; Zubiri, I.; de la Cuesta, F.; Mourino-Alvarez, L.; Barderas, M.G.; Heredero, A.; Aldamiz-Echevarría, G.; et al. Cytoskeleton deregulation and impairment in amino acids and energy metabolism in early atherosclerosis at aortic tissue with reflection in plasma. *Biochim. Biophys. Acta* **2016**, *1862*, 725–732. [CrossRef] [PubMed]
14. Callesen, K.T.; Yuste-Montalvo, A.; Poulsen, L.K.; Jensen, B.M.; Esteban, V. In Vitro Investigation of Vascular Permeability in Endothelial Cells from Human Artery, Vein and Lung Microvessels at Steady-State and Anaphylactic Conditions. *Biomedicines* **2021**, *9*, 439. [CrossRef] [PubMed]
15. Martin-Lorenzo, M.; Gonzalez-Calero, L.; Martinez, P.J.; Baldan-Martin, M.; Lopez, J.A.; Ruiz-Hurtado, G.; de la Cuesta, F.; Segura, J.; Vazquez, J.; Vivanco, F.; et al. Immune system deregulation in hypertensive patients chronically RAS suppressed developing albuminuria. *Sci. Rep.* **2017**, *7*, 8894. [CrossRef]
16. Kattoor, A.J.; Pothineni, N.V.K.; Palagiri, D.; Mehta, J.L. Oxidative Stress in Atherosclerosis. *Curr. Atheroscler. Rep.* **2017**, *19*, 42. [CrossRef] [PubMed]
17. Yang, X.; Li, Y.; Li, Y.; Ren, X.; Zhang, X.; Hu, D.; Gao, Y.; Xing, Y.; Shang, H. Oxidative Stress-Mediated Atherosclerosis: Mechanisms and Therapies. *Front. Physiol.* **2017**, *8*, 600. [CrossRef]
18. Eisenberg, T.; Abdellatif, M.; Schroeder, S.; Primessnig, U.; Stekovic, S.; Pendl, T.; Harger, A.; Schipke, J.; Zimmermann, A.; Schmidt, A.; et al. Cardioprotection and lifespan extension by the natural polyamine spermidine. *Nat. Med.* **2016**, *22*, 1428–1438. [CrossRef]

19. Krüger, A.; Vowinckel, J.; Mülleder, M.; Grote, P.; Capuano, F.; Bluemlein, K.; Ralser, M. Tpo1-mediated spermine and spermidine export controls cell cycle delay and times antioxidant protein expression during the oxidative stress response. *EMBO Rep.* **2013**, *14*, 1113–1119. [CrossRef]

20. Durante, W. The Emerging Role of l-Glutamine in Cardiovascular Health and Disease. *Nutrients* **2019**, *11*, 2092. [CrossRef]

21. Ellis, A.C.; Patterson, M.; Dudenbostel, T.; Calhoun, D.; Gower, B. Effects of 6-month supplementation with β-hydroxy-β-methylbutyrate, glutamine and arginine on vascular endothelial function of older adults. *Eur. J. Clin. Nutr.* **2016**, *70*, 269–273. [CrossRef]

22. Martinez, P.J.; Agudiez, M.; Molero, D.; Martin-Lorenzo, M.; Baldan-Martin, M.; Santiago-Hernandez, A.; García-Segura, J.M.; Madruga, F.; Cabrera, M.; Calvo, E.; et al. Urinary metabolic signatures reflect cardiovascular risk in the young, middle-aged, and elderly populations. *J. Mol. Med. Berl. Ger.* **2020**, *98*, 1603–1613. [CrossRef]

23. Vieira-Alves, I.; Coimbra-Campos, L.M.C.; Sancho, M.; da Silva, R.F.; Cortes, S.F.; Lemos, V.S. Role of the α7 Nicotinic Acetylcholine Receptor in the Pathophysiology of Atherosclerosis. *Front. Physiol.* **2020**, *11*, 621769. [CrossRef] [PubMed]

24. Li, D.-J.; Fu, H.; Tong, J.; Li, Y.-H.; Qu, L.-F.; Wang, P.; Shen, F.-M. Cholinergic anti-inflammatory pathway inhibits neointimal hyperplasia by suppressing inflammation and oxidative stress. *Redox Biol.* **2018**, *15*, 22–33. [CrossRef] [PubMed]

25. Seldin, M.M.; Meng, Y.; Qi, H.; Zhu, W.; Wang, Z.; Hazen, S.L.; Lusis, A.J.; Shih, D.M. Trimethylamine N-Oxide Promotes Vascular Inflammation Through Signaling of Mitogen-Activated Protein Kinase and Nuclear Factor-κB. *J. Am. Heart Assoc.* **2016**, *5*, e002767. [CrossRef]

26. Würtz, P.; Havulinna, A.S.; Soininen, P.; Tynkkynen, T.; Prieto-Merino, D.; Tillin, T.; Ghorbani, A.; Artati, A.; Wang, Q.; Tiainen, M.; et al. Metabolite profiling and cardiovascular event risk: A prospective study of 3 population-based cohorts. *Circulation* **2015**, *131*, 774–785. [CrossRef] [PubMed]

27. Cavus, E.; Karakas, M.; Ojeda, F.M.; Kontto, J.; Veronesi, G.; Ferrario, M.M.; Linneberg, A.; Jørgensen, T.; Meisinger, C.; Thorand, B.; et al. Association of Circulating Metabolites With Risk of Coronary Heart Disease in a European Population: Results From the Biomarkers for Cardiovascular Risk Assessment in Europe (BiomarCaRE) Consortium. *JAMA Cardiol* **2019**, *4*, 1270–1279. [CrossRef]

Article

Aminoguanidine Prevents the Oxidative Stress, Inhibiting Elements of Inflammation, Endothelial Activation, Mesenchymal Markers, and Confers a Renoprotective Effect in Renal Ischemia and Reperfusion Injury

Consuelo Pasten [1,2], Mauricio Lozano [1], Jocelyn Rocco [1], Flavio Carrión [3], Cristobal Alvarado [4,5], Jéssica Liberona [6], Luis Michea [6,7] and Carlos E. Irarrázabal [1,2,*]

[1] Laboratorio de Fisiología Integrativa y Molecular, Programa de Fisiología, Centro de Investigación e Innovación Biomédica, Universidad de los Andes, Santiago 7620157, Chile; mpasten@uandes.cl (C.P.); mlozano@uandes.cl (M.L.); joce.valeska.roccog@gmail.com (J.R.)

[2] Facultad de Medicina, Universidad de los Andes, Santiago 7620157, Chile

[3] Facultad de Ciencias de la Salud, Universidad del Alba, Santiago 7620157, Chile; flavio.carrion@udalba.cl

[4] Clinical Research Unit, Hospital Las Higueras, Talcahuano 4260000, Chile; cristobalalvarado@ucsc.cl

[5] Department of Basic Sciences, School of Medicine, Universidad Católica de la Santísima Concepción, Concepción 4030000, Chile

[6] Instituto de Ciencias Biomédicas, School of Medicine, Universidad de Chile, Santiago 7620157, Chile; jliberona@clinicauandes.cl (J.L.); lmichea@med.uchile.cl (L.M.)

[7] Millennium Institute on Immunology and Immunotheraphy, Santiago 762015, Chile

* Correspondence: cirarrazabal@uandes.cl; Tel.: +56-2-4129607

Citation: Pasten, C.; Lozano, M.; Rocco, J.; Carrión, F.; Alvarado, C.; Liberona, J.; Michea, L.; Irarrázabal, C.E. Aminoguanidine Prevents the Oxidative Stress, Inhibiting Elements of Inflammation, Endothelial Activation, Mesenchymal Markers, and Confers a Renoprotective Effect in Renal Ischemia and Reperfusion Injury. *Antioxidants* **2021**, *10*, 1724. https://doi.org/10.3390/antiox10111724

Academic Editors: Soliman Khatib and Dana Atrahimovich Blatt

Received: 31 August 2021
Accepted: 6 October 2021
Published: 28 October 2021

Publisher's Note: MDPI stays neutral with regard to jurisdictional claims in published maps and institutional affiliations.

Abstract: Oxidative stress produces macromolecules dysfunction and cellular damage. Renal ischemia-reperfusion injury (IRI) induces oxidative stress, inflammation, epithelium and endothelium damage, and cessation of renal function. The IRI is an inevitable process during kidney transplantation. Preliminary studies suggest that aminoguanidine (AG) is an antioxidant compound. In this study, we investigated the antioxidant effects of AG (50 mg/kg, intraperitoneal) and its association with molecular pathways activated by IRI (30 min/48 h) in the kidney. The antioxidant effect of AG was studied measuring GSSH/GSSG ratio, GST activity, lipoperoxidation, iNOS, and Hsp27 levels. In addition, we examined the effect of AG on elements associated with cell survival, inflammation, endothelium, and mesenchymal transition during IRI. AG prevented lipid peroxidation, increased GSH levels, and recovered the GST activity impaired by IRI. AG was associated with inhibition of iNOS, Hsp27, endothelial activation (VE-cadherin, PECAM), mesenchymal markers (vimentin, fascin, and HSP47), and inflammation (IL-1β, IL-6, Foxp3, and IL-10) upregulation. In addition, AG reduced kidney injury (NGAL, clusterin, Arg-2, and TFG-β1) and improved kidney function (glomerular filtration rate) during IRI. In conclusion, we found new evidence of the antioxidant properties of AG as a renoprotective compound during IRI. Therefore, AG is a promising compound to treat the deleterious effect of renal IRI.

Keywords: aminoguanidine; antioxidants; oxidative stress; ischemia-reperfusion injury; renal protection

1. Introduction

Aminoguanidine (AG) is a small, nontoxic molecule with different biological actions. It has been shown that AG inhibits the formation of highly reactive advanced glycosylation end products (AGEs) associated with diabetes mellitus (DM) and decreases the complications related to proteinuria [1], retinopathy [2], and neuropathy [3]. Due to these effects, AG was proposed as a therapeutic agent inhibiting AGEs formation [4]. However, in a phase III clinical trial in type 1 diabetic patients, aminoguanidine reduced proteinuria and retinopathy, whereas the progression to overt nephropathy was not statistically improved. In addition, a high dose of aminoguanidine provoked abnormalities in liver function and other side effects associated with flu-like symptoms, gastrointestinal alterations, rare vasculitis, and anemia [5]. However, the beneficial therapeutic properties of low doses of AG in other kidney diseases are poorly understood.

In another view, AG is a potent antioxidant compound and free radical scavenger. High doses (AG $\geq$ 5 mM) have an inhibitory effect on the xanthine oxidase (XO) activity [6]. AG also is an effective in vitro quencher of reactive carbonyl species (RCS) [7]. The radical scavenging properties of AG may contribute to protective effects during glycation and explain the prevention of diabetic complications. In a model of streptozotocin-induced diabetes, AG demonstrated antioxidant effects decreasing plasma lipid hydroperoxide and thiobarbituric acid (TBARS) [8]. AG also showed antioxidant action against brain injury induced by doxorubicin (DOX), decreasing the levels of MDA (malondialdehyde) and GPx (Glutathione peroxidase) and increasing the GST (glutathione S-transferase) activity [9]. Similarly, in a model of intestinal IRI, administration of AG reduced the intestinal mucosal injury and decreased MDA, protein carbonyl (PC), SOD (superoxide dismutase), and GPx levels [10].

Renal IRI induces hypoxia, oxidative stress, vasoconstriction, destruction of the tubular cells, endothelial cell damage, inflammation, cell necrosis, and cessation of renal function [11]. The reperfusion phase results in more production of reactive oxygen species (ROS) [12–14], including, among others, superoxide anion (O_2^-), hydrogen peroxide (H_2O_2), and hydroxyl radicals (OH·), all of which have chemical properties that confer reactivity to different biological targets [15,16]. Further, IRI injury led to a significant increase in NO synthesis, and high concentrations produce peroxynitrite (ONOO-), which interacts with lipids, DNA, and proteins, leading to cell damage [17]. We previously described a murine model of IRI and observed increased levels of oxidative stress and kidney disfunction (glomerular filtration rate, neutrophil gelatinase-associated lipocalin, and clusterin) [18]. Previously, the literature described that co-treatment with α-tocopherol and aminoguanidine during unilateral IRI (I/R: 2 h/24 h) prevented the formation of MDA and nitrite/nitrate in plasma [19]. Besides, in a bilateral renal IRI model (I/R 60 min/24 h), AG (100 mg /kg, i.p.) decreased the MDA levels but did not attenuate the histological damage [20], indicating an antioxidant effect of AG during renal IR. In addition, the post-treatment with AG (50 mg/kg, i.p.) in a model of bilateral ischemia (IR 60 min/24 h) significantly reduced serum urea and creatinine levels and improved histopathological lesions observed by IRI (45 min/24 h) [21]. Therefore, the role of AG during renal IRI is not very clearly understood.

Interestingly, AG has been described as an effective inhibitor of NOS, specifically the inducible NO synthase isoform (iNOS). AG suppresses the NO production after the reperfusion phase in experimental models of cerebral [22] or myocardial [23] IRI. AG suppresses iNOS activity in mice with brain ischemia to a level equivalent to those seen in iNOS knockout mice [24]. In the kidney, there are three different isoforms of nitric oxide synthase (NOS)—neuronal NOS (nNOS), endothelial NOS (eNOS), and inducible NOS (iNOS). While nNOS and eNOS are constitutively expressed, iNOS is a calcium-independent synthase whose expression is induced by cytokines, oxidative stress, and transcription factors such as NF-κb [25]. We previously described that the pharmacological inhibition of iNOS inhibition (l-NIL) prevented oxidative stress and improved renal function during IRI [18]. On the other hand, the Hsp27 is a stress protein that shows an early and transient increase after acute ischemia [26], inhibiting apoptosis by decreasing intracellular reactive oxygen species and the mitochondrial caspase-dependent apoptotic pathway [27]. Selective renal overexpression of Hsp27 in mice through lentiviral gene delivery protects against ischemic renal injury [28]. Arginase-2 (Arg-2) and transforming growth factor-beta 1 (TGF-β1) are two markers of kidney damage by renal IR. Arg2 plays a significant role in renal fibrosis via its action on NO and mitochondrial function [29], and TGF-β is recognized as a central mediator of renal tubulointerstitial fibrosis [30]. Finally, new evidence suggests that the development of vascular rarefaction and fibrosis after an injury is dependent on the differentiation of renal endothelial and tubular cells into mesenchymal cells by IRI.

These processes are called epithelial-to-mesenchymal transition (EMT) and endothelial-to-mesenchymal transition (EndMT), respectively [31,32]. EMT is reversible and leads to the loss of cellular polarity, migratory capacity development, resistance to apoptosis, and the induction of extracellular matrix synthesis [33]. EndMT resembles and leads to loss of cellular adhesion molecules, cytoskeletal reorganization, and change of the compacted cobblestone-like into spindle-shaped phenotype without polarity, associated with reduced expression of endothelial markers (e.g., vascular endothelial-cadherin, CD31/PECAM-1, and von Willebrand factor) [34]. Both EMT and EndMT show increased mesenchymal cell markers such as vimentin, fascin1, and Hsp47 [35,36].

This study aimed to investigate the antioxidant property of AG in the kidney and its potential renoprotective effect during renal IRI. We studied the effect of intraperitoneal injection of AG (50 mg/kg) before the IRI on reduced and oxidized glutathione ratio (GSH/GSSG), glutathione S-transferase (GST) activity, lipoperoxidation, iNOS, and Hsp27. In addition, we identified new molecular pathways of the antioxidant effect of AG implicated in inflammation, cell injury, endothelial activation, and mesenchymal transition to provide novel evidence of the renoprotective effect of AG during renal IRI.

2. Materials and Methods

2.1. Animals

Previously, we described an IRI model using male BALB/c mice [18]. The mice (20–25 g) were housed in a 12 h light/dark cycle. Animals were maintained at the University de los Andes Animal Care Facility with food and water ad libitum. All experimental procedures followed institutional and international standards for humane care and use of laboratory animals (Animal Welfare Assurance Publication A5427-01, Office for Protection from Research Risks, Division of Animal Welfare. The National Institutes of Health). All procedures were approved by the Committee on the Ethics of Animal Experiments of the University de los Andes, Chile.

2.2. Ischemia-Reperfusion (IR) Experiments

The animals were anesthetized with 25 mg/kg i.p. ketamine/15 mg/kg i.p. xylazine (Drag Pharma, Santiago, Chile) and maintained on a 37 °C blanket during the surgical procedure. A flank incision exposed both kidneys, and the renal pedicle was occluded for 30 min with a non-traumatic vascular clamp (cat N° 18055-02 Fine Science Tools). Renal blood flow was re-established (reperfusion phase) by clamp removal, and both incisions were sutured. Sham animals did not undergo renal pedicle occlusion [18,37]. To study the effects of AG, mice were treated intraperitoneally (i.p.) with either vehicle or 50 mg/kg of aminoguanidine (aminoguanidine hydrochloride, TOCRIS TO.0787/100) [21] before ischemia. Then, the animals were subjected to 48 h of reperfusion according to our previous publication [18]. Following protocols, mice were euthanized with CO_2, and kidneys were dissected and processed for analysis.

2.3. Oxidative Stress Experiments

Kidney samples were homogenized in a Dounce homogenizer in 0.15 M potassium chloride (KCl). After measuring the total protein concentration [38], aliquots were deproteinized with cold trichloroacetic acid 30% v/v for 10 min, and the pellet was removed by centrifugation. Oxidative stress was evaluated in kidney homogenates as changes of the reduced glutathione (GSH) pool assayed in deproteinized kidney homogenates and lipid peroxidation, measured with the thiobarbituric acid reactive substances (TBARS) assays [39]. In addition, total enzymatic glutathione S-transferase (GST) activity was assayed in homogenates using 1-chloro-2,4-dinitrobenzene (CDNB) as a substrate and GSH as a cofactor [40].

2.4. Quantification of Neutrophil Gelatinase-Associated Lipocalin

Neutrophil gelatinase-associated lipocalin (NGAL) was quantified in plasma samples using a commercial ELISA kit (Bioporto) according to the manufacturer's instructions.

2.5. Glomerular Filtration Rate (GFR) Measurements

GFR was determined in conscious unrestrained mice with 48 h of reperfusion using the excretion kinetics of an intravenous bolus of FITC-sinistrin (6 mg/100 g body weight; dissolved in sodium chloride (NaCl) 0.9% Fresenius Kabi, GmgH Estermannstraße 17 A-4020 Linz, Austria) using a miniaturized fluorometric detector (NIC-Kidney excitation 480 nm/emission 521 nm; Mannheim Pharma & Diagnostics; MediBeacon St. Louis, MO, USA). GFR was calculated using the half-life derived from the rate constant of the single exponential phase of the FITC-sinistrin excretion curve as previously described [41].

2.6. Real-Time PCR

We studied the cortex and the medulla separately to understand the effect of AG in both kidney regions. Total RNA was isolated using an RNeasy Mini Kit (#74014, Qiagen, Germantown, MD, USA) according to the manufacturer's directions. Extracted RNA was quantified at 260 nm in a NanoDrop 2000 Spectrophotometer (NanoDrop Technologies, Madison, WI, USA), and the integrity of the RNA was assessed by agarose gel electrophoresis. cDNA was prepared from total RNA (1 µg) using a reverse transcription system (random hexamers (C-1118), Improm II Reverse Transcriptase System (A-3802) from Promega). Then, PCR was duplicated for each experiment (HotStartTaq DNA polymerase from Qiagen or BRILLIANT III ULTRA-FAST SYBR GREEN QPCR (Stratagene). Amplicons were detected for real-time fluorescence detection (Rotor-Gene Q, Qiagen, Germantown, MD, USA). The primers used are detailed in Table S1. Relative mRNA abundance was calculated using Ct values and normalized to the relative abundance of each transcript.

2.7. Western Blot Assay

We studied the cortex and the medulla separately to understand the effect of AG in both kidney regions. Western blot was realized as was previously published with some modifications [18]. Briefly, the renal cortex and the medulla were dissected and homogenized with an Ultra-Turrax homogenizer (Model PRO-200, PRO Scientific, Vernon Hills, IL, USA) in ice-cooled 10 mM Tris·HCl buffer at pH 7.4, supplemented with 1 mM EDTA, 1 mM EGTA, 0.25 M sucrose, 1% vol/vol Triton X-100, and a protease inhibitor cocktail (Complete Mini, Roche Applied Science # 5892970001). Tissue homogenates were subject to centrifugation. Step one was $900\times g$ by 10 min. Next, the tissue was sonicated for 30 s on ice, vortexed, and centrifuged again by $2400\times g$ and $17{,}000\times g$ by 5 min. All procedures were held at 4 °C. Total proteins in supernatants were measured using the BCA Protein Assay Kit (ThermoFisher Scientific, Rockford, IL, USA), and samples were stored at −80 °C. The antibodies were: anti-HSP27 (sc-13132; Santa Cruz, OR, USA) and mouse anti-β-actin (A5441; Sigma, St. Louis, MO, USA) antibodies. Secondary antibodies were anti-mouse (A-21257) or anti-rabbit (A-21039) IgG conjugated with Alexa Fluor-750 (Thermo Scientific, South San Francisco, CA, USA). The resulting band intensities were quantified using Odyssey equipment and Image Studio Lite software (version 5.25; Li-Cor, Lincoln, NE, USA).

2.8. Histochemical Analysis and Tissue Damage Determination

Kidneys were fixed in 10% formalin included in paraffin. The kidneys were fixed in 10% buffered formalin, embedded in paraffin, sectioned, dewaxed, rehydrated, rinsed in water, and stained with hematoxylin-eosin (HE). The morphologic analysis was carried out in a blinded manner as detailed previously by a physician/pathologist [18]. The cortex and the medulla were evaluated for epithelial necrosis, loss of brush border, tubular dilation, and tubular congestion, among other kidney alterations observed in AKI response.

2.9. Statistical Analysis

Differences between groups were analyzed using one-way ANOVA, nonparametric test, and Kruskal–Wallis test using GraphPad Prism Software. The levels of significance were represented by * $p < 0.05$ or ** $p < 0.005$. Data were the mean ± standard error of the mean (SEM).

3. Results

3.1. AG Prevented Kidney Injury due to Renal Ischemia and Reperfusion (IR)

Kidneys from Balb/c adult mice were subjected to 30 min ischemia and 48 h of reperfusion as described previously [18]. We studied kidney damage by measuring glomerular filtration rate (GFR), acute kidney injury biomarker (neutrophil gelatinase-associated lipocalin, NGAL), and histological analysis. The GFR decreased by approximately 50% in mice subjected to renal IR compared with sham (476.7 ± 56.0 and 869.2 ± 62.0 µL/min/100 g body for IR and sham, respectively; $p < 0.05$). In addition, the blood NGAL concentration was significantly higher (3.5-fold) from mice subjected to the IR compared with sham mice (64.1 ± 37.1 and 18.5 ± 5.7 pg/mL for IR and sham, respectively; $p < 0.05$). Interestingly, the AG administration (50 mg/kg) 30 min before IR protocol prevented the GFR downregulation (918.4 ± 79.0 µL/min/100 g body weight; $p < 0.05$) and avoided the NGAL upregulation (16.8 ± 16.8 pg/mL; $p < 0.05$) induced by IR (Figure 1). The kidney morphology was analyzed earlier (at 30 min of reperfusion) than 48 h of reperfusion. The data revealed normal kidney morphology in the cortex and the medulla of the control animals (sham group). Glomeruli and tubules had normal structures (Figure 1C). In contrast, acute tubular necrosis (ATN) was present in the kidney in the IR group, characterized by loss of the brush border, intratubular cellular detritus, loss of the nucleus, and vascular congestion (Figure 1C, arrows). In contrast, no signal of ATN was observed in the kidneys from sham and IR mice treated with AG. Altogether, these results suggested the renoprotective effect of AG against ischemia and reperfusion injury.

In addition, inflammation is one of the most critical stages in renal injury after ischemia. We investigated the mRNA expression of two pro-inflammatory cytokines (IL-1β and IL-6) and two regulatory factors of inflammation (IL-10 and Foxp3). We did not find a significant upregulation of IL-1β by IRI in the cortex or the medulla. However, we observed a significant upregulation of IL-6 mRNA in the renal medulla (but not in the cortex) by IR compared to the sham group. Besides, renal IR upregulated the Foxp3 mRNA expression in the kidney cortex but not in the medulla (Figure 2C,D). In addition, the IL-10 mRNA was not significantly upregulated by IR in the cortex or the medulla (Figure 2). Interestingly, AG reduced IL-1β and IL-6 mRNA levels in the kidney medulla sections, preventing the Foxp3 upregulation in the cortex. These data suggest that AG prevents the upregulation of inflammatory elements observed during renal IRI.

Besides, clusterin (CLU) and klotho are two markers of kidney protection by renal IR. We previously reported that CLU was increased in kidneys subjected to IRI, and it was downregulated by l-NIL (a specific iNOS inhibitor) [18]. Here, we also showed that mRNA levels of CLU were significantly increased in both cortex and medulla in the IRI animals compared with sham. Remarkably, the AG administration decreased the upregulation of CLU observed by IRI (Figure 3A,B). On another side, we studied the expression of klotho (a renoprotective and anti-aging gene highly expressed in the kidney), and we did not observe changes in the mRNA klotho expression in either cortex or medulla kidney sections by IRI or AG.

Finally, we showed that Arg-2 and TGF-β1 mRNA levels were significantly increased in the medulla in the IRI animals compared with sham. In the cortex, we observed similar results, but it was not statistically significant. Remarkably, the AG administration decreased the upregulation of Arg-2 and TGF-β1 observed by IRI (Figure 4)

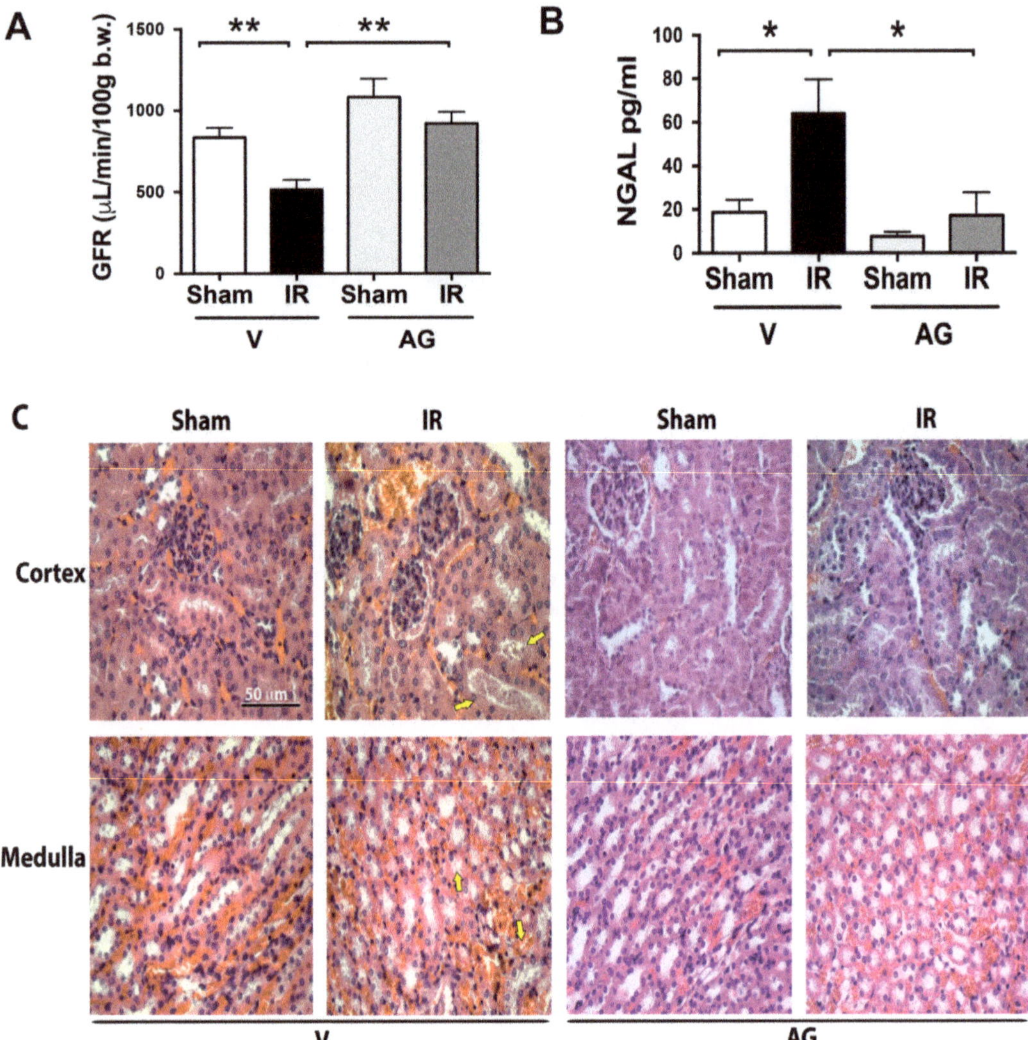

Figure 1. AG prevented the kidney damage induced by renal IRI. BALB/c mice were treated with either vehicle or AG (50 mg/kg i.p) before sham or ischemia-reperfusion (IR) surgery and subjected to 30 min of ischemia and 48 h of reperfusion. (**A**) GFR (μL/min/100 g body weight) was determined in sham (n = 5), IR (n = 6), AG-sham (n = 5), and AG-IR (n = 6). (**B**) Neutrophil gelatinase-associated lipocalin (NGAL) was measured by ELISA in blood samples in sham (n = 9), IR (n = 6), AG-sham (n = 5), and AG-IR (n = 6). The bar graphs represent mean $\pm$ SEM, and the data were analyzed using ANOVA and non-parametric Kruskal–Wallis test. * $p < 0.05$ and ** $p < 0.005$. (**C**) Histological analysis: representative hematoxylin/eosin (H/E) staining of the cortex and the medulla kidney sections. Yellow arrows indicate areas with evident kidney acute tubular necrosis (ATN) characterized by loss of nucleus in tubules or intratubular cellular detritus. Three kidneys for each protocol were analyzed. Representative images correspond to 400X scale bar = 50 μm.

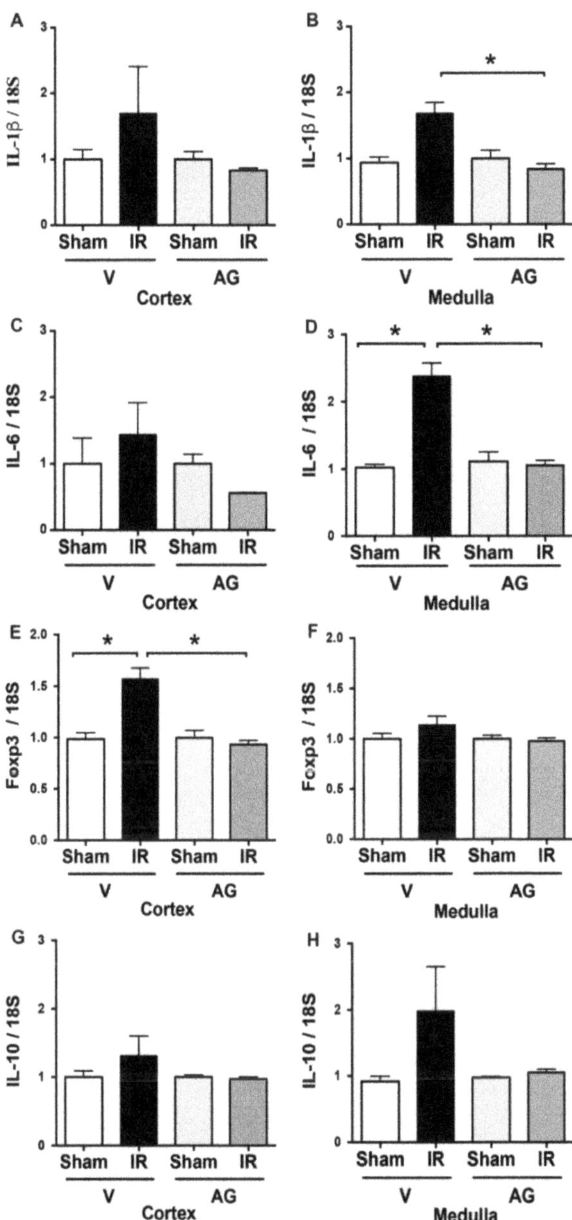

Figure 2. AG inhibited the IL-1β, IL-6, Foxp3, and IL-10 mRNA upregulation observed during IR. BALB/c mice were treated with either vehicle or AG (50 mg/kg i.p) before sham or subjected to 30 min of ischemia and 48 h of reperfusion (IR). Cytokine expression was determined by qPCR in sham (n = 5), IR (n = 5), AG-sham (n = 6), and AG IR (n = 6). (**A**) IL-1β mRNA in the cortex. (**B**) IL-1β mRNA in the medulla. (**C**) IL-6 mRNA in the cortex. (**D**) IL-6 mRNA in the medulla. (**E**) Foxp3 mRNA in the cortex. (**F**) Foxp3 mRNA in the medulla. (**G**) IL-10 mRNA in the cortex. (**H**) IL-10 mRNA in the medulla. The bar graphs represent mean ± SEM, and the data were analyzed using ANOVA and non-parametric Kruskal–Wallis test. * $p < 0.05$.

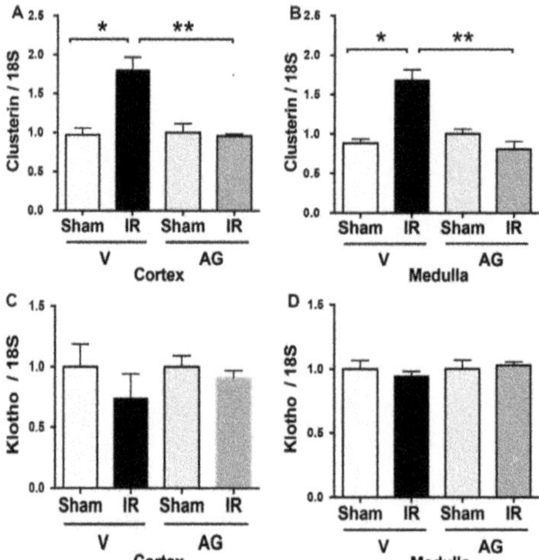

Figure 3. AG inhibited the upregulation of clusterin mRNA observed during IR, but not for klotho. BALB/c mice were treated with either vehicle or AG (50 mg/kg i.p) before sham or subjected to 30 min of ischemia and 48 h of reperfusion (IR). The clusterin and the klotho mRNA expressions were determined by qPCR in sham (n = 5), IR (n = 5), AG-sham (n = 6), and AG-IR (n = 6). (**A**) Clusterin mRNA in the cortex. (**B**) Clusterin mRNA in the medulla. (**C**) Klotho mRNA in the cortex (**D**) Klotho mRNA in the medulla. The bar graphs represent mean ± SEM, and the data were analyzed using ANOVA and non-parametric Kruskal–Wallis test. * $p < 0.05$, ** $p < 0.005$.

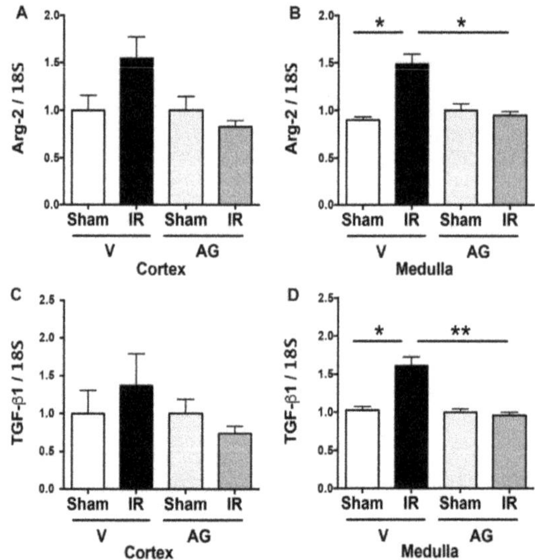

Figure 4. AG inhibited the upregulation of Arg-2 and TGF-β1 mRNA observed during IR. BALB/c mice were treated with either vehicle or AG (50 mg/kg i.p) before sham or subjected to 30 min of ischemia and 48 h of reperfusion (IR). The Arg-2 and the TGF-β1 mRNA expressions were determined by qPCR in sham (n = 5), IR (n = 5), AG-sham (n = 6), and AG-IR (n = 6). (**A**) Arg-2 mRNA in the cortex. (**B**) Arg-2 mRNA in the medulla. (**C**) TGF-β1 mRNA in the cortex (**D**) TGF-β1 mRNA in the medulla. The bar graphs represent mean ± SEM, and the data were analyzed using ANOVA and Kruskal–Wallis test. * $p < 0.05$, ** $p < 0.005$.

3.2. AG Prevented Oxidative Stress

We selected a panel of oxidative stress markers according to our previous publication [18]. We studied the oxidative stress in the whole kidney, assaying the GSH-to-GSSG ratio as a measure of the soluble antioxidant status (Figure 5A), the glutathione S-transferase (GST) activity as an antioxidant mechanism (Figure 5B), and the TBARS as a readout for lipid peroxidation (Figure 5C). Mice exposed to renal IR had 3.8-fold reduced levels of GSH:GSSG ratio compared to the sham group (15.5 ± 4 and 57.3 ± 1 for IR and sham, respectively; $p < 0.05$. Figure 5A). Besides, the GST activity was deeply downregulated by renal IR (4.0 ± 0.5 and 79.9 ± 12.5 μmol/min/g tissue for IR and sham, respectively; $p < 0.05$. Figure 5B). In consequence, the renal lipid peroxidation increased by renal IR (345.3 ± 15 and 175.8 ± 1.7 nmol/g-tissue for IR and sham, respectively; $p < 0.05$. Figure 5C). Noteworthy, AG treatment before renal IR increased the GSH:GSSG (33.4 ± 3.5 for AG-IR) ratio, recovered the GST activity (63.1 ± 3.0 μmol/min/g tissue for AG-IR), and reduced the oxidative damage (TBARS: 180.3 ± 3.8 nmol/g tissue for AG-IR) levels. All these data suggested that AG prevented oxidative stress in kidneys exposed to IRI.

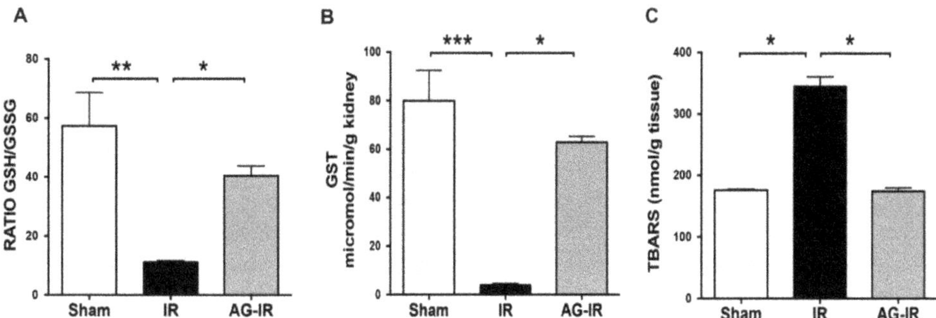

Figure 5. AG ameliorated the oxidative stress induced by renal IR. BALB/c mice were treated with either vehicle or AG (50 mg/kg i.p) before sham or IR surgery and subjected to 30 min of ischemia and 48 h of reperfusion. GSH:GSSG ratio, GST activity, and TBARS (lipoperoxidation) were determined in whole kidneys in sham (n = 5), IR (n = 5), and AG-IR (n = 6) (**A**) GSH:GSSG ratio. (**B**) Glutathione S-transferase (GST) activity. (**C**) Thiobarbituric acid reactive substances (TBARS) levels. Bar graph represents mean ± SEM, and data were analyzed by ANOVA non-parametric Kruskal–Wallis analysis, * $p < 0.05$, ** $p < 0.005$, *** $p < 0.0005$.

3.3. Molecular Sensors of AG Antioxidant Effect

During renal IRI, the iNOS activity induces oxidative stress [42–44], and Hsp27 prevents it [45,46]. Here, we observed an upregulation of iNOS mRNA by renal IR in the cortex and the medulla (Figure 6A,B). Noteworthy, AG significantly prevented the upregulation of iNOS by renal IR. On the other hand, Hsp27 protein expression was significantly upregulated by IR in the cortex kidney, and AG avoided its expression. These data suggested that the antioxidant effect of AG prevented the upregulation of iNOS and Hsp27 (Figure 6C,D) involved in the oxidant response in the kidney by IRI.

3.4. Effect of AG on Endothelial Markers Expression

To examine the mechanisms involved in the protection induced by AG treatment against IRI, we studied the mRNA expression of endothelial markers VE-cadherin and PECAM-1 (platelet/endothelial cell adhesion molecule 1 or CD31) [32]. The results demonstrated that VE-cadherin mRNAs were significantly upregulated in the renal medulla in mice subjected to 48 h of reperfusion. The PECAM mRNA also increased, but it was not statistically significant. The administration of AG before IR significantly reduced VE-cadherin and PECAM mRNA (Figure 7). The findings suggested that AG prevented the endothelial dysfunction induced by IRI, most likely through its antioxidant effects.

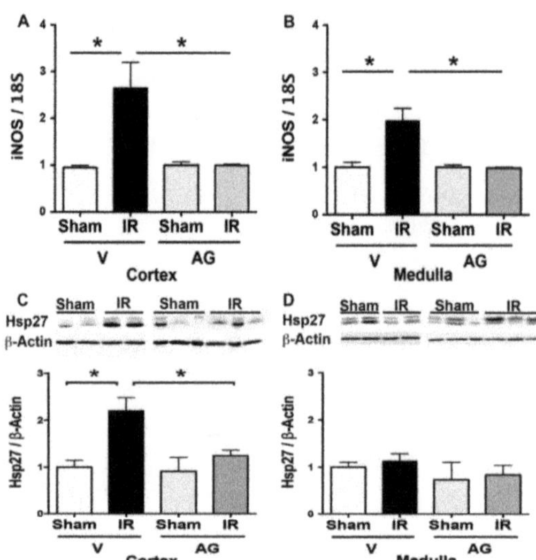

Figure 6. AG inhibited upregulation of iNOS and Hsp27 observed during IR. BALB/c mice were treated with AG (50 mg/kg i.p.) before sham or subjected to 30 min of ischemia and 48 h of reperfusion (IR). The iNOS mRNA expression was determined by qPCR and Hsp27 protein by Western blot in sham (n = 5), IR (n = 5), AG-sham (n = 6), and AG-IR (n = 6). (**A**) INOS mRNA in the cortex. (**B**) INOS mRNA in the medulla. (**C**) Hsp27 protein in the cortex. (**D**) Hsp27 protein in the medulla. The bar graph represents mean ± SEM, and the data were analyzed by using ANOVA and non-parametric Kruskal–Wallis test. * $p < 0.05$.

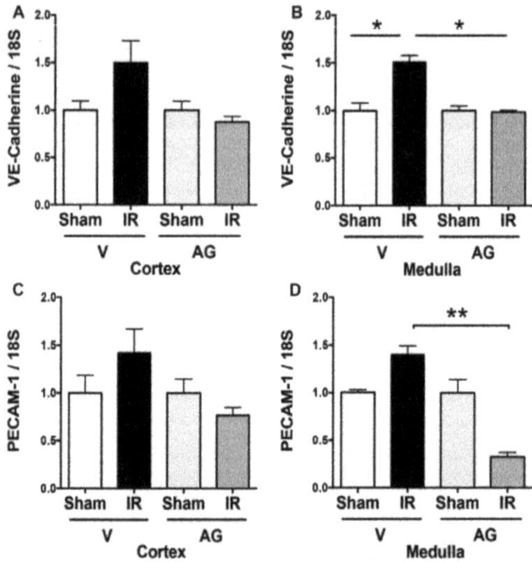

Figure 7. AG prevented VE-cadherin and PECAM upregulation induced by IR. BALB/c mice were treated with either vehicle or AG (50 mg/kg i.p) before sham or subjected to 30 min of ischemia and 48 h of reperfusion. VE-cadherin and PECAM-1 expressions were determined by qPCR in sham (n = 5), IR (n = 5), AG-sham (n = 6), and AG-IR (n = 6). (**A**) VE-cadherin mRNA in the cortex. (**B**) VE-cadherin mRNA in the medulla. (**C**) PECAM-1 mRNA in the cortex. (**D**) PECAM-1 mRNA in the medulla. The bar graph represents mean ± SEM using ANOVA non-parametric Kruskal–Wallis test. * $p < 0.05$, ** $p < 0.005$.

3.5. Effect of AG on Mesenchymal Markers Expression: Vimentin, Fascin, and Hsp47

Vimentin is an intermediate filament used as a marker for the state of differentiation and is expressed in kidney mesenchymal cells but not in tubular epithelial cells. On the other hand, fascin-1 is an actin-bundling protein involved in cell migration and expressed in the mesenchymal phenotype. The Hsp47 is a pro-fibrotic protein required for procollagen biosynthesis and is also suggested to be increased by mesenchymal phenotype. The mRNA expressions of vimentin, fascin, and Hsp47 were upregulated in the cortex and the medulla kidney in mice subjected to IR (Figure 8). Notably, AG treatment significantly prevented vimentin, fascin, and Hsp47 upregulation observed by IR in both kidney sections. These data suggested that the antioxidant effect of aminoguanidine prevented the expression of mesenchymal transition markers in the kidney induced by IRI.

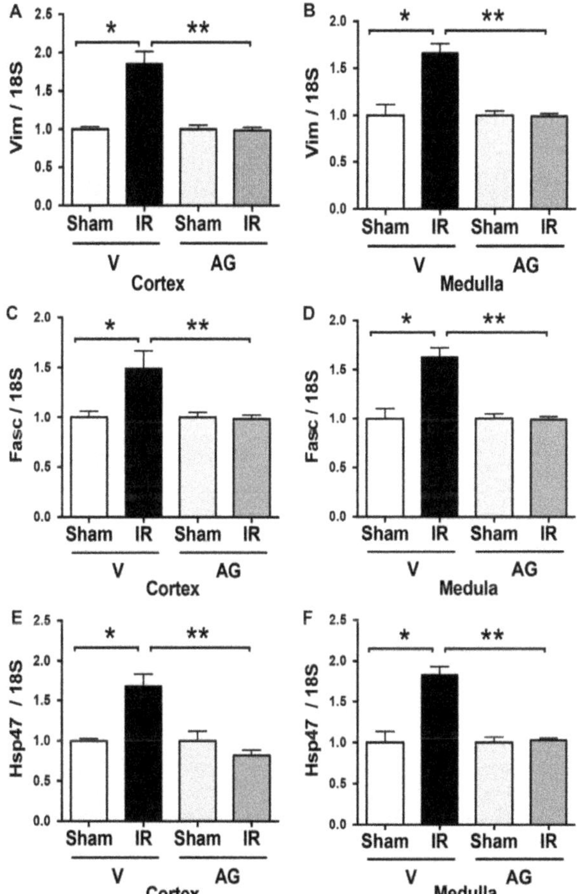

Figure 8. AG inhibited vimentin, fascin, and Hsp47 mRNA upregulation during IR. BALB/c mice were treated with either vehicle or AG (50 mg/kg i.p) before sham or subjected to 30 min of ischemia and 48 h of reperfusion (IR). Vimentin, fascin, and Hsp47 mRNA expression were determined by qPCR in sham (n = 5), IR (n = 5), AG-sham (n = 6), and AG-IR (n = 6). (**A**) Vimentin mRNA in the cortex. (**B**) Vimentin mRNA in the medulla (**C**) Fascin mRNA in the cortex. (**D**) Fascin mRNA in the medulla. (**E**) Hsp47 mRNA in the cortex. (**F**) Hsp47 mRNA in the medulla. The bar graph represents mean + SEM, and the data were analyzed using ANOVA non parametric Kruskal–Wallis test. * $p < 0.05$, ** $p < 0.005$.

4. Discussion

We found an antioxidant effect of AG in the murine model of ischemic (30 min) and reperfusion (48 h) injury. The major findings of this study are: (1) the AG pretreatment recovered GSH levels and GST activity in the kidney during renal IRI, therefore reducing the lipoperoxidation levels in the whole kidney. (2) The upregulation of iNOS (cortex and medulla) and Hsp27 (cortex) by IRI was prevented by AG pretreatment. (3) The upregulation of clusterin, Arg-2, and TGF-β1 by IRI was also prevented by AG. (4) The kidney damage biomarkers (GFR, NGAL, and histology) observed by IRI were prevented by AG. (5) AG inhibited IL-1β, IL-6, and Foxp3 mRNA upregulation induced by IRI. (6) The AG prevented the endothelial dysfunction (VE-cadherin and PECAM) and the upregulation of mesenchymal transition markers (vimentin, fascin1, and Hsp47) observed by IRI.

4.1. Antioxidant Effect of AG and Prosurvival Genes

In the kidney, after injury, NADPH oxidase, mitochondria, and inducible nitric oxide synthase are the primary sources of oxidative stress [14,15]. The glutathione S-transferase (GST) is an isozyme whose primary function is to detoxify and neutralize a wide variety of electrophilic molecules by mediating their conjugation with reduced glutathione [47]. Low levels of GST increase the presence of oxidative stress and reduce the GSH:GSSG ratio, therefore increasing levels of lipoperoxidation. We previously described the downregulation of GST activity by renal IRI in mice [18]. Our results indicated that the administration of AG before renal IR increased the levels of the GSH, recovered the GST activity, and reduced oxidative damage. To our knowledge, this is the first report showing that, during renal IRI in mice, AG recovered GSH levels and GST activity, reducing the lipoperoxidation.

A high level of oxidative stress can lead to the induction of different pro-survival signals. Here, we showed increased oxidative stress and upregulation of Hsp27 only in the cortex during IRI. During oxidative stress, Hsp27 functions as an antioxidant in cells, lowering ROS levels by reducing intracellular iron levels and raising intracellular levels of reduced glutathione [48]. Hsp27 confers cytoprotection from ischemia [49] by stabilizing the actin cytoskeleton, displaying an anti-apoptotic function, reducing ATP depletion, and inhibiting the reactive oxygen species [50]. The antioxidant effect of Hsp27 was described in the heart and the brain [51], and, together with the A1 adenosine receptor (AR), it works as a renal protector against IRI [52]. Therefore, the present data showed that AG inhibited the upregulation of Hsp27 during renal IRI, which could be a consequence of the antioxidant effect of this compound.

Clusterin (CLU) is an extremely sensitive biosensor to exogenous or endogenous stress, particularly free radicals and their derivatives. Several studies demonstrated that CLU has a cytoprotective role against the deleterious effects of oxidants, working as an antioxidant protein [53]. Recently, it was observed that glomerular CLU is upregulated in diabetic nephropathy and may have a protective effect against oxidative stress-induced apoptosis in podocytes [54]. Induction of CLU during IRI was proposed as a survival function in response to high levels of oxidative stress through a mechanism involved in survival autophagy [55]. In addition, autophagy protects renal tubular cells against ischemia [56]. The prevention of Hsp47, Arg-2, and TGF-β1 mRNA levels upregulation by the AG treatment suggested inhibition of the kidney injury process associated with the development of fibrosis. The present work provided the first evidence that AG inhibits CLU, Arg-2, TGF-β1, and Hsp27 upregulation by IRI (Figures 3, 4 and 6, respectively), suggesting that the antioxidant effect of AG alleviates the injury induced by IR.

4.2. Antioxidant Effect of AG and Endothelial Injury

During renal IRI, endothelial damage is observed. Previously, it was described that, in a murine model of IRI, the renal VE-cadherin mRNA levels were downregulated in the first hour of reperfusion. However, after 24 h of reperfusion, the VE-cadherin mRNA levels were significantly upregulated compared to the sham group as a compensatory effect [57]. We found that VE-cadherin and PECAM-1 mRNA were upregulated after 48 h of reperfusion compared with the sham group. Remarkably, the antioxidant effect of AG prevented this phenomenon. Interestingly, PECAM could result in the transmission of a pro-survival signal suppressing the mitochondria-dependent Bax-mediated intrinsic apoptotic pathway or inhibiting cytochrome c release from mitochondria [58]. These data suggested that the antioxidant effect of AG had a protective effect on renal endothelium during IRI.

Additionally, we noted upregulation of vimentin, fascin, and Hsp47 after IRI. Previously, Xu-Dubois et al. demonstrated increased expression of these markers in early posttransplant biopsies, associated with poor renal graft recovery and/or late graft dysfunction and endothelium injury [36]. Therefore, our data suggested that AG could have a renoprotective effect during kidney transplants, reducing endothelium dysfunction. Future studies are needed to determine the role of AG during kidney transplantation.

4.3. AG and Oxidative Stress in the Kidney

The pathophysiology of renal IR is a complex process regulated by several intracellular pathways, in which reactive oxygen and nitrogen species (ROS/RNS) appear to be central mediators of detrimental effects. During the renal IR, the reperfusion phase results in increased ROS production [12,15]. After the injury, the primary source of oxidative stress in the kidney comes from NADPH oxidase, mitochondria, and inducible nitric oxide synthase [14,15]. Overproduction of inflammatory mediators and ROS can lead to perturbations of microcirculation and tissue oxygenation [59]. On another note, NO produced by iNOS enhances oxidative stress and impairs renal function [60]. The NO combines with the superoxide radical and forms the cytotoxic metabolite, peroxynitrite, which causes cell membrane damage through protein nitration [61]. Thus, an increase in nitrotyrosine levels in patients with acute kidney injury is associated with overall mortality [62]. Therefore, ROS and NO can inhibit the compensatory changes in renal hemodynamics produced by IR after renal ischemia, therefore decreasing tubular function [63,64]. Our animal model of renal IRI produced elevated oxidative stress (inactivation of GST, increased GSSG levels, and elevated lipid peroxidation), decreased the glomerular filtration rate, and increased kidney damage (NGAL and clusterin). Noteworthy, the pretreatment of AG 30 min before IRI reduced the oxidative stress, and the renoprotective effect was observed.

Besides ischemia, other pathophysiological mechanisms can contribute to kidney damage, such as antibiotics aminoglycosides and chemotherapeutic agents. Data on the role of AG in kidney toxicity have been limited to studies with rats. In a cyclophosphamide (CP) induced nephrotoxicity model, AG protects the kidney from oxidative stress. The CP administration (50 mg/kg body weight, i.p.) induced oxidative stress in the kidney by increase of lipoperoxidation, protein oxidation, depletion of GSH, as well as loss of catalase, glutathione peroxidase (GPx), myeloperoxidase (MPO), and GST. The pretreatment with AG (200 mg/kg body weight, i.p. 1 h before CP administration) prevented CP-induced oxidative stress, decreased lipoperoxidation, protein oxidation, GSH levels, and recovered GPx catalase, GST, and MPO activities [65]. Another study explored the protective effect of AG on nephrotoxicity induced by gentamicine (GEN). The GEN administration to control group rats increased renal MDA and NO levels but decreased GPx, SOD, CAT activities, and GSH content. The AG administration with GEN injection significantly decreased MDA and NO generation and increased GPx, SOD, CAT activities, and GSH abundance [66]. Using a mice model of IR, our finding reproduced the antioxidant effect of AG observed by cyclophosphamide and gentamicine oxidative stress in rats. In our case, we used a lower dose of AG (50 mg/kg body weight, i.p), and we found that the antioxidant effect of AG (recovered GSH, reduced lipoperoxidation, and increased GST activity) was associated with inhibition of iNOS, Hsp27, endothelial activation (VE-cadherin and PECAM) and mesenchymal markers (vimentin, fascin1, and Hsp47). Besides, we studied the mRNA of two pro-inflammatory (IL-1β and IL-6) and two anti-inflammatories (Foxp3 and IL-10) elements in the cortex and the medulla. After 48 h of reperfusion, we found significant upregulation of Foxp3 mRNA in the cortex. On the other hand, IL-1b and IL-6 were upregulated by IR (48 h reperfusion) in the medulla but not the anti-inflammatory elements (Foxp3 and IL-10). Remarkably, the pro-inflammation–anti-inflammatory equilibrium was entirely abolished by AG. Finally, AG improved the glomerular filtration rate and reduced the tubular epithelial damage (NGAL, clusterin, arg-2, and TFG-β1). Therefore, the antioxidant properties of AG prevented acute kidney injury markers and improved kidney function. Consequently, AG is a promising pharmacological compound to treat the deleterious effect of acute renal ischemia and reperfusion injury.

5. Conclusions

The present study demonstrated the beneficial effect of AG on renal IRI through the suppression of oxidative stress, iNOS, Hsp27, Arg-2, and TFG-β1. In addition, AG prevented endothelium activation mesenchymal transition markers and inflammation elements. Moreover, AG reduced kidney injury and improved kidney function. These results may be of potential clinical relevance, and the protective effect of AG as a therapeutic strategy may be of value in the future.

Supplementary Materials: The following are available online at https://www.mdpi.com/article/10.3390/antiox10111724/s1. Table S1. Primers for Real-Time quantitative PCR.

Author Contributions: Conceptualization, C.E.I., methodology, C.P., M.L., J.R., C.A., J.L., L.M. and C.E.I., software, L.M., J.L., M.L., C.E.I. and L.M., validation, C.E.I.; formal analysis, L.M., J.L., M.L., F.C., C.E.I.; investigation, C.P., C.E.I., resources; L.M., C.E.I., F.C., data curation, C.A., J.L., L.M. and C.E.I., writing—original draft preparation, C.P., C.E.I., writing—review and editing, C.P., C.E.I., visualization, C.E.I., supervision, C.E.I., project administration, C.P., funding acquisition, C.E.I.; J.L.; and C.P. All authors have read and agreed to the published version of the manuscript.

Funding: This research was funded by: FONDECYT-1151157, FONDECYT-21150304 and FONDECYT-1211949 (supported by ANID, Agencia Nacional de Investigación y Desarrollo, Chile); and FAI-2019 and FAI-Puente 2019 (supported by Fondo de Ayuda a la Investigación, Universidad de los Andes, Chile).

Institutional Review Board Statement: The study was conducted according to the guidelines of the Declaration of Helsinki and approved by the Institutional Review Board of the Committee on the Ethics of Animal Experiments of the Universidad de los Andes, Chile.

Informed Consent Statement: Not applicable.

Data Availability Statement: Data is contained within the article and supplementary materials.

Conflicts of Interest: The authors declare no conflict of interest.

References

1. Itakura, M.; Yoshikawa, H.; Bannai, C.; Kato, M.; Kunika, K.; Kawakami, Y.; Yamaoka, T.; Yamashita, K. Aminoguanidine decreases urinary albumin and high-molecular-weight proteins in diabetic rats. *Life Sci.* **1991**, *49*, 889–897. [CrossRef]
2. Luo, D.; Fan, Y.; Xu, X. The effects of aminoguanidine on retinopathy in STZ-induced diabetic rats. *Bioorg. Med. Chem. Lett.* **2012**, *22*, 4386–4390. [CrossRef]
3. Yagihashi, S.; Kamijo, M.; Baba, M.; Yagihashi, N.; Nagai, K. Effect of aminoguanidine on functional and structural abnormalities in peripheral nerve of STZ-induced diabetic rats. *Diabetes* **1992**, *41*, 47–52. [CrossRef] [PubMed]
4. Yamagishi, S.; Nakamura, N.; Suematsu, M.; Kaseda, K.; Matsui, T. Advanced Glycation End Products: A Molecular Target for Vascular Complications in Diabetes. *Mol. Med.* **2015**, *21*, S32–S40. [CrossRef] [PubMed]
5. Schalkwijk, C.G.; Miyata, T. Early- and advanced non-enzymatic glycation in diabetic vascular complications: The search for therapeutics. *Amino Acids* **2012**, *42*, 1193–1204. [CrossRef]
6. Dobšák, P.; Courderot-Masuyer, C.; Siegelová, J.; Svačinová, H.; Jančík, J.; Vergely-Vanriessen, C.; Rochette, L. Antioxidant properties of aminoguanidine: A Paramagnetic Resonance test. *Scr. Med. Fac. Med. Univ. Brun. Masaryk.* **2001**, *74*, 41–50.
7. Colzani, M.; De Maddis, D.; Casali, G.; Carini, M.; Vistoli, G.; Aldini, G. Reactivity, Selectivity, and Reaction Mechanisms of Aminoguanidine, Hydralazine, Pyridoxamine, and Carnosine as Sequestering Agents of Reactive Carbonyl Species: A Comparative Study. *ChemMedChem* **2016**, *11*, 1778–1789. [CrossRef] [PubMed]
8. Ihm, S.H.; Yoo, H.J.; Park, S.W.; Ihm, J. Effect of aminoguanidine on lipid peroxidation in streptozotocin-induced diabetic rats. *Metabolism* **1999**, *48*, 1141–1145. [CrossRef]
9. Abd El-Gawad, H.M.; El-Sawalhi, M.M. Nitric oxide and oxidative stress in brain and heart of normal rats treated with doxorubicin: Role of aminoguanidine. *J. Biochem. Mol. Toxicol.* **2004**, *18*, 699–777. [CrossRef]

10. Tunc, T.; Kesik, V.; Demirin, H.; Ersoz, N.; Vurucu, S.; Kul, M.; Uysal, B.; Sadir, S.; Guven, A.; Oztas, E. Effects of aminoguanidine and melatonin on intestinal ischemia/reperfusion injury in rats: An assessor-blinded, controlled experimental study. *Curr. Ther. Res. Clin. Exp.* **2009**, *6*, 449–459. [CrossRef]

11. Esson, M.L.; Schrier, R.W. Diagnosis and treatment of acute tubular necrosis. *Ann. Intern. Med.* **2002**, *9*, 744–752. [CrossRef]

12. Noiri, E.; Nakao, A.; Uchida, K.; Tsukahara, H.; Ohno, M.; Fujita, T.; Brodsky, S.; Goligorsky, M.S. Oxidative and nitrosative stress in acute renal ischemia. *AJP-Ren. Physiol.* **2001**, *281*, F948–F957. [CrossRef] [PubMed]

13. Paller, M.S.; Hoidal, J.R.; Ferris, T.F. Oxygen free radicals in ischemic acute renal failure in the rat. *J. Clin. Investig.* **1984**, *74*, 1156–1164. [CrossRef] [PubMed]

14. Goligorsky, M.S.; Brodsky, S.V.; Noiri, E. Nitric oxide in acute renal failure: NOS versus NOS. *Kidney Int.* **2002**, *61*, 855–861. [CrossRef] [PubMed]

15. Ratliff, B.B.; Abdulmahdi, W.; Pawar, R.; Wolin, M.S. Oxidant mechanisms in renal injury and disease. *Antioxid. Redox Signal.* **2016**, *25*, 119–146. [CrossRef]

16. Ozbek, E. Induction of oxidative stress in kidney. *Int. J. Nephrol.* **2012**, *2012*, 465897. [CrossRef] [PubMed]

17. Dounousi, E.; Papavasiliou, E.; Makedou, A.; Ioannou, K.; Katopodis, K.P.; Tselepis, A.; Siamopoulos, K.C.; Tsakiris, D. Oxidative Stress Is Progressively Enhanced with Advancing Stages of CKD. *Am. J. Kidney Dis.* **2006**, *48*, 752–760. [CrossRef]

18. Pasten, C.; Alvarado, C.; Rocco, J.; Contreras, L.; Aracena, P.; Liberona, J.; Suazo, C.; Michea, L.F.; Irarrázabal, C.E. L-NIL prevents the ischemia and reperfusion injury involving TRL4, GST, clusterin and NFAT5 in mice. *AJP Ren. Physiol.* **2018**, *316*, F624–F634. [CrossRef]

19. Fatemikia, H.; Ketabchi, F.; Karimi, Z.; Moosavi, S.M. Distant effects of unilateral renal ischemia/reperfusion on contralateral kidney but not lung in rats: The roles of ROS and iNOS. *Can. J. Physiol. Pharmacol.* **2016**, *94*, 447–487. [CrossRef]

20. Sahna, E.; Parlakpinar, H.; Cihan, O.F.; Turkoz, Y.; Acet, A. Effects of aminoguanidine against renal ischaemia-reperfusion injury in rats. *Cell Biochem. Funct.* **2006**, *24*, 137–141. [CrossRef]

21. Onem, Y.; Ipcioglu, O.M.; Haholu, A.; Sen, H.; Aydinoz, S.; Suleymanoglu, S.; Bilgi, O.; Akyol, I. Posttreatment with aminoguanidine attenuates renal ischemia/reperfusion injury in rats. *Ren. Fail.* **2009**, *31*, 50–53. [CrossRef] [PubMed]

22. Tsuji, M.; Higuchi, Y.; Shiraishi, K.; Kume, T.; Akaike, A.; Hattori, H. Protective effect of aminoguanidine on hypoxic-ischemic brain damage and temporal profile of brain nitric oxide in neonatal rat. *Pediatric Res.* **2000**, *47*, 79–83. [CrossRef]

23. Parlakpinar, H.; Ozer, M.K.; Acet, A. Effect of aminoguanidine on ischemia-reperfusion induced myocardial injury in rats. *Mol. Cell Biochem.* **2005**, *277*, 137–142. [CrossRef] [PubMed]

24. Sugimoto, K.; Iadecola, C. Effects of aminoguanidine on cerebral ischemia in mice: Comparison between mice with and without inducible nitric oxide synthase gene. *Neurosci. Lett.* **2002**, *331*, 25–28. [CrossRef]

25. Mattson, D.L.; Wu, F. Nitric oxide synthase activity and isoforms in rat renal vasculature. *Hypertension* **2000**, *35*, 337–341. [CrossRef] [PubMed]

26. Guo, Q.; Du, X.; Zhao, Y.; Zhang, D.; Yue, L.; Wang, Z. Ischemic postconditioning prevents renal ischemia reperfusion injury through the induction of heat shock proteins in rats. *Mol. Med. Rep.* **2014**, *10*, 2875–2881. [CrossRef] [PubMed]

27. Chebotareva, N.; Bobkova, I.; Shilov, E. Heat shock proteins and kidney disease: Perspectives of HSP therapy. *Cell Stress Chaperones* **2017**, *22*, 319–343. [CrossRef]

28. Kim, M.; Park, S.W.; Kim, M.; Chen, S.W.; Gerthoffer, W.T.; D'Agati, V.D.; Lee, H.T. Selective renal overexpression of human heat shock protein 27 reduces renal ischemia-reperfusion injury in mice. *Am. J. Physiol. Ren. Physiol.* **2010**, *299*, F347–F358. [CrossRef]

29. Wetzel, M.D.; Stanley, K.; Wang, W.W.; Maity, S.; Madesh, M.; Reeves, W.B.; Awad, A.S. Selective inhibition of arginase-2 in endothelial cells but not proximal tubules reduces renal fibrosis. *JCI Insight* **2020**, *5*, e142187. [CrossRef]

30. Chung, S.; Overstreet, J.M.; Li, Y.; Wang, Y.; Niu, A.; Wang, S.; Fan, X.; Sasaki, K.; Jin, G.N.; Khodo, S.N.; et al. TGF-β promotes fibrosis after severe acute kidney injury by enhancing renal macrophage infiltration. *JCI Insight* **2018**, *3*, e123563. [CrossRef]

31. Cruz-Solbes, A.S.; Youker, K. Epithelial to Mesenchymal Transition (EMT) and Endothelial to Mesenchymal Transition (EndMT): Role and Implications in Kidney Fibrosis. *Results Probl. Cell Differ.* **2017**, *60*, 345–372. [CrossRef] [PubMed]

32. Kovacic, J.C.; Mercader, N.; Torres, M.; Boehm, M.; Fuster, V. Epithelial-to-mesenchymal and endothelial-to-mesenchymal transition from cardiovascular development to disease. *Circulation* **2012**, *125*, 1795–1808. [CrossRef] [PubMed]

33. Lamouille, S.; Xu, J.; Derynck, R. Molecular mechanisms of epithelial-mesenchymal transition. *Nat. Rev. Mol. Cell Biol.* **2014**, *15*, 178–196. [CrossRef]

34. Ranchoux, B.; Antigny, F.; Rucker-Martin, C.; Hautefort, A.; Péchoux, C.; Bogaard, H.J.; Dorfmüller, P.; Remy, S.; Lecerf, F.; Planté, S.; et al. Endothelial-to-mesenchymal transition in pulmonary hypertension. *Circulation* **2015**, *131*, 1006–1018. [CrossRef]

35. LeBleu, V.S.; Taduri, G.; O'Connell, J.; Teng, Y.; Cooke, V.G.; Woda, C.; Sugimoto, H.; Kalluri, R. Origin and function of myofibroblasts in kidney fibrosis. *Nat. Med.* **2013**, *19*, 1047–1053. [CrossRef]

36. Xu-Dubois, Y.C.; Ahmadpoor, P.; Brocheriou, I.; Kevin, L.; Arzouk Snanoudj, N.; Rouvier, P.; Taupin, J.-L.; Corchia, A.; Galichon, P.; Barrou, B.; et al. Microvasculature partial endothelial mesenchymal transition in early posttransplant biopsy with acute tubular necrosis identifies poor recovery renal allografts. *Am. J. Transplant.* **2020**, *20*, 2400–2412. [CrossRef]

37. Wei, Q.; Dong, Z. Mouse model of ischemic acute kidney injury: Technical notes and tricks. *Am. J. Physiol. Renal. Physiol.* **2012**, *303*, F1487–F1494. [CrossRef] [PubMed]

38. Lowry, O.H.; Rosebrough, N.J.; Farr, A.L.; Randall, R.J. Protein measurement with the Folin phenol reagent. *J. Biol. Chem.* **1951**, *193*, 265. [CrossRef]

39. Ohkawa, H.; Ohishi, N.; Yagi, K. Assay for lipid peroxides in animal tissues by thiobarbituric acid reaction. *Anal. Biochem.* **1979**, *95*, 351–358. [CrossRef]

40. Tietze, F. Enzymic method for quantitative determination of nanogram amounts of total and oxidized glutathione: Applications to mammalian blood and other tissues. *Anal. Biochem.* **1969**, *27*, 502–522. [CrossRef]

41. Giani, J.F.; Bernstein, K.E.; Janjulia, T.; Han, J.; Toblli, J.E.; Shen, X.Z.; Rodriguez-Iturbe, B.; McDonough, A.A.; Gonzalez-Villaloboset, G.A. Salt Sensitivity in Response to Renal Injury Requires Renal Angiotensin-Converting Enzyme. *Hypertension* **2015**, *66*, 534. [CrossRef]

42. Lieberthal, W. Biology of ischemic and toxic renal tubular cell injury: Role of nitric oxide and the inflammatory response. *Curr. Opin. Nephrol. Hypertens.* **1998**, *7*, 289–295. [CrossRef]

43. Taylor, E.L.; Megson, I.L.; Haslett, C.; Rossi, A.G. Nitric oxide: A key regulator of myeloid inflammatory cell apoptosis. *Cell Death Differ.* **2003**, *10*, 418–430. [CrossRef]

44. Mark, L.A.; Robinson, A.V.; Schulak, J.A. Inhibition of nitric oxide synthase reduces renal ischemia/reperfusion injury. *J. Surg. Res.* **2005**, *129*, 236–241. [CrossRef]

45. Vidyasagar, A.; Reese, S.; Acun, Z.; Hullett, D.; Djamali, A. HSP27 is involved in the pathogenesis of kidney tubulointerstitial fibrosis. *Am. J. Physiol. Ren. Physiol.* **2008**, *295*, F707–F716. [CrossRef] [PubMed]

46. Park, S.W.; Chen, S.W.; Kim, M.; D'Agati, V.D.; Lee, H.T. Human heat shock protein 27-overexpressing mice are protected against acute kidney injury after hepatic ischemia and reperfusion. *Am. J. Physiol. Ren. Physiol.* **2009**, *297*, F885–F894. [CrossRef]

47. Chatterjee, A.; Gupta, S. The multifaceted role of glutathione S-transferases in cancer. *Cancer Lett.* **2018**, *433*, 33–42. [CrossRef]

48. Arrigo, A.P.; Virot, S.; Chaufour, S.; Firdaus, W.; Kretz-Remy, C.; Diaz-Latoud, C. Hsp27 consolidates intracellular redox homeostasis by upholding glutathione in its reduced form and by decreasing iron intracellular levels. *Antioxid. Redox Signal.* **2005**, *7*, 414–422. [CrossRef] [PubMed]

49. Jin, C.; Cleveland, J.C.; Ao, L.; Li, J.; Zeng, Q.; Fullerton, D.A.; Meng, X. Human myocardium releases heat shock protein 27 (HSP27) after global ischemia: The proinflammatory effect of extracellular HSP27 through toll-like receptor (TLR)-2 and TLR4. *Mol. Med.* **2014**, *20*, 280–289. [CrossRef] [PubMed]

50. Stetler, R.A.; Gao, Y.; Signore, A.P.; Cao, G.; Chen, J. HSP27: Mechanisms of cellular protection against neuronal injury. *Curr. Mol. Med.* **2007**, *9*, 863–872. [CrossRef]

51. Efthymiou, C.A.; Mocanu, M.M.; de Belleroche, J.; Wells, D.J.; Latchmann, D.S.; Yellon, D.M. Heat shock protein 27 protects the heart against myocardial infarction. *Basic Res. Cardiol.* **2004**, *99*, 392–394. [CrossRef]

52. Xiong, B.; Li, M.; Xiang, S.; Han, L. A1AR-mediated renal protection against ischemia/reperfusion injury is dependent on HSP27 induction. *Int. Urol. Nephrol.* **2018**, *50*, 1355–1363. [CrossRef] [PubMed]

53. Trougakos, I.P. The molecular chaperone apolipoprotein J/clusterin as a sensor of oxidative stress: Implications in therapeutic approaches—A mini-review. *Gerontology* **2013**, *59*, 514–523. [CrossRef]

54. He, J.; Dijkstra, K.L.; Bakker, K.; Bus, P.; Bruijn, J.A.; Scharpfenecker, M.; Baelde, H.J. Glomerular clusterin expression is increased in diabetic nephropathy and protects against oxidative stress-induced apoptosis in podocytes. *Sci. Rep.* **2020**, *10*, 14888. [CrossRef]

55. Guan, X.; Qian, Y.; Shen, Y.; Zhang, L.; Du, Y.; Dai, H.; Qian, J.; Yan, Y. Autophagy protects renal tubular cells against ischemia/reperfusion injury in a time-dependent manner. *Cell Physiol. Biochem.* **2015**, *36*, 285–298. [CrossRef] [PubMed]

56. Kaushal, G.P. Autophagy protects proximal tubular cells from injury and apoptosis. *Kidney Int.* **2012**, *82*, 1250–1253. [CrossRef]

57. Paulus, P.; Rupprecht, K.; Baer, P.; Obermüller, N.; Penzkofer, D.; Reissig, C.; Scheller, B.; Holfeld, J.; Zacharowski, K.; Dimmeler, S.; et al. The early activation of toll-like receptor (TLR)-3 initiates kidney injury after ischemia and reperfusion. *PLoS ONE* **2014**, *9*, e94366. [CrossRef] [PubMed]

58. Newman, P.J.; Newman, D.K. Signal transduction pathways mediated by PECAM-1: New roles for an old molecule in platelet and vascular cell biology. *Arterioscler. Thromb. Vasc. Biol.* **2003**, *23*, 953–964. [CrossRef] [PubMed]

59. Eleftheriadis, T.; Pissas, G.; Filippidis, G.; Liakopoulos, V.; Stefanidis, I. Reoxygenation induces reactive oxygen species production and ferroptosis in renal tubular epithelial cells by activating aryl hydrocarbon receptor. *Mol. Med. Rep.* **2021**, *23*, 41. [CrossRef]

60. Chatterjee, P.K.; Patel, N.S.A.; Kvale, E.O.; Cuzzocrea, S.; Brown, P.A.J.; Stewart, K.N.; Mota-Filipe, H.; Thiemermann, C. Inhibition of inducible nitric oxide synthase reduces renal ischemia/reperfusion injury. *Kidney Int.* **2002**, *61*, 862–871. [CrossRef]

61. Radi, R.; Beckman, J.S.; Bush, K.M.; Freeman, B.A. Peroxynitrite-induced membrane lipid peroxidation: The cytotoxic potential of superoxide and nitric oxide. *Arch. Biochem. Biophys.* **1991**, *288*, 481–487. [CrossRef]

62. Qian, J.; You, H.; Zhu, Q.; Ma, S.; Zhou, Y.; Zheng, Y.; Liu, J.; Kuang, D.; Gu, Y.; Hao, C.; et al. Nitrotyrosine level was associated with mortality in patients with acute kidney injury. *PLoS ONE* **2013**, *8*, E7996. [CrossRef] [PubMed]

63. Lee, J.U. Nitric oxide in the kidney: Its physiological role and pathophysiological implications. *Electrolyte Blood Press* **2008**, *6*, 27–34. [CrossRef] [PubMed]

64. Legrand, M.; Almac, E.; Mik, E.G.; Johannes, T.; Kandil, A.; Bezemer, R.; Payen, D.; Ince, C. L-NIL prevents renal microvascular hypoxia and increase of renal oxygen consumption after ischemia-reperfusion in rats. *AJP Ren. Physiol.* **2009**, *296*, F1109–F1117. [CrossRef] [PubMed]

65. Abraham, P.; Rabi, S. Protective effect of aminoguanidine against cyclophosphamide-induced oxidative stress and renal damage in rats. *Redox Rep.* **2011**, *16*, 8–14. [CrossRef] [PubMed]

66. Polat, A.; Parlakpinar, H.; Tasdemir, S.; Colak, C.; Vardi, N.; Ucar, M.; Emre, M.H.; Acet, A. Protective role of aminoguanidine on gentamicin-induced acute renal failure in rats. *Acta Histochem.* **2006**, *108*, 365–371. [CrossRef]

 *antioxidants*

MDPI

Communication

Antioxidant-Based Therapy Reduces Early-Stage Intestinal Ischemia-Reperfusion Injury in Rats

Gaizka Gutiérrez-Sánchez [1], Ignacio García-Alonso [1,2], Jorge Gutiérrez Sáenz de Santa María [1], Ana Alonso-Varona [3] and Borja Herrero de la Parte [1,2,*]

[1]   Department of Surgery and Radiology and Physical Medicine, University of The Basque Country, ES48940 Leioa, Biscay, Spain; gaizka_gutierrez@hotmail.es (G.G.-S.); ignacio.galonso@ehu.eus (I.G.-A.); jorge.gssm@gmail.com (J.G.S.d.S.M.)

[2]   Interventional Radiology Research Group, Biocruces Bizkaia Health Research Institute, ES48903 Barakaldo, Biscay, Spain

[3]   Department of Cell Biology and Histology, University of The Basque Country, ES48940 Leioa, Biscay, Spain; ana.alonsovarona@ehu.eus

*   Correspondence: borja.herrero@ehu.eus

Citation: Gutiérrez-Sánchez, G.; García-Alonso, I.; Gutiérrez Sáenz de Santa María, J.; Alonso-Varona, A.; Herrero de la Parte, B. Antioxidant-Based Therapy Reduces Early-Stage Intestinal Ischemia-Reperfusion Injury in Rats. *Antioxidants* **2021**, *10*, 853. https://doi.org/10.3390/antiox10060853

Academic Editors: Soliman Khatib, Dana Atrahimovich Blatt and Jeffrey B. Blumberg

Received: 23 April 2021
Accepted: 24 May 2021
Published: 27 May 2021

Publisher's Note: MDPI stays neutral with regard to jurisdictional claims in published maps and institutional affiliations.

Abstract: Intestinal ischemia-reperfusion injury (i-IRI) is a rare disorder with a high mortality rate, resulting from the loss of blood flow to an intestinal segment. Most of the damage is triggered by the restoration of flow and the arrival of cytokines and reactive oxygen species (ROS), among others. Inactivation of these molecules before tissue reperfusion could reduce intestinal damage. The aim of this work was to analyze the preventive effect of allopurinol and nitroindazole on intestinal mucosal damage after i-IRI. Wag/RijHsd rats were subjected to i-IRI by clamping the superior mesenteric artery (for 1 or 2 h) followed by a 30 min period of reperfusion. Histopathological intestinal damage (HID) was assessed by microscopic examination of histological sections obtained from injured intestine. HID was increased by almost 20% by doubling the ischemia time (from 1 to 2 h). Nitroindazole reduced HID in both the 1 and 2 h period of ischemia by approximately 30% and 60%, respectively ($p < 0.001$). Our preliminary results demonstrate that nitroindazole has a preventive/protective effect against tissue damage in the early stages of i-IRI. However, to better understand the molecular mechanisms underlying this phenomenon, further studies are needed.

Keywords: antioxidant treatment; intestinal ischemia-reperfusion injury; allopurinol; nitroindazole; animal model

1. Introduction

Intestinal ischemia-reperfusion injury (i-IRI) is an uncommon disorder that affects 0.09 to 0.2% of hospitalized patients. It involves decreased intestinal blood flow followed by a local inflammatory response that leads to necrosis of the intestinal wall, and in the absence of correct and quick management, may lead to patient death (the mortality rate is reported to be as high as 90%). The acute onset results from an abrupt and spontaneous interruption of intestinal blood flow due to occlusion of a venous or arterial vessel, or to non-occlusive causes (NOMI). NOMI comprises approximately 20% of cases of intestinal ischemia, which are commonly related to superior mesenteric artery (SMA) vasoconstriction or poor cardiac performance that triggers hypoperfusion of the intestine [1–4].

This blood deprivation hinders the aerobic metabolism of enterocytes, compromising mitochondrial oxidative phosphorylation, forcing the cells to perform anaerobic metabolism for energy production [5]. Moreover, the inhibition of the Na^+/K^+-ATPase pump activity alters the transmembrane potential and the electrolyte balance of the intra- and extracellular media, generating edema that is of special interest at the time of reperfusion [6]. In addition, the different intermediate active metabolites generated during the different cellular reactions cannot be secreted into the systemic circulation for renal excretion, thus, they accumulate inside the cells and contribute to cell damage and death [5]. However, the greatest damage during i-IRI does not occur in the ischemia period, but rather, during reperfusion of the ischemic tissue with the arrival of oxygen and the action of cytokines, the complement system, reactive oxygen species (ROS), neutrophils and the alteration of capillary permeability, which, in addition to promoting expansion to the rest of the organism, leads to bacterial translocation [7].

Damage at the systemic level as a consequence of i-IRI is mainly mediated by toxic metabolites produced during the ischemia period and the affinity and high reactivity of ROS for other tissues. These toxic metabolites generate an imbalance between the intracellular and extracellular environment, which triggers fluid sequestration at the extracellular level, the creation of a third-space and a decrease in effective circulating volume. This manifests clinically as hypotension, shock, or even systemic inflammatory response syndrome (SIRS), and multiple organ dysfunction syndrome (MODS), which also perpetuates the low cardiac output situation [8] (Figure 1).

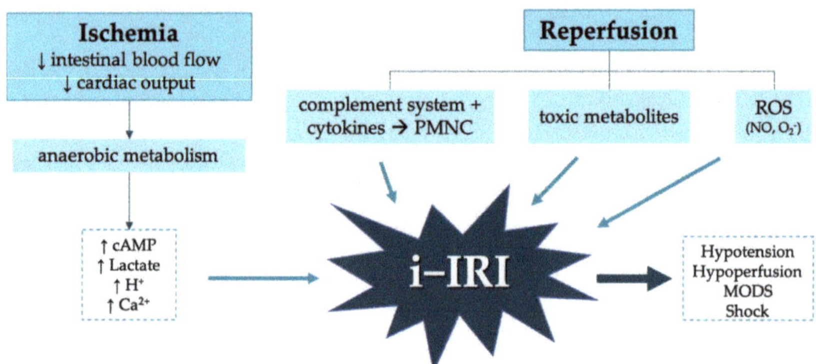

Figure 1. Pathophysiology of mesenteric ischemia reperfusion syndrome (i-IRI). cAMP (cyclic adenosine monophosphate), PMNC (polymorphonuclear cells), ROS (reactive oxygen species). The upward pointing arrow (↑) signifies increase, and The downward pointing arrow (↓) signifies decrease.

The therapeutic management of NOMI-related i-IRI must be individualized, based on the underlying precipitating cause. The main therapeutic approaches include fluid resuscitation, optimization of cardiac output, and elimination of vasopressors; furthermore, vasodilators or antispasmodics may also be used. In those patients in whom ischemia has resulted in severe infarction of an intestinal segment, surgical resection of the affected portion is necessary; however, when this resection involves an important segment of intestine, it can trigger short bowel syndrome [4].

These approaches are mainly aimed at restoring intestinal flow as soon as possible [3]. However, another interesting and complementary approach is the management of oxidative stress damage by decreasing/inhibiting the action of the ROS. Xanthine oxidase (XO) and nitric oxide synthase (NOS) play a key role in ROS production. There are compounds such as allopurinol, which is used in the management of chronic gout that inhibit XO, preventing the generation of ROS by this pathway. The prophylactically administration of this compound in animal models or in clinical trials of cardiac, hepatic or renal ischemia-reperfusion injury (IRI) has demonstrated its protective effect, acting at the level of vascular permeability, polymorphonuclear cells infiltration, bacterial translocation, chemokine signaling, motility and mortality derived from this situation [9–14]. On the other hand, nitroindazole, a compound of the imidazole and indazole family, inhibits the neuronal nitric oxide synthase isoform (nNOS), which induces a decrease in nitric oxide (NO) synthesis during the early stages of reperfusion with no alterations observed during the period of ischemia or late reperfusion [15,16]. Moreover, it has been shown that this molecule contributes to protein degradation and cell damage during IRI [17]. Therefore, it is expected that blocking or decreasing the activity of XO and NOS could prevent or minimize reperfusion-associated damage in the organism as a whole.

This study analyzes the individualized effect of the prophylactic administration of these two compounds, due to their recognized antioxidant capacity, as therapeutic tools for the treatment of i-IRI by analyzing the response at the anatomopathological level.

2. Materials and Methods

All procedures were carried out in accordance with current legislation and were approved by the Animal Experimentation Ethics Committee (CEEA) (reference M20/2019/207) and the Biological Agents Research Ethics Committee (CEIAB) of the University of The Basque Country (reference M30/2019/208).

Sixty male 3-months-old WAG/RijHsd rats were induced to develop i-IRI following a 1 or 2 h period of ischemia and 30 min of reperfusion, and were randomized into 10 groups (Table 1). The animals were maintained in 12 h light/dark cycles with food and water ad libitum. Another 6 animals were used as a control.

Table 1. Experimental groups.

Group	Ischemia	Reperfusion	Treatment	N° of Animals
0	no	no	no	6
1			no	6
2			Saline	6
3	1 h	30′	ClinOleic® 20% [1]	6
4			Allopurinol	6
5			Nitroindazole	6
6			no	6
7			Saline	6
8	2 h	30′	ClinOleic® 20% [1]	6
9			Allopurinol	6
10			Nitroindazole	6

[1] ClinOleic® 20% (Baxter S.L., Valencia, Spain).

2.1. Surgical Procedure

Under isoflurane anesthesia (1.5%), a middle laparotomy was performed to locate the SMA (Figure 2a). Briefly, part of the small intestine was pulled out over moistened gauze and the superior mesenteric artery was dissected, and then clamped with a Yasargil microvascular clamp (Figure 2b). Once the absence of arterial pulse in the mesenteric region was observed, the bowel was reintroduced and the abdominal muscles were sutured with a running suture and the skin was sutured with single stitches. Prior to the animal's recovery, a single dose of meloxicam (2 mg/kg, sc) was administered.

When the ischemia period was over (1 or 2 h), the animals were re-anesthetized, the laparotomy was reopened, and the Yasargil clip was removed. The wound was closed as previously described and 30 min of reperfusion time was allowed.

2.2. Experimental Treatment

Either vehicle (saline or ClinOleic® 20% (Baxter S.L., Valencia, Spain)) or experimental drug treatment were administered through the femoral vein 30 min before the end of the ischemia period. Allopurinol (50 mg/kg) was suspended in saline, and nitroindazole (10 mg/kg) in absolute ethanol and then dissolved in ClinOleic®, at a ratio of 1:7.

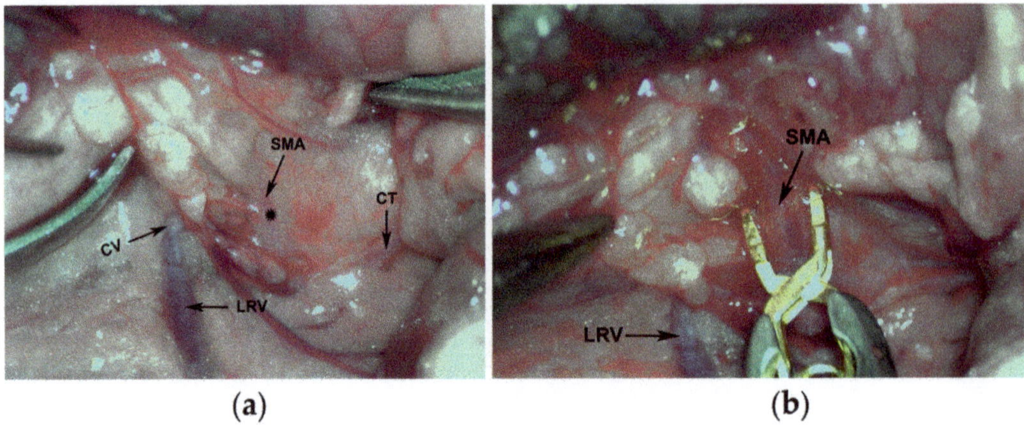

Figure 2. View of the surgical field: (**a**) anatomical location of the main structures: CV (cava vein), LRV (left renal vein), CT (celiac trunk), and SMA (superior mesenteric artery), the asterisk indicates the site of placement of the microvascular clip; (**b**) detail of the dissected SMA with the microvascular clip.

For this purpose, under 1.5% isoflurane anesthesia, an incision was made in the skin of the inguinal area and the femoral vein was exposed, and the corresponding vehicle or drug was administered using a 27G needle, in accordance with the experimental group (Table 1). To avoid bleeding at the puncture site, direct hemostasis was performed for 1 to 2 min, using a cotton swab. Then, the incision was closed with single stitches.

2.3. Tissue Sample Collection and Histological Examination

At the end of the reperfusion period, the animals were anesthetized and 6 cm of terminal ileum (measured from the ileocecal valve) was removed and immersed in tempered saline to gently wash the specimen. Then, the piece was fixed to a plastic guide to keep it stretched and immersed for 24 h in 4% formaldehyde. Each piece of intestine was transversely split into 4 fragments (each 1 cm long), which were embedded in the same paraffin mold. Three paraffin-embedded slices (5 μm) were obtained from each intestine sample (12 sections of intestine from each animal) and stained with hematoxylin/eosin.

The histological injury degree (HID) score was assigned according to the scale summarized in Table 2 (adapted from Chiu et al. [18]). Four quadrants were defined in each histological section and the HID was assigned to each of them. Subsequently, the HID of each section was calculated by summing the index assigned to each quadrant. The HID of each animal was calculated as the mean of the 12 analyzed sections.

Table 2. Criteria to assess the corresponding histological injury degree (HID), according to Chiu et al. [18].

Grade	Description
0	Normal mucosal villi; no histological changes
1	Epithelium of the villi is almost fully preserved; development of Gruenhagen's subepithelial spaces (normally located in the apex); capillary congestion
2	Extension of the subepithelial spaces with moderate lifting of the epithelial layer of the lamina propria
3	Preserved villous structure with almost complete loss of epithelium (preservation > 50%); presence of intraluminal hemorrhage.
4	Destructuring of the villi, mostly denuded (preservation < 50%)
5	Loss of villi, disintegration of the lamina propria; hemorrhage and ulceration

2.4. Statistical Analyses

All analyses were performed with the GraphPad Prism 6 (GraphPad Software, San Diego, CA, USA), and the minimum significance level was set as $p < 0.05$. The quantitative variables described in this piece of work were represented by the mean and standard deviation. Statistical treatment of the data was performed by analysis of variance (ANOVA). Additionally, once the significant differences between the groups were demonstrated, comparisons of the different groups were performed using Tukey's multiple comparison test.

3. Results

The surgical model used to emulate i-IRI induced the development of i-IRI in 100% of the animals, without resulting in the death of any animal. Overall, the procedure was well tolerated.

Histological samples obtained from animals subjected to 1 and 2 h of ischemia showed high intestinal mucosal damage. After 1 h of ischemia and 30 min of reperfusion there was major destruction of the villi (Figure 3a) with the accumulation of inflammatory cells and wide Gruenhagen's spaces in those villi that had preserved their structure. When doubling the ischemia period (Figure 3d), the damage was more pronounced with the complete loss of the villus structure and fragments of detached villi appearing in the lumen of the intestine. In addition, intraluminal hemorrhage was evidenced by the presence of erythrocytes in the intestinal lumen.

Allopurinol treatment of those animals that underwent 1 h of ischemia resulted in a slight reduction in tissue damage (Figure 3b). Nonetheless, there were evident signs of mucosal injury, such as inflammation (edema of the lamina propria and submucosal layers and presence of inflammatory cells), erythrocyte extravasation and loss of epithelial goblet cells. In the case of 2 h of ischemia (Figure 3e), treatment with allopurinol also resulted in a less severe tissue damage compared to the untreated animals, although extensive epithelial lifting, inflammatory cells presence and edema were still evident.

When nitroindazole treatment was applied, we found a striking improvement in the histological structure of the intestines subjected to ischemia. Thus, after 1 h of ischemia, the structure of intestinal mucosa was almost totally preserved, and the only significant change was the presence of small Gruenhagen's spaces located at the tip of some villi (Figure 3c). However, nitroindazole was not as effective when administered to animals in which the arterial blood flow of the intestine was occluded for 2 h (Figure 3f). Even though the structure of the villi was better preserved than in the non-treated animals, almost complete loss of the epithelium at the tip of some villi was observed. In addition, evident signs of erythrocyte starvation and inflammatory cell infiltration were found.

Regarding HID (Figure 4), we observed that the degree of injury after 1 h of ischemia was 15 times higher than that of the control (not subjected to ischemia), 1.06 ± 0.65 vs. 16.3 ± 0.96, respectively ($p < 0.001$). When the ischemia time was doubled to 2 h, although HID did not increase in the same order of magnitude, it was significantly higher (19.10 ± 0.77; $p < 0.001$) than that observed after 1 h. Regarding the vehicles used for the drugs (saline and ClinOleic®), neither of them modified the HID ($p < 0.05$; Table 2).

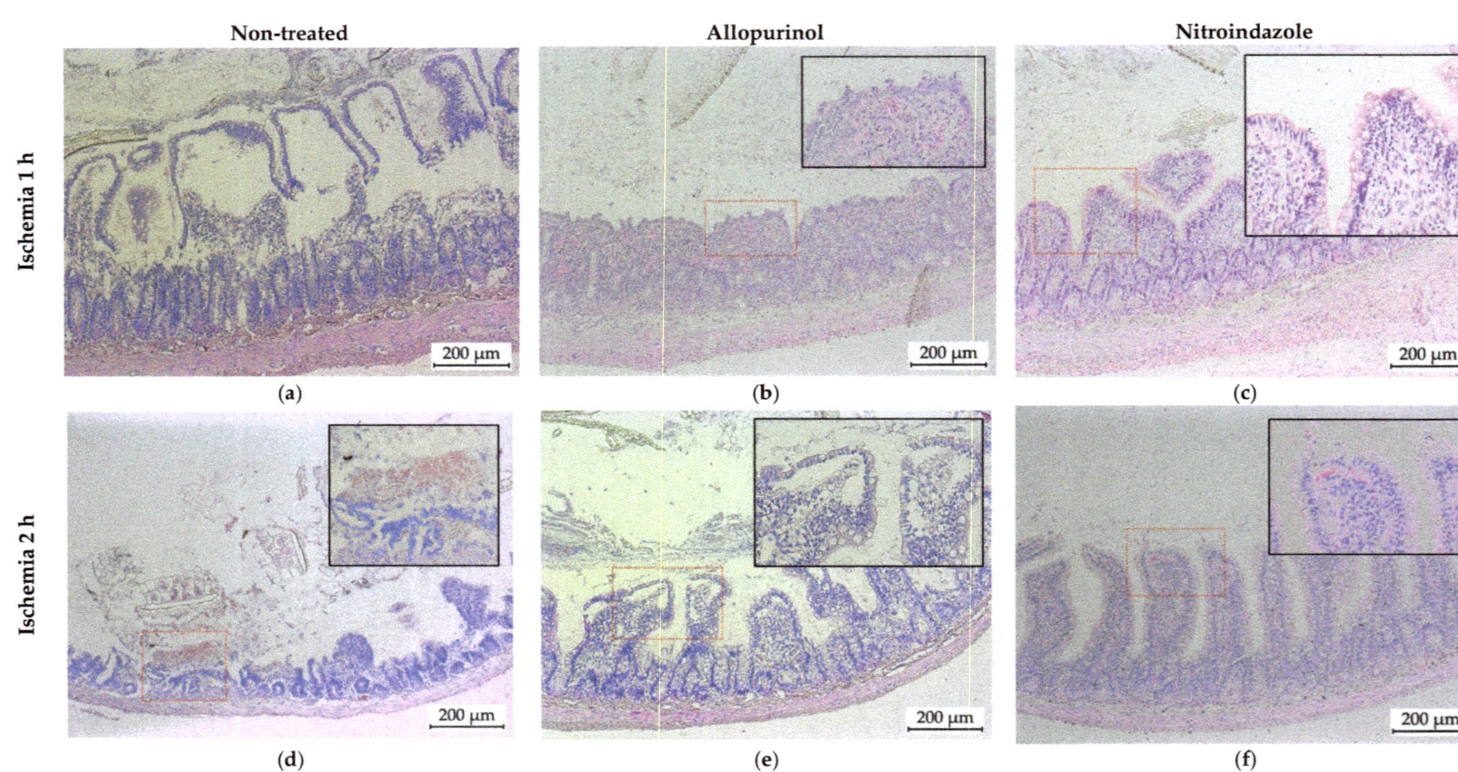

Figure 3. Representative photomicrographs of small intestine sections stained with hematoxylin/eosin. Animals subjected to a period of 1 (**a**–**c**) or 2 h (**d**–**f**) of ischemia, followed by 30 min of reperfusion. The photographs on the left (**a**,**c**) correspond to histological sections obtained from untreated animals, the photographs in the center (**b**,**d**) to animals treated with allopurinol and the photographs on the right (**c**,**f**) to animals treated with nitroindazole.

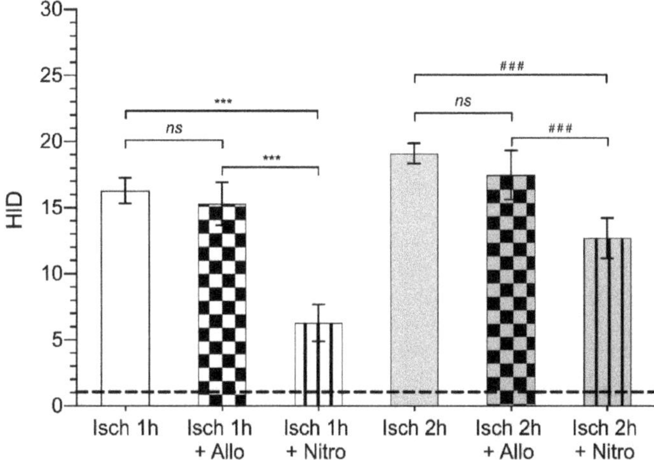

Figure 4. Pharmacological modulation of the histological injury degree (HID) after treatment with allopurinol (grid pattern) or nitroindazole (line pattern), after 1 (white) or 2 h (grey) of ischemia. ***: $p < 0.001$; ###: $p < 0.001$; ns: $p > 0.05$. The dashed line indicates the HID score of the control non-ischemic tissue. Non-relevant data (vehicle groups) have been excluded from the figure.

Treatment with allopurinol did not have a beneficial effect on the histological damage induced by 1 h or 2 h of ischemia (15.3 ± 1.64 and 17.49 ± 1.86, respectively; $p > 0.05$). Nevertheless, treatment with nitroindazole did substantially modify the HID observed in the intestinal mucosa. Following 1 h ischemia, HID was reduced by more than half in animals treated with 10 mg/kg nitroindazole (16.3 ± 0.96 vs. 6.3 ± 1.4, $p > 0.001$). In animals subjected to 2 h of ischemia, nitroindazole also induced a significant reduction in HID, however the reduction observed was smaller, decreasing by 1.5-fold (19.10 ± 0.77 vs. 12.69 ± 1.52; $p < 0.001$) (Figure 4 and Table 3).

Table 3. Histological injury degree (HID) following a period of ischemia of 1 or 2 h and 30 min of reperfusion in control- or treated-animals (vehicles, allopurinol or nitroindazole). The statistical significance shown corresponds to the HID of non-treated animals subjected to i-IRI.

Group	Control	Saline	Allopurinol	ClinOleic®	Nitroindazole
Ischemia 1 h	16.3 ± 0.96	15.2 ± 1.91 *ns* [1]	15.3 ± 1.64 *ns* [1]	14.4 ± 0.98 *ns* [1]	6.3 ± 1.4 $p < 0.001$
Ischemia 2 h	19.10 ± 0.77	18.1 ± 0.9 *ns* [1]	17.49 ± 1.86 *ns* [1]	17.9 ± 0.67 *ns* [1]	12.69 ± 1.52 $p < 0.001$

[1] *ns*: not significant ($p > 0.05$).

4. Discussion

The effectiveness of antioxidant treatments on IRI has been demonstrated by a large number of previous studies [19–24]. Our results are in accordance with this and show that nitroindazole reduces intestinal mucosal damage after i-IRI. SMA clamping is a simple, reproducible procedure with low mortality that allows the simulation of i-IRI for a low output situation (NOMI). In our study, 100% of the animals survived, and we were always able to induce comparable intestinal damage.

We analyzed the relationship between ischemia time and HID, and in contrast to the study by Da Costa et al. [25] where no direct proportionality between both factors was observed, our study shows a significant increase in intestinal mucosal damage as the period of blood flow deprivation increases. These different results could be explained by the type of ischemia induction model that was used; our model recreates a low flow condition when the SMA is clamped, which allows minimal flow because the collateral circulation remains intact. However, in the previously cited model of Da Costa, the aorta is clamped prior to the birth of the renal arteries; in the absence of collateral supply, such a lot of damage occurs in the first minutes of ischemia that it does not change significantly when this period is increased.

As for the relationship between HID and reperfusion time, some studies have shown that the greatest damage, from a biochemical point of view, occurs in the first moments of reperfusion (even in the first 15 min) when those molecules involved in the pathophysiology of i-IRI reach the area deprived of blood flow [24,26–28]. In addition, Clark et al. [11] have shown that two 15 min periods of ischemia are more harmful than a period of half an hour. This is explained by the fact that there is a physiological adaptation to time-limited and not very long periods of oxidative stress, which endogenous defense mechanisms are able to cope with [29]. However, if the damage is too severe and prolonged over time, this endogenous system loses its capacity to cope with the oxidative agents [30].

The analysis of the effect of the antioxidant agents tested in this piece of work shows that the prophylactic use of allopurinol during the ischemia period does not achieve a statistically significant reduction in local intestinal damage, regardless of the ischemia time. The literature reviewed showed that several authors have reported results in agreement with those described here. They conclude that when allopurinol is administered, the effect on the prevention of IRI is only significant in those cases in which this treatment is associated with ischemic postconditioning, that is, periods of arterial reocclusion of a few minutes duration applied at the beginning of tissue reperfusion and which have shown a decrease in the damage associated with ROS [31,32]. In contrast, there are also studies in which its oral administration prior to ischemia instauration does demonstrate an antioxidant effect, as evidenced by a reduction in HID and a decrease in the production of oxidative stress markers such as malondialdehyde (MDA) [33]. Other studies performed in a renal IRI mode have shown the beneficial effect of XO inhibition [34–38]. After prophylactic administration of allopurinol, a decrease of up to 75% in the damage was noticed, both biochemical and anatomopathological. In addition to renal IRI models, in studies of this syndrome in upper limbs, allopurinol demonstrated an increase of up to 70% in muscle viability after 6 h of reperfusion [39]. In any case, the pathophysiological mechanism of XO-mediated damage found in rodents and felines, but absent in humans, casts doubt on the already questionable beneficial effect of allopurinol in the management of i-IRI [7].

Regarding the use of nitroindazole in reperfusion injury, most studies have been performed in experimental models of cerebral ischemia. It has been shown that various markers of damage due to oxidative stress, such as MDA, glutathione (GSH) or lipid peroxidation, are greatly reduced with nitroindazole, and reached levels close to those of control groups [27,40,41]. It has been proposed that the prophylactic effect of this compound involves more than its action as an NOS inhibitor, since it is also involved in other enzymatic pathways. Hirabayashi et al. aimed to study the effect of nitroindazole administration on the role of nNOS, in the early phase of cerebral ischemia-reperfusion in mice [42]. Following a 2 h ischemia period and 30 min of reperfusion, they found that nitrotyrosine (NO2-Tyr) formation, a well-known marker for NO-related cytotoxicity, was reduced by 45% and 100% after 25 mg/kg or 50 mg/kg of nitroindazole, respectively. These findings are in accordance to those previously reported by Sorrenti et al. [43] and Liu [44], where nitroindazole increased c-fos mRNA levels, which is related to the recovery of cellular function after cerebral injury. In a murine model of retinal IRI, San Cristobal et al. also reported that prophylactic administration of nitroindazole (10 mg/k) reduced retinal histological damage, the effect being nearly as strong as that achieved by folic acid [45].

It has been proven that the solvent used in the preparation of the nitroindazole solution, ClinOleic® modulates excessive inflammatory reactions [46]. However, in our study we did not found any significant effect in those animals that received ClinOleic® as a transporter for nitroindazole. This emulsion contains 80% refined olive oil and 20%

refined soybean oil, and is rich in a large variety of monounsaturated fatty acids (MUFAs) (up to 65%), mainly in the form of oleic acid (C18:1n-9) (found in the olive oil), and lower amounts of polyunsaturated fatty acids (PUFAs) (20%) and triglycerides [47]. According to the United States Food and Drug Administration (FDA), the total amount of both refined olive and soybean oil in ClinOleic® is around 200 g/L [48].

Huang et al. [46] have reported the protective effect of ClinOleic®, by decreasing the intensity of the cytokine storm and apoptosis in an acute lung injury model. These results may seem opposed to those found in our work, however, we must take into account that in their study ClinOleic doses of 6 g/kg/day were administered for 7 consecutive days before the induction of lung damage (total dosage over 10 mL), while we administered a single dose of 0.43 mL of ClinOleic®.

Furthermore, the concentration of the antioxidants included in ClinOleic® are quite different from those used in therapeutic studies. For example, Farías et al. [49] used 225 g rats that received daily doses of 135 mg of PUFAs over 8 weeks, while our animals received just one dose of 13.76 mg of PUFAs. A similar disparity can be observed with the experiments of Sukhotnik et al. [50].

ClinOleic® also has tiny quantities of α-tocopherol, an active isomer of vitamin E with high antioxidant capacity [47]. Its presence could explain the better results obtained in animals treated with nitroindazole since it acts at different levels and in a synergistic manner. However, the total amount of α-tocopherol administered to each animal was below 0.04 mg/kg. This amount is more than 500 times lower than the dose of this compound administered when used for therapeutic purposes [51–53], so the effect of the α-tocopherol contained in ClinOleic® would be minimal. In any case, no benefit was observed when it was administered alone, without nitroindazole. Moreover, when α-tocopherol was tested in several models of IRI, it was shown that it has an enhancing effect over other treatments, such as allopurinol [54], vitamin C [28], or ischemic preconditioning [55]. However, Catilayan et al. [19] studied its isolated effect on i-IRI and found no statistically significant differences compared to the non-treated groups.

5. Conclusions

In our i-IRI experimental setting, we observed that the duration of the ischemia period has a directly proportional effect on intestinal mucosa damage. The use of allopurinol achieved a discrete but not significant reduction in intestinal damage, ruling out its applicability in this model of i-IRI. On the other hand, nitroindazole administered prophylactically during ischemia significantly reduced the histological injury, with its effect being more striking when ischemia was limited to 1 h.

Author Contributions: Conceptualization, B.H.d.l.P. and G.G.-S.; surgical procedures, G.G.-S. and J.G.S.d.S.M.; histological analyses, G.G.-S.; contributed to the interpretation of the results, B.H.d.l.P., G.G.-S., J.G.S.d.S.M., A.A.-V. and I.G.-A.; verified the analytical methods, B.H.d.l.P., A.A.-V. and I.G.-A.; B.H.d.l.P. wrote the manuscript with support from G.G.-S. and A.A.-V. All authors discussed the results and contributed to the final manuscript. All authors have read and agreed to the published version of the manuscript.

Funding: This research received funding from the University of The Basque Country UPV/EHU (grant reference GIU19/088).

Institutional Review Board Statement: All procedures were carried out in accordance with current legislation and were approved by the Animal Experimentation Ethics Committee (CEEA) (reference M20/2019/207) and the Biological Agents Research Ethics Committee (CEIAB) of the UPV/EHU (reference M30/2019/208).

Informed Consent Statement: Not applicable.

Data Availability Statement: The data that support the findings of this study are available from the corresponding author, (B.H.P.), upon reasonable request.

Acknowledgments: G. Gutiérrez-Sánchez is grateful to the Department of Education of the Basque Government for his undergraduate fellowship for undergraduate university students through the program Ikasiker (ref. number IkasC_2019_1_0224).

Conflicts of Interest: The authors declare no conflict of interest.

References

1. Montoro, M.Á.; García Egea, J.; Fabregat, G. Isquemia intestinal. In *Gastroenterología y Hepatología*; Montoro, M.A., García Pagán, J.C., Eds.; Jarpyo Editores, S.A: Madrid, Spain, 2012; pp. 383–410.
2. Wyers, M.C.; Michelle, M.C. Acute Mesenteric Arterial Disease. In *Rutherford's Vascular Surgery and Endovascular Therapy*; Sidawy, A.N., Perler, B.A., Eds.; Elsevier Inc.: Amsterdam, The Netherlands, 2019; pp. 1754–1770.
3. Tendler, D.A.; Lamont, J.T. Nonocclusive Mesenteric Ischemia. Available online: https://www.uptodate.com/contents/nonocclusive-mesenteric-ischemia (accessed on 4 May 2021).
4. Bala, M.; Kashuk, J.; Moore, E.E.; Kluger, Y.; Biffl, W.; Gomes, C.A.; Ben-Ishay, O.; Rubinstein, C.; Balogh, Z.J.; Civil, I.; et al. Acute mesenteric ischemia: Guidelines of the World Society of Emergency Surgery. *World J. Emerg. Surg.* **2017**, *12*, 38. [CrossRef]
5. Cerqueira, N.F.; Hussni, C.A.; Yoshida, W.B. Pathophysiology of mesenteric ischemia/reperfusion: A review. *Acta Cir. Bras.* **2005**, *20*, 333–343. [CrossRef]
6. Eltzschig, H.K.; Collard, C.D. Vascular ischaemia and reperfusion injury. *Br. Med. Bull.* **2004**, *70*, 71–86. [CrossRef]
7. Gonzalez, L.M.; Moeser, A.J.; Blikslager, A.T. Animal models of ischemia-reperfusion-induced intestinal injury: Progress and promise for translational research. *Am. J. Physiol. Gastrointest. Liver Physiol.* **2015**, *308*, G63–G75. [CrossRef]
8. Yasuhara, H. Acute Mesenteric Ischemia: The Challenge of Gastroenterology. *Surg. Today* **2005**, *33*, 185–195. [CrossRef]
9. Pacher, P.; Nivorozhkin, A.; Szabó, C. Therapeutic effects of xanthine oxidase inhibitors: Renaissance half a century after the discovery of allopurinol. *Pharmacol. Rev.* **2006**, *58*, 87–114. [CrossRef]
10. Shadid, M.; Van Bel, F.; Steendijk, P.; Dorrepaal, C.A.; Moison, R.; Van der Velde, E.T.; Baan, J. Pretreatment with allopurinol in cardiac hypoxic-ischemic reperfusion injury in newborn lambs exerts its beneficial effect through afterload reduction. *Basic Res. Cardiol.* **1999**, *94*, 23–30. [CrossRef]
11. Guan, W.; Osanai, T.; Kamada, T.; Hanada, H.; Ishizaka, H.; Onodera, H.; Iwasa, A.; Fujita, N.; Kudo, S.; Ohkubo, T.; et al. Effect of allopurinol pretreatment on free radical generation after primary coronary angioplasty for acute myocardial infarction. *J. Cardiovasc. Pharmacol.* **2003**, *41*, 699–705. [CrossRef] [PubMed]
12. Riaz, A.A.; Schramm, R.; Sato, T.; Menger, M.D.; Jeppsson, B.; Thorlacius, H. Oxygen radical-dependent expression of CXC chemokines regulate ischemia/reperfusion-induced leukocyte adhesion in the mouse colon. *Free Radic. Biol. Med.* **2003**, *35*, 782–789. [CrossRef]
13. Kulah, B.; Besler, H.T.; Akdag, M.; Oruc, T.; Altinok, G.; Kulacoglu, H.; Ozmen, M.M.; Coskun, F. The effects of verapamil vs. allopurinol on intestinal ischemia/reperfusion injury in rats. "An experimental study". *Hepatogastroenterology* **2004**, *51*, 401–407.
14. Kakita, T.; Suzuki, M.; Takeuchi, H.; Unno, M.; Matsuno, S. Significance of xanthine oxidase in the production of intracellular oxygen radicals in an in-vitro hypoxia-reoxygenation model. *J. Hepatobiliary Pancreat. Surg.* **2002**, *9*, 249–255. [CrossRef] [PubMed]
15. Southan, G.J.; Szabó, C. Selective pharmacological inhibition of distinct nitric oxide synthase isoforms. *Biochem. Pharmacol.* **1996**, *51*, 383–394. [CrossRef]
16. San Cristóbal Epalza, J. Modulación Farmacológica del Síndrome de Isquemia-Reperfusión Retiniana en la Rata. Available online: https://addi.ehu.es/bitstream/handle/10810/23865/TESIS_SANCRISTOBAL_EPALZA_JUAN.pdf?sequence=1&isAllowed=y (accessed on 3 May 2021).
17. Rivera, L.R.; Pontell, L.; Cho, H.J.; Castelucci, P.; Thacker, M.; Poole, D.P.; Frugier, T.; Furness, J.B. Knock out of neuronal nitric oxide synthase exacerbates intestinal ischemia/reperfusion injury in mice. *Cell Tissue Res.* **2012**, *349*, 565–576. [CrossRef]
18. Chiu, C.J.; McArdle, A.H.; Brown, R.; Scott, H.J.; Gurd, F.N. Intestinal mucosal lesion in low-flow states I. A morphological, hemodynamic, and metabolic reappraisal. *Arch. Surg.* **1970**, *101*, 478–483. [CrossRef]
19. Catilayan, F. Treatment of Intestinal Reperfusion Using Antioxidative Agents. *J. Pediatr. Surg.* **1998**, *33*, 1536–1539.
20. Castelo Branco Couto de Miranda, C.T.; Fagundes, D.J. The role of ischemic preconditioning in gene expression related to inflammation in a rat model of intestinal. *Acta Cir. Bras.* **2018**, *33*, 1095–1102. [CrossRef]
21. Drucker, N.A.; Jensen, A.R.; Winkel, J.P.; Markel, T.A. ScienceDirect Association for Academic Surgery Hydrogen Sulfide Donor GYY4137 Acts Through Endothelial Nitric Oxide to Protect Intestine in Murine Models of Necrotizing Enterocolitis and Intestinal Ischemia. *J. Surg. Res.* **2018**, *234*, 294–302. [CrossRef] [PubMed]
22. Treskes, N.; Persoon, A.M.; Zanten, A.R.H. Van Diagnostic accuracy of novel serological biomarkers to detect acute mesenteric ischemia: A systematic review and meta-analysis. *Intern. Emerg. Med.* **2017**, *12*, 821–836. [CrossRef]
23. Mohamed, D.A.; Almallah, A.A. Possible protective role of verapamil on ischemia/reperfusion induced changes in the jejunal mucosa of adult male albino rat: Histological and biochemical study. *J. Innov. Pharm. Biol. Sci.* **2018**, *5*, 92–102.
24. Yagmurdur, M.C.; Topaloglu, S.; Kilinc, K. Effects of alpha tocopherol and verapamil on liver and small bowel following mesenteric ischemia-reperfusion. *Turk. J. Gastroenterol.* **2002**, *13*, 40–46.
25. Da Costa Rocha, B. Experimental model of mesenteric ischemia reperfusion by abdominal aorta clamping in Wistar rats. *Rev. Col. Bras. Cir.* **2012**, *39*, 207–210. [CrossRef]

26. Liu, K.-X.; Li, Y.-S.; Huang, W.-Q.; Chen, S.-Q.; Wang, Z.-X.; Liu, J.-X.; Xia, Z. Immediate postconditioning during reperfusion attenuates intestinal injury. *Intensive Care Med.* **2009**, *35*, 933. [CrossRef]

27. Chalimoniuk, M.; Strosznajder, J. NMDA Receptor-Dependent Nitric Oxide and cGMP Synthesis in Brain Hemispheres and Cerebellum During Reperfusion After Transient Forebrain Ischemia in Gerbils: Effect of 7-Nitroindazole. *J. Neurosci. Res.* **1998**, *690*, 681–690. [CrossRef]

28. Rhee, J.; Jung, S.; Shin, S.; Suh, G.; Noh, D.; Youn, Y.; Oh, S.; Choe, K. The Effects of Antioxidants and Nitric Oxide Modulators on Hepatic Ischemic-Reperfusion Injury in Rats. *J. Korean Med. Sci.* **2002**, *107*, 502–506. [CrossRef]

29. Clark, E.T.; Gewertz, B.L. Intermittent ischemia potentiates intestinal reperfusion injury. *J. Vasc. Surg.* **1991**, *13*, 601–606. [CrossRef]

30. Ristow, M.; Schmeisser, S. Extending life span by increasing oxidative stress. *Free Radic. Biol. Med.* **2011**, *51*, 327–336. [CrossRef]

31. Genki, H.; Akahane, K.; Gomes, R.Z.; Paludo, K.S.; Linhares, F. The influence of allopurinol and post-conditioning on lung injuries induced by lower-limb ischemia and reperfusion in. *Acta Cir. Bras.* **2017**, *32*, 746–754.

32. Brandão, R.I.; Gomes, R.Z.; Lopes, L.; Linhares, F.S.; Carlos, J.; Vellosa, R.; Paludo, K.S. Remote post-conditioning and allopurinol reduce ischemia-reperfusion injury in an infra-renal ischemia model. *Ther. Adv. Cardiovasc. Dis.* **2018**, *12*, 341–349. [CrossRef]

33. Sapalidis, K.; Papavramidis, T.S.; Gialamas, E.; Deligiannidis, N.; Tzioufa, V.; Papavramidis, S. The role of allopurinol's timing in the ischemia reperfusion injury of small intestine. *J. Emerg. Trauma Shock* **2013**, *6*, 203–208. [CrossRef] [PubMed]

34. Tani, T.; Okamoto, K.; Fujiwara, M.; Katayama, A.; Tsuruoka, S. Metabolomics analysis elucidates unique influences on purine/pyrimidine metabolism by xanthine oxidoreductase inhibitors in a rat model of renal ischemia-reperfusion injury. *Mol. Med.* **2019**, *25*, 1–13. [CrossRef] [PubMed]

35. Zhou, J.Q.; Qiu, T.; Zhang, L.; Chen, Z.B.; Wang, Z.S.; Ma, X.X.; Li, D. Allopurinol preconditioning attenuates renal ischemia/reperfusion injury by inhibiting HMGB1 expression in a rat model. *Acta Cir. Bras.* **2016**, *31*, 176–182. [CrossRef] [PubMed]

36. Keel, C.E.; Wang, Z.; Colli, J.; Grossman, L.; Majid, D.; Lee, B.R. Protective effects of reducing renal ischemia-reperfusion injury during renal hilar clamping: Use of allopurinol as a nephroprotective agent. *Urology* **2013**, *81*, 210.e5–210.e10. [CrossRef]

37. Choi, E.K.; Jung, H.; Kwak, K.H.; Yeo, J.; Yi, S.J.; Park, C.Y.; Ryu, T.H.; Jeon, Y.H.; Park, K.M.; Lim, D.G. Effects of Allopurinol and Apocynin on Renal Ischemia-Reperfusion Injury in Rats. *Transplant. Proc.* **2015**, *47*, 1633–1638. [CrossRef] [PubMed]

38. Gomes, R.Z.; Romanek, G.M.M.; Przybycien, M.; Amaral, D.C.; Akahane, H.G.K. Evaluation of the effect of allopurinol as a protective factor in post ischemia and reperfusion inflammation in Wistar rats. *Acta Cir. Bras.* **2016**, *31*, 127–132. [CrossRef] [PubMed]

39. Bánfi, A.; Tiszlavicz, L.; Székely, E.; Peták, F.; Tóth-Szüki, V.; Baráti, L.; Bari, F.; Novak, Z. Development of bronchus-associated lymphoid tissue hyperplasia following lipopolysaccharide-induced lung inflammation in rats. *Exp. Lung Res.* **2009**, *35*, 186–197. [CrossRef]

40. Chi, O.Z.; Rah, K.H.; Barsoum, S.; Liu, X.; Weiss, H.R. Inhibition of neuronal nitric oxide synthase improves microregional O2 balance in cerebral ischemia–reperfusion. *Neurol. Res.* **2015**, 1–8. [CrossRef]

41. Vitcheva, V.; Simeonova, R.; Kondeva-burdina, M.; Mitcheva, M. Selective Nitric Oxide Synthase Inhibitor 7-Nitroindazole Protects against Cocaine-Induced Oxidative Stress in Rat Brain. *Oxid. Med. Cell. Longev.* **2015**, *2015*, 157876. [CrossRef]

42. Hirabayashi, H.; Takizawa, S.; Fukuyama, N.; Nakazawa, H.; Shinohara, Y. 7-Nitroindazole attenuates nitrotyrosine formation in the early phase of cerebral ischemia-reperfusion in mice. *Neurosci. Lett.* **1999**, *268*, 111–113. [CrossRef]

43. Sorrenti, V.; Di Giacomo, C.; Campisi, A.; Perez-Polo, J.R.; Vanella, A. Nitric Oxide Synthetase Activity in Cerebral Post-Ischemic Reperfusion and Effects of L-NG-Nitroarginine and 7-Nitroindazole on the Survival. *Neurochem. Res.* **1999**, *24*, 861–866. [CrossRef]

44. Liu, P.K. Ischemia-Reperfusion-Related Repair Deficit after Oxidative Stress: Implications of Faulty Transcripts in Neuronal Sensitivity after Brain Injury. *J. Biomed. Sci.* **2003**, *10*, 4–13. [CrossRef]

45. San Cristobal, J.; Garcia-Alonso, I.; Herrero, B. Pharmacological modulation of retinal ischemia—Reperfusion syndrome in rats. *Investig. Ophthalmol. Vis. Sci.* **2018**, *59*, 5287.

46. Huang, L.-M.; Hu, Q.; Huang, X.; Qian, Y.; Lai, X.-H. Preconditioning rats with three lipid emulsions prior to acute lung injury affects cytokine production and cell apoptosis in the lung and liver. *Lipids Health Dis.* **2020**, *19*, 1–9. [CrossRef]

47. Thomas-Gibson, S.; Jawhari, A.; Atlan, P.; Le Brun, A.L.A.; Farthing, M.; Forbes, A. Safe and efficacious prolonged use of an olive oil-based lipid emulsion (ClinOleic) in chronic intestinal failure. *Clin. Nutr.* **2004**, *23*, 697–703. [CrossRef] [PubMed]

48. Center for Drug Evaluation and Research (CDER); United States Food and Drug Administration. Available online: https://www.accessdata.fda.gov/drugsatfda_docs/nda/2013/204508Orig1s000MedR.pdf (accessed on 13 May 2021).

49. Farías, J.G.; Carrasco-Pozo, C.; Carrasco Loza, R.; Sepúlveda, N.; Álvarez, P.; Quezada, M.; Quiñones, J.; Molina, V.; Castillo, R.L. Polyunsaturated fatty acid induces cardioprotection against ischemia-reperfusion through the inhibition of NF-kappaB and induction of Nrf2. *Exp. Biol. Med.* **2017**, *242*, 1104–1114. [CrossRef]

50. Sukhotnik, I.; Slijper, N.; Pollak, Y.; Chemodanov, E.; Shaoul, R.; Coran, A.G.; Mogilner, J.G. Parenteral omega-3 fatty acids (Omegaven) modulate intestinal recovery after intestinal ischemia-reperfusion in a rat model. *J. Pediatr. Surg.* **2011**, *46*, 1353–1360. [CrossRef]

51. European Union. *Commission Implementing Regulation (EU) No 1348/2013, of 16 December 2013 Amending Regulation (EEC) No 2568/91 on the Characteristics of Olive Oil and Olive-Residue Oil and on the Relevant Methods of Analysis*; Official Journal of the European Union: Luxembourg, 2013.

52. Tubert-Brohman, I.; Sherman, W.; Repasky, M.; Beuming, T. Improved docking of polypeptides with Glide. *J. Chem. Inf. Model.* **2013**, *53*, 1689–1699. [CrossRef]

53. Jimenez-Lopez, C.; Carpena, M.; Lourenço-Lopes, C.; Gallardo-Gomez, M.; Lorenzo, J.M.; Barba, F.J.; Prieto, M.A.; Simal-Gandara, J. Bioactive Compounds and Quality of Extra Virgin Olive Oil. *Foods* **2020**, *9*, 1014. [CrossRef]
54. Augustin, A.J.; Lutz, J. Influence of Perflubron on the Creation of Oxygen Free Radical Products in Mesenteric Artery Occlusion Shock. *Artif. Cells Blood Substit. Biotechnol.* **1994**, *22*, 1223–1230. [CrossRef]
55. Demirkan, A.; Savas, B.; Melli, M. Endotoxin level in ischemia–reperfusion injury in rats: Effect of glutamine pretreatment on endotoxin levels and gut morphology. *Nutrition* **2010**, *26*, 106–111. [CrossRef]

Perspective

The Influence of Oxidative Stress on Thyroid Diseases

Joanna Kochman [1], Karolina Jakubczyk [1,*], Piotr Bargiel [2] and Katarzyna Janda-Milczarek [1]

[1] Department of Human Nutrition and Metabolomics, Pomeranian Medical University, 24 Broniewskiego Street, 71-460 Szczecin, Poland; kochmaan@gmail.com (J.K.); katarzyna.janda-milczarek@pum.edu.pl (K.J.-M.)

[2] Clinic of Plastic, Endocrine and General Surgery, Pomeranian Medical University, 2 Siedlecka Street, 72-010 Police, Poland; bergiel87@gmail.com

* Correspondence: jakubczyk.kar@gmail.com; Tel.: +48-790-233-164

Abstract: Thyroid diseases, including neoplasms, autoimmune diseases and thyroid dysfunctions, are becoming a serious social problem with rapidly increasing prevalence. The latter is increasingly linked to oxidative stress. There are many methods for determining the biomarkers of oxidative stress, making it possible to evaluate the oxidative profile in patients with thyroid diseases compared to the healthy population. This opens up a new perspective for investigating the role of elevated parameters of oxidative stress and damage in people with thyroid diseases, especially of neoplastic nature. An imbalance between oxidants and antioxidants is observed at different stages and in different types of thyroid diseases. The organ, which is part of the endocrine system, uses free radicals (reactive oxygen species, ROS) to produce hormones. Thyroid cells release enzymes that catalyse ROS generation; therefore, a key role is played by the internal defence system and non-enzymatic antioxidants that counteract excess ROS not utilised to produce thyroid hormones, acting as a buffer to neutralise free radicals and ensure whole-body homeostasis. An excess of free radicals causes structural cell damage, undermining genomic stability. Looking at the negative effects of ROS accumulation, oxidative stress appears to be implicated in both the initiation and progression of carcinogenesis. The aim of this review is to investigate the oxidation background of thyroid diseases and to summarise the links between redox imbalance and thyroid dysfunction and disease.

Keywords: oxidative stress; ROS; thyroid diseases; antioxidants

Citation: Kochman, J.; Jakubczyk, K.; Bargiel, P.; Janda-Milczarek, K. The Influence of Oxidative Stress on Thyroid Diseases. *Antioxidants* **2021**, *10*, 1442. https://doi.org/10.3390/antiox10091442

Academic Editors: Soliman Khatib and Dana Atrahimovich Blatt

Received: 28 July 2021
Accepted: 6 September 2021
Published: 10 September 2021

Publisher's Note: MDPI stays neutral with regard to jurisdictional claims in published maps and institutional affiliations.

1. Introduction

Reactive oxygen species (ROS) are molecules capable of independent existence, which contain an oxygen atom and unpaired electrons [1]. ROS arise mainly as by-products in a series of bioenergetic processes of ATP synthesis in mitochondrial respiratory chains [2,3]. Inflammatory processes are an additional source of ROS [1,4]. The most common reactive oxygen species include radicals derived from the electron reduction of molecular oxygen–superoxide anion ($O_2^{\bullet-}$), hydrogen peroxide (H_2O_2) and the more reactive hydroxyl radical ($HO^\bullet$), released in reactions involving metal ions [5].

The body's antioxidant defence against the negative effects of ROS works across a number of different platforms. It involves preventing the formation of radicals, scavenging them and repairing ROS-induced damage. The leading role in the body's defence system is played by antioxidant enzymes, breaking down ROS molecules and thus protecting cells from excessive exposure to ROS [6–8]. The repair system of ROS-induced damage partly relies on autophagy and apoptosis processes, eliminating damaged cells [9–11]. In spite of the range of internal mechanisms of enzymatic regulation, the antioxidant defence system should also be supported by non-enzymatic mechanisms. The latter include the action of molecules with powerful antioxidant properties, notably including glutathione, coenzyme Q10, as well as exogenous substances—polyphenolic compounds, ascorbic acid, retinol, β-carotene and tocopherol. Exogenous substances with confirmed antioxidant properties reinforce the body antioxidant defence, increasing total antioxidant capacity [7,12–14].

Oxidative stress is an effect of redox imbalance between reactive oxygen species and antioxidant defence [9,15]. It may be caused both by the excessive production of ROS and by an inefficient antioxidant system, resulting in molecular damage [16]. Additionally, ROS generation in different subcellular compartments likely involves a positive feedback mechanism, creating a vicious circle of pathological conditions related to oxidative stress [17–19]. Redox homeostasis requires an equilibrium of ROS production and scavenging [20]. Even though the concept of oxidative stress was introduced in the 1980s, its definition and scope of research have been continually elaborated and expanded [6].

Thyroid diseases are a common health problem worldwide, especially among women. The occurrence of subclinical thyroid disorders, which often remain undiagnosed, is also significant [21–24]. Thyroid diseases are increasingly linked to oxidative stress [25–28]. It has been shown that thyroid dysfunction can co-occur with metabolic disorders, including obesity [29–31]. Obesity is a metabolic disease involving mitochondrial dysfunction and chronic oxidative stress, as in several metabolic disorders [32–38]. Since the incidence of thyroid diseases is increased in individuals with increased body weight, the related substrate of metabolic disorders and thyroid dysfunction seems relevant [30,31,39]. However, current reports do not distinguish between the causes and consequences of metabolic abnormalities, so there is a need to develop research on the pathogenesis of thyroid disorders.

2. Physiological Redox Signalling and the Role of ROS in Thyroid Function

Signalling functions in immune responses are initiated when molecular oxygen is oxidised to the reactive superoxide anion radical by the NADPH oxidase (NOX) complex, itself an additional source of ROS [4]. Subsequently, the superoxide is converted by superoxide dismutase (SOD) to H_2O_2. Hydrogen peroxide is associated with a signalling function regulating cellular processes, due to its capacity to reversibly modify cysteine residues [20]. The process alters redox signalling [17]. Accumulation of excessive concentrations of H_2O_2 activates thiolate anion (Cys-S-) oxidation pathways. This is an irreversible process, resulting in permanent protein damage [40]. Antioxidant systems serve a protective function, preventing intracellular accumulation of ROS by reversing the modification of cysteine residues [20].

The role (physiological or pathological) played by ROS depends largely on their concentration and the conditions accompanying biochemical transformations. The initial concentration dictates downstream responses [7]. Excessive amounts of ROS at the subcellular level activates pathways leading to damage in particularly susceptible cell structures or apoptosis [40]. In turn, at low physiological levels, ROS play a signalling role, essential for normal cellular processes [8,41]. Reactive oxygen species also serve as intracellular mediators produced in phagocytic cells, controlling the inflammatory response and antimicrobial defence [4].

ROS play an important role in normal thyroid function. Thyroid cells release oxidases, which catalyse ROS production [42–44]. Inositols are also involved in thyroid hormone synthesis and normal thyroid function, activating a cascade of processes including regulating TSH-dependent signalling (as a TSH transmitter) and generating H_2O_2 production used for iodination and coupling of iodotyrosine and iodothyronine [45–48]. Inositol deficiency or impairment of inositol cascades may result in insufficient synthesis of thyroid hormones, leading to hypothyroidism, which may be further compounded by an increased need for inositols in response to high TSH levels [45,48]. Myoinositol supplementation in hypothyroid patients effectively lowers TSH levels. Its effect has been demonstrated in combination with metformin and selenium compared to treatment without inositol [49,50].

The synthesis of thyroxine (T4) and triiodothyronine (T3) catalysed by thyroid peroxidase (TPO) in thyroid follicles is a very complex process involving ROS, notably, H_2O_2 (Figure 1) [51]. ROS are already essential in the initial stages of thyroid hormone production, during iodide oxidation [52]. Additionally, thyroid hormones perform a metabolic regulatory function by affecting mitochondrial activity [53]. Because of the reliance on ROS in its function, the thyroid is particularly exposed to oxidative damage [54]. Therefore, the antioxidant defence system of the thyroid must effectively regulate ROS production and scavenging [26,55].

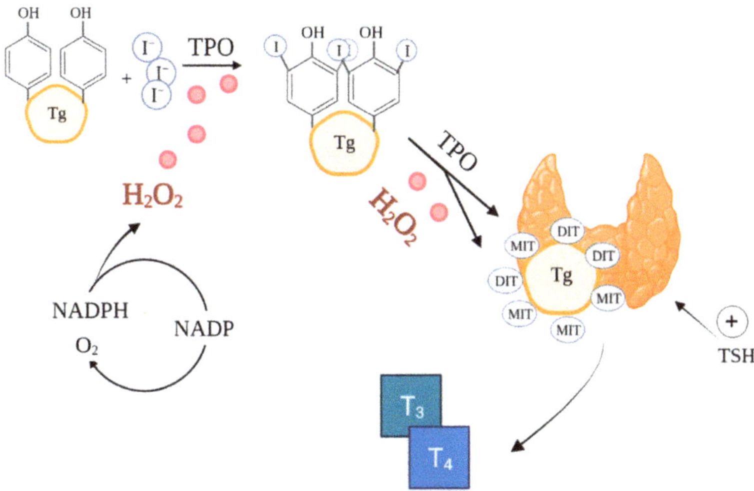

Figure 1. Role of ROS in thyroid hormones synthesis. Based on [47,56]. Created with BioRender.com.(accessed on 26/08/2021) I—iodine, TPO—thyroid peroxidase, Tg—thyroglobulin, MIT—monoiodotyrosine, DIT—diiodotyrosine, T_3—triiodothyronine, T_4—thyroxine.

3. Biomarkers of Oxidative Stress in Thyroid Diseases

Enzymatic mechanisms of antioxidant defence constitute the internal system for maintaining ROS homeostasis (Figure 2). Superoxide dismutases (SOD1, SOD2, SOD3) are antioxidant enzymes, neutralising $O_2^{\bullet-}$ [17,57]. The key enzyme responsible for neutralising hydrogen peroxide is catalase (CAT), which converts it to water and oxygen [58]. Likewise, glutathione peroxidase (GPX) scavenges and detoxifies H_2O_2 [20]. Glutathione serves as an intracellular buffer against oxidation. In response to excessive ROS release, it forms an oxidised dimer structure by bridging two glutathione molecules. Glutathione reductase (GR) then restores the reduced form of glutathione, lowering its reactivity [59]. Measurement of antioxidant enzyme activity in serum makes it possible to evaluate the condition of the antioxidant defence system. Lower levels of this activity, compared to the control, may be a sign of inadequate defence against free radicals [60].

Biomarkers of oxidative stress also include prooxidant enzymes—NADPH oxidases (NOX), which are an endogenous source of ROS, especially in thyroid tissue [46]. Their increased activity is associated with elevated concentrations of reactive oxygen species in pathological conditions. Direct measurement of ROS concentrations may be a helpful marker in the evaluation of medical conditions, yet its utility may be limited given the short half-life of these molecules [15,18].

Malondialdehyde (MDA) is a product of lipid peroxidation by ROS. The marker can be used to evaluate oxidative damage and measure whole-body or tissue-specific oxidative stress [61,62]. Advanced glycation end products (AGE) are believed to be associated with the onset and progression of metabolic disorders, notably diabetes and obesity, due to their formation both through lipid peroxidation and glycoxidation reactions; that is, in response to an increased intake of simple carbohydrates [15,63]. Elevated levels are observed in ROS-damaged tissues, as the final product of peroxidation, making them markers of oxidative stress in the body [64]. Among DNA bases, guanine is the most easily oxidised, due to its relatively low redox potential. Its oxidised form (8-oxo-2'-deoxyguanosine) may therefore serve as a measurement of DNA damage in cells exposed to oxidative stress and in carcinogenesis. 8-oxo-2'-deoxyguanosine has mutagenic potential [9,65].

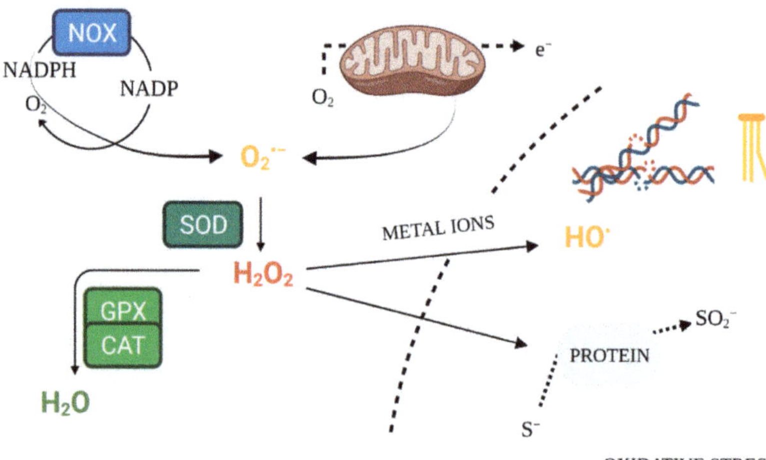

Figure 2. Free Radical Physiology. Created with BioRender.com. (accessed on 26 July 2021).

Total antioxidant capacity (TAC) is a parameter indicative of the body's overall ability to neutralise oxidants. It takes into account all the antioxidants contained in bodily fluids, including exogenous and endogenous compounds [15]. In turn, total oxidant status (TOS) is based on the oxidation of ferrous ion to ferric ion in the presence of various oxidants. It reflects the oxidation state of bodily fluids, represented by the level of radicals [66]. Oxidative stress index (OSI) is a measure of oxidative stress, calculated as the ratio of total oxidant status to total antioxidant status and therefore represents the overall oxidation state of the body [67].

All the biomarkers employed in the determination of the role of oxidative stress in thyroid diseases in this review are listed in Table 1.

Table 1. Biomarkers of oxidative stress used in thyroid disease research [15].

Biomarkers	Mechanism of Development, Role	References
ROS	Energy metabolism in mitochondria	[68]
MDA, HNE	Lipid peroxidation products	[62]
AGE, ALE	Protein oxidation products; Advanced peroxidation end products	[64]
SOD, CAT, GPX, GR	Antioxidant enzymes	[62,68,69]
NOX, DUOX	ROS-generating enzymes	[70]
GSH/GSSG	Reduced/oxygenated glutathione	[69]
TAC, TOS	Number of moles of oxidants neutralised by one litre of body fluid; total oxidative status;	[71,72]

ROS—reactive oxygen species, MDA—malondialdehyde, HNE—hydroxynonenal, AGE-advanced glycation end products, ALE—advanced lipoxidation end products, SOD—superoxide dismutase, CAT—catalase, GPX—glutathione peroxidase, GR—glutathione reductase, NOX—NADPH oxidases, DUOX—dual oxidase, GSH/GSSG—the reduced glutathione/oxidized glutathione ratio, TAC—total antioxidant capacity, TOS—total oxidant status.

4. Relationship between Oxidative Stress, ROS and Thyroid Diseases

4.1. Thyroid Disorders

4.1.1. Underactive Thyroid (Hypothyroidism)

Ref [61] in hypothyroidism, including its subclinical form, elevated levels of MDA have been noted, compared to healthy individuals. Apart from inadequate antioxidant defence, this may be related to altered lipid metabolism in thyroid cells [61]. The treatment of hypothyroidism, despite lowering lipid peroxidation levels, does not bring serum MDA concentrations down to the levels observed in healthy individuals, but it may significantly boost SOD activity [73]. The relationship between hypothyroidism and oxidative stress is probably based on the lower activity of the internal antioxidant system, which does not provide adequate protection to cells against free radical accumulation, leading to oxidative damage [74]. Similarly, a mutation in the gene encoding NOX activity may contribute to excessive stimulation of ROS production. Accumulation of oxygen free radicals may inhibit TPO activity, consequently interfering with thyroid hormone production and leading to the development of hypothyroidism [46,75].

4.1.2. Overactive Thyroid Gland (Hyperthyroidism)

Thyroid hormones also stimulate mitochondrial respiration, leading to an increase in ROS release in the respiratory chain. Overproduction of thyroid hormones therefore causes oxidative stress through the overproduction of free radicals, unlike in hypothyroidism, where redox imbalance can be attributed to an inefficient antioxidant defence system [74]. Consequently, overproduction of thyroid hormones (hyperthyroidism) may be associated with oxidative damage to cell structures. Individuals with hyperthyroidism present higher rates of lipid peroxidation than euthyroid individuals, which is indicative of oxidative damage to membrane lipids [76,77]. In addition, in a study investigating the effects of lead exposure on the parameters of thyroid function and antioxidant markers, thyroid hormones were shown to be positively correlated with MDA, with a positive relationship between TSH and glutathione. These findings suggest a close relationship between hyperthyroidism and the progression of oxidative stress [27].

4.1.3. Thyroid Multinodules Goitre and Nodules

Elevated MDA levels were observed in tissues collected from patients with toxic and non-toxic multinodular goitre, with reduced activity of SOD, GPx and selenium content, compared to adjacent, non-pathologic tissue. Patients did not unequivocally demonstrate hyperthyroidism before surgery, as their thyroid parameters were stabilized in a euthyroid state before sampling [62]. Moreover, tissues of benign thyroid nodules show significantly reduced TAS and reduced OSI [71]. In addition, it was demonstrated that the size of thyroid nodules may decrease as a result of supplementation with extracts of plants with powerful antioxidant and anti-inflammatory properties [78]. The presence of elevated oxidative stress parameters and levels of SOD and CAT activities in toxic multinodular goitre with hyperthyroidism and decreased plasma GPx and GR activities, compared with the control group, were also demonstrated [68]. These findings suggest an impaired redox status and antioxidant defence in patients with thyroid nodules and nodular goitre.

4.1.4. Autoimmune Thyroid Diseases

Chronic lymphocytic thyroiditis, also known as Hashimoto's thyroiditis, is an autoimmune thyroid disease which presents with inflammatory cell infiltration of the thyroid gland and is characterised by the production of autoantibodies to thyroglobulin (anti-TG) and thyroperoxidase (anti-TPO) [79,80]. Inflammatory lesions in the thyroid gland result in the destruction of follicular cells and fibrosis, leading to hypothyroidism [67]. NOX participation in the production of hydrogen peroxide for the purposes of thyroid hormone synthesis may be associated with the pathophysiology of autoimmune thyroid diseases, through interacting with thyroperoxidase and thyroglobulin (TG) and altering their activity, promoting immunogenicity [75,81]. Excessive iodine intake is regarded as an additional risk factor for the development of autoimmune thyroid disease due to enhancing ROS production and, at the same time, reducing internal antioxidant levels. Anti-TPO antibodies show a dependence on glutathione levels, demonstrating an inverse relationship in individuals with Hashimoto's thyroiditis. Additionally, both antibodies (anti-TG and anti-TPO) show a positive correlation with TOS and OSI. Decreased glutathione levels appear to be a distinctive parameter related to the activation and development of oxidative stress in Hashimoto's thyroiditis, as oxidative stress is associated with thyroid hormone deficiency, inflammation and autoimmune parameters. Patients also present with elevated AGE levels. In addition, increased TOS and OSI parameters were shown to precede findings of hypothyroidism in autoimmune thyroiditis and could therefore be treated as predictors of thyroid cell damage [25,64,67,69,72,82].

Graves' disease (GD) is the most common cause of hyperthyroidism and oxidative DNA damage appears to play an important role in its pathogenesis [83,84]. Enhanced inflammatory response modulates the upregulation of autoimmune response [85]. Oxidative stress, in inducing and augmenting inflammation in the thyroid, disrupts self-tolerance, consequently leading to autoimmune thyroid dysfunction. The antibodies found in GD (TSAb, thyroid stimulating antibodies) are involved in oxidation processes. The degree of DNA damage in individuals with untreated GD was shown to be significantly higher than in patients with toxic nodular goitre and individuals without thyroid dysfunction. At the same time, lipid peroxidation markers were higher than in the control. The above-mentioned parameters of oxidative stress, as well as prooxidant enzyme activity, showed a positive correlation with TSAb, suggesting their involvement in the disruption of redox homeostasis [86].

4.1.5. Thyroid Cancer

Oxidative genetic damage caused by the interaction between ROS and DNA, disrupting genomic integrity, leads to mutagenesis. Thus, oxidative stress may cause DNA damage, initiating neoplastic processes [26,87]. A simplified chart of the mechanisms of carcinogenesis, including the free-radical background, is presented in Figure 3. In murine models, oxidative damage is observed much more often in the thyroid gland than in other organs [88]. Patients with different thyroid conditions, in particular neoplasms, present higher baseline genome damage compared with healthy controls [56,89].

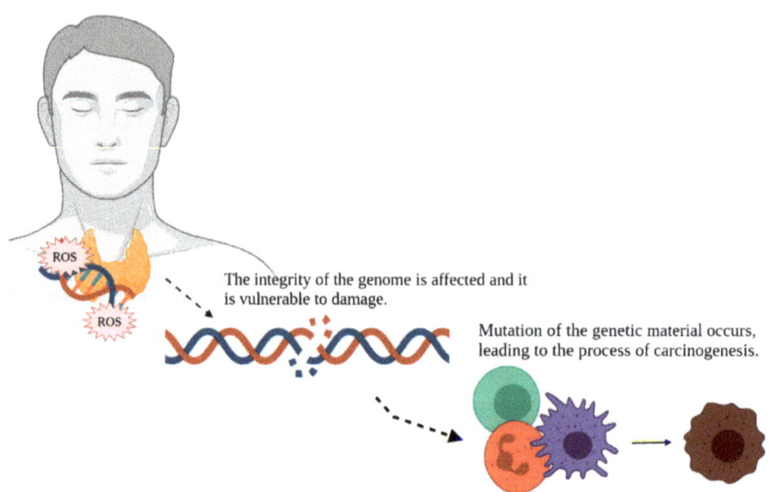

Figure 3. Free radical-mediated carcinogenesis in thyroid cancer. Created with BioRender.com. (accessed on 5 July 2021).

Patients with different types of thyroid cancer have higher serum ROS levels than healthy individuals. Apart from increased whole-body oxidation, they also present with lower activity of internal antioxidants belonging to the antioxidant defence system [60,76,89,90]. Because of the reduced activity of antioxidant enzymes in thyroid cancer cells, the inefficient defence system is not able to neutralise ROS overproduction, resulting in oxidative stress [91]. In a study evaluating the change in biomarkers of oxidative stress in individuals with thyroid cancer before and after thyroidectomy, a significant difference was demonstrated between the study and the control group in terms of glutathione peroxidase activity and MDA levels. Surgical removal of the thyroid had a significant effect on the parameters under analysis, improving oxidative status in favour of antioxidants; however, lipid peroxidation levels remained significantly higher than in healthy individuals [92]. In addition, thyroid tissues in cancer patients have altered metabolic pathways, aimed at improving cancer cell adaptation to unfavourable conditions. Metabolic pathways are shifted to promote glycolysis, more resistant to the conditions of high oxidative stress in cells. This might be an additional target for therapies aimed at processes related to cancer cell metabolism [91]. Apart from higher rates of oxidative processes in cancer patients compared to healthy individuals, those with papillary thyroid cancer had a worse oxidative profile than patients with autoimmune thyroid disease [28]. Obese patients were also found to be at an increased risk for thyroid cancer [93]. There are many reports identifying metabolic links between obesity and mitochondrial dysfunction, excessive ROS generation and oxidative stress [74,94–97]. The relationship between the development of thyroid diseases and obesity, as well as the mechanisms involved, are nevertheless unclear and require in-depth analysis and more detailed research.

5. Conclusions

It is most likely that many of the mechanisms participating in the development of thyroid pathologies are still unknown. However, there is a notable connection of increased ROS generation and findings of oxidative damage with the development of thyroid cancer and other diseases described here. In addition, thyroid disorders may also initiate or increase ROS release and oxidative stress, enhancing oxidative damage. The most recent studies suggest a close link between thyroid diseases and oxidative stress.

Taking into consideration research findings to date, it would appear that preventive nutrition therapy against redox imbalance, in enriching the daily diet in products with a high antioxidant value and supporting the internal antioxidant defence systems, may constitute a promising approach to preventing the development of many chronic thyroid diseases. This creates a prospect for developing measures precisely targeted at the free-radical background, which can be used in the treatment and prevention of thyroid diseases as well as other oxidative diseases.

Author Contributions: Conceptualization, J.K., K.J.; investigation, J.K.; writing—original draft preparation, J.K.; writing—review and editing, K.J., P.B., J.K.; visualization, J.K.; supervision, K.J.; project administration, K.J.-M.; funding acquisition, P.B. All authors have read and agreed to the published version of the manuscript.

Funding: This research was funded by the Pomeranian Medical University in Szczecin.

Institutional Review Board Statement: Not applicable.

Informed Consent Statement: Not applicable.

Data Availability Statement: Not applicable.

Conflicts of Interest: The authors declare no conflict of interest.

References

1. Jakubczyk, K.; Dec, K.; Kałduńska, J.; Kawczuga, D.; Kochman, J.; Janda, K. Reactive Oxygen Species—Sources, Functions, Oxidative Damage. *Pol. Merkur. Lek. Organ Pol. Tow. Lek.* **2020**, *48*, 124–127.
2. Tan, B.L.; Norhaizan, M.E.; Liew, W.-P.-P. Nutrients and Oxidative Stress: Friend or Foe? *Oxid. Med. Cell. Longev.* **2018**, *2018*. [CrossRef] [PubMed]
3. Yang, S.; Lian, G. ROS and Diseases: Role in Metabolism and Energy Supply. *Mol. Cell. Biochem.* **2020**, *467*, 1–12. [CrossRef] [PubMed]
4. Shekhova, E. Mitochondrial Reactive Oxygen Species as Major Effectors of Antimicrobial Immunity. *PLoS Pathog.* **2020**, *16*, e1008470. [CrossRef]
5. Yun, H.R.; Jo, Y.H.; Kim, J.; Shin, Y.; Kim, S.S.; Choi, T.G. Roles of Autophagy in Oxidative Stress. *Int. J. Mol. Sci.* **2020**, *21*, 3289. [CrossRef] [PubMed]
6. Sies, H. Oxidative Stress: A Concept in Redox Biology and Medicine. *Redox Biol.* **2015**, *4*, 180–183. [CrossRef] [PubMed]
7. Di Marzo, N.; Chisci, E.; Giovannoni, R. The Role of Hydrogen Peroxide in Redox-Dependent Signaling: Homeostatic and Pathological Responses in Mammalian Cells. *Cells* **2018**, *7*, 156. [CrossRef]

8. Sies, H.; Berndt, C.; Jones, D.P. Oxidative Stress. *Annu. Rev. Biochem.* **2017**, *86*, 715–748. [CrossRef]

9. Filomeni, G.; De Zio, D.; Cecconi, F. Oxidative Stress and Autophagy: The Clash between Damage and Metabolic Needs. *Cell Death Differ.* **2015**, *22*, 377–388. [CrossRef]

10. Gu, Y.; Han, J.; Jiang, C.; Zhang, Y. Biomarkers, Oxidative Stress and Autophagy in Skin Aging. *Ageing Res. Rev.* **2020**, *59*, 101036. [CrossRef]

11. Vostrikova, S.M.; Grinev, A.B.; Gogvadze, V.G. Reactive Oxygen Species and Antioxidants in Carcinogenesis and Tumor Therapy. *Biochem. Mosc.* **2020**, *85*, 1254–1266. [CrossRef]

12. Mahdavi, A.; Naeini, A.A.; Najafi, M.; Maracy, M.; Ghazvini, M.A. Effect of Levetiracetam Drug on Antioxidant and Liver Enzymes in Epileptic Patients: Case-Control Study. *Afr. Health Sci.* **2020**, *20*, 984–990. [CrossRef] [PubMed]

13. Jakubczyk, K.; Kałduńska, J.; Dec, K.; Kawczuga, D.; Janda, K. Antioxidant Properties of Small-Molecule Non-Enzymatic Compounds. *Pol. Merkur. Lek. Organ Pol. Tow. Lek.* **2020**, *48*, 128–132.

14. Kowalska, K.; Brodowski, J.; Pokorska-Niewiada, K.; Szczuko, M. The Change in the Content of Nutrients in Diets Eliminating Products of Animal Origin in Comparison to a Regular Diet from the Area of Middle-Eastern Europe. *Nutrients* **2020**, *12*, 2986. [CrossRef] [PubMed]

15. Marrocco, I.; Altieri, F.; Peluso, I. Measurement and Clinical Significance of Biomarkers of Oxidative Stress in Humans. *Oxid. Med. Cell. Longev.* **2017**, *2017*, 6501046. [CrossRef] [PubMed]

16. Sies, H.; Jones, D.P. Reactive Oxygen Species (ROS) as Pleiotropic Physiological Signalling Agents. *Nat. Rev. Mol. Cell Biol.* **2020**, *21*, 363–383. [CrossRef]

17. Fukai, T.; Ushio-Fukai, M. Cross-Talk between NADPH Oxidase and Mitochondria: Role in ROS Signaling and Angiogenesis. *Cells* **2020**, *9*, 1849. [CrossRef]

18. Kim, Y.-M.; Kim, S.-J.; Tatsunami, R.; Yamamura, H.; Fukai, T.; Ushio-Fukai, M. ROS-Induced ROS Release Orchestrated by Nox4, Nox2, and Mitochondria in VEGF Signaling and Angiogenesis. *Am. J. Physiol. Cell Physiol.* **2017**, *312*, C749–C764. [CrossRef]

19. Aldosari, S.; Awad, M.; Harrington, E.O.; Sellke, F.W.; Abid, M.R. Subcellular Reactive Oxygen Species (ROS) in Cardiovascular Pathophysiology. *Antioxid. Basel Switz.* **2018**, *7*, 14. [CrossRef]

20. Irazabal, M.V.; Torres, V.E. Reactive Oxygen Species and Redox Signaling in Chronic Kidney Disease. *Cells* **2020**, *9*, 1342. [CrossRef] [PubMed]

21. Garmendia Madariaga, A.; Santos Palacios, S.; Guillén-Grima, F.; Galofré, J.C. The Incidence and Prevalence of Thyroid Dysfunction in Europe: A Meta-Analysis. *J. Clin. Endocrinol. Metab.* **2014**, *99*, 923–931. [CrossRef] [PubMed]

22. Canaris, G.J.; Manowitz, N.R.; Mayor, G.; Ridgway, E.C. The Colorado Thyroid Disease Prevalence Study. *Arch. Intern. Med.* **2000**, *160*, 526–534. [CrossRef] [PubMed]

23. Kasagi, K.; Takahashi, N.; Inoue, G.; Honda, T.; Kawachi, Y.; Izumi, Y. Thyroid Function in Japanese Adults as Assessed by a General Health Checkup System in Relation with Thyroid-Related Antibodies and Other Clinical Parameters. *Thyroid* **2009**, *19*, 937–944. [CrossRef] [PubMed]

24. Empson, M.; Flood, V.; Ma, G.; Eastman, C.J.; Mitchell, P. Prevalence of Thyroid Disease in an Older Australian Population. *Intern. Med. J.* **2007**, *37*, 448–455. [CrossRef]

25. Rostami, R.; Aghasi, M.R.; Mohammadi, A.; Nourooz-Zadeh, J. Enhanced Oxidative Stress in Hashimoto's Thyroiditis: Inter-Relationships to Biomarkers of Thyroid Function. *Clin. Biochem.* **2013**, *46*, 308–312. [CrossRef]

26. Ameziane El Hassani, R.; Buffet, C.; Leboulleux, S.; Dupuy, C. Oxidative Stress in Thyroid Carcinomas: Biological and Clinical Significance. *Endocr. Relat. Cancer* **2019**, *26*, R131–R143. [CrossRef]

27. Fahim, Y.A.; Sharaf, N.E.; Hasani, I.W.; Ragab, E.A.; Abdelhakim, H.K. Assessment of Thyroid Function and Oxidative Stress State in Foundry Workers Exposed to Lead. *J. Health Pollut.* **2020**, *10*, 200903. [CrossRef]

28. Lassoued, S.; Mseddi, M.; Mnif, F.; Abid, M.; Guermazi, F.; Masmoudi, H.; El Feki, A.; Attia, H. A Comparative Study of the Oxidative Profile in Graves' Disease, Hashimoto's Thyroiditis, and Papillary Thyroid Cancer. *Biol. Trace Elem. Res.* **2010**, *138*, 107–115. [CrossRef]

29. Mehran, L.; Amouzegar, A.; Rahimabad, P.K.; Tohidi, M.; Tahmasebinejad, Z.; Azizi, F. Thyroid Function and Metabolic Syndrome: A Population-Based Thyroid Study. *Horm. Metab. Res.* **2017**, *49*, 192–200. [CrossRef]

30. Du, F.-M.; Kuang, H.-Y.; Duan, B.-H.; Liu, D.-N.; Yu, X.-Y. Effects of Thyroid Hormone and Depression on Common Components of Central Obesity. *J. Int. Med. Res.* **2019**, *47*, 3040–3049. [CrossRef]

31. Song, R.-H.; Wang, B.; Yao, Q.-M.; Li, Q.; Jia, X.; Zhang, J.-A. The Impact of Obesity on Thyroid Autoimmunity and Dysfunction: A Systematic Review and Meta-Analysis. *Front. Immunol.* **2019**, *10*, 2349. [CrossRef] [PubMed]

32. Heinonen, S.; Buzkova, J.; Muniandy, M.; Kaksonen, R.; Ollikainen, M.; Ismail, K.; Hakkarainen, A.; Lundbom, J.; Lundbom, N.; Vuolteenaho, K.; et al. Impaired Mitochondrial Biogenesis in Adipose Tissue in Acquired Obesity. *Diabetes* **2015**, *64*, 3135–3145. [CrossRef]

33. Parra, M.D.; Martínez de Morentin, B.E.; Martínez, J.A. Postprandial Insulin Response and Mitochondrial Oxidation in Obese Men Nutritionally Treated to Lose Weight. *Eur. J. Clin. Nutr.* **2005**, *59*, 334–340. [CrossRef]

34. Anderson, E.J.; Lustig, M.E.; Boyle, K.E.; Woodlief, T.L.; Kane, D.A.; Lin, C.-T.; Price, J.W.; Kang, L.; Rabinovitch, P.S.; Szeto, H.H.; et al. Mitochondrial H_2O_2 Emission and Cellular Redox State Link Excess Fat Intake to Insulin Resistance in Both Rodents and Humans. *J. Clin. Investig.* **2009**, *119*, 573–581. [CrossRef]

35. Saraf-Bank, S.; Ahmadi, A.; Paknahad, Z.; Maracy, M.; Nourian, M. Effects of Curcumin Supplementation on Markers of Inflammation and Oxidative Stress among Healthy Overweight and Obese Girl Adolescents: A Randomized Placebo-Controlled Clinical Trial. *Phytother. Res.* **2019**, *33*, 2015–2022. [CrossRef]
36. Yin, X.; Lanza, I.R.; Swain, J.M.; Sarr, M.G.; Nair, K.S.; Jensen, M.D. Adipocyte Mitochondrial Function Is Reduced in Human Obesity Independent of Fat Cell Size. *J. Clin. Endocrinol. Metab.* **2014**, *99*, E209–E216. [CrossRef]
37. Fischer, B.; Schöttl, T.; Schempp, C.; Fromme, T.; Hauner, H.; Klingenspor, M.; Skurk, T. Inverse Relationship between Body Mass Index and Mitochondrial Oxidative Phosphorylation Capacity in Human Subcutaneous Adipocytes. *Am. J. Physiol. Endocrinol. Metab.* **2015**, *309*, E380–E387. [CrossRef]
38. Christe, M.; Hirzel, E.; Lindinger, A.; Kern, B.; von Flüe, M.; Peterli, R.; Peters, T.; Eberle, A.N.; Lindinger, P.W. Obesity Affects Mitochondrial Citrate Synthase in Human Omental Adipose Tissue. *ISRN Obes.* **2013**, *2013*, 826027. [CrossRef] [PubMed]
39. Schmid, D.; Ricci, C.; Behrens, G.; Leitzmann, M.F. Adiposity and Risk of Thyroid Cancer: A Systematic Review and Meta-Analysis. *Obes. Rev. Off. J. Int. Assoc. Study Obes.* **2015**, *16*, 1042–1054. [CrossRef] [PubMed]
40. Schieber, M.; Chandel, N.S. ROS Function in Redox Signaling and Oxidative Stress. *Curr. Biol. CB* **2014**, *24*, R453–R462. [CrossRef]
41. Sies, H. Hydrogen Peroxide as a Central Redox Signaling Molecule in Physiological Oxidative Stress: Oxidative Eustress. *Redox Biol.* **2017**, *11*, 613–619. [CrossRef]
42. Ameziane-El-Hassani, R.; Schlumberger, M.; Dupuy, C. NADPH Oxidases: New Actors in Thyroid Cancer? *Nat. Rev. Endocrinol.* **2016**, *12*, 485–494. [CrossRef] [PubMed]
43. Cardoso, L.C.; Martins, D.C.; Figueiredo, M.D.; Rosenthal, D.; Vaisman, M.; Violante, A.H.; Carvalho, D.P. Ca^{2+}/Nicotinamide Adenine Dinucleotide Phosphate-Dependent H_2O_2 Generation Is Inhibited by Iodide in Human Thyroids. *J. Clin. Endocrinol. Metab.* **2001**, *86*, 4339–4343. [CrossRef] [PubMed]
44. Dupuy, C.; Virion, A.; Ohayon, R.; Kaniewski, J.; Dème, D.; Pommier, J. Mechanism of Hydrogen Peroxide Formation Catalyzed by NADPH Oxidase in Thyroid Plasma Membrane. *J. Biol. Chem.* **1991**, *266*, 3739–3743. [CrossRef]
45. Piras, C.; Pibiri, M.; Leoni, V.P.; Balsamo, A.; Tronci, L.; Arisci, N.; Mariotti, S.; Atzori, L. Analysis of Metabolomics Profile in Hypothyroid Patients before and after Thyroid Hormone Replacement. *J. Endocrinol. Investig.* **2021**, *44*, 1309–1319. [CrossRef]
46. Ohye, H.; Sugawara, M. Dual Oxidase, Hydrogen Peroxide and Thyroid Diseases. *Exp. Biol. Med. Maywood NJ* **2010**, *235*, 424–433. [CrossRef]
47. Benvenga, S.; Nordio, M.; Laganà, A.S.; Unfer, V. The Role of Inositol in Thyroid Physiology and in Subclinical Hypothyroidism Management. *Front. Endocrinol.* **2021**, *12*, 662582. [CrossRef]
48. Grasberger, H.; Van Sande, J.; Hag-Dahood Mahameed, A.; Tenenbaum-Rakover, Y.; Refetoff, S. A Familial Thyrotropin (TSH) Receptor Mutation Provides in Vivo Evidence That the Inositol Phosphates/Ca^{2+} Cascade Mediates TSH Action on Thyroid Hormone Synthesis. *J. Clin. Endocrinol. Metab.* **2007**, *92*, 2816–2820. [CrossRef]
49. Morgante, G.; Musacchio, M.C.; Orvieto, R.; Massaro, M.G.; De Leo, V. Alterations in Thyroid Function among the Different Polycystic Ovary Syndrome Phenotypes. *Gynecol. Endocrinol.* **2013**, *29*, 967–969. [CrossRef]
50. Pace, C.; Tumino, D.; Russo, M.; Le Moli, R.; Naselli, A.; Borzì, G.; Malandrino, P.; Frasca, F. Role of Selenium and Myo-Inositol Supplementation on Autoimmune Thyroiditis Progression. *Endocr. J.* **2020**, *67*, 1093–1098. [CrossRef]
51. Thanas, C.; Ziros, P.G.; Chartoumpekis, D.V.; Renaud, C.O.; Sykiotis, G.P. The Keap1/Nrf2 Signaling Pathway in the Thyroid—2020 Update. *Antioxidants* **2020**, *9*, 1082. [CrossRef] [PubMed]
52. Massart, C.; Hoste, C.; Virion, A.; Ruf, J.; Dumont, J.E.; Van Sande, J. Cell Biology of H_2O_2 Generation in the Thyroid: Investigation of the Control of Dual Oxidases (DUOX) Activity in Intact Ex Vivo Thyroid Tissue and Cell Lines. *Mol. Cell. Endocrinol.* **2011**, *343*, 32–44. [CrossRef] [PubMed]
53. Venditti, P.; Puca, A.; Di Meo, S. Effects of Thyroid State on H_2O_2 Production by Rat Heart Mitochondria: Sites of Production with Complex I- and Complex II-Linked Substrates. *Horm. Metab. Res.* **2003**, *35*, 55–61. [CrossRef] [PubMed]
54. Paunkov, A.; Chartoumpekis, D.V.; Ziros, P.G.; Chondrogianni, N.; Kensler, T.W.; Sykiotis, G.P. Impact of Antioxidant Natural Compounds on the Thyroid Gland and Implication of the Keap1/Nrf2 Signaling Pathway. *Curr. Pharm. Des.* **2019**, *25*, 1828–1846. [CrossRef] [PubMed]
55. Poncin, S.; Gérard, A.-C.; Boucquey, M.; Senou, M.; Calderon, P.B.; Knoops, B.; Lengelé, B.; Many, M.-C.; Colin, I.M. Oxidative Stress in the Thyroid Gland: From Harmlessness to Hazard Depending on the Iodine Content. *Endocrinology* **2008**, *149*, 424–433. [CrossRef] [PubMed]
56. Szanto, I.; Pusztaszeri, M.; Mavromati, M. H_2O_2 Metabolism in Normal Thyroid Cells and in Thyroid Tumorigenesis: Focus on NADPH Oxidases. *Antioxidants* **2019**, *8*, 126. [CrossRef]
57. Eleutherio, E.C.A.; Magalhães, R.S.S.; de Araújo Brasil, A.; Neto, J.R.M.; de Holanda Paranhos, L. SOD1, More than Just an Antioxidant. *Arch. Biochem. Biophys.* **2021**, *697*. [CrossRef]
58. Sepasi Tehrani, H.; Moosavi-Movahedi, A.A. Catalase and Its Mysteries. *Prog. Biophys. Mol. Biol.* **2018**, *140*, 5–12. [CrossRef]
59. Couto, N.; Wood, J.; Barber, J. The Role of Glutathione Reductase and Related Enzymes on Cellular Redox Homoeostasis Network. *Free Radic. Biol. Med.* **2016**, *95*, 27–42. [CrossRef]
60. Metere, A.; Frezzotti, F.; Graves, C.E.; Vergine, M.; De Luca, A.; Pietraforte, D.; Giacomelli, L. A Possible Role for Selenoprotein Glutathione Peroxidase (GPx1) and Thioredoxin Reductases (TrxR1) in Thyroid Cancer: Our Experience in Thyroid Surgery. *Cancer Cell Int.* **2018**, *18*, 7. [CrossRef]

61. Torun, A.N.; Kulaksizoglu, S.; Kulaksizoglu, M.; Pamuk, B.O.; Isbilen, E.; Tutuncu, N.B. Serum Total Antioxidant Status and Lipid Peroxidation Marker Malondialdehyde Levels in Overt and Subclinical Hypothyroidism. *Clin. Endocrinol.* **2009**, *70*, 469–474. [CrossRef] [PubMed]

62. Erdamar, H.; Cimen, B.; Gülcemal, H.; Saraymen, R.; Yerer, B.; Demirci, H. Increased Lipid Peroxidation and Impaired Enzymatic Antioxidant Defense Mechanism in Thyroid Tissue with Multinodular Goiter and Papillary Carcinoma. *Clin. Biochem.* **2010**, *43*, 650–654. [CrossRef] [PubMed]

63. Loomis, S.J.; Chen, Y.; Sacks, D.B.; Christenson, E.S.; Christenson, R.H.; Rebholz, C.M.; Selvin, E. Cross-Sectional Analysis of AGE-CML, SRAGE, and EsRAGE with Diabetes and Cardiometabolic Risk Factors in a Community-Based Cohort. *Clin. Chem.* **2017**, *63*, 980–989. [CrossRef] [PubMed]

64. Ruggeri, R.M.; Giovinazzo, S.; Barbalace, M.C.; Cristani, M.; Alibrandi, A.; Vicchio, T.M.; Giuffrida, G.; Aguennouz, M.H.; Malaguti, M.; Angeloni, C.; et al. Influence of Dietary Habits on Oxidative Stress Markers in Hashimoto's Thyroiditis. *Thyroid Off. J. Am. Thyroid Assoc.* **2021**, *31*, 96–105. [CrossRef] [PubMed]

65. Kasai, H. Analysis of a Form of Oxidative DNA Damage, 8-Hydroxy-2′-Deoxyguanosine, as a Marker of Cellular Oxidative Stress during Carcinogenesis. *Mutat. Res. Mutat. Res.* **1997**, *387*, 147–163. [CrossRef]

66. Rovcanin, B.R.; Gopcevic, K.R.; Kekic, D.L.; Zivaljevic, V.R.; Diklic, A.D.; Paunovic, I.R. Papillary Thyroid Carcinoma: A Malignant Tumor with Increased Antioxidant Defense Capacity. *Tohoku J. Exp. Med.* **2016**, *240*, 101–111. [CrossRef] [PubMed]

67. Ates, I.; Arikan, M.F.; Altay, M.; Yilmaz, F.M.; Yilmaz, N.; Berker, D.; Guler, S. The Effect of Oxidative Stress on the Progression of Hashimoto's Thyroiditis. *Arch. Physiol. Biochem.* **2018**, *124*, 351–356. [CrossRef]

68. Bednarek, J.; Wysocki, H.; Sowinski, J. Oxidation Products and Antioxidant Markers in Plasma of Patients with Graves' Disease and Toxic Multinodular Goiter: Effect of Methimazole Treatment. *Free Radic. Res.* **2004**, *38*, 659–664. [CrossRef]

69. Rostami, R.; Nourooz-Zadeh, S.; Mohammadi, A.; Khalkhali, H.R.; Ferns, G.; Nourooz-Zadeh, J. Serum Selenium Status and Its Interrelationship with Serum Biomarkers of Thyroid Function and Antioxidant Defense in Hashimoto's Thyroiditis. *Antioxidants* **2020**, *9*, 1070. [CrossRef]

70. Fortunato, R.S.; Braga, W.M.O.; Ortenzi, V.H.; Rodrigues, D.C.; Andrade, B.M.; Miranda-Alves, L.; Rondinelli, E.; Dupuy, C.; Ferreira, A.C.F.; Carvalho, D.P. Sexual Dimorphism of Thyroid Reactive Oxygen Species Production Due to Higher NADPH Oxidase 4 Expression in Female Thyroid Glands. *Thyroid Off. J. Am. Thyroid Assoc.* **2013**, *23*, 111–119. [CrossRef]

71. Faam, B.; Ghadiri, A.A.; Ghaffari, M.A.; Totonchi, M.; Khorsandi, L. Comparing Oxidative Stress Status Among Iranian Males and Females with Malignant and Non-Malignant Thyroid Nodules. *Int. J. Endocrinol. Metab.* **2021**, *19*, e105669. [CrossRef]

72. Ates, I.; Yilmaz, F.M.; Altay, M.; Yilmaz, N.; Berker, D.; Güler, S. The Relationship between Oxidative Stress and Autoimmunity in Hashimoto's Thyroiditis. *Eur. J. Endocrinol.* **2015**, *173*, 791–799. [CrossRef]

73. Baskol, G.; Atmaca, H.; Tanrıverdi, F.; Baskol, M.; Kocer, D.; Bayram, F. Oxidative Stress and Enzymatic Antioxidant Status in Patients with Hypothyroidism before and after Treatment. *Exp. Clin. Endocrinol. Diabetes* **2007**, *115*, 522–526. [CrossRef] [PubMed]

74. Mancini, A.; Di Segni, C.; Raimondo, S.; Olivieri, G.; Silvestrini, A.; Meucci, E.; Currò, D. Thyroid Hormones, Oxidative Stress, and Inflammation. *Mediators Inflamm.* **2016**, *2016*, 6757154. [CrossRef] [PubMed]

75. Fortunato, R.S.; Ferreira, A.C.F.; Hecht, F.; Dupuy, C.; Carvalho, D.P. Sexual Dimorphism and Thyroid Dysfunction: A Matter of Oxidative Stress? *J. Endocrinol.* **2014**, *221*, R31–R40. [CrossRef] [PubMed]

76. Wang, D.; Feng, J.-F.; Zeng, P.; Yang, Y.-H.; Luo, J.; Yang, Y.-W. Total Oxidant/Antioxidant Status in Sera of Patients with Thyroid Cancers. *Endocr. Relat. Cancer* **2011**, *18*, 773–782. [CrossRef] [PubMed]

77. Piazera, B.K.L.; Gomes, D.V.; Vigário, P.; Salerno, V.P.; Vaisman, M. Evaluation of Redox Profiles in Exogenous Subclinical Hyperthyroidism at Two Different Levels of TSH Suppression. *Arch. Endocrinol. Metab.* **2018**, *62*, 545–551. [CrossRef]

78. Stancioiu, F.; Mihai, D.; Papadakis, G.Z.; Tsatsakis, A.; Spandidos, D.A.; Badiu, C. Treatment for Benign Thyroid Nodules with a Combination of Natural Extracts. *Mol. Med. Rep.* **2019**, *20*, 2332–2338. [CrossRef]

79. Burek, C.L.; Rose, N.R. Autoimmune Thyroiditis and ROS. *Autoimmun. Rev.* **2008**, *7*, 530–537. [CrossRef]

80. Laganà, A.S.; Santoro, G.; Triolo, O.; Giacobbe, V.; Certo, R.; Palmara, V. Hashimoto Thyroiditis Onset after Laparoscopic Removal of Struma Ovarii: An Overview to Unravel a Rare and Intriguing Finding. *Clin. Exp. Obstet. Gynecol.* **2015**, *42*, 673–678.

81. Duthoit, C.; Estienne, V.; Giraud, A.; Durand-Gorde, J.M.; Rasmussen, A.K.; Feldt-Rasmussen, U.; Carayon, P.; Ruf, J. Hydrogen Peroxide-Induced Production of a 40 KDa Immunoreactive Thyroglobulin Fragment in Human Thyroid Cells: The Onset of Thyroid Autoimmunity? *Biochem. J.* **2001**, *360*, 557–562. [CrossRef] [PubMed]

82. Baser, H.; Can, U.; Baser, S.; Yerlikaya, F.H.; Aslan, U.; Hidayetoglu, B.T. Assesment of Oxidative Status and Its Association with Thyroid Autoantibodies in Patients with Euthyroid Autoimmune Thyroiditis. *Endocrine* **2015**, *48*, 916–923. [CrossRef] [PubMed]

83. Zarković, M. The Role of Oxidative Stress on the Pathogenesis of Graves' Disease. *J. Thyroid Res.* **2012**, *2012*, 302537. [CrossRef] [PubMed]

84. De Leo, S.; Lee, S.Y.; Braverman, L.E. Hyperthyroidism. *Lancet Lond. Engl.* **2016**, *388*, 906–918. [CrossRef]

85. Rasool, M.; Malik, A.; Saleem, S.; Ashraf, M.A.B.; Khan, A.Q.; Waquar, S.; Zahid, A.; Shaheen, S.; Abu-Elmagd, M.; Gauthaman, K.; et al. Role of Oxidative Stress and the Identification of Biomarkers Associated With Thyroid Dysfunction in Schizophrenics. *Front. Pharmacol.* **2021**, *12*, 646287. [CrossRef]

86. Diana, T.; Daiber, A.; Oelze, M.; Neumann, S.; Olivo, P.D.; Kanitz, M.; Stamm, P.; Kahaly, G.J. Stimulatory TSH-Receptor Antibodies and Oxidative Stress in Graves Disease. *J. Clin. Endocrinol. Metab.* **2018**, *103*, 3668–3677. [CrossRef]

87. Nakashima, M.; Suzuki, K.; Meirmanov, S.; Naruke, Y.; Matsuu-Matsuyama, M.; Shichijo, K.; Saenko, V.; Kondo, H.; Hayashi, T.; Ito, M.; et al. Foci Formation of P53-Binding Protein 1 in Thyroid Tumors: Activation of Genomic Instability during Thyroid Carcinogenesis. *Int. J. Cancer* **2008**, *122*, 1082–1088. [CrossRef]

88. Maier, J.; van Steeg, H.; van Oostrom, C.; Karger, S.; Paschke, R.; Krohn, K. Deoxyribonucleic Acid Damage and Spontaneous Mutagenesis in the Thyroid Gland of Rats and Mice. *Endocrinology* **2006**, *147*, 3391–3397. [CrossRef]

89. Gerić, M.; Domijan, A.M.; Gluščić, V.; Janušić, R.; Šarčević, B.; Garaj-Vrhovac, V. Cytogenetic Status and Oxidative Stress Parameters in Patients with Thyroid Diseases. *Mutat. Res. Toxicol. Environ. Mutagen.* **2016**, *810*, 22–29. [CrossRef]

90. Ramli, N.S.F.; Mat Junit, S.; Leong, N.K.; Razali, N.; Jayapalan, J.J.; Abdul Aziz, A. Analyses of Antioxidant Status and Nucleotide Alterations in Genes Encoding Antioxidant Enzymes in Patients with Benign and Malignant Thyroid Disorders. *PeerJ* **2017**, *5*. [CrossRef]

91. Metere, A.; Graves, C.E.; Chirico, M.; Caramujo, M.J.; Pisanu, M.E.; Iorio, E. Metabolomic Reprogramming Detected by 1H-NMR Spectroscopy in Human Thyroid Cancer Tissues. *Biology* **2020**, *9*, 112. [CrossRef]

92. Akinci, M.; Kosova, F.; Çetin, B.; Sepici, A.; Altan, N.; Aslan, S.; Çetin, A. Oxidant/Antioxidant Balance in Patients with Thyroid Cancer. *Acta Cirúrgica Bras.* **2008**, *23*, 551–554. [CrossRef] [PubMed]

93. Oberman, B.; Khaku, A.; Camacho, F.; Goldenberg, D. Relationship between Obesity, Diabetes and the Risk of Thyroid Cancer. *Am. J. Otolaryngol.* **2015**, *36*, 535–541. [CrossRef]

94. Kanikowska, D.; Kanikowska, A.; Swora-Cwynar, E.; Grzymisławski, M.; Sato, M.; Bręborowicz, A.; Witowski, J.; Korybalska, K. Moderate Caloric Restriction Partially Improved Oxidative Stress Markers in Obese Humans. *Antioxidants* **2021**, *10*, 1018. [CrossRef]

95. Włodarczyk, M.; Nowicka, G. Obesity, DNA Damage, and Development of Obesity-Related Diseases. *Int. J. Mol. Sci.* **2019**, *20*, 1146. [CrossRef] [PubMed]

96. Lahera, V.; de Las Heras, N.; López-Farré, A.; Manucha, W.; Ferder, L. Role of Mitochondrial Dysfunction in Hypertension and Obesity. *Curr. Hypertens. Rep.* **2017**, *19*, 11. [CrossRef]

97. Zaki, M.; Basha, W.; El-Bassyouni, H.T.; El-Toukhy, S.; Hussein, T. Evaluation of DNA Damage Profile in Obese Women and Its Association to Risk of Metabolic Syndrome, Polycystic Ovary Syndrome and Recurrent Preeclampsia. *Genes Dis.* **2018**, *5*, 367–373. [CrossRef] [PubMed]

 *antioxidants*

Review

Risk Assessment of Oxidative Stress Induced by Metal Ions Released from Fixed Orthodontic Appliances during Treatment and Indications for Supportive Antioxidant Therapy: A Narrative Review

Jasmina Primožič [1], Borut Poljšak [2,*], Polona Jamnik [3], Vito Kovač [2], Gordana Čanadi Jurešić [4] and Stjepan Spalj [5]

[1] Department of Orthodontics and Jaw Orthopedics, Medical Faculty, University of Ljubljana, Vrazov trg 2, SI-1000 Ljubljana, Slovenia; jasmina.primozic@mf.uni-lj.si

[2] Laboratory of Oxidative Stress Research, Faculty of Health Sciences, University of Ljubljana, Zdravstvena pot 5, SI-1000 Ljubljana, Slovenia; kovacv@zf.uni-lj.si

[3] Biotechnical Faculty, University of Ljubljana, SI-1000 Ljubljana, Slovenia; polona.jamnik@bf.uni-lj.si

[4] Department of Medical Chemistry, Biochemistry and Clinical Chemistry, Faculty of Medicine, University of Rijeka, 51000 Rijeka, Croatia; gordanacj@medri.uniri.hr

[5] Department of Orthodontics, Faculty of Dental Medicine, University of Rijeka, 51000 Rijeka, Croatia; stjepan.spalj@fdmri.uniri.hr

* Correspondence: borut.poljsak@zf.uni-lj.si

Citation: Primožič, J.; Poljšak, B.; Jamnik, P.; Kovač, V.; Čanadi Jurešić, G.; Spalj, S. Risk Assessment of Oxidative Stress Induced by Metal Ions Released from Fixed Orthodontic Appliances during Treatment and Indications for Supportive Antioxidant Therapy: A Narrative Review. *Antioxidants* **2021**, *10*, 1359. https://doi.org/10.3390/antiox10091359

Academic Editors: Soliman Khatib and Dana Atrahimovich Blatt

Received: 2 August 2021
Accepted: 25 August 2021
Published: 26 August 2021

Publisher's Note: MDPI stays neutral with regard to jurisdictional claims in published maps and institutional affiliations.

Abstract: The treatment with fixed orthodontic appliances could have an important role in the induction of oxidative stress and associated negative consequences. Because of the simultaneous effects of corrosion, deformation, friction, and mechanical stress on fixed orthodontic appliances during treatment, degradation of orthodontic brackets and archwires occurs, causing higher concentrations of metal ions in the oral cavity. Corroded appliances cause the release of metal ions, which may lead to the increased values of reactive oxygen species (ROS) due to metal-catalyzed free radical reactions. Chromium, iron, nickel, cobalt, titanium, and molybdenum all belong to the group of transition metals that can be subjected to redox reactions to form ROS. The estimation of health risk due to the amount of heavy metals released and the level of selected parameters of oxidative stress generated for the time of treatment with fixed orthodontic appliances is presented. Approaches to avoid oxidative stress and recommendations for the preventive use of topical or systemic antioxidants during orthodontic treatment are discussed.

Keywords: fixed orthodontic appliances; oxidative stress; metal ions; risk assessment; antioxidant therapy

1. Transition Metal Ions, Fenton-like Chemistry and Oxidative Stress

Oxidative stress was first explicated by Sies [1] as "a disturbance in the balance between pro-oxidants and antioxidants in favor of the former, leading to potential damage". Oxidative stress occurs in case of significant imbalance between the production and degradation of reactive oxygen species (ROS) with prominent pro-oxidants, generating possible oxidative damage. ROS comprise both free radicals and non-free radical oxygen-containing molecules, such as hydrogen peroxide (H_2O_2), superoxide ($O_2{}^-$), singlet oxygen ($1/2O_2$), and the hydroxyl radical (HO^-) [2]. The cell undergoes the oxidative stress, which is the result of the activity of the ROS generating reactions as well as the efficiency of the ROS scavenger system composed of endogenous and exogenous antioxidants. Endogenous defense system includes enzymatic and nonenzymatic antioxidants, such as superoxide dismutase (Mn-SOD, Cu/Zn-SOD, and extracellular (EC)-SOD), glutathione peroxidase, catalase, peroxiredoxins, glutathione (GSH), thioredoxin, uric acid, and a system for repairing oxidative damages of molecules [3]. On the other hand, the antioxidant defense can be raised through ingested food. Exogenous antioxidants contain vitamin E and C, polyphenols, carotenoids and minerals, such as zinc and selenium. They all play a significant role in the preservation of intracellular redox status and avoidance of oxidative stress-induced damage [4]. The elevated ROS production is due to endogenous (increased mitochondrial leakage, increased O_2 concentration, and inflammation) and exogenous causes (pollution of the environment, physically demanding sport activities, diet, chronic inflammation, smoking, psychological and emotional stress, and other) [3,5,6]. Free metal ions are yet another source of increased oxidative stress since certain metals (e.g., chromium, iron, copper, cobalt, and vanadium) have the potential of redox cycling. During this process, the metal ion can accept or donate a single electron. This process catalyzes reactions in which reactive radicals are generated and reactive oxygen species can be formed. Reduced forms of redox-active metal ions play a part in the Fenton reaction where the hydroxyl radical (HO) is produced from hydrogen peroxide.

$$\text{Metal}^{(n+1)} + H_2O_2 \rightarrow \text{Metal}^{(n+1)+} + HO^{\cdot} + HO^{-} \qquad \text{(1 Fenton Reaction)} \quad (1)$$

Additionally, the Haber–Weiss reaction between the oxidized forms of the redox-active metal ions and the superoxide anion produces the reduced form of the metal ion, which can be coupled with Fenton chemistry to produce the hydroxyl radical.

$$\text{Metal}^{(n+1)+} + O_2{}^{\cdot -} \rightarrow \text{Metal}^{(n+1)} + O_2 \qquad \text{(2 Haber-Weiss Reaction)} \quad (2)$$

Both $O_2{}^{-}$ and H_2O_2 by themselves are not overly reactive and are consequently not particularly harmful in physiological concentrations. Nonetheless, their reactions with poorly bound iron species can cause the formation of hydroxyl radicals that are exceptionally harmful and are the prime source of cellular oxidative damage [7]. In the net reaction, a molecule of hydrogen peroxide undergoes a conversion to a hydroxyl radical and water in the presence of metal ion as a catalyst. Since the free metal concentration in biological systems is frequently very low, the crucial component for the activity of Fenton chemistry in biological systems is a functional metal redox cycle mechanism [2]. It can be concluded that any increase in the content of $O_2{}^{-}$, H_2O_2, or the redox active metal ions will most probably lead to the elevation levels of HO by the abovementioned chemical mechanisms.

Metal ions (e.g., Fe, Cr, Ni, Co, Ti, and Mo ions), released from corroded orthodontic appliances, all fall into the category of transition metals and could be subjected to redox cycling reactions as well as cause oxidative stress [8] in their vicinity. Metal ions are released from various types of alloys present in different components, including wires, brackets, or molar bands of fixed orthodontic appliances. Indeed, nickel, iron, and chromium can each intensify free radical production through a Fenton-like reaction [9–11], while cadmium, mercury, nickel, and lead can trigger ROS formation via indirect mechanism [12]. Although orthodontic alloys form an oxide layer that makes them corrosion resistant, these biocompatible metal materials tend to corrode locally in the harsh, ever-changing oral environment and degrade over time, releasing metal ions into the oral cavity [13,14], contributing to the general toxicity of fixed orthodontic appliances [15]. Moreover, due to frequent mechanical loading of the appliances in the oral cavity, the protective oxide layer is being constantly disrupted, accelerating wear as well as electrochemical corrosion. In further discussions, we will attempt to answer the specific research question: do fixed orthodontic appliances pose a health risk to patients due to corrosion and biodegradation?

2. Fixed Orthodontic Appliances

Fixed orthodontic appliances consist of archwires, brackets, and ligatures. During treatment, the archwires are ligated into the slot of the bracket [16]. The most commonly used materials from which parts of fixed appliances are manufactured are alloys of stainless steel (SS) and nickel-nitanium (NiTi) because of their advantageous mechanical properties. Less often, archwires made of other materials (e.g., cobalt-chromium alloy, ß-titanium (ß-Ti) alloy) are used. Orthodontic fixed appliances are usually in the oral cavity for extended time periods and are, therefore, exposed to degradation processes. The degradation process of dental metal alloys is the subject of several in vitro researches in which different parameters and their possible synergistic effects are analyzed, although a complete mapping of the oral cavity is hardly achieved due to the complex intraoral processes [17]. As a result of the ongoing temperature and pH changes, nutrition, breakdown of food and cells, varying oral flora, and the presence of plaque, various products are processed that can destroy the surface of fixed orthodontic appliances [18–21]. Metal alloys in the oral cavity are simultaneously subjected to corrosion and mechanical stress, e.g., due to masticatory forces, and in orthodontic treatments also due to the sliding of the archwire along the bracket slot. Therefore, the contact of the bracket, archwire, and ligature in the electrolytic fluid (saliva) leads to a faster and more intensive corrosion process on the metal surface [22]. Simultaneous exposure of fixed orthodontic appliances to corrosion, deformation, friction, and mechanical stress during treatment results in degradation of orthodontic brackets and archwires, which can result in elevated concentrations of metal ions in the oral cavity [23,24]. Mucosal erythema, allergic contact dermatitis, contact stomatitis, periodontitis, gingival hyperplasia, glossitis, gingivitis, and multiform erythema were all noticed during orthodontic treatment, which could be generated also by the toxic action of metal ions relieved by fixed orthodontic appliances [25,26]. As reported by the American Board of Orthodontics, treatment with fixed appliances continues for about 24 months [27]. Therefore, the safety of such appliances should be well studied as metal ions can induce either minor or major toxic effects locally (e.g., on oral cavity tissues) [28] or even systemic effects when they are absorbed and enter the systemic bloodstream [29,30].

3. Metal Ions Release during Treatment with Fixed Orthodontic Appliances

Nearly all fixed orthodontic appliances are constructed of metal alloys, which over certain period of time release these metal ions into the environment [31]. Among metal alloys, stainless steel, cobalt-chromium, NiTi, and β-titanium alloys are the most frequently used orthodontic materials [32]. The type of alloy, the exposure time and the environment are factors that impact the amount of metal ions released. Electrochemical corrosion, oxidation, and friction caused by continual sliding of the wires in the bracket slots impact the degradation of surface (tribocorrosion) and, subsequently, elevate the amount of ions that are released into saliva from the fixed appliances [33].

The study by Kovac et al. analyzed the amount of the metal ions released from various segments of orthodontic appliances (brackets, wires, and molar bands) composed of diverse alloys (SS, NiTi, β-Ti, and Co-Cr-Ni) over a three-month incubation period in an artificial oral environment. The results of the study show that the amounts of metal ions released from orthodontic appliances in an artificial oral environment were at a lower level than the upper intake limits. Not even one of the studied released metal ions exceeded the recommended daily dose concentrations. Molar bands and stainless-steel brackets discharged quadruple the number of ions in comparison to the stainless-steel wires. It was also observed that Ni-Ti wire released larger amount of Ni than the wires manufactured from stainless steel. Additionally, Ti-Mo wires released the lowest amount of metal ions. In terms of kinetics, metal release rates over 90 days were nonlinear, with significantly higher release in the first seven days of exposure as compared to the final seven-day exposure period.

Mikulewicz et al. [34] reported that the concentration of metal ions in artificial saliva reached 573 for Ni, 101 for Cr, 68 for Mn, and 2382 for Fe (in micrograms per liter). The increased amounts of specific metals could be attributed to the fact that Mikulewicz et al. study design constituted continuous stirring of the samples at 120 rpm during a one-month period while the samples in the study by Kovac et al. (2021) were only slightly shaken before sampling.

The release of Ni ions from different types of archwires used in orthodontics containing NiTi, SS, Cu-NiTi, and ion-implanted NiTi was studied by Charles et al. [35]. The amount of Ni ions released in all test solutions did not exceed the critical level that may trigger allergy and is found to be lower than the amount in the average daily intake. The study by Hussain et al. [36] compared nickel release from SS and NiTi archwires in artificial saliva over the period of three months using simulated fixed orthodontic appliances. The nickel release from NiTi archwires was 4.85 ng/mL and was higher in comparison to SS archwires, in which 0.41 ng/mL Ni ion release was detected. Hwang et al. [37] and Kutha et al. [38] corroborated the findings of Mikulewicz et al. for Ni release, while Barrett et al. [39] and Park et al. [40] observed significantly increased Ni levels. On the contrary, Gürsoy et al. [41] identified a lower concentration of Ni in the range of 20 µg/L.

Ortiz et al. [25] and Kuhta et al. [38] compared a NiTi archwire with an SS archwire and reported similar results on Ti release as reported by Kovac et al. Both studies found that Ti-containing alloys were more biocompatible than others. Kuhta et al. found that the difference in pH had a significant impact on ion release and that ion release was conditioned by archwire composition. It is, however, not in proportion to the metal composition of the archwire. Hwang et al. [37] determined metal release from fixed orthodontic appliances soaked in artificial saliva and noted that the values of Fe and Cr ions released from SS orthodontic appliances were even above the values reported by Kovac et al. Besides, a decrease in metal release with increasing immersion time was noted after a three-month period. Similarly, Kuhta et al. measured the highest amount of ions released in the first seven-day immersion of appliance. Hussain et al. [42] observed that ion release depends more on the type of alloy and the process of its manufacture than on the amount of ions in the alloy. According to He et al., saliva proteins influence the alloys metal ion release and have also an important role in corrosion resistance improvement [43].

There is a strong evidence that the results from in vitro studies investigating the release of metal ions over 30 days vary widely, e.g., some orthodontic alloys could not be ascertained or were as high as 7000 ng/mL [44]. The observed differences in the results regarding the quantity of metal ions released are due to different study designs, material analyzed, electroplated coatings, and different compositions of the alloys, the immersion media and different techniques and methods used at various times. On the other hand, metal ion concentrations from in vivo experiments were much lower and with less pronounced release variability, in the range between 4 and 30 ng/mL [45]. However, it should be emphasized that the metal ion concentration closer to the orthodontic alloy may be much higher than that measured in saliva, due to the constant saliva flow.

4. Cytotoxicity

Various methodological approaches have been used to treat cells with heavy metals. Namely, immersing orthodontic components in medium allowing immediate contact with the cells or incubating the orthodontic components in medium (e.g., artificial saliva) and dissolving out metal ions have been used to treat the cells. Another approach was reported using ingredients, such as metal chlorides, in which all metal ions acting on an orthodontic alloy are released identically, allowing the correct metal ion concentrations to be selected.

Additionally, in vitro toxicity of archwires have been performed on different cell cultures including fibroblasts [46–49], osteoblasts [46,47], smooth muscle cells [50,51], and neurons [52]. Results on SS, Ni, and Cr treated cells were generally toxic [48,52], while the results on NiTi archwires were inconsistent, with indications that it is both toxic [50,51] and non-toxic [46,47]. In the study of Spalj et al. [53], NiTi orthodontic archwire induced the lowest viability of mouse fibroblast cells, while SS archwires revealed the highest viability. The lowest inhibition of cell growth was observed in SS and TiMo, lower than in NiTi, CoCr, and the positive control.

The toxicity of metal ions employed in dental alloys was evaluated in the Saccharomyces cerevisaie yeast cultivated in various media. Metal toxicity was lower in the nutrient rich media (containing yeast extract and peptone) compared to synthetic complete media. In synthetic media, toxicity was the highest for HG, followed by Ag and Au, while Cu, Ni, Co, and Zn exhibited the lowest toxicity. The authors attributed this reduced toxicity to the sequestration of metal ions by various components in the yeast growth media [54].

Metal mixtures, simulating different orthodontic alloys and prepared in four different concentrations, were employed in the treatment of yeast cells for the period of 24 h in the study of Kovac et al. [32]. At 1000 μM concentrations, each simulated orthodontic alloy was cytotoxic, and in the instance of the CoCr alloy, even the 10-fold lower concentration was cytotoxic to S. cerevisiae cells. Mutant yeast cells showed a negative effect at the 100 μM concentration of all metal ion combinations, with the exception of SS. Cytotoxicity of orthodontic materials (brackets, archwires, resin, elastomers, and silver solder) using multiple yeast mutants either with antioxidant defense or DNA repair deficiency as a model organism was assessed also by Limberger et al. [55]. The results revealed significant cytotoxicity of silver solder, possibly induced by oxidative stress.

On the contrary, no major distinction in cell viability was identified among ceramic, metal, and polymeric conventional brackets on murine fibroblast cells L929 in the period following the in vitro exposure to four types of self-ligating and three types of conventional brackets [56]. Fixed orthodontic appliances (a combination of brackets and archwires) were examined in buccal mucosal cells. Decreased cellular viability, induced DNA damage, and elevated Ni and Cr levels were observed three months after the treatment [57]. Cicedo et al. [58] studied how metals (iron, chromium, aluminum, cobalt, copper, molybdenum, nickel, vanadium, niobium, and zirconium) released from implants effect human CD4+ T lymphocytes. All the examined metals caused T cell apoptosis at doses that were below the level causing the DNA damage. Ni induced apoptosis at 100 μM, while Co inhibited viability at a concentration of 500 and proliferation at 100 μM. An in vitro evaluation of the cytotoxicity and genotoxicity of a commercial Ti alloy for dental implantology was performed by Velasco-Ortega et al. [59]. The results showed that the Ti alloy (Ti-6Al-4V) was not cytotoxic or genotoxic in none of the conducted tests. Comparable data were obtained by Kovac et al. [32] on Ti-containing Ni-Ti and β-Ti ion mixtures, which showed no evidence of toxicity and oxidative stress formation.

NiTi shape memory alloys used in orthodontics were assessed by electrochemical assays in artificial saliva and in vitro biological tests with L132 cells and HEPM cells. The results proved the corrosiveness and cytotoxicity of Ni, as 425 μM of pure Ni ions resulted in 50% viability in human epithelial embryotic cell lines (L132). On the contrary, Ti ions and the commixture of both did not affect cell lines up to 3750 μM concentrations [60]. Issa et al. [61] measured the citotoxicity induced by changing concentrations of several metal salts on human oligodendrocyte (MO3.13) and human gingival fibroblasts (HGF). The greatest impact on survivability (50%) was observed in Co concentrations, succeeded by the Ni and Cr concentrations. Cd revealed the most noticeable cytotoxic effects on MO3.13 cells while Hg demonstrated the most significant cytotoxic effect on HGF in comparison to other tested metals.

5. Oxidative Stress and Oxidative Damage during the Treatment with Fixed Orthodontic Appliances

There are not many studies that investigated oxidative stress in the treatment with fixed orthodontic devices. Until recently, only one study determined the oxidative stress induced by orthodontic treatment at the systemic level [30], while the oxidative stress parameters were evaluated during orthodontic treatment either in saliva [62–65] or the gingival crevicular fluid [64]. Systemic oxidative stress parameters in the study of Kovac et al. [30] were evaluated in patients in the first seven-day period of treatment with fixed orthodontic appliances. The reactive oxygen species formation (ROS) as well as the antioxidant defense potential (AD) were assessed in the capillary blood samples from fifty-four male patients with malocclusion undergoing orthodontic treatment and untreated control subjects. Twenty-four hours after orthodontic treatment with fixed appliances, the ROS level was markedly elevated in the treatment group in comparison to the control group. Moreover, 24 h after archwire insertion, a significantly higher ROS/AD ratio was identified in the treatment group than in the control group. The authors concluded that the treatment with fixed orthodontic devices could cause systemic oxidative stress in the short run as the values of ROS and ROS/AD normalize during the period of one week following archwire insertion. Indeed, as revealed by Buczko et al. [62], the treatment with fixed orthodontic appliances alters the oxidant-antioxidant balance in saliva of individuals who are otherwise clinically healthy. Highest oxidative stress parameters (tiobarbituric acid reactive substance and total oxidant status) were detected in saliva within a week following the insertion of orthodontic appliance. Decrease in superoxide dismutase (SOD1) and catalase (CAT) was observed in stimulated saliva one week after treatment, while peroxidase (Px) was increased. It was noted that the total antioxidant status (TAS) was decreased twenty-four weeks after orthodontic treatment. The oxidative status index (OSI) increased a week after the treatment. The levels of selected markers of oxidative stress (malondialdehyde, ceruloplasmin, hydrogen donors) in the saliva of the treated patients prior to and after the beginning of treatment were additionally studied by Olteanu et al. [63]. Additionally, the current research shows increased markers of oxidative stress in the 24 h period, while seven days after appliance use the concentrations were close to baseline values. In contrast, Atuğ Oezcan et al. [64] observed no increase in oxidative stress markers and damage in saliva and gingival-crevicular fluid samples of patients with fixed orthodontic appliances when comparing the pretreatment results and the results at the sixth month of orthodontic treatment. Similarly, orthodontic treatment with self-ligating metal brackets did not statistically significantly affect salivary antioxidant parameters in the period of the first ten weeks of treatment [65].

During the treatment with fixed orthodontic appliances, metal ions are released, which could potentially induce oxidative stress and have a local oxidative effect, as observed in in vitro studies. Systemically, increased oxidative stress was observed only in the first seven-day period after the implementation of fixed orthodontic appliances, most likely due to the activation of endogenous adaptive stress responses and the induction of antioxidant endogenous defenses. Numerous in vitro studies have shown that orthodontic brackets as well as archwires cause oxidative stress linked to the release of heavy metals. All types of orthodontic brackets tested, disregarding the material components, are the origin of oxidative stress in vitro, as indicated by an increased concentration of the oxidative stress marker 8-hydroxy-2′-deoxyguanosine (8-OHdG) in the DNA of murine fibroblast cells L929. The highest stress levels were observed in the all-metal and polyurethane brackets [56]. In addition, all orthodontic archwires tested induced oxidative stress in vitro by measuring (8-OHdG) in DNA of murine fibroblast cells L929. Standard NiTi archwires produced the highest oxidative stress, while TiMo and SS triggered the lowest stress [53].

Different mixtures of Cr, Fe, Co, Ni, Ti, and Mo metal ions simulating either NiTi, SS, β-Ti, or CoCr orthodontic alloys were used in testing the capacity of metal-ions as oxidative stress-inducing agents using wild-type yeast Saccharomyces cerevisiae and two mutant ΔSod1 and ΔCtt1 as model organisms, to determine whether the lack of defense system from superoxide anions and H_2O_2 contributes to metal ion-induced toxicity [32]. Indeed, mutants lacking the Sod1-Ctt1 gene should be more sensitive to transition metal ion-induced ROS cytosolic SOD (Sod1) scavenges O_2^- and converts it to H_2O_2, which is further degraded to H_2O and O_2 by cytosolic catalase (Ctt1) [66]. A 1000 µM metal ion treatment with SS, Co-Cr-Ni resulted in significantly higher ROS values in all yeast strains than in the untreated control group, but the treatment using equal and lower concentrations of Ti-Mo did not significantly affect the ROS value. While SS, Co-Cr-Ni showed a significant increase in oxidative lipid damage in all yeast strains at the concentration of 1000 µM in comparison to the control group, and in the mutants (ΔSod1 and ΔCtt1) lipid oxidation presented already at 100 µM. Yeast mutants without the Sod1 gene have greater intracellular ROS levels than the Ctt1-lacking mutants although the same amount of lipid oxidation was observed in both mutants. Lipid oxidation and intracellular formation of ROS was almost twice as high in the mutants compared to the wild type.

Stainless steel bands, with (SSB) or without (NSB) silver soldered joints, were investigated for the cytostatic, cytotoxic, genotoxic, and DNA damage-inducing effects on the

HepG2 and HOK cell lines [67]. After performing MTT reduction assay, alkaline and modified comet assay, cytokinesis block micronucleus assay, cytostasis assay and cytotoxicity assay, the results showed higher cytotoxicity and genotoxicity of SSB eluates in comparison to NSB samples. Ni and Fe were found in the SSB as well as NSB medium samples, while Cd, Cr, Ag, Cu, and Zn were found only in the SSB medium samples.

It can be concluded that oxidative stress during orthodontic treatment can be caused by exposure to heavy metals (e.g., Cr, Fe, Co, Ni, and Ti), both locally and systemically. It can be induced also by many other factors, including aseptic inflammation in the periodontal ligament by cause of mechanical force and inflammation of periodontal tissues as a result of improper oral hygiene [30]. We would also like to draw attention to the occurrence of a synergistic effect when cells are simultaneously exposed to two or more metal ions. This is because their toxic effect not only adds up with simultaneous exposure, but it can even be multiplied or potentiated. According to Rinčić Mlinarić et al. (2019), the collective concentration of Ni and Ti in concentrations of at least 162 μg/L is necessary for synergism, provoking moderate to strong cytotoxicity and a notable induction of free radicals [68]. Terpilovska and Siwicki reported that Cr^{3+} and Fe^{3+} show synergistic effects in cytotoxicity, genotoxicity, and mutagenicity assays, but in some other combinations metal ions can act antagonistically—Cr^{3+} and Mo^{3+} or Cr^{3+} and Ni^{2+} show antagonistic effect—Cr^{3+} protects from Ni^{2+} or Mo^{3+} toxicity [69].

6. Risk Assessment

The Acceptable Daily Intake (ADI), also known as the Tolerable Daily Intake (TDI), is a "measure of the amount of a particular substance in food or drinking water that can be ingested daily (orally) over a lifetime without posing a significant health risk" [70]. The ADI is based on the NOAEL (no-observed-adverse-effect level) calculated from animal studies and observations in humans. NOAEL is a dose defined on experiments and observations where no statistically or biologically significant adverse effects have been found. When a NOAEL cannot be determined experimentally, the term "lowest-observed-adverse-effect level (LOAEL)" is used [71]. The NOAEL is numerically divided by a safety factor (SF) to explicate and justify differences found in experimental animals and humans (factor of 10) and potential dissimilarities in sensitivity among humans (factor of 10). An additional factor of 10 may be considered if the NOAEL has not been established because adverse effects are detected at all dose levels tested. Instead of the NOAEL, the lowest observed adverse effect level (LOAEL) is employed to determine the ADI divided by 10-fold safety factor. The safety factor of 10 shall also apply in cases where no satisfactory results of chronic toxicological tests are available, and the NOAEL is consequently determined on the basis of subchronic studies [71]. The level of the safety factor used shall depend on the reliability of the data collected and the severity of the adverse reactions observed.

$$ADI \text{ (human dose)} = NOAEL \text{ (experimental dose)}/SF \qquad (3)$$

NOAEL and LOAEL values derived from toxicological studies on metal ions are shown in Table 1. It should be noted that Table 1 does not include studies investigating the leaching of metal ions from orthodontic appliances, but all kind of studies investigating the toxicity of metal ions.

Table 1. NOAEL *, LOAEL * values, tested concentration range, type of study (chronic vs. acute) and model organism tested to determine the toxicity of specific metal ions.

Source of Metal Ions	NOAEL	LOAEL	Concentration (Dose)	Study Type	Model Organism	Ref.
			Fe			
$FeSO_4$	10 μM	250 μM	10–1000 μM	acute toxicity	human leukocytes	[72]
	50 μM	100 μM	0, 10, 25, 50, 100 μM	acute toxicity	microglia cells	[73]
$FeCl_3 \cdot 6H_2O$	3.2 μM	11.1 μM	3.2, 11.1, 111, 1111 μM	chronic toxicity	rats	[74] *
	100 μM	200 μM	0–1400 μM	acute toxicity	BALB/3T3, HepG2 cells	[75]
iron sucrose injection	1790 μM	3580 μM	360, 890, 1790, 3580, 8950 μM	acute toxicity	hemodialysis patients	[76] *
			Ni			
	27.2 mg/kg	54.4 mg/kg	3.4, 6.8, 13.6, 27.2, 54.4, 108.8 mg/kg	acute toxicity	mice	[77]
$NiCl_2$	100 μM	250 μM	0, 100, 250, 500, 600, 750, 1000 μM	acute toxicity	human HeLa cells	[78]
	10 μM	100 μM	0–10,000 μM	acute toxicity	human keratinocytes	[79]
$NiCl_2 \cdot 6H_2O$	100 μM	200 μM	0–1400 μM	acute toxicity	BALB/3T3 cells	[75]
$NiSO_4$	1.25 mg/kg/day	2.5 mg/kg/day	1.25, 2.5, 5 mg/kg/day	sub-acute toxicity	rats	[80]
$Ni(NO_3)_2$	150 μM	300 μM	0, 30, 75, 150, 300, 600 μM	acute toxicity	chromatin from rat liver cells	[81]
			Cr			
$K_2Cr_2O_7$	1 μM	10 μM	0–1000 μM	acute toxicity	human keratinocytes	[79]
	50 μM	400 μM	50, 100, 200, 400, 600, 1000 μM	acute toxicity	lymphocytes	[82]
$CrCl_3$	400 μM	600 μM	50, 100, 200, 400, 600, 1000 μM	acute toxicity	lymphocytes	[82]
$CrCl_3 \cdot 6H_2O$	100 μM	200 μM	0–1400 μM	acute toxicity	BALB/3T3 cells	[75]
			Mo			
MoO_3	100 μM	200 μM	0–1400 μM	acute toxicity	BALB/3T3, HepG2 cells	[75]
Na_2MoO_4	60.7 μM	121.4 μM	0, 60.7, 121.4, 242.7, 485.4, 970.9 μM	Sub-acute toxicity	mice	[83] *
$MoCl_5$	100 μM	500 μM	50, 100, 500, 1000, 5000 μM	acute toxicity	human CD4þ T lymphocytes	[58] *
			Ti			
Ti particles	13 μM	26 μM	13, 26, 52, 104, 209 μM	acute toxicity	Osteoblasts MC3T3-E1	[84] *
			Co			
$CoCl_2$	50 μM	100 μM	50, 100, 500, 1000, 5000 μM	acute toxicity	human CD4þ T lymphocytes	[58] *
$CoCl_2$	/	1 μM	1–100 μM	acute toxicity	Balb/3T3 mouse fibroblasts	[85]

* Abbreviations: NOAEL no-observed-adverse-effect level; LOAEL lowest-observed-adverse-effect level.

Similarly, the U.S. EPA has introduced the Reference Dose (RfD), which by definition is the maximum acceptable oral dose of a toxic substance derived from the formula: RfD = NOAEL/UF (uncertainty factor) [86]. A useful additional measure in risk assessment is also the exposure margin (MOE), which presents the amount by which the NOAEL of the critical toxic effect exceeds the estimated exposure dose (EED), both expressed in the same units: MOE = NOAEL (experimental dose)/EED (human dose) [86]. According to Food and Nutrition Board from the Institute of Medicine, National Academy of Sciences defines Dietary Reference Intakes (DRI) as the general term for a set of reference values used to plan and evaluate the nutrient intake of healthy people. These values, which vary by age and gender, additionally include [87,88].

Recommended Dietary Allowance (RDA): "average daily intake sufficient to meet the nutrient requirements of almost all (97–98%) healthy people".

Adequate Intake (AI): "a value based on observed or experimentally determined approximations of the nutrient intake of a group (or groups) of healthy people-used when an RDA cannot be determined".

Tolerable Upper Intake Level (UL): "the highest daily nutrient intake that is unlikely to pose a risk of adverse health effects for almost all individuals in the general population. As intake increases above the UL, the risk of adverse effects increases".

The derived no objection level (DNEL) is the level of exposure to a substance to which people should not be exposed [89].

Table 2 shows the limits of intake for each element, expressed in units of mass per day. Table 3 includes orthodontic studies showing the leaching of metal ions mainly in artificial saliva. The results are presented in different units (ppm, ng/mL, g/L, ng/cm^2) and the observation period included different time intervals (1 day–90 days). The studies also differed in the fact that some examined leaching from whole orthodontic appliances, while others examined only individual parts (archwire/bands/brackets). These differences restrict the possibility of comparison of individual studies and extrapolate the data for human safety as well as the comparison of data with the limits presented in Table 2. In addition, it should be noted that the leaching of metals over time is not simply linear.

Table 2. Limit intake levels for selected metals.

Metal	Threshold		References
	Men	**Women**	
Iron	RDA * = 8 mg/day	RDA = 18 mg/dayRDA = 8 mg/day (postmenopause)	[90]
	UL * = 45 mg/day	UL = 45 mg/day	[90]
	RDA = 17 mg/day	RDA = 17 mg/day	[91]
	AI * = 11 mg/day	AI = 16 mg/dayAI = 11 mg/day (postmenopause)	[92]
	RDA = 10 mg/day	RDA = 10–15 mg/day	[93]
	DNEL * = 710 µg/kg bw/day (chronic exposure)	DNEL * = 710 µg/kg bw/day (chronic exposure)	[94]
Nickel	UL = 1.0 mg/day	UL = 1.0 mg/day	[95]
	GL = 0.26 mg/day	GL = 0.26 mg/day	[91]
	DNEL = 11 µg/kg bw/day (chronic exposure)	DNEL = 11 µg/kg bw/day (chronic exposure)	[96]
Chromium	AI = 35 µg/day	AI = 25 µg/day	[97]
	RDA = 30–100 µg/day	RDA = 30–100 µg/day	[93]
	GL = 10 mg/day	GL = 10 mg/day	[91]
Titanium	DNEL = 350 mg/kg bw/day (chronic exposure)	DNEL = 350 mg/kg bw/day (chronic exposure)	[98]
Molybdenum	AI = 45 µg/day	AI = 45 µg/day	[99]
	RDA = 65 µg/day	RDA = 65 µg/day	[100]
	RDA = 50–100 µg/day	RDA = 50–100 µg/day	[93]
	DNEL = 3.4 mg/kg bw/day	DNEL = 3.4 mg/kg bw/day	[101]
Cobalt	GL = 1.4 mg/day	GL = 1.4 mg/day	[91]
	DNEL = 29.8 µg/kg bw/day (chronic exposure)	DNEL = 29.8 µg/kg bw/day (chronic exposure)	[102]

* Abbreviations: RDA Recommended Dietary Allowance; UL Tolerable Upper Intake Level; AI Adequate Intake; DNEL The derived no objection level; ADI Acceptable Daily Intake.

The highest quantities of released metal ions were observed in the study of Mikulewicz et al. Metal concentrations observed in saliva compared to the maximum acceptable concentrations of metal ions in drinking water (80/778EEC/; 98/83/ES; 2020/2184) [103,104] should be of concern. For example, the set levels for Ni (573 µg/L) and Cr (101 µg/L) were exceeded 11 times and twice, respectively. Additionally, the concentration of Mn was (68 µg/L) 1.4 times higher and Fe (2382 µg/L) concentration was 12 times higher than the maximum acceptable values for drinking water as set in the EU Drinking Water Directive 98/83/EC. Following American standards, the concentration of Ni ions was 5.5 times higher, while other elements were at an acceptable level [105]. The volume of artificial saliva (in milliliters) that the recommended daily doses of these elements would provide, would amount to 61 for Ni, suggesting that the only possible exposure risk to orthodontic patients is from it [34], suggesting that the latter presents the sole potential exposure risk to the treated patients. Nevertheless, the amount of ions released by orthodontic appliances may induce delayed allergic reactions, which should be considered when selecting the type of archwire, with special attention to patients with metal hypersensitivity or limited oral hygiene [38]. Finally, it should be noted that the individual reaction to metal ions may occur in some individuals at much lower concentrations than indicated in the legal limits for the general population.

Table 3. Average concentrations of leached metal ions from orthodontic appliances.

| Metal Ion Release from Orthodontic Appliances | | | | | | | | |
Fe	Ni	Cr	Ti	Mo	Co	Alloy Type	Time	Ref.
310 ± 123.1 ppb	1292 ± 437.5 ppb	0 ppb	/	/	4 ± 5.477 ppb	SS wire + SS brackets + SS band		
450 ± 222.8 ppb	1864 ± 600.2 ppb	18 ± 34.9 ppb	/	/	10 ± 7.071 ppb	NiTi wire + SS brackets + SS band	90 days	[106]
614 ± 531.7 ppb	2466 ± 867.8 ppb	30 ± 51.9 ppb	/	/	12 ± 4.47 ppb	Co-NiTi wire + SS brackets + SS band		
534 ± 558 ppb	2132 ± 1143 ppb	20 ± 44.72 ppb	/	/	12 ± 4.472 ppb	Elgiloy wire + SS brackets + SS band		
2382 ppb	573 ppb	101 ppb	/	/	/	SS wires, SS brackets, SS bands, SS metal ligatures	30 days	[34]
1107 ppb	15 ppb	8 ppb	/	/	/	control		
/	18.5 ± 13.1 ppb	2.6 ± 1.6 ppb	/	/	/	SS brackets and bands; SS + NiTi archwires	16 ± 2 months	[29]
/	11.9 ± 11.4 ppb	2.2 ± 1.8 ppb	/	/	/	control		
10.78 ± 17.66 µg	28.33 ± 29.19 µg	12.66 ± 4.61 µg	/	133.33 ± 57.7 µg	/	NiTi archwire + Dentaurum brackets		
21.33 ± 8.73 µg	17.66 ± 11/59 µg	<10 ± 0 µg	/	110 ± 17.32 µg	/	NiTi archwire + American Ortho brackets		
14.33 ± 5.33 µg	76.66 ± 62.52 µg	16.66 ± 11.54 µg	/	<130 ± 0 µg	/	NiTi archwire + Shinye brackets		
<10 ± 0 µg	22 ± 7.54 µg	30 ± 0 µg	/	120 ± 17.32 µg	/	NiTi archwire + ORJ brackets	28 days	[107]
16 ± 5.29 µg	37 ± 20.042 µg	26 ± 6.92 µg	/	110 ± 17.32 µg	/	SS archwire + Dentaurum brackets		
11 ± 1.73 µg	45 ± 31 µg	26 ± 6.92 µg	/	0 ± <100 µg	/	SS archwire + American Ortho brackets		
<10 ± 0 µg	293.2 ± 365.6 µg	15.33 ± 4.61 µg	/	0 ± <100 µg	/	SS archwire + Shinye brackets		
<10 ± 0 µg	44.66 ± 35.50 µg	15.33 ± 4.61 µg	/	110 ± 17.32 µg	/	SS archwire + ORJ brackets		
67.7 ± 9.6 ppb	10 ± 6.3 ppb	4.5 ± 0.5 ppb	2 ± 0.6 ppb	/	/	NiTi New brackets + new archwire		
64.3 ± 8 ppb	15 ± 3.3 ppb	4 ± 0.7 ppb	2 ± 0.7 ppb	/	/	NiTi Recycled brackets + new archwire	45 days	[41]
115 ± 11.9 ppb	20 ± 6.5 ppb	7.3 ± 0.6 ppb	3.5 ± 1.3 ppb	/	/	NiTi New brackets + recycled archwire		
140 ± 9.4 ppb	20 ± 3.3 ppb	7.5 ± 0,6 ppb	4 ± 0.7 ppb	/	/	NiTi Recycled brackets + recycled archwire		
/	2 ppb	/	/	/	/	NiTi archwire		
/	/	/	/	/	/	TMO archwire	28 days	[108]
289 ppb	6 ppb	11 ppb	/	/	/	SS archwire		
/	4 ppb	3 ppb	/	/	/	Cu-NiTi archwire		
/	67 ± 10.8 ppb	30.8 ± 4.3 ppb	/	/	/	SS brackets, bands NiTi & SS archwires	1.5 years	[31]
/	5.02 ± 0.001 ppb	1.27 ± 0.09 ppb	/	/	/	control		
/	0.93 ± 0.04 µg	/	/	/	/	NiTi archwires		
/	0.66 ± 0.02 µg	/	/	/	/	SS archwires	21 days	[35]
/	0.67 ± 0.02 µg	/	/	/	/	CuNiTi archwires		
/	0.77 ± 0.05 µg	/	/	/	/	Ion implanted NiTi archwires		
/	125 ± 22 µg	112 ± 18 µg	/	/	/	SS $\frac{1}{4}$ archwire + bands + brackets	12 days	[40]

Table 3. *Cont.*

| Metal Ion Release from Orthodontic Appliances | | | | | | Alloy Type | Time | Ref. |
Fe	Ni	Cr	Ti	Mo	Co			
96.06 ± 57.4 ppb/day	41.66 ± 33.99 ppb/day	33.43 ± 24.05 ppb/day	0 ± 0	/	/		1st day	
25.55 ± 10.00 ppb/day	10.21 ± 2.68 ppb/day	3.83 ± 1.93 ppb/day	0 ± 0	/	/	SS archwire	6th day	[38]
11.08 ± 5.89 ppb/day	5.28 ± 1.87 ppb/day	1.43 ± 0.69 ppb/day	0 ± 0	/	/		7th day	
5.62 ± 1.47 ppb/day	3.84 ± 0.86 ppb/day	0.70 ± 0.10 ppb/day	0 ± 0	/	/		14th day	
38.47 ± 15.67 ppb/day	11.77 ± 2.84 ppb/day	10.49 ± 3.90 ppb/day	0.14 ± 0.04 ppb/day	/	/		1st day	
26.93 ± 5.44 ppb/day	10.83 ± 3.49 ppb/day	3.30 ± 0.95 ppb/day	0.01 ± 0 ppb/day	/	/	NiTi archwire	6th day	[38]
13.07 ± 4.01 ppb/day	6.13 ± 1.39 ppb/day	1.76 ± 0.34 ppb/day	0.01 ± 0.01 ppb/day	/	/		7th day	
6.81 ± 1.70 ppb/day	3.38 ± 1.67 ppb/day	1.06 ± 0.21 ppb/day	0.01 ± 0 ppb/day	/	/		14th day	
21.18 ± 6,43 ppb/day	7.12 ± 1.33 ppb/day	4.39 ± 1.99 ppb/day	0.12 ± 0.02 ppb/day	/	/		1st day	
19.11 ± 6.28 ppb/day	7.26 ± 1.10 ppb/day	2.10 ± 0.84 ppb/day	0.01 ± 0 ppb/day	/	/	Termo NiTi archwire	6th day	[38]
8.79 ± 3.79 ppb/day	5.06 ± 1.57 ppb/day	1.23 ± 0.80 ppb/day	0.01 ± 0 ppb/day	/	/		7th day	
2.50 ± 0.59 ppb/day	2.33 ± 0.77 ppb/day	0.40 ± 0.15 ppb/day	0 ± 0	/	/		14th day	
42.33 ± 27.06 ppb	0.00 ± 0.00 ppb	0.36 ± 0.33 ppb	6.64 ± 2.00 ppb	0.80 ± 0.27 ppb	0.00 ± 0.00 ppb	control		
520.0 ± 210.1 ppb	416.9 ± 133.5 ppb	8.97 ± 6.73 ppb	9.03 ± 2.08 ppb	3.11 ± 1.47 ppb	1.38 ± 1.25 ppb	SS brackets + SS bands	30 days	[25]
49.67 ± 29.29 ppb	0.00 ± 0.00 ppb	0.06 ± 0.05 ppb	5.35 ± 3.96 ppb	0.11 ± 0.07 ppb	0.03 ± 0.01 ppb	Ti brackets + Ti bands		
179.49 ± 99.1 ppb	5.79 ± 2.07 ppb	0.91 ± 0.28 ppb	3.79 ± 3.11 ppb	0.60 ± 0.21 ppb	0.00 ± 0.00 ppb	Ni-free brackets + bands		

7. Indications for Antioxidant Therapy: Pros and Cons

The awareness that leaching of metal ions occurs in the treatment with fixed orthodontic appliances, and the evidence of the resulting elevated oxidative stress and oxidative impairment require a careful consideration of the antioxidant application in patients during treatment. Based on the present literature review and risk assessment, an increased intake of antioxidants one week before and one week (as well as after each reactivation of the device/wire change) after treatment with fixed orthodontic appliances would be advocated, while prolonged intake of synthetic antioxidants is not recommended for several reasons (reviewed in [6,109]). Although several in vitro studies indicate that supplementation with synthetic antioxidants or plant extracts reduces oxidative stress and oxidative damage (reviewed in [110]), the results of epidemiological and human studies are conflicting [111–114]. Moreover, the effect of antioxidants as free radical scavengers in vitro is well studied, but their effect on ROS -induced damage prevention in vivo is still controversial. ROS effects various vital physiological processes in a human, including pathogen defense, stress response induction, regulation of glucose, cellular growth and proliferation, and systemic signaling [115]. Both excessive "antioxidant/reductive stress" and oxidative stresses can be harmful to organisms because antioxidants have no capacity to discriminate among physiologically beneficial and harmful radicals, which induce oxidative damage to biomolecules [6]. Antioxidants can cause similar damage as oxidants when present in large concentrations or in the presence of redox cycling metal ions that can induce the creation of highly reactive free radicals through the Fenton-like reaction. For example, vitamins C and E reduce iron, which generate free radicals through the Fenton-like reaction [116,117].

$$2\,Fe^{3+} + Ascorbate \rightarrow 2\,Fe^{2+} + Dehydroascorbate \tag{4}$$

$$2\,Fe^{2+} + 2\,H_2O_2 \rightarrow 2\,Fe^{3+} + 2\,OH\cdot + 2\,OH^- \tag{5}$$

The estimated half-life of the same ROS is very short (e.g., for OH is 10^{-9} s), which means that low weight synthetic antioxidants (e.g., vitamins C, E) are unable to specifically scavenge such ROS in vivo due to their extreme reactivity [118,119]. In contrast, enzymatic antioxidants have the ability to accelerate chemical reactions. This ability to eliminate ROS is, therefore, much higher than that of low molecular weight antioxidants ingested through the diet. Therefore, activation of enzymatic antioxidant defenses in vivo is more relevant than radical scavenging by exogenous antioxidants consumed through the diet [120].

Although there are no official guidelines regarding the necessary quantity of antioxidants [121], it is recommended to primarily follow a diverse and balanced diet. Daily consumption of at least 400 g of fruits and vegetables is recommended by the World Health Organization to gain adequate protective compounds from the normal diet to scavenge ROS.

Since preventing the formation of ROS is considered to be more effective in reducing oxidative stress than the suppression of already created free radicals with antioxidants, metal chelating antioxidants (e.g., transferrin, albumin, ceruloplasmin) could be used to prevent radical formation by inhibiting the Fenton reaction catalyzed by transition metal ions. In addition, various flavonoids act as chelators of transition metal ions involved in Fenton chemistry [122–124].

In any case, the justification and appropriateness of supplemental intake of antioxidants or topical application of antioxidants should be confirmed by the results of human studies that would establish the most effective antioxidants, the most appropriate concentration, combination and timing of intake of antioxidants in patients treated with non-removable orthodontic appliances. Antioxidant therapy during dental treatment proved to be effective in the reduction of oxidative stress and cytotoxicity during tooth whitening and in restorative dentistry (reviewed in [125]).

Another approach that may be effective in preventing oxidative stress in the oral cavity is the direct application of antioxidants to metal parts of orthodontic appliances, which could intercept free radicals directly in the vicinity of their formation before they can damage surrounding tissues. On the other hand, the application of antioxidants to the metal parts of orthodontic appliances may lead to the damage of the protective layer, corrosion, and increased leaching of metal ions. Further research into these domains is, therefore, to be undertaken.

8. Conclusions

The broad topic was considered from many different perspectives, as the narrative review was divided into various sections. To summarize: The results of in vitro and in vivo studies that examined the release of metal ions show differences in the amount of metal ions released due to different study designs. Metal ions released from corroded orthodontic appliances could be exposed to redox cycling reactions and cause oxidative stress. Therefore, the safety of such appliances should be investigated, the risk assessed, and the justification and appropriateness of additional antioxidant intake confirmed by the results of future human studies to demonstrate the efficacy of antioxidant therapy in reducing oxidative stress during dental treatment. The consumer growing awareness, interest and knowledge impels the manufacturing industry to confront the challenges of choosing adequate orthodontic appliance materials to minimize their impact on health.

Although Žukowski et al. [126] reported some beneficial effects of generating ROS in the oral cavity during dental treatment (e.g., stimulation of immune response and wound healing, cytotoxic effect on pathogenic bacteria), the metal ions released into the oral cavity from fixed orthodontic appliances due to corrosion and biodegradation pose a health risk to patients. This risk is, however, a minor one, as only very high metal concentrations induce cytotoxicity and oxidative stress, which was found in the studies on cell cultures. On the other hand, several studies reported that orthodontic treatment can induce transient imbalance in the ratio between oxidants and antioxidants in saliva as well as at the systemic level of clinically healthy subjects.

Nevertheless, the increased ROS levels can occur at a local level in the oral cavity, which could pose a problem to patients with lower efficiency of endogenous antioxidant defense systems, metal ions hypersensitive individuals, or patients with decreased salivary antioxidant power [127]. This topic should be further investigated, since there is currently only scarce scientific research on the effect of orthodontic materials on the formation of oxidative stress.

Funding: The authors acknowledge the financial support from the Slovenian Research Agency (Research Core Funding No. P3-0388, J3-2520) and Croatian Research Agency (Research Funding No. IPS-2020-01).

Institutional Review Board Statement: Not applicable.

Informed Consent Statement: Not applicable.

Acknowledgments: We would like to thank Tina Skorjanc for her help with the figures.

Conflicts of Interest: The authors declare no conflict of interest.

References

1. Sies, H. Oxidative stress: From basic research to clinical application. *Am. J. Med.* **1991**, *91*, 31–38. [CrossRef]
2. Deguillaume, L.; Leriche, M.; Chaumerliac, N. Impact of radical versus non-radical pathway in the Fenton chemistry on the iron redox cycle in clouds. *Chemosphere* **2005**, *60*, 718–724. [CrossRef]
3. Poljsak, B. Strategies for reducing or preventing the generation of oxidative stress. *Oxidative Med. Cell. Longev.* **2011**, *2011*, 194586. [CrossRef]
4. Nuran Ercal, B.S.P.; Hande Gurer-Orhan, B.S.P.; Nukhet Aykin-Burns, B.S.P.; Ercal, N.; Gurer-Orhan, H.; Aykin-Burns, N. Toxic metals and oxidative stress. Part I: Mechanisms involved in metal induced oxidative damage. *Curr. Top. Med. Chem.* **2001**, *1*, 529–539. [CrossRef] [PubMed]
5. Poljšak, B.; Jamnik, P.; Raspor, P.; Pesti, M. Oxidation-Antioxidation-Reduction Processes in the Cell: Impacts of Environmental Pollution. In *Encyclopedia of Environmental Health*; Elsevier Inc.: Amsterdam, The Netherlands, 2011; pp. 300–306, ISBN 9780444522726.
6. Poljsak, B.; Milisav, I. The neglected significance of "antioxidative stress". *Oxidative Med. Cell. Longev.* **2012**, *2012*, 480895. [CrossRef]
7. Kell, D.B. Iron behaving badly: Inappropriate iron chelation as a major contributor to the aetiology of vascular and other progressive inflammatory and degenerative diseases. *BMC Med. Genom.* **2009**, *2*, 2. [CrossRef]
8. Kanti Das, T.; Wati, M.R.; Fatima-Shad, K. Oxidative Stress Gated by Fenton and Haber Weiss Reactions and Its Association With Alzheimer's Disease. *Arch. Neurosci.* **2014**, *2*, e60038. [CrossRef]
9. Torreilles, J.; Guerin, M.C.; Slaoui-Hasnaoui, A. Nickel (II) complexes of histidyl-peptides as fenton-reaction catalysts. *Free Radic. Res.* **1990**, *11*, 159–166. [CrossRef] [PubMed]
10. Halliwell, B.; Gutteridge, J.M.C. Biologically relevant metal ion-dependent hydroxyl radical generation An update. *FEBS Lett.* **1992**, *307*, 108–112. [CrossRef]
11. Sugden, K.D.; Geer, R.D.; Rogers, S.J. Oxygen Radical-Mediated DNA Damage by Redox-Active Cr(III) Complexes. *Biochemistry* **1992**, *31*, 11626–11631. [CrossRef]
12. Glaser, V.; Leipnitz, G.; Straliotto, M.R.; Oliveira, J.; dos Santos, V.V.; Wannmacher, C.M.D.; de Bem, A.F.; Rocha, J.B.T.; Farina, M.; Latini, A. Oxidative stress-mediated inhibition of brain creatine kinase activity by methylmercury. *Neurotoxicology* **2010**, *31*, 454–460. [CrossRef]
13. Sifakakis, I.; Eliades, T. Adverse reactions to orthodontic materials. *Aust. Dent. J.* **2017**, *62*, 20–28. [CrossRef] [PubMed]
14. Lucchese, A.; Carinci, F.; Brunelli, G.; Monguzzi, R. An in vitro study of resistance to corrosion in brazed and laser welded orthodontic appliances. *Eur. J. Inflamm.* **2011**, *9*, 67–72.
15. Bandeira, A.M.; Ferreira Martinez, E.; Dias Demasi, A.P. Evaluation of toxicity and response to oxidative stress generated by orthodontic bands in human gingival fibroblasts. *Angle Orthod.* **2020**, *90*, 285–290. [CrossRef] [PubMed]
16. Andrews, L.F. The straight-wire appliance. *Br. J. Orthod.* **1979**, *6*, 125–143. [CrossRef]

17. Wichelhaus, A.; Geserick, M.; Hibst, R.; Sander, F.G. The effect of surface treatment and clinical use on friction in NiTi orthodontic wires. *Dent. Mater.* **2005**, *21*, 938–945. [CrossRef]
18. Eliades, T.; Bourauel, C. Intraoral aging of orthodontic materials: The picture we miss and its clinical relevance. *Am. J. Orthod. Dentofac. Orthop.* **2005**, *127*, 403–412. [CrossRef]
19. Eliades, T.; Athanasiou, A.E. In Vivo Aging of Orthodontic Alloys: Implications for Corrosion Potential, Nickel Release, and Biocompatibility. *Angle Orthod.* **2002**, *72*, 222–237.
20. Cortizo, M.C.; De Mele, M.F.L.; Cortizo, A.M. Metallic dental material biocompatibility in osteoblastlike cells: Correlation with metal ion release. *Biol. Trace Elem. Res.* **2004**, *100*, 151–168. [CrossRef]
21. Kao, C.T.; Ding, S.J.; Min, Y.; Hsu, T.C.; Chou, M.Y.; Huang, T.H. The cytotoxicity of orthodontic metal bracket immersion media. *Eur. J. Orthod.* **2007**, *29*, 198–203. [CrossRef]
22. Drescher, D.; Bourauel, C.; Schumacher, H.A. Frictional forces between bracket and arch wire. *Am. J. Orthod. Dentofac. Orthop.* **1989**. [CrossRef]
23. Landolt, D.; Mischler, S.; Stemp, M. Electrochemical methods in tribocorrosion: A critical appraisal. *Electrochim. Acta* **2001**, *46*, 3913–3929. [CrossRef]
24. Močnik, P.; Kosec, T.; Kovač, J.; Bizjak, M. The effect of pH, fluoride and tribocorrosion on the surface properties of dental archwires. *Mater. Sci. Eng. C* **2017**, *78*, 682–689. [CrossRef] [PubMed]
25. Ortiz, A.J.; Fernández, E.; Vicente, A.; Calvo, J.L.; Ortiz, C. Metallic ions released from stainless steel, nickel-free, and titanium orthodontic alloys: Toxicity and DNA damage. *Am. J. Orthod. Dentofac. Orthop.* **2011**, *140*, 115–122. [CrossRef]
26. Keinan, D.; Mass, E.; Zilberman, U. Absorption of Nickel, Chromium, and Iron by the Root Surface of Primary Molars Covered with Stainless Steel Crowns. *Int. J. Dent.* **2010**, *2010*, 326124. [CrossRef]
27. Moresca, R. Orthodontic treatment time: Can it be shortened? *Dent. Press J. Orthod.* **2018**, *23*, 90–105. [CrossRef]
28. Wataha, J.C. Biocompatibility of dental casting alloys: A review. *J. Prosthet. Dent.* **2000**, *83*, 223–234. [CrossRef]
29. Amini, F.; Jafari, A.; Amini, P.; Sepasi, S. Metal ion release from fixed orthodontic appliances—An in vivo study. *Eur. J. Orthod.* **2012**, *34*, 126–130. [CrossRef]
30. Kovac, V.; Poljsak, B.; Perinetti, G.; Primozic, J.; Reis, F.S. Systemic Level of Oxidative Stress during Orthodontic Treatment with Fixed Appliances. *Biomed. Res. Int.* **2019**, *2019*, 5063565. [CrossRef]
31. Quadras, D.; Nayak, U.; Kumari, N.; Priyadarshini, H.; Gowda, S.; Fernandes, B. In vivo study on the release of nickel, chromium, and zinc in saliva and serum from patients treated with fixed orthodontic appliances. *Dent. Res. J.* **2019**, *16*, 209–215. [CrossRef]
32. Kovač, V.; Poljšak, B.; Primožič, J.; Jamnik, P. Are metal ions that make up orthodontic alloys cytotoxic, and do they induce oxidative stress in a yeast cell model? *Int. J. Mol. Sci.* **2020**, *21*, 7993. [CrossRef]
33. Greenwald, A.S. *Biological Performance of Materials: Fundamentals of Biocompatibility*, 3rd ed.; Taylor Francis Group: Abingdon, UK, 2001; Volume 83.
34. Mikulewicz, M.; Chojnacka, K.; Woźniak, B.; Downarowicz, P. Release of metal ions from orthodontic appliances: An in vitro study. *Biol. Trace Elem. Res.* **2012**, *146*, 272–280. [CrossRef]
35. Charles, A.; Gangurde, P.; Jacob, S.; Jatol-Tekade, S.; Senkutvan, R.; Vadgaonkar, V. Evaluation of nickel ion release from various orthodontic arch wires: An in vitro study. *J. Int. Soc. Prev. Community Dent.* **2014**, *4*, 12. [CrossRef] [PubMed]
36. Hussain, H.D.; Ajith, S.D.; Goel, P. Nickel release from stainless steel and nickel titanium archwires—An in vitro study. *J. Oral Biol. Craniofacial Res.* **2016**, *6*, 213–218. [CrossRef] [PubMed]
37. Hwang, C.J.; Shin, J.S.; Cha, J.Y. Metal release from simulated fixed orthodontic appliances. *Am. J. Orthod. Dentofac. Orthop.* **2001**, *120*, 383–391. [CrossRef]
38. Kuhta, M.; Pavlin, D.; Slaj, M.M.; Varga, S.; Lapter-Varga, M.; Slaj, M.M. Type of archwire and level of acidity: Effects on the release of metal ions from orthodontic appliances. *Angle Orthod.* **2009**, *79*, 102–110. [CrossRef] [PubMed]
39. Barrett, R.D.; Bishara, S.E.; Quinn, J.K. Biodegradation of orthodontic appliances. Part I. Biodegradation of nickel and chromium in vitro. *Am. J. Orthod. Dentofac. Orthop.* **1993**, *103*, 8–14. [CrossRef]
40. Park, H.Y.; Shearer, T.R. In vitro release of nickel and chromium from simulated orthodontic appliances. *Am. J. Orthod.* **1983**, *84*, 156–159. [CrossRef]
41. Gürsoy, S.; Acar, A.G.; Şeşen, Ç. Comparison of Metal Release from New and Recycled Bracket-Archwire Combinations. *Angle Orthod.* **2005**, *75*, 92–94.
42. Hussain, S.; Asshaari, A.; Osman, B.; AL-Bayaty, F. In Vitro-Evaluation of Biodegradation of Different Metallic Orthodontic Brackets. *J. Int. Dent. Med. Res.* **2017**, *7*, 76–83.
43. He, L.; Cui, Y.; Zhang, C. Effect of Protein and Mechanical Strain on the Corrosion Resistance and Cytotoxicity of the Orthodontic Composite Arch Wire. *ACS Omega* **2020**, *5*, 8992–8998. [CrossRef] [PubMed]
44. Bhaskar, V.; Subba Reddy, V. Biodegradation of nickel and chromium from space maintainers: An in vitro study. *J. Indian Soc. Pedod. Prev. Dent.* **2010**, *28*, 6. [CrossRef]
45. Mikulewicz, M.; Chojnacka, K. Trace Metal release from orthodontic appliances by in vivo studies: A systematic literature review. *Biol. Trace Elem. Res.* **2010**, *137*, 127–138. [CrossRef] [PubMed]
46. Ryhä, J.; Nen, R.; Niemi, E.; Serlo, W.; Niemelä, E.N.; Sandvik, P.; Pernu, H.; Salo, T. Biocompatibility of nickel-titanium shape memory metal and its corrosion behavior in human cell cultures. *J. Biomed. Mater. Res.* **1997**, *35*, 451–457.

47. Wever, D.J.; Veldhuizen, A.G.; Sanders, M.M.; Schakenraad, J.M.; Van Horn, J.R. Cytotoxic, allergic and genotoxic activity of a nickel-titanium alloy. *Biomaterials* **1997**, *18*, 1115–1120. [CrossRef]
48. Rose, E.C.; Jonas, I.E.; Kappert, H.F. In vitro investigation into the biological assessment of orthodontic wires. *J. Orofac. Orthop.* **1998**, *59*, 253–264. [CrossRef]
49. Es-Souni, M.; Fischer-Brandies, H.; Es-Souni, M. In-vitro-bioverträglichkeit von Elgiloy®, einer kobalt-basislegierung, im vergleich zu zwei titanlegierungen. *J. Orofac. Orthop.* **2003**, *64*, 16–26. [CrossRef]
50. Shih, C.-C.; Lin, S.-J.; Chen, Y.-L.; Su, Y.-Y.; Lai, S.-T.; Wu, G.J.; Kwok, C.-F.; Chung, K.-H. The cytotoxicity of corrosion products of nitinol stent wire on cultured smooth muscle cells. *J. Biomed. Mater. Res.* **2000**, *52*, 395–403. [CrossRef]
51. Shih, C.-C.; Shih, C.-M.; Chen, Y.-L.; Su, Y.-Y.; Shih, J.-S.; Kwok, C.-F.; Lin, S.-J. Growth inhibition of cultured smooth muscle cells by corrosion products of 316 L stainless steel wire. *J. Biomed. Mater. Res.* **2001**, *112*, 200–207. [CrossRef]
52. David, A.; Lobner, D. In vitro cytotoxicity of orthodontic archwires in cortical cell cultures. *Eur. J. Orthod.* **2004**, *26*, 421–426. [CrossRef]
53. Spalj, S.; Mlacovic Zrinski, M.; Tudor Spalj, V.; Ivankovic Buljan, Z. In-vitro assessment of oxidative stress generated by orthodontic archwires. *Am. J. Orthod. Dentofac. Orthop.* **2012**, *141*, 583–589. [CrossRef]
54. Yang, H.C.; Pon, L.A. Toxicity of metal ions used in dental alloys: A study in the yeast Saccharomyces cerevisiae. *Drug Chem. Toxicol.* **2003**, *26*, 75–85. [CrossRef]
55. Limberger, K.M.; Westphalen, G.H.; Menezes, L.M.; Medina-Silva, R. Cytotoxicity of orthodontic materials assessed by survival tests in Saccharomyces cerevisiae. *Dent. Mater.* **2011**, *27*, e81–e86. [CrossRef]
56. Buljan, Z.I.; Ribaric, S.P.; Abram, M.; Ivankovic, A.; Spalj, S. In vitro oxidative stress induced by conventional and self-ligating brackets. *Angle Orthod.* **2012**, *82*, 340–345. [CrossRef]
57. Hafez, H.S.; Selim, E.M.N.; Kamel Eid, F.H.; Tawfik, W.A.; Al-Ashkar, E.A.; Mostafa, Y.A. Cytotoxicity, genotoxicity, and metal release in patients with fixed orthodontic appliances: A longitudinal in-vivo study. *Am. J. Orthod. Dentofac. Orthop.* **2011**, *140*, 298–308. [CrossRef]
58. Caicedo, M.; Jacobs, J.J.; Reddy, A.; Hallab, N.J. Analysis of metal ion-induced DNA damage, apoptosis, and necrosis in human (Jurkat) T-cells demonstrates Ni^{2+} and V^{3+} are more toxic than other metals: Al^{3+}, Be^{2+}, Co^{2+}, Cr^{3+}, Cu^{2+}, Fe^{3+}, Mo^{5+}, Nb^{5+}, Zr^{2+}. *J. Biomed. Mater. Res. Part A* **2008**, *86*, 905–913. [CrossRef] [PubMed]
59. Velasco-Ortega, E.; Jos, A.; Cameán, A.M.; Pato-Mourelo, J.; Segura-Egea, J.J. In vitro evaluation of cytotoxicity and genotoxicity of a commercial titanium alloy for dental implantology. *Mutat. Res. Genet. Toxicol. Environ. Mutagen.* **2010**, *702*, 17–23. [CrossRef]
60. El Medawar, L.; Rocher, P.; Hornez, J.C.; Traisnel, M.; Breme, J.; Hildebrand, H.F. Electrochemical and cytocompatibility assessment of NiTiNOL memory shape alloy for orthodontic use. In Proceedings of the Biomolecular Engineering; Elsevier: Amsterdam, The Netherlands, 2002; Volume 19, pp. 153–160.
61. Issa, Y.; Brunton, P.; Waters, C.M.; Watts, D.C. Cytotoxicity of metal ions to human oligodendroglial cells and human gingival fibroblasts assessed by mitochondrial dehydrogenase activity. *Dent. Mater.* **2008**, *24*, 281–287. [CrossRef] [PubMed]
62. Buczko, P.; Knaś, M.; Grycz, M.; Szarmach, I.; Zalewska, A. Orthodontic treatment modifies the oxidant–antioxidant balance in saliva of clinically healthy subjects. *Adv. Med. Sci.* **2017**, *62*, 129–135. [CrossRef] [PubMed]
63. Olteanu, C.; Muresan, A.; Daicoviciu, D.; Tarmure, V.; Olteanu, I.; Irene, K.M.L.W. Variations of some saliva markers of the oxidative stress in patients with orthodontic appliances. *Fiziologia* **2009**, *19*, 27–29.
64. Atuğ Özcan, S.S.; Ceylan, I.; Özcan, E.; Kurt, N.; Dağsuyu, I.M.; Çanakçi, C.F. Evaluation of oxidative stress biomarkers in patients with fixed orthodontic appliances. *Dis. Markers* **2014**, *2014*, 1–7. [CrossRef] [PubMed]
65. Portelli, M.; Militi, A.; Cervino, G.; Lauritano, F.; Sambataro, S.; Mainardi, A.; Nucera, R. Oxidative Stress Evaluation in Patients Treated with Orthodontic Self-ligating Multibracket Appliances: An Case-Control Study. *Open Dent. J.* **2017**, *11*, 257–265. [CrossRef] [PubMed]
66. Farrugia, G.; Balzan, R.; Madeo, F.; Breitenbach, M. *Oxidative Stress and Programmed Cell Death in Yeast*; Frontiers Media SA: Lausanne, Switzerland, 2012; pp. 1–21.
67. Gonçalves, T.S.; de Menezes, L.M.; Trindade, C.; Machado, M. da S.; Thomas, P.; Fenech, M.; Henriques, J.A.P. Cytotoxicity and genotoxicity of orthodontic bands with or without silver soldered joints. *Mutat. Res. Genet. Toxicol. Environ. Mutagen.* **2014**, *762*, 1–8. [CrossRef]
68. Rincic Mlinaric, M.; Durgo, K.; Katic, V.; Spalj, S. Cytotoxicity and oxidative stress induced by nickel and titanium ions from dental alloys on cells of gastrointestinal tract. *Toxicol. Appl. Pharmacol.* **2019**, *383*. [CrossRef] [PubMed]
69. Terpilowska, S.; Siwicki, A.K. Interactions between chromium(III) and iron(III), molybdenum(III) or nickel(II): Cytotoxicity, genotoxicity and mutagenicity studies. *Chemosphere* **2018**, *201*, 780–789. [CrossRef] [PubMed]
70. The International Programme on Chemical Safety. *Principles for the Safety Assessment of Food Additives and Contaminants in Food*; WHO: Geneva, Switzerland, 1987.
71. Faustman, E.M.; Omenn, B.S. Risk assessment. In *Casarett & Doull's Toxicology: The Basic Science of Poisons*; Klaassen, C.D., Ed.; McGraw: New York, NY, USA, 2001; pp. 92–94.
72. Park, J.H.; Park, E. Influence of iron-overload on DNA damage and its repair in human leukocytes in vitro. *Mutat. Res. Genet. Toxicol. Environ. Mutagen.* **2011**, *718*, 56–61. [CrossRef]
73. Yauger, Y.J.; Bermudez, S.; Moritz, K.E.; Glaser, E.; Stoica, B.; Byrnes, K.R. Iron accentuated reactive oxygen species release by NADPH oxidase in activated microglia contributes to oxidative stress in vitro. *J. Neuroinflamm.* **2019**, *16*, 1–15. [CrossRef]

74. Ceylan, H.; Budak, H.; Kocpinar, E.F.; Baltaci, N.G.; Erdogan, O. Examining the link between dose-dependent dietary iron intake and Alzheimer's disease through oxidative stress in the rat cortex. *J. Trace Elem. Med. Biol.* **2019**, *56*, 198–206. [CrossRef]

75. Terpilowska, S.; Siwicki, A.K. Cell cycle and transmembrane mitochondrial potential analysis after treatment with chromium(iii), iron(iii), molybdenum(iii) or nickel(ii) and their mixtures. *Toxicol. Res.* **2019**, *8*, 188–195. [CrossRef]

76. Kuo, K.L.; Hung, S.C.; Wei, Y.H.; Tarng, D.C. Intravenous iron exacerbates oxidative DNA damage in peripheral blood lymphocytes in chronic hemodialysis patients. *J. Am. Soc. Nephrol.* **2008**, *19*, 1817–1826. [CrossRef]

77. Danadevi, K.; Rozati, R.; Banu, B.S.; Grover, P. In vivo genotoxic effect of nickel chloride in mice leukocytes using comet assay. *Food Chem. Toxicol.* **2004**, *42*, 751–757. [CrossRef]

78. Dally, H.; Hartwig, A. Induction and repair inhibition of oxidative DNA damage by nickel(II) and cadmium(II) in mammalian cells. *Carcinogenesis* **1997**, *18*, 1021–1026. [CrossRef]

79. Curtis, A.; Morton, J.; Balafa, C.; MacNeil, S.; Gawkrodger, D.J.; Warren, N.D.; Evans, G.S. The effects of nickel and chromium on human keratinocytes: Differences in viability, cell associated metal and IL-1α release. *Toxicol. Vitr.* **2007**, *21*, 809–819. [CrossRef]

80. Su, L.; Deng, Y.; Zhang, Y.; Li, C.; Zhang, R.; Sun, Y.; Zhang, K.; Li, J.; Yao, S. Protective effects of grape seed procyanidin extract against nickel sulfate-induced apoptosis and oxidative stress in rat testes. *Toxicol. Mech. Methods* **2011**, *21*, 487–494. [CrossRef]

81. Rabbani-Chadegani, A.; Fani, N.; Abdossamadi, S.; Shahmir, N. Toxic effects of lead and nickel nitrate on rat liver chromatin components. *J. Biochem. Mol. Toxicol.* **2011**, *25*, 127–134. [CrossRef]

82. Blasiak, J.; Kowalik, J. A comparison of the in vitro genotoxicity of tri- and hexavalent chromium. *Mutat. Res. Genet. Toxicol. Environ. Mutagen.* **2000**, *469*, 135–145. [CrossRef]

83. Zhai, X.W.; Zhang, Y.L.; Qi, Q.; Bai, Y.; Chen, X.L.; Jin, L.J.; Ma, X.G.; Shu, R.Z.; Yang, Z.J.; Liu, F.J. Effects of molybdenum on sperm quality and testis oxidative stress. *Syst. Biol. Reprod. Med.* **2013**, *59*, 251–255. [CrossRef]

84. Qiu, S.; Zhao, F.; Tang, X.; Pei, F.; Dong, H.; Zhu, L.; Guo, K. Type-2 cannabinoid receptor regulates proliferation, apoptosis, differentiation, and OPG/RANKL ratio of MC3T3-E1 cells exposed to Titanium particles. *Mol. Cell. Biochem.* **2015**, *399*, 131–141. [CrossRef] [PubMed]

85. Ponti, J.; Sabbioni, E.; Munaro, B.; Broggi, F.; Marmorato, P.; Franchini, F.; Colognato, R.; Rossi, F. Genotoxicity and morphological transformation induced by cobalt nanoparticles and cobalt chloride: An in vitro study in Balb/3T3 mouse fibroblasts. *Mutagenesis* **2009**, *24*, 439–445. [CrossRef] [PubMed]

86. United States Environmental Protection Agency. Reference Dose (RfD): Description and Use in Health Risk Assessments. Available online: https://www.epa.gov/iris/reference-dose-rfd-description-and-use-health-risk-assessments (accessed on 7 June 2021).

87. Institute of Medicine (US) Food and Nutrition Board. *Dietary Reference Intakes*; National Academies Press: Washington, DC, USA, 1998; ISBN 978-0-309-06348-7.

88. Institute of Medicine (US) Food and Nutrition Board. What are Dietary Reference Intakes? In *Dietary Reference Intakes: A Risk Assessment Model for Establishing Upper Intake Levels for Nutrients*; National Academies Press: Washington, DC, USA, 1998.

89. European Food Safety Authority. Trusted science for safe food. Available online: https://www.efsa.europa.eu/en (accessed on 9 June 2021).

90. Institute of Medicine (US) Panel on Micronutrients. Iron. In *Dietary Reference Intakes for Vitamin A, Vitamin K, Arsenic, Boron, Chromium, Copper, Iodine, Iron, Manganese, Molybdenum, Nickel, Silicon, Vanadium, and Zinc*; National Academies Press: Washington, DC, USA, 2001; pp. 290–394.

91. Gaby, A.R. "Safe Upper Levels" for Nutritional Supplements: One Giant Step Backward. *J. Orthomol. Med.* **2003**, *18*, 126–130.

92. Bresson, J.L.; Burlingame, B.; Dean, T.; Fairweather-Tait, S.; Heinonen, M.; Hirsch-Ernst, K.I.; Mangelsdorf, I.; McArdle, H.; Naska, A.; Neuhäuser-Berthold, M.; et al. Scientific Opinion on Dietary Reference Values for iron. *EFSA J.* **2015**, *13*. [CrossRef]

93. Inštitut za Varovanje Zdravja Republike Slovenije. *Referenčne Vrednosti za Vnos Vitaminov in Mineralov—Tabelarična Priporočila za Otroke, Mladostnike, Odrasle in Starejše*; Inštitut za varovanje zdravja Republike Slovenije: Ljubljana, Slovenia, 2013.

94. Iron—Registration Dossier—ECHA. Available online: https://echa.europa.eu/sl/registration-dossier/-/registered-dossier/15 429 (accessed on 7 June 2021).

95. Institute of Medicine (US) Panel on Micronutrients. Arsenic, Boron, Nickel, Silicon, and Vanadium. In *Dietary Reference Intakes for Vitamin A, Vitamin K, Arsenic, Boron, Chromium, Copper, Iodine, Iron, Manganese, Molybdenum, Nickel, Silicon, Vanadium, and Zinc*; National Academies Press: Washington, DC, USA, 2001; pp. 502–554.

96. Nickel—Registration Dossier—ECHA. Available online: https://echa.europa.eu/sl/registration-dossier/-/registered-dossier/15544/1/2 (accessed on 7 June 2021).

97. Institute of Medicine (US) Panel on Micronutrients. Chromium. In *Dietary Reference Intakes for Vitamin A, Vitamin K, Arsenic, Boron, Chromium, Copper, Iodine, Iron, Manganese, Molybdenum, Nickel, Silicon, Vanadium, and Zinc*; National Academies Press: Washington, DC, USA, 2001; pp. 197–224.

98. Titanium—Registration Dossier—ECHA. Available online: https://echa.europa.eu/sl/registration-dossier/-/registered-dossier/15537 (accessed on 7 June 2021).

99. Institute of Medicine (US) Panel on Micronutrients. Molybdenum. In *Dietary Reference Intakes for Vitamin A, Vitamin K, Arsenic, Boron, Chromium, Copper, Iodine, Iron, Manganese, Molybdenum, Nickel, Silicon, Vanadium, and Zinc*; National Academies Press: Washington, DC, USA, 2001; pp. 420–442.

100. Agostoni, C.; Berni Canani, R.; Fairweather-Tait, S.; Heinonen, M.H.K.; La Vieille, S.; Marchelli, R.; Martin, A.; Naska, A.; Neuhäuser-Berthold, M.G.; Nowicka, Y.S.; et al. Scientific Opinion on Dietary Reference Values for molybdenum. *EFSA J.* **2013**, *11*. [CrossRef]

101. Molybdenum—Registration Dossier—ECHA. Available online: https://echa.europa.eu/sl/registration-dossier/-/registered-dossier/15524/11/?documentUUID=d7d1afd6-94e2-45d8-90fe-71fc15de0917 (accessed on 7 June 2021).

102. Cobalt—Registration Dossier—ECHA. Available online: https://echa.europa.eu/sl/registration-dossier/-/registered-dossier/15506/11 (accessed on 7 June 2021).

103. Council Directive from 15th July 1980 Relating the Quality of Water Intended for Human Consumption (80/778/EEC). Available online: https://eur-lex.europa.eu/legal-content/EN/TXT/?uri=OJ:L:1980:229:TOC (accessed on 7 June 2021).

104. Directive (EU) 2020/2184 of the European Parliament and of the Council of 16 December 2020 on The Quality of Water Intended for Human Consumption (Recast) (Text with EEA Relevance). Available online: https://eur-lex.europa.eu/eli/dir/2020/2184/oj (accessed on 7 June 2021).

105. Smith, P.G.P.G.; Scott, J.G.J.G. *Dictionary of Water and Waste Management*; Elsevier: Oxford, UK, 2005.

106. Karnam, K.S.; Reddy, A.N.; Manjith, C. Comparison of Metal Ion Release from Different Bracket Archwire Combinations: An in vitro Study. *J. Contemp. Dent. Pract.* **2012**, *13*, 376–381. [CrossRef]

107. Tahmasbi, S.; Ghorbani, M.; Sheikh, T.; Yaghoubnejad, Y. Galvanic Corrosion and Ion Release from Different Orthodontic Brackets and Wires in Acidic Artificial Saliva. *J. Dent. Sch. Shahid Beheshti Univ. Med. Sci.* **2019**, *32*, 37–44. [CrossRef]

108. Gopikrishnan, S.; Melath, A.; Ajith, V.V.; Mathews, N.B. A comparative study of bio degradation of various orthodontic arch wires: An in vitro study. *J. Int. Oral Health* **2015**, *7*, 12–17.

109. Poljsak, B.; Šuput, D.; Milisav, I. Achieving the balance between ROS and antioxidants: When to use the synthetic antioxidants. *Oxidative Med. Cell. Longev.* **2013**, *2013*, 956792. [CrossRef]

110. Poljšak, B.; Fink, R. The protective role of antioxidants in the defence against ROS/RNS-mediated environmental pollution. *Oxidative Med. Cell. Longev.* **2014**, *2014*, 671539. [CrossRef]

111. Bjelakovic, G.; Nikolova, D.; Simonetti, R.G.; Gluud, C. Antioxidant supplements for prevention of gastrointestinal cancers: A systematic review and meta-analysis. *Lancet* **2004**, *364*, 1219–1228. [CrossRef]

112. Miller, E.R.; Pastor-Barriuso, R.; Dalal, D.; Riemersma, R.A.; Appel, L.J.; Guallar, E. Meta-analysis: High-dosage vitamin E supplementation may increase all-cause mortality. *Ann. Intern. Med.* **2005**, *142*, 37–46. [CrossRef]

113. Vivekananthan, D.P.; Penn, M.S.; Sapp, S.K.; Hsu, A.; Topol, E.J. Use of antioxidant vitamins for the prevention of cardiovascular disease: Meta-analysis of randomised trials. *Lancet* **2003**, *361*, 2017–2023. [CrossRef]

114. Caraballoso, M.; Sacristan, M.; Serra, C.; Cosp, X.B. Drugs for preventing lung cancer in healthy people. *Cochrane Database Syst. Rev.* **2003**, *2*. [CrossRef]

115. Rhee, S.G. Redox signaling: Hydrogen peroxide as intracellular messenger. *Exp. Mol. Med.* **1999**, *31*, 53–59. [CrossRef]

116. Poljšak, B.; Raspor, P. The antioxidant and pro-oxidant activity of vitamin C and trolox in vitro: A comparative study. *J. Appl. Toxicol.* **2008**, *28*, 183–188. [CrossRef]

117. Poljšak, B.; Gazdag, Z.; Jenko-Brinovec, Š.; Fujs, Š.; Pesti, M.; Bélagyi, J.; Plesničar, S.; Raspor, P. Pro-oxidative vs antioxidative properties of ascorbic acid in chromium(VI)-induced damage: An in vivo and in vitro approach. *J. Appl. Toxicol.* **2005**, *25*, 535–548. [CrossRef] [PubMed]

118. Cheeseman, K.H.; Slater, T.F. An introduction to free radical biochemistry. *Br. Med. Bull.* **1993**, *49*, 481–493. [CrossRef] [PubMed]

119. Halliwell, B.; Gutteridge, J.M.C. Oxygen toxicity, oxygen radicals, transition metals and disease. *Biochem. J.* **1984**, *219*, 1–14. [CrossRef]

120. Forman, H.J.; Davies, K.J.A.; Ursini, F. How do nutritional antioxidants really work: Nucleophilic tone and para-hormesis versus free radical scavenging in vivo. *Free Radic. Biol. Med.* **2014**, *66*, 24–35. [CrossRef] [PubMed]

121. Argüelles, S.; Gómez, A.; Machado, A.; Ayala, A. A preliminary analysis of within-subject variation in human serum oxidative stress parameters as a function of time. *Rejuvenation Res.* **2007**, *10*, 621–636. [CrossRef]

122. Mira, L.; Fernandez, M.T.; Santos, M.; Rocha, R.; Florêncio, M.H.; Jennings, K.R. Interactions of flavonoids with iron and copper ions: A mechanism for their antioxidant activity. *Free Radic. Res.* **2002**, *36*, 1199–1208. [CrossRef] [PubMed]

123. Bhuiyan, M.N.I.; Mitsuhashi, S.; Sigetomi, K.; Ubukata, M. Quercetin inhibits advanced glycation end product formation via chelating metal ions, trapping methylglyoxal, and trapping reactive oxygen species. *Biosci. Biotechnol. Biochem.* **2017**, *81*, 882–890. [CrossRef]

124. Zhang, W.; Chen, C.; Shi, H.; Yang, M.; Liu, Y.; Ji, P.; Chen, H.; Tan, R.X.; Li, E. Curcumin is a biologically active copper chelator with antitumor activity. *Phytomedicine* **2016**, *23*, 1–8. [CrossRef] [PubMed]

125. Zieniewska, I.; Maciejczyk, M.; Zalewska, A. The effect of selected dental materials used in conservative dentistry, endodontics, surgery, and orthodontics as well as during the periodontal treatment on the redox balance in the oral cavity. *Int. J. Mol. Sci.* **2020**, *21*, 9684. [CrossRef] [PubMed]

126. Żukowski, P.; Maciejczyk, M.; Waszkiel, D. Sources of free radicals and oxidative stress in the oral cavity. *Arch. Oral Biol.* **2018**, *92*, 8–17. [CrossRef] [PubMed]

127. Tartaglia, G.M.; Gagliano, N.; Zarbin, L.; Tolomeo, G.; Sforza, C. Antioxidant capacity of human saliva and periodontal screening assessment in healthy adults. *Arch. Oral Biol.* **2017**, *78*, 34–38. [CrossRef] [PubMed]

Systematic Review

Meta-Analysis and Systematic Review of the Association between a Hypoactive *NCF1* Variant and Various Autoimmune Diseases

Liang Zhang [1], Jacqueline Wax [1], Renliang Huang [2], Frank Petersen [1] and Xinhua Yu [1,*]

[1] Priority Area Chronic Lung Diseases, Research Center Borstel, Member of the German Center for Lung Research (DZL), 23845 Borstel, Germany
[2] Hainan Women and Children's Medical Center, Haikou 571100, China
*   Correspondence: xinhuayu@fz-borstel.de

Abstract: Genetic association studies have discovered the *GTF2I-NCF1* intergenic region as a strong susceptibility locus for multiple autoimmune disorders, with the missense mutation *NCF1* rs201802880 as the causal polymorphism. In this work, we aimed to perform a comprehensive meta-analysis of the association of the *GTF2I-NCF1* locus with various autoimmune diseases and to provide a systemic review on potential mechanisms underlying the effect of the causal *NCF1* risk variants. The frequencies of the two most extensively investigated polymorphisms within the locus, *GTF2I* rs117026326 and *NCF1* rs201802880, vary remarkably across the world, with the highest frequencies in East Asian populations. Meta-analysis showed that the *GTF2I-NCF1* locus is significantly associated with primary Sjögren's syndrome, systemic lupus erythematosus, systemic sclerosis, and neuromyelitis optica spectrum disorder. The causal *NCF1* rs201802880 polymorphism leads to an amino acid substitution of p.Arg90His in the p47phox subunit of the phagocyte NADPH oxidase. The autoimmune disease risk His90 variant results in a reduced ROS production in phagocytes. Clinical and experimental evidence shows that the hypoactive His90 variant might contribute to the development of autoimmune disorders via multiple mechanisms, including impairing the clearance of apoptotic cells, regulating the mitochondria ROS-associated formation of neutrophil extracellular traps, promoting the activation and differentiation of autoreactive T cells, and enhancing type I IFN responses. In conclusion, the identification of the association of *NCF1* with autoimmune disorders demonstrates that ROS is an essential regulator of immune tolerance and autoimmunity mediated disease manifestations.

Keywords: neutrophil cytosolic factor 1; reactive oxygen species; genetic association

Citation: Zhang, L.; Wax, J.; Huang, R.; Petersen, F.; Yu, X. Meta-Analysis and Systematic Review of the Association between a Hypoactive *NCF1* Variant and Various Autoimmune Diseases. *Antioxidants* **2022**, *11*, 1589. https://doi.org/10.3390/antiox11081589

Academic Editors: Soliman Khatib and Dana Atrahimovich Blatt

Received: 12 July 2022
Accepted: 12 August 2022
Published: 16 August 2022

Publisher's Note: MDPI stays neutral with regard to jurisdictional claims in published maps and institutional affiliations.

1. Introduction

Genome-wide association studies (GWAS) have revolutionized the dissection of the genetic basis of autoimmune disorders. GWAS often lead to the discovery of a large number of susceptibility loci, each of which shows a mild contribution to the disease. For example, multiple GWAS in the last 15 years have uncovered more than 200 genetic loci that are independently associated with multiple sclerosis, while none of them show a odds ratio (OR) of more than 1.5 [1,2]. In 2013, a strong susceptibility loci within the *GTF2I–NCF1* intergenic region at 7q11.23 was identified for primary Sjögren's syndrome (pSS), with the *GTF2I* rs117026326 as the most significant polymorphism (OR = 2.20, $p = 1.31 \times 10^{-53}$) [3]. Subsequently, this strong association has been confirmed [4] and extended to other autoimmune disorders, including systemic lupus erythematosus (SLE) [5], systemic sclerosis (SSc) [6], and neuromyelitis optica spectrum disorder (NMOSD) [7].

Since the *GTF2I* rs117026326 is an intronic SNP, efforts have been made to identify the causal variant of this novel susceptibility locus for multiple autoimmune diseases. In 2017, two studies reported that a missense mutation in neutrophil cytosolic factor 1 (*NCF1*)

within the locus is associated with SLE, pSS and rheumatoid arthritis (RA) [8,9]. The *NCF1* rs201802880 G > A polymorphism that is associated disease susceptibility, age at diagnosis but not disease activity in SLE leads to a shift from Arg to His at position 90 which is a evolutionarily conserved residue in the p47phox subunit of the phagocyte NADPH oxidase complex [8,10]. The disease risk His90 variant reduces the production of reactive oxygen species (ROS) and increases the expression of type 1 interferon (IFN-I)-regulated genes [8,9], suggesting it is a putative causal variant of the *GTF2I–NCF1* intergenic susceptibility locus. This notion was verified by experimental evidence obtained from NCF1-His90 knock-in (KI) mice [11]. Compared to wild type (WT) littermate controls, NCF1-His90 KI mice show a reduced ROS production, elevated type IFN-I scores, splenomegaly, and increased germinal center B cells and plasma cells. Moreover, NCF1-His90 KI mice but not WT littermate controls develop autoantibodies and SLE-like kidney pathology after challenge with pristane [11]. In the current study, we performed a comprehensive meta-analysis to evaluate the association of the *GTF2I–NCF1* intergenic susceptibility locus with various autoimmune diseases. In addition, we aimed to provide an overview of potential mechanisms underlying the role of the causal NCF1-His90 variant in autoimmune conditions.

2. Methods

2.1. Identification of Eligible Studies

To obtain an overview of the association between the *GTF2I–NCF1* locus and autoimmune diseases, we carried out a comprehensive meta-analysis. A search of the Medline database (https://www.ncbi.nlm.nih.gov/pubmed (accessed on 30 March 2022)) was performed to identify eligible studies. First of all, the keyword 'rs117026326' and '(Arg90His) OR (rs201802880)' were used for the search without any limitations. In a second step, the full text of all articles from step one were reviewed to identify eligible studies. The following studies were excluded: (a) review articles or comment, (b) non-genetic association studies, and (c) genetic studies for non-autoimmune diseases.

2.2. Data Extraction

Data extraction was conducted as described previously [12]. Briefly, the following information was collected from each study: first author's name, year of publication, the population of origin, type of autoimmune disease, the number of cases and controls, and frequencies of the rs117026326 and rs201802880 allele in both cases and controls. For studies including several case–control populations, each case–control population was extracted separately. Two participants searched whole the articles that needed to be extracted independently, and the extracted data were checked by a third participant.

2.3. Data Evaluation and Statistical Analysis

Cochran's Q-statistics was applied to evaluate the heterogeneity across studies. The random-effect model was used for meta-analysis when heterogeneity was indicated by a significant Q-statistic ($p < 0.10$), otherwise the common/fixed effect model was used. By comparing frequencies of alleles, the odds ratio (OR), 95% confidence intervals (CI), and p values were estimated for each individual case–control study. Meta-analysis was performed to calculate the pooled OR, 95% CI and p values. The presence of publication bias was evaluated by examining the asymmetry of the funnel plot using Egger's regression test. All statistical analyses were performed using the R software (version 4.1.2) and Comprehensive Meta-Analysis computer program (Biosta, Englewood, NJ, USA).

3. Results

3.1. Frequency Distribution of the GTF2I and NCF1 Polymorphisms across Populations

Notably, although the *GTF2I-NCF1* locus at 7q11.23 is strongly associated with pSS in Chinese [3], such an association has not been observed in Caucasians [13], suggesting that susceptibility variants within the locus might be remarkably less prevalent in Caucasians than in Chinese. This notion is supported by the frequency distribution of the *GTF2I* rs117026326 and *NCF1* rs201802880 polymorphisms across populations.

The allele frequency of the *GTF2I* rs117026326 polymorphism varies considerably across the world (Figure 1). The highest frequency (17.6%) was reported in Changchun, a city in Northeast China [7]. The frequency decreases toward the south and west China, showing 14.9% and 13.9% in Beijing [5,6], and 11.9% in Shanghai [9], 11.9% and 9.4% in Chengdu [14,15], 11.5% in Xiamen [4], and 10.4% in Taiwan [16]. This polymorphism is also prevalent in other Eastern Asia populations, with allele frequencies of 11.4% in Korean [9] and 8.0% in Japanese [17]. In line with this finding, the frequency in Chinese American in Los Angeles was reported to be 11.4% [9]. By contrast, those values in European American and African American are only 0.7% and 0.0%, respectively [9].

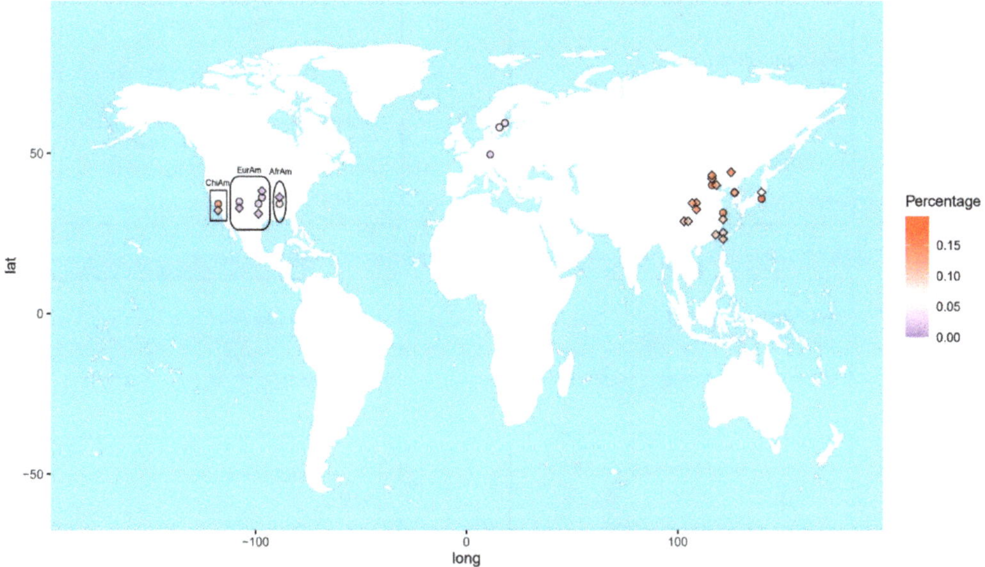

Figure 1. Heat map of the allele frequencies of the *GTF2I* rs117026326 (diamond) and the *NCF1* rs201802880 (circle) across the world. Each color dot represents frequency of the *GTF2I* rs117026326 or the *NCF1* rs201802880 in a single population. ChiAm: Chinese American; EurAm: European American; AfrAm: African American.

Due to highly homologous sequences among *NCF1* and its two pseudogenes, *NCF1B* and *NCF1C*, genotyping of the *NCF1* rs201802880 polymorphism is not possible with conventional methods [9]. To obtain the correct genotypes of the polymorphism, a NCF1-specific genomic fragment needs to be amplified and, subsequently, subjected to sequencing or TaqMan SNP Genotyping Assay [8,9,17]. The highest allele frequency of the *NCF1* rs201802880 polymorphism was reported in Japan (19.6%) [17], followed by Beijing, China (18.3%); Korea (18.1%); and Shanghai, China (16.7%). The frequency in Chinese American (15.6%) is considerably higher than in African American (8.2%) and European American (2.3%). As expected, the frequencies in Swedish (2.45%) and German (1.8%) populations are comparable to that in European American [8,18] (Figure 1).

Taken together, allele frequencies of *GTF2I* rs117026326 and *NCF1* rs201802880 polymorphisms are population dependent, with higher frequencies in East Asian populations than in others, which explains the strong association of the *GTF2I-NCF1* locus with autoimmune diseases in China, Korea, and Japan.

3.2. Meta-Analysis for the Association of the GTF2I-NCF1 Locus with Autoimmune Diseases

The keyword 'rs117026326´ and '(Arg90His) OR (rs201802880)' were used for the search without any limitations, which identified 11 and 8 articles, respectively. Further review of the 19 articles led to the identification of 11 eligible studies for the meta-analysis (Figure 2). To access potential publication bias, the asymmetry of the funnel plot was examined. All *p* values of Egger's regression test were >0.05, suggesting that there was no publication bias for the association of the two polymorphisms within the *GTF2I-NCF1* loci with pSS or SLE (Figure 3).

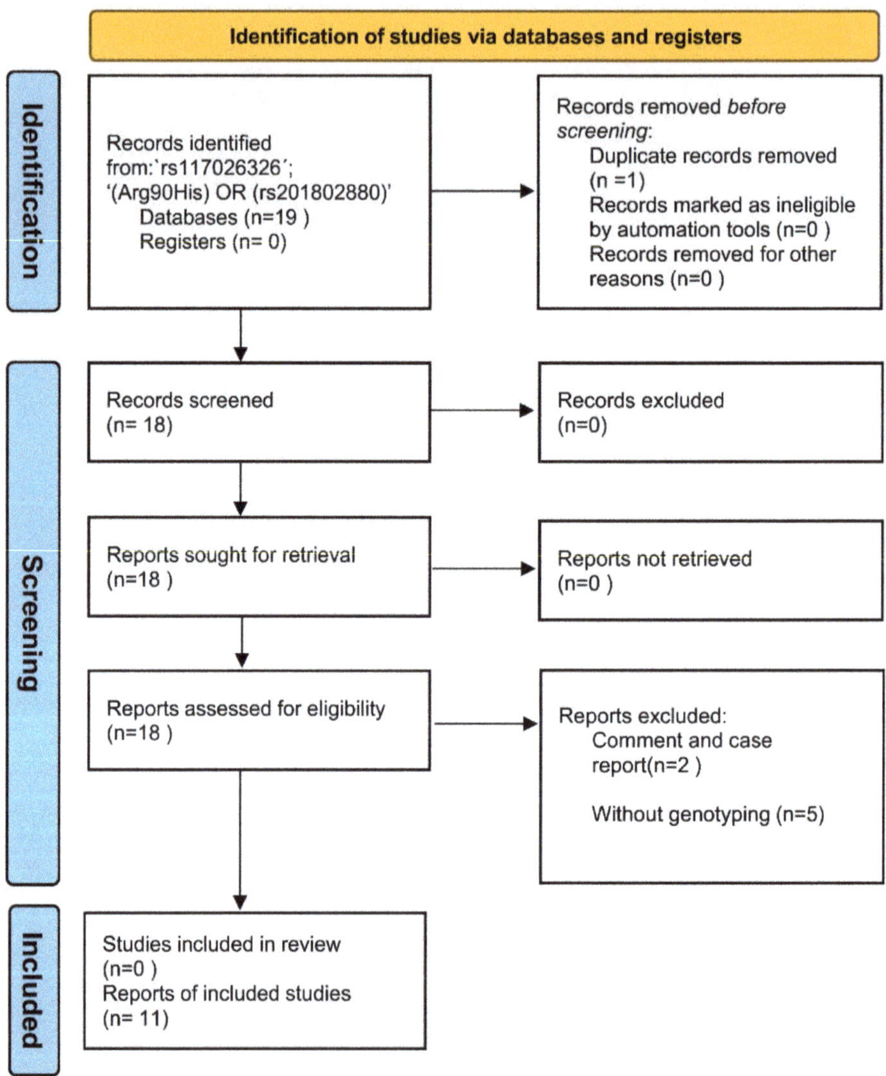

Figure 2. The PRISMA flow diagram of the meta-analysis.

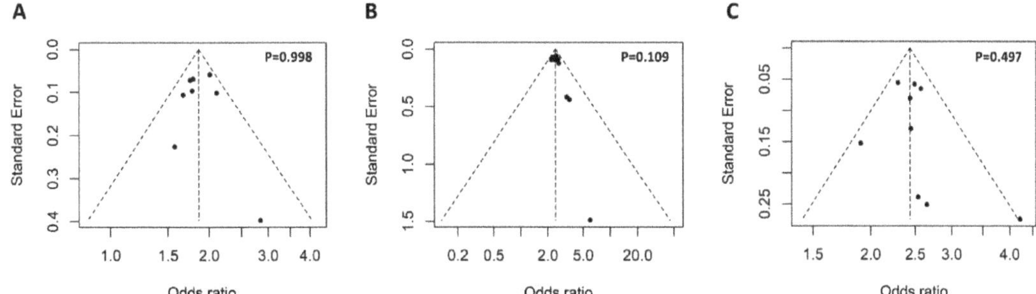

Figure 3. Symmetrical funnel plot used to assess publication bias. The funnel plot of the association of the *GTF2I* rs117026326 polymorphism with pSS (**A**) and SLE (**B**), as well as association of the *NCF1* rs201802880 polymorphism with SLE (**C**). Egger's regression test was used to assess potential publication bias via funnel plot asymmetry, and a *p*-value less than 0.05 indicates presence publication bias.

3.2.1. Meta-Analysis for the Association of the GTF2I rs117026326 Polymorphism with Autoimmune Diseases

Since the *GTF2I* rs117026326 polymorphism exerts the strongest association with pSS in the discovery study of the susceptibility locus [3], this SNP has been extensively studied for its association with various autoimmune diseases, including pSS, SLE, SSc, NMOSD, multiple sclerosis (MS), and antineutrophil cytoplasmic antibody (ANCA)-associated vasculitides (AAV) [3–5,7,9,14–17]. The association of the *GTF2I* rs117026326 polymorphism with pSS, SLE, SSc and NMOSD has been investigated in at least two case–control studies, making them applicable for meta-analysis (Figure 4).

For the meta-analysis of the association between the *GTF2I* rs117026326 polymorphism and pSS, 8 case–control studies were recruited. All of them showed that the minor allele is associated with an increased risk of pSS [3,4,9,16]. As expected, our meta-analysis confirmed a highly significant association between the *GTF2I* rs117026326 polymorphism and pSS (OR = 2.17; 95% CI = 2.00–2.34, $p < 1 \times 10^{-16}$) (Figure 4A). The association between the GTF2I rs117026326 polymorphism and SLE has been investigated in 9 case–control studies, and 8 of them showed a significant association [5,9,14,17]. Our meta-analysis showed a strong association of the *GTF2I* rs117026326 polymorphism with SLE, where the minor allele conferred an approximate 3-fold risk for the disease (OR = 3.04; 95% CI = 2.81–3.30, $p < 1 \times 10^{-16}$) (Figure 4B). Meta-analysis also revealed a significant association between the *GTF2I* rs117026326 polymorphism and SSc with two case–control studies [6,16] (OR = 1.77; 95% CI = 1.49–2.11, $p = 9.36 \times 10^{-11}$) (Figure 4C). The association between the *GTF2I* rs117026326 polymorphism and NMOSD has been investigated in two case–control studies where both have shown a significant association [7,15]. The meta-analysis with the two studies revealed a marginally significant association of the *GTF2I* rs117026326 polymorphism with NMOSD (OR = 1.92; 95% CI = 0.99–3.73, $p = 0.054$) (Figure 4D).

Unlike the four autoimmune diseases mentioned above, MS and AAV are reported not to be associated with the *GTF2I* rs117026326 polymorphism [7,17]. The difference between two autoimmune demyelinating diseases of the central nervous system, MS and NMOSD, in their association with *GTF2I* is supported by our recent studies reporting that the *GTF2I* rs73366469, another Taq-SNP of the intergenic locus, is associated with NMOSD, but not with MS, myelin oligodendrocyte glycoprotein-associated disorders or anti-*N*-Methyl-D-Aspartate receptor encephalitis [19,20].

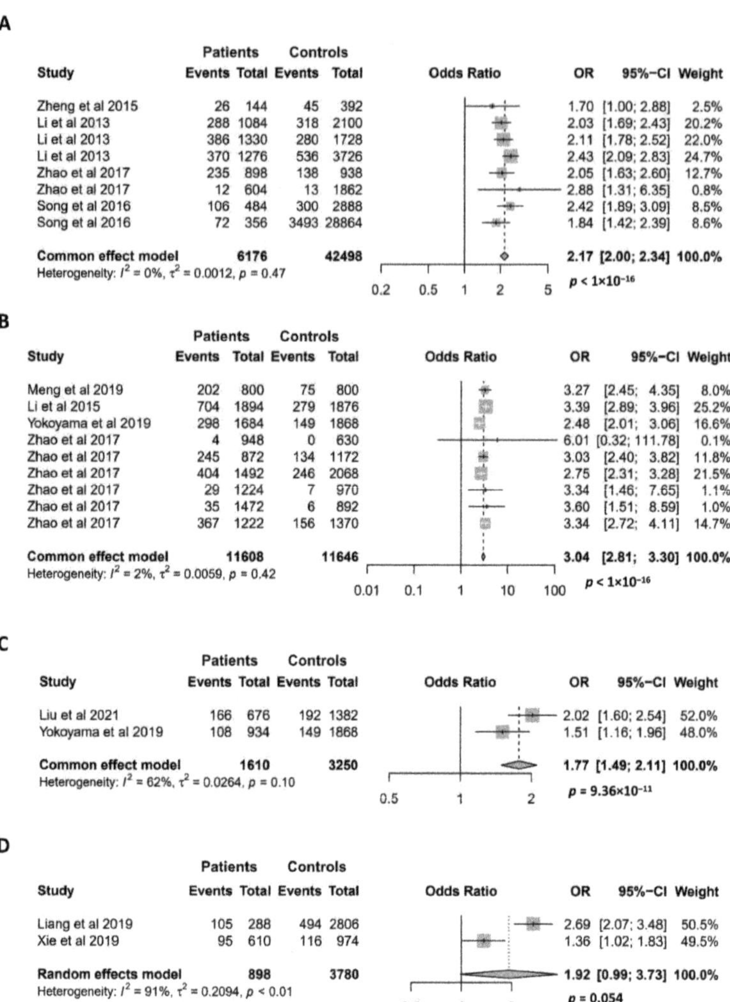

Figure 4. Association of the GTF2I rs117026326 polymorphism with pSS (**A**), SLE (**B**), SSc (**C**), and NMOSD (**D**). Data derived from the meta-analysis are presented as forest plot. Odds ratios (OR) and 95% confidence intervals (CI) of individual studies are represented by squares and horizontal lines, respectively. Size of the square represents the weight of the individual study in the meta-analysis. Diamonds represent the OR (center line of the respective diamond) and 95% CI (lateral tips of the respective diamond). The OR, 95% CI and *p* values for the meta-analysis are indicated by bold text. Information regarding individual case–control studies, including the first author name and year of publication, are indicated [3–7,9,14–17].

3.2.2. Meta-Analysis for the Association of the *NCF1* rs201802880 Polymorphism with Autoimmune Diseases

Since it was discovered in 2017 [9], the missense mutation *NCF1* rs201802880 has been investigated for its association with 5 autoimmune conditions, namely SLE, pSS, RA, SSc, and AAV [8,9,17]. As shown in Figure 5A, the association of the *NCF1* rs201802880 polymorphism with SLE has been investigated in 9 case–control studies, and all of them reported that the minor allele is significantly associated with an increased risk of SLE [8,9,17]. Accordingly, our meta-analysis showed a strong association of the *NCF1* rs201802880 polymorphism with SLE, where the minor allele conferred a more than 3-fold risk of developing the disease (OR = 3.20; 95% CI = 2.79–3.66, $p < 1 \times 10^{-16}$). Meta-analysis of the association between the *NCF1* rs201802880 polymorphism and pSS was performed with two case–control studies [9]. The result showed a strong association of the mutant allele with an increased risk for pSS (OR = 2.63; 95% CI = 2.16–3.20, $p < 1 \times 10^{-16}$) (Figure 5B). Association of the *NCF1* rs201802880 polymorphism with other three autoimmune diseases was investigated in only one case–control study, where SSc and RA, but not AAV, are reported to be associated with the missense mutation [17].

A

Study	Patients Events	Total	Controls Events	Total	Odds Ratio	OR	95%–CI	Weight
Zhao et al 2017	149	948	52	630		2.07	[1.49; 2.89]	9.3%
Zhao et al 2017	332	872	183	1172		3.32	[2.70; 4.09]	14.2%
Zhao et al 2017	621	1492	345	2068		3.56	[3.05; 4.16]	16.8%
Zhao et al 2017	570	1222	248	1370		3.96	[3.31; 4.72]	15.7%
Zhao et al 2017	67	1224	20	970		2.75	[1.66; 4.57]	5.4%
Zhao et al 2017	88	1472	21	892		2.64	[1.63; 4.28]	5.7%
Yokoyama et al 2019	744	1652	344	1752		3.35	[2.88; 3.91]	16.9%
Olsson et al 2017	132	1206	96	2148		2.63	[2.00; 3.45]	11.5%
Olsson et al 2017	88	852	15	613		4.59	[2.63; 8.02]	4.6%
Random effects model		**10940**		**11615**		**3.20**	**[2.79; 3.66]**	**100.0%**

Heterogeneity: $I^2 = 55\%$, $\tau^2 = 0.0222$, $p = 0.02$

0.2 0.5 1 2 5 $\qquad p < 1\times10^{-16}$

B

Study	Patients Events	Total	Controls Events	Total	Odds Ratio	OR	95%–CI	Weight
Zhao et al 2017	339	898	172	938		2.70	[2.18; 3.34]	84.6%
Zhao et al 2017	29	604	41	1862		2.24	[1.38; 3.64]	15.4%
Common effect model		**1502**		**2800**		**2.63**	**[2.16; 3.20]**	**100.0%**

Heterogeneity: $I^2 = 0\%$, $\tau^2 = 0$, $p = 0.49$

0.5 1 2 $\qquad p < 1\times10^{-16}$

Figure 5. Association of the *NCF1* rs201802880 polymorphism with SLE (**A**) and pSS (**B**). Data derived from the meta-analysis are presented as forest plot. Odds ratios (OR) and 95% confidence intervals (CI) of individual studies are represented by squares and horizontal lines, respectively. Size of the square represents the weight of the individual study in the meta-analysis. Diamonds represent the OR (center line of the respective diamond) and 95% CI (lateral tips of the respective diamond). The OR, 95% CI and *p* values for the meta-analysis are indicated by bold text. Information regarding individual case–control studies, including the first author name and year of publication, are indicated [8,9,17].

4. Discussion

In the present study, we performed a meta-analysis demonstrating that the *GTF2I-NCF1* intergenic locus is associated with multiple autoimmune diseases, including pSS, SLE, SSc, and NMOSD. A major limitation of this study is the relative small number of studies retrieved for the analysis. Since the causal mutation within the *GTF2I-NCF1* locus was identified in 2017, its association with autoimmune diseases has not been extensively investigated. It is conceivable that some other autoimmune disorders are also associated with the susceptibility locus, and examination of the pattern of the association will help to understand the underlying mechanisms of the association.

Both clinical and experimental evidence have shown that the causal variation is the *NCF1* rs201802880 polymorphism which leads to the p.Arg90His substitution. *NCF1* encodes the p47phox which is one of five subunits of the NADPH oxidase 2 (NOX2), and the p.Arg90His residue is located in the phox domain which mediates binding of p47phox to the cellular membrane. It has been shown that the evolutionary conserved Arg90 residue is essentially involved in the binding to the cellular membrane and the NAPDH oxidase activity in response to various stimuli [21]. Compared to Arg90, the

His90 variant reduces the ROS production and leads to an increased risk of developing multiple autoimmune diseases including SLE, pSS, SSc, and RA [8,9,22]. In accordance with these findings, *NCF1* variants that reduce NOX2-derived ROS production in rat and mice have also been shown to promote multiple experimental autoimmune diseases, such as pristane-induced arthritis, experimental autoimmune encephalomyelitis, and SLE-like disease [23–25]. The causal relationship between the human and rodent hypoactive *NCF1* mutations and various autoimmune disorders suggests an essential role of NOX2-derived ROS in the regulation of development of autoimmune disorders [26]. Clinical and experimental evidence have demonstrated that the NOX2-derived ROS might contribute to the pathogenesis of autoimmune diseases via multiple mechanisms.

4.1. Regulation of Apoptotic Cell Clearance

The phagocyte NOX2 complex is composed of two catalytic transmembrane subunits and three regulatory cytosolic subunits, including the p47phox encoded by *NCF1* [27]. ROS-reducing *NCF1* variants that lead to hypofunction of the NOX2 might impair the physiological function of phagocytes. Very recently, Geng and colleagues reported that bone marrow-derived macrophages from *NCF1* His90-KI mice result in an impaired efferocytosis [11]. Compared to WT controls, macrophages from the His90-KI mice show a decreased ROS production, reduced Hv1-dependent acidification, impaired maturation and proteolysis of phagosomes, which lead to a slower digestion of apoptotic cells [11]. This observation is in line with the finding that macrophages from mice carrying loss-of-function mutation in *gp91phox* encoding another NOX2 subunit show an impaired ability to clear apoptotic cells [28].

Under physiological condition, most tissue undergoes routine turnover of cells which mainly die via apoptosis, and the rapid removal of those apoptotic cells by phagocytes via efferocytosis is an essential step for the maintenance of tissue homeostasis. Defective clearance of apoptotic cells is associated with many pathological conditions, including autoimmune diseases, such as SLE [29]. For example, it has been shown that macrophages from patients with SLE are featured by a decreased phagocytic clearance of apoptotic cells [30,31]. Moreover, mice deficient in genes encoding molecules involved in apoptotic cell recognition and binding on the cell membrane of phagocytes develop spontaneously lupus-like disease [32–34]. Therefore, it is conceivable that the ROS-reducing *NCF1* variant causes a defective efferocytosis and thus impairs clearance of apoptotic cells, which further contributes to the development of SLE. This concept is supported by the observation that mice carrying hypoactive *NCF1* variant produce SLE-associated autoantibodies and develop chronic kidney inflammation spontaneously or after application of pristane [11,25]. In line with these findings, macrophages from SLE patients carrying the His90 variant show a decreased phagocytic ability compared to those from Arg90-carrying patients with SLE [11].

4.2. Regulation of Mitochondrial ROS-Associated NET Formation

As a crucial component in the first line of defense against micro-organisms, neutrophils are equipped with multiple anti-infective strategies, including phagocytosis, generation of ROS, proteases, and formation of neutrophil extracellular traps (NETs) [35]. Apart from their essential role in infection, neutrophils have been emerged as a player in the development of various autoimmune disorders [36].

In SLE, neutrophils from patients are characterized with multiple abnormalities in their functions, such as increased apoptosis, elevated release of NETs, aberrant NETosis, impaired phagocytosis, and enriched circulating low-density granulocytes [37–40], supporting a crucial role of these cells in the disease pathogenesis. Among the different neutrophil functions, the role of NET formation in the pathogenesis of SLE has been extensively investigated. As complex three-dimensional structures committed to trap circulating micropathogens, NETs are composed of chromosome DNA, histone, and cytoplasmic granule proteins, providing a potential resource of SLE autoantigens [41,42]. Despite extensive investigation, the role of NETs in SLE remains elucidative. On the one hand, increased NET release and impaired clearance of NET components have been observed in SLE [43,44] and immune complexes formed by autoantibodies and NET-derived antigen can stimulate the production of IFN-α [40] and activate autoreactive B cells [45], favoring disease-promoting role of NETs in SLE. On the other hand, studies using murine models of SLE have shown that NET deficiency can promote [46], inhibit [47] or have no impact [48] on lupus-like disease, suggesting that NETs are not an indispensible component but may contribute to the development of SLE depending on the respective pathological condition.

Neutrophils from patients with CGD are unable to produce NETs [49], highlighting an essential role of NOX2 activity in NET formation. Recently, Linge et al. reported that human neutrophils with the hypoactive *NCF1* His90 variant show a decreased NET formation in response to PMA stimulation, compared to Arg90 neutrophils [10]. Moreover, an increased dependence on mtROS in PMA-mediated NET formation has been observed in His90 neutrophils [10]. Apart from ROS produced via NOX2, neutrophil activation also leads to the generation of mitochondria-derived ROS (mtROS) [50]. When NOX2-derived ROS are impaired, mtROS are likely involved in the formation of NETs [50]. By contrast to the anti-inflammatory role of the NOX2-derived ROS, mtROS are potentially inflammatory because it causes oxidation of mitochondrial DNA [51–53], implicating that imbalance in the usage of NOX2-derived ROS and mtROS might affect autoimmune inflammatory processes. Therefore, it is likely that the involvement of mtROS in the formation of NETs in His90 neutrophils makes the NETs pathogenic. This concept is partially supported by evidence that mtROS-associated NETs containing mtDNA aggravates lupus-like disease in mice [51]. However, it needs to be further validated in future studies.

4.3. Regulation of T Cell Responses

Dysregulated autoreactive T cell responses have been observed in mice carrying His90 allele or other ROS-reducing variants [11,23,54], suggesting a role of autoimmune disease risk *NCF1* variants in the regulation of T cell homeostasis. Although T cells do not exert oxidative burst, multiple evidence show that *NCF1* mutation are able to regulate T cell homeostasis via antigen presenting cells (APCs).

As compared with WT mice, His90 KI mice are featured by increased ratios of splenic follicular T helper 2 (Tfh2) to either T follicular regulatory (Tfr) or Tfh1 cells [11], suggesting a role of the hypoactive His90 variant and reduced ROS in the regulation of differentiation of follicular T helpers. In line with this experimental observation, SLE patients carrying His90 variant have increased frequencies of circulating Tfh and Tfh2 cells but decreased frequencies of Tfh1 and Tfr cells [11]. Expanded Tfh2 cell populations are likely caused by the interaction between T cells and APCs with defective efferocytosis. Mechanistically, the His90 variant causes defective efferocytosis and persistence of apoptotic cell proteins within phagosomes of APCs, which leads to the enhanced presentation of an apoptotic cell-associated antigen with increased CD40 expression and consequent expansion of Tfh and Tfh2 cells [11,55].

Hypoactive *NCF1* variants also promote the expansion of autoreactive effector T helpers. In 2017, Sarelia and colleagues reported that KI mice carrying a ROS-reducing *NCF1* variant originated from rat, develop severe collagen induced arthritis (CIA) while mice expressing functional *NCF1* develop only mild symptoms [54], demonstrating a protective role of ROS in the development of autoimmune arthritis. Notably, the hypoactive *NCF1* variant promotes the expansion of immunodominant collagen II-specific Th17 cells [54] which are a major contributor to the pathogenesis of CIA [56]. The effect of promoting the expansion of collagen II-specific CD4 effector T cells has also been reported in mice expressing another hypoactive *NCF1* variant [23]. A possible mechanism underlying the role of reduced ROS in regulation of CD4 effector T cells has been proposed by Gelderman and colleagues [57,58]. According to their hypothesis, APCs with lower capacity to produce ROS are associated with an increased number of reduced thiol groups (-SH) on membrane surfaces of T cells, which lowers the threshold for T cell reactivity and enhances proliferative responses [57,58].

In addition to the regulation of T helper cells, NOX2-derieved ROS is also involved in the induction of regulatory T cells (Tregs). Compared to controls with functional *NCF1*, animals carrying hypoactive *NCF1* variant show similar levels of naïve Tregs but a decreased levels of induced Tregs, suggesting that NOX2-derived ROS is require for the induction of Tregs [59]. This notion is supported by in vitro evidence that macrophages induce CD4$^+$CD25$^+$FoxP3$^+$ Tregs in a ROS-dependent manner [59]. Taken together, the hypoactive *NCF1* variant might contribute to T cell autoimmunity by promoting autoreactive T helpers and inhibiting the induction of Tregs.

4.4. Regulation of Type 1 IFN Signaling

In 2014, Kelkka and colleagues reported that both chronic granulomatous disease (CGD) patients and mice lacking functional NOX2 complex exhibit a prominent type I interferon (IFN) response signature and elevated levels of autoantibodies [25], suggesting for the first time that type I IFN signaling as a potential mediator connecting ROS deficiency to autoimmunity. In line with this finding, the hypoactive *NCF1* His90 variant is associated with an increased expression of type 1 interferon-regulated genes in patients with autoimmune diseases, such as RA and SLE [8,10]. Furthermore, *NCF1* His90 KI mice show an elevated type I IFN response and SLE-associated autoantibodies compared to WT littermate controls [11]. Therefore, the prominent type I IFN response might be a potential contributor to the His90 variant-caused autoimmune conditions. The type I IFN response is committed to combat viral infection, and it can be induced by both environmental and endogenous factors [60,61]. By investigating germ-free *NCF1* mutated mice, Kelkka and colleagues showed that the upregulated type I IFN response associated with reduced ROS is of endogenous origin [25].

There is an increasing body of evidences that type I IFN response is associated with autoimmune conditions, especially autoimmune rheumatoid diseases, such as SLE, pSS, and RA [62,63]. Elevated levels of circulating IFN-α have been observed in patients with SLE and they are associated with disease activity and clinical manifestations [64]. In murine models of SLE, increased type I IFN response has been observed [65], and application of IFN-α increases immune complex deposition in the kidneys and consequently exacerbates the disease [66,67]. In addition to in SLE, unregulated expression of type I IFN and IFN-stimulated genes has been observed in pSS [68,69]. In addition, elevated IFN signature is regarded as a biomarker for disease activity and response to therapy in RA [70,71]. Type I IFNs are capable to promote multiple immunological processes, including maturation of myeloid DCs (mDCs), CD4 T helper cell differentiation, B cell activation, plasma cell differentiation, antibody production, and Ig class switching [72–74]. Specifically with regard to defective NCF1-caused type I IFN responses, it has been shown that hypoactive *NCF1* variant leads to elevated type I IFN responses and the expansion of germinal center (GC) B cells in mice [11]. Given that type I IFN is capable to induce B cells to express CD38 [75] that prevents apoptosis of GC B cells [76], and that B cells from CDG patients expressed higher levels of CD38 [25], it is conceivable that type I IFN response contributes to the NCF1-associated autoimmunity via inhibiting apoptosis of autoreactive GC B cells.

5. Conclusions

The meta-analysis in this study demonstrates that the *GTFI-NCF1* susceptibility locus is associated with multiple autoimmune diseases. The causal *NCF1* His90 variant leads to an impaired NOX2 activity and confers a risk for various autoimmune conditions. Experimental and clinical evidence suggests that the hypoactive *NCF1* His90 variant might contribute to autoimmune conditions via multiple pathways, including impairing apoptotic cell clearance, regulating the formation of N.

ETs, promoting the differentiation of autoreactive T cells, and increasing type I IFN responses. In addition, since NOX2-derived ROS play an important role in autophagy and aberrations of autophagy has been observed in autoimmune diseases, such as SLE [77,78], a potential mechanism underlying hypoactive *NCF1* variant-caused autoimmune conditions might be regulating the homeostasis of autophagy. Of note, the *NCF1* His90 variant might contribute to different autoimmune diseases via different mechanisms, e.g., impairment of apoptotic cell clearance likely contributes to SLE. Since NOX2 is expressed in all phagocytes, the hypoactive His90 variant might contribute to autoimmune diseases via other mechanisms which need to be explored in future studies. In addition, given that susceptibility polymorphisms may impact the treatment of disease [79], it is interesting to explore the association of *NCF1* His90 variant with response to treatment in autoimmune diseases. In conclusion, the identification of the autoimmune disease risk *NCF1* His90 variant demonstrates an essential role of NOX2-derived ROS in the development of autoimmune diseases.

Author Contributions: Study design: X.Y.; Data acquisition: L.Z., J.W. and R.H.; Data analysis and interpretation: L.Z.; Writing: X.Y. and F.P. All authors have read and agreed to the published version of the manuscript.

Funding: This study was supported by Deutsche Forschungsgemeinschaft (DFG) project YU 142/1-3 (No. 272606465) and by funding of German Center for Lung Research (DZL), Airway Research Center North (ARCN).

Institutional Review Board Statement: Not applicable.

Informed Consent Statement: Not applicable.

Data Availability Statement: Not applicable.

Conflicts of Interest: The authors declare no conflict of interest.

References

1. Baranzini, S.E.; Oksenberg, J.R. The Genetics of Multiple Sclerosis: From 0 to 200 in 50 Years. *Trends Genet.* **2017**, *33*, 960–970. [CrossRef] [PubMed]
2. Sawcer, S.; Franklin, R.J.; Ban, M. Multiple sclerosis genetics. *Lancet Neurol.* **2014**, *13*, 700–709. [CrossRef]
3. Li, Y.; Zhang, K.; Chen, H.; Sun, F.; Xu, J.; Wu, Z.; Li, P.; Zhang, L.; Du, Y.; Luan, H.; et al. A genome-wide association study in Han Chinese identifies a susceptibility locus for primary Sjögren's syndrome at 7q11.23. *Nat. Genet.* **2013**, *45*, 1361–1365. [CrossRef] [PubMed]
4. Zheng, J.; Huang, R.; Huang, Q.; Deng, F.; Chen, Y.; Yin, J.; Chen, J.; Wang, Y.; Shi, G.; Gao, X.; et al. The GTF2I rs117026326 polymorphism is associated with anti-SSA-positive primary Sjögren's syndrome. *Rheumatology* **2015**, *54*, 562–564. [CrossRef]
5. Li, Y.; Li, P.; Chen, S.; Wu, Z.; Li, J.; Zhang, S.; Cao, C.; Wang, L.; Liu, B.; Zhang, F.; et al. Association of GTF2I and GTF2IRD1 polymorphisms with systemic lupus erythematosus in a Chinese Han population. *Clin. Exp. Rheumatol.* **2015**, *33*, 632–638.
6. Liu, C.; Yan, S.; Chen, H.; Wu, Z.; Li, L.; Cheng, L.; Li, H.; Li, Y. Association of GTF2I, NFKB1, and TYK2 Regional Polymorphisms with Systemic Sclerosis in a Chinese Han Population. *Front. Immunol.* **2021**, *12*, 640083. [CrossRef]
7. Liang, H.; Gao, W.; Liu, X.; Liu, J.; Mao, X.; Yang, M.; Long, X.; Zhou, Y.; Zhang, Q.; Zhu, J.; et al. The GTF2I rs117026326 polymorphism is associated with neuromyelitis optica spectrum disorder but not with multiple sclerosis in a Northern Han Chinese population. *J. Neuroimmunol.* **2019**, *337*, 577045. [CrossRef]
8. Olsson, L.M.; Johansson, C.; Gullstrand, B.; Jönsen, A.; Saevarsdottir, S.; Rönnblom, L.; Leonard, D.; Wetterö, J.; Sjöwall, C.; Svenungsson, E.; et al. A single nucleotide polymorphism in the *NCF1* gene leading to reduced oxidative burst is associated with systemic lupus erythematosus. *Ann. Rheum. Dis.* **2017**, *76*, 1607–1613. [CrossRef]
9. Zhao, J.; Ma, J.; Deng, Y.; Kelly, J.; Kim, K.; Bang, S.-Y.; Lee, H.-S.; Li, Q.-Z.; Wakeland, Q.-Z.L.E.K.; Qiu, R.; et al. A missense variant in NCF1 is associated with susceptibility to multiple autoimmune diseases. *Nat. Genet.* **2017**, *49*, 433–437. [CrossRef]
10. Linge, P.; Arve, S.; Olsson, L.M.; Leonard, D.; Sjöwall, C.; Frodlund, M.; Gunnarsson, I.; Svenungsson, E.; Tydén, H.; Jönsen, A.; et al. NCF1-339 polymorphism is associated with altered formation of neutrophil extracellular traps, high serum interferon activity and antiphospholipid syndrome in systemic lupus erythematosus. *Ann. Rheum. Dis.* **2020**, *79*, 254–261. [CrossRef]
11. Geng, L.; Zhao, J.; Deng, Y.; Molano, I.; Xu, X.; Xu, L.; Ruiz, P.; Li, Q.; Feng, X.; Zhang, M.; et al. Human SLE variant *NCF1-R90H* promotes kidney damage and murine lupus through enhanced Tfh2 responses induced by defective efferocytosis of macrophages. *Ann. Rheum. Dis.* **2022**, *81*, 255–267. [CrossRef]
12. Chen, Y.; Li, S.; Huang, R.; Zhang, Z.; Petersen, F.; Zheng, J.; Yu, X. Comprehensive meta-analysis reveals an association of the HLA-DRB1*1602 allele with autoimmune diseases mediated predominantly by autoantibodies. *Autoimmun. Rev.* **2020**, *19*, 102532. [CrossRef]

13. Lessard, C.J.; Registry, F.U.P.S.S.; Li, H.; Adrianto, I.; A Ice, J.; Rasmussen, A.; Grundahl, K.M.; Kelly, J.; Dozmorov, M.; Miceli-Richard, C.; et al. Variants at multiple loci implicated in both innate and adaptive immune responses are associated with Sjögren's syndrome. *Nat. Genet.* **2013**, *45*, 1284–1292. [CrossRef]
14. Meng, Y.; He, Y.; Zhang, J.; Xie, Q.; Yang, M.; Chen, Y.; Wu, Y. Association of GTF2I gene polymorphisms with renal involvement of systemic lupus erythematosus in a Chinese population. *Medicine* **2019**, *98*, e16716. [CrossRef]
15. Zhou, H.-Y.; Xie, J.-L.; Liu, J.; Lian, Z.-Y.; Chen, H.-X.; Shi, Z.-Y.; Zhang, Q.; Feng, H.-R.; Du, Q.; Miao, X.-H. Association of GTF2IRD1–GTF2I polymorphisms with neuromyelitis optica spectrum disorders in Han Chinese patients. *Neural Regen. Res.* **2019**, *14*, 346–353. [CrossRef]
16. Song, I.W.; Chen, H.C.; Lin, Y.F.; Yang, J.-H.; Chang, C.-C.; Chou, C.-T.; Lee, M.-T.M.; Chou, Y.-C.; Chen, C.-H.; Chen, Y.-T.; et al. Identification of susceptibility gene associated with female primary Sjögren's syndrome in Han Chinese by genome-wide association study. *Hum. Genet.* **2016**, *135*, 1287–1294. [CrossRef]
17. Yokoyama, N.; Kawasaki, A.; Matsushita, T.; Furukawa, H.; Kondo, Y.; Hirano, F.; Sada, K.-E.; Matsumoto, I.; Kusaoi, M.; Amano, H.; et al. Association of NCF1 polymorphism with systemic lupus erythematosus and systemic sclerosis but not with ANCA-associated vasculitis in a Japanese population. *Sci. Rep.* **2019**, *9*, 16366. [CrossRef]
18. Löhr, S.; Ekici, A.B.; Uebe, S.; Büttner, C.; Köhm, M.; Behrens, F.; Böhm, B.; Sticherling, M.; Schett, G.; Simon, D.; et al. Analyses of association of psoriatic arthritis and psoriasis vulgaris with functional NCF1 variants. *Rheumatology* **2019**, *58*, 915–917. [CrossRef]
19. Shu, Y.; Guo, J.; Ma, X.; Yan, Y.; Wang, Y.; Chen, C.; Sun, X.; Wang, H.; Yin, J.; Long, Y.; et al. Anti-N-methyl-D-aspartate receptor (NMDAR) encephalitis is associated with IRF7, BANK1 and TBX21 polymorphisms in two populations. *Eur. J. Neurol.* **2021**, *28*, 595–601. [CrossRef]
20. Shu, Y.; Ma, X.; Chen, C.; Wang, Y.; Sun, X.; Zhang, L.; Lu, Z.; Petersen, F.; Qiu, W.; Yu, X. Myelin oligodendrocyte glycoprotein-associated disease is associated with BANK1, RNASET2 and TNIP1 polymorphisms. *J. Neuroimmunol.* **2022**, *577937, in press.* [CrossRef]
21. Li, X.J.; Marchal, C.C.; Stull, N.D.; Stahelin, R.V.; Dinauer, M.C. p47phox Phox Homology Domain Regulates Plasma Membrane but Not Phagosome Neutrophil NADPH Oxidase Activation. *J. Biol. Chem.* **2010**, *285*, 35169–35179. [CrossRef]
22. Olsson, L.M.; Nerstedt, A.; Lindqvist, A.-K.; Johansson, C.; Medstrand, P.; Olofsson, P.; Holmdahl, R. Copy Number Variation of the Gene *NCF1* Is Associated with Rheumatoid Arthritis. *Antioxid. Redox Signal.* **2012**, *16*, 71–78. [CrossRef]
23. Hultqvist, M.; Olofsson, P.; Holmberg, J.; Backstrom, B.T.; Tordsson, J.; Holmdahl, R. Enhanced autoimmunity, arthritis, and encephalomyelitis in mice with a reduced oxidative burst due to a mutation in the Ncf1 gene. *Proc. Natl. Acad. Sci. USA* **2004**, *101*, 12646–12651. [CrossRef]
24. Olofsson, P.; Holmberg, J.; Tordsson, J.; Lu, S.; Akerström, B.; Holmdahl, R. Positional identification of Ncf1 as a gene that regulates arthritis severity in rats. *Nat. Genet.* **2003**, *33*, 25–32. [CrossRef]
25. Kelkka, T.; Kienhöfer, D.; Hoffmann, M.; Linja, M.; Wing, K.; Sareila, O.; Hultqvist, M.; Laajala, E.; Chen, Z.; Vasconcelos, J.; et al. Reactive Oxygen Species Deficiency Induces Autoimmunity with Type 1 Interferon Signature. *Antioxid. Redox Signal.* **2014**, *21*, 2231–2245. [CrossRef]
26. Scherlinger, M.; Tsokos, G.C. Reactive oxygen species: The Yin and Yang in (auto-)immunity. *Autoimmun. Rev.* **2021**, *20*, 102869. [CrossRef]
27. Vermot, A.; Petit-Härtlein, I.; Smith, S.; Fieschi, F. NADPH Oxidases (NOX): An Overview from Discovery, Molecular Mechanisms to Physiology and Pathology. *Antioxidants* **2021**, *10*, 890. [CrossRef]
28. Fernandez-Boyanapalli, R.F.; Frasch, S.C.; McPhillips, K.; Vandivier, R.W.; Harry, B.L.; Riches, D.W.H.; Henson, P.M.; Bratton, D.L. Impaired apoptotic cell clearance in CGD due to altered macrophage programming is reversed by phosphatidylserine-dependent production of IL-4. *Blood* **2009**, *113*, 2047–2055. [CrossRef]
29. Petersen, F.; Yue, X.; Riemekasten, G.; Yu, X. Dysregulated homeostasis of target tissues or autoantigens—A novel principle in autoimmunity. *Autoimmun. Rev.* **2017**, *16*, 602–611. [CrossRef]
30. Ren, Y.; Tang, J.; Mok, M.Y.; Chan, A.W.K.; Wu, A.; Lau, C.S. Increased apoptotic neutrophils and macrophages and impaired macrophage phagocytic clearance of apoptotic neutrophils in systemic lupus erythematosus. *Arthritis Rheum.* **2003**, *48*, 2888–2897. [CrossRef]
31. Tas, S.W.; Quartier, P.; Botto, M.; Fossati-Jimack, L. Macrophages from patients with SLE and rheumatoid arthritis have defective adhesion in vitro, while only SLE macrophages have impaired uptake of apoptotic cells. *Ann. Rheum. Dis.* **2006**, *65*, 216–221. [CrossRef] [PubMed]
32. Cohen, P.L.; Caricchio, R.; Abraham, V.; Camenisch, T.D.; Jennette, J.C.; Roubey, R.A.; Earp, H.S.; Matsushima, G.; Reap, E.A. Delayed Apoptotic Cell Clearance and Lupus-like Autoimmunity in Mice Lacking the c-mer Membrane Tyrosine Kinase. *J. Exp. Med.* **2002**, *196*, 135–140. [CrossRef] [PubMed]
33. Hanayama, R.; Tanaka, M.; Miyasaka, K.; Aozasa, K.; Koike, M.; Uchiyama, Y.; Nagata, S. Autoimmune Disease and Impaired Uptake of Apoptotic Cells in MFG-E8-Deficient Mice. *Science* **2004**, *304*, 1147–1150. [CrossRef] [PubMed]
34. Ramirez-Ortiz, Z.G.; Iii, W.F.P.; Prasad, A.; Byrne, M.H.; Iram, T.; Blanchette, C.J.; Luster, A.D.; Hacohen, N.; El Khoury, J.; Means, T.K.; et al. The scavenger receptor SCARF1 mediates the clearance of apoptotic cells and prevents autoimmunity. *Nat. Immunol.* **2013**, *14*, 917–926. [CrossRef]
35. Kaplan, M.J. Neutrophils in the pathogenesis and manifestations of SLE. *Nat. Rev. Rheumatol.* **2011**, *7*, 691–699. [CrossRef]
36. Kaplan, M.J. Role of neutrophils in systemic autoimmune diseases. *Arthritis Res. Ther.* **2013**, *15*, 219. [CrossRef]

37. Brandt, L.; Hedberg, H. Impaired Phagocytosis by Peripheral Blood Granulocytes in Systemic Lupus Erythematosus. *Scand. J. Haematol.* **1969**, *6*, 348–353. [CrossRef]
38. Courtney, P.A.; Crockard, A.D.; Williamson, K.; Irvine, A.E.; Kennedy, R.J.; Bell, A.L. Increased apoptotic peripheral blood neutrophils in systemic lupus erythematosus: Relations with disease activity, antibodies to double stranded DNA, and neutropenia. *Ann. Rheum. Dis.* **1999**, *58*, 309–314. [CrossRef]
39. Denny, M.F.; Yalavarthi, S.; Zhao, W.; Thacker, S.G.; Anderson, M.; Sandy, A.R.; Joseph McCune, W.; Kaplan, M.J. A distinct subset of proinflammatory neutrophils isolated from patients with systemic lupus erythematosus induces vascular damage and synthesizes type I IFNs. *J. Immunol.* **2010**, *184*, 3284–3297. [CrossRef]
40. Garcia-Romo, G.S.; Caielli, S.; Vega, B.; Connolly, J.; Allantaz, F.; Xu, Z.; Punaro, M.; Baisch, J.; Guiducci, C.; Coffman, R.L.; et al. Netting Neutrophils Are Major Inducers of Type I IFN Production in Pediatric Systemic Lupus Erythematosus. *Sci. Transl. Med.* **2011**, *3*, 73ra20. [CrossRef]
41. Bosch, X. Systemic Lupus Erythematosus and the Neutrophil. *N. Engl. J. Med.* **2011**, *365*, 758–760. [CrossRef]
42. Lee, K.H.; Kronbichler, A.; Park, D.D.-Y.; Park, Y.; Moon, H.; Kim, H.; Choi, J.H.; Choi, Y.; Shim, S.; Lyu, I.S.; et al. Neutrophil extracellular traps (NETs) in autoimmune diseases: A comprehensive review. *Autoimmun. Rev.* **2017**, *16*, 1160–1173. [CrossRef]
43. Hakkim, A.; Fürnrohr, B.G.; Amann, K.; Laube, B.; Abed, U.A.; Brinkmann, V.; Herrmann, M.; Voll, R.E.; Zychlinsky, A. Impairment of neutrophil extracellular trap degradation is associated with lupus nephritis. *Proc. Natl. Acad. Sci. USA* **2010**, *107*, 9813–9818. [CrossRef]
44. Leffler, J.; Martin, M.; Gullstrand, B.; Tydén, H.; Lood, C.; Truedsson, L.; Bengtsson, A.A.; Blom, A.M. Neutrophil Extracellular Traps That Are Not Degraded in Systemic Lupus Erythematosus Activate Complement Exacerbating the Disease. *J. Immunol.* **2012**, *188*, 3522–3531. [CrossRef]
45. Gestermann, N.; Di Domizio, J.; Lande, R.; Demaria, O.; Frasca, L.; Feldmeyer, L.; Di Lucca, J.; Gilliet, M. Netting Neutrophils Activate Autoreactive B Cells in Lupus. *J. Immunol.* **2018**, *200*, 3364–3371. [CrossRef]
46. Kienhöfer, D.; Hahn, J.; Stoof, J.; Csepregi, J.Z.; Reinwald, C.; Urbonaviciute, V.; Johnsson, C.; Maueröder, C.; Podolska, M.J.; Biermann, M.; et al. Experimental lupus is aggravated in mouse strains with impaired induction of neutrophil extracellular traps. *JCI Insight* **2017**, *2*, e92920. [CrossRef]
47. Knight, J.S.; Subramanian, V.; A O'Dell, A.; Yalavarthi, S.; Zhao, W.; Smith, C.K.; Hodgin, J.B.; Thompson, P.R.; Kaplan, M.J. Peptidylarginine deiminase inhibition disrupts NET formation and protects against kidney, skin and vascular disease in lupus-prone MRL/*lpr* mice. *Ann. Rheum. Dis.* **2015**, *74*, 2199–2206. [CrossRef]
48. Gordon, R.; Herter, J.M.; Rosetti, F.; Campbell, A.; Nishi, H.; Kashgarian, M.; Bastacky, S.I.; Marinov, A.; Nickerson, K.; Mayadas, T.N.; et al. Lupus and proliferative nephritis are PAD4 independent in murine models. *JCI Insight* **2017**, *2*, e92926. [CrossRef]
49. Fuchs, T.A.; Abed, U.; Goosmann, C.; Hurwitz, R.; Schulze, I.; Wahn, V.; Weinrauch, Y.; Brinkmann, V.; Zychlinsky, A. Novel cell death program leads to neutrophil extracellular traps. *J. Cell Biol.* **2007**, *176*, 231–241. [CrossRef]
50. Vorobjeva, N.; Prikhodko, A.; Galkin, I.; Pletjushkina, O.; Zinovkin, R.; Sud'Ina, G.; Chernyak, B.; Pinegin, B. Mitochondrial reactive oxygen species are involved in chemoattractant-induced oxidative burst and degranulation of human neutrophils in vitro. *Eur. J. Cell Biol.* **2017**, *96*, 254–265. [CrossRef]
51. Lood, C.; Blanco, L.P.; Purmalek, M.M.; Carmona-Rivera, C.; De Ravin, S.S.; Smith, C.K.; Malech, H.L.; A Ledbetter, J.; Elkon, K.B.; Kaplan, M.J. Neutrophil extracellular traps enriched in oxidized mitochondrial DNA are interferogenic and contribute to lupus-like disease. *Nat. Med.* **2016**, *22*, 146–153. [CrossRef] [PubMed]
52. Wirestam, L.; Arve, S.; Linge, P.; Bengtsson, A.A. Neutrophils—Important Communicators in Systemic Lupus Erythematosus and Antiphospholipid Syndrome. *Front. Immunol.* **2019**, *10*, 2734. [CrossRef] [PubMed]
53. Barrera, M.-J.; Aguilera, S.; Castro, I.; Carvajal, P.; Jara, D.; Molina, C.; González, S.; González, M.-J. Dysfunctional mitochondria as critical players in the inflammation of autoimmune diseases: Potential role in Sjögren's syndrome. *Autoimmun. Rev.* **2021**, *20*, 102867. [CrossRef] [PubMed]
54. Sareila, O.; Hagert, C.; Kelkka, T.; Linja, M.; Xu, B.; Kihlberg, J.; Holmdahl, R. Reactive Oxygen Species Regulate Both Priming and Established Arthritis, but with Different Mechanisms. *Antioxid. Redox Signal.* **2017**, *27*, 1473–1490. [CrossRef] [PubMed]
55. Akiba, H.; Takeda, K.; Kojima, Y.; Usui, Y.; Harada, N.; Yamazaki, T.; Ma, J.; Tezuka, K.; Yagita, H.; Okumura, K. The Role of ICOS in the CXCR5+ Follicular B Helper T Cell Maintenance In Vivo. *J. Immunol.* **2005**, *175*, 2340–2348. [CrossRef]
56. Kelchtermans, H.; Schurgers, E.; Geboes, L.; Mitera, T.; Van, D.J.; Van, S.J.; Uyttenhove, C.; Matthys, P. Effector mechanisms of interleukin-17 in collagen-induced arthritis in the absence of interferon-gamma and counteraction by interferon-gamma. *Arthritis Res. Ther.* **2009**, *11*, R122. [CrossRef] [PubMed]
57. Gelderman, K.A.; Hultqvist, M.; Holmberg, J.; Olofsson, P.; Holmdahl, R. T cell surface redox levels determine T cell reactivity and arthritis susceptibility. *Proc. Natl. Acad. Sci. USA* **2006**, *103*, 12831–12836. [CrossRef]
58. Gelderman, K.A.; Hultqvist, M.; Pizzolla, A.; Zhao, M.; Nandakumar, K.S.; Mattsson, R.; Holmdahl, R. Macrophages suppress T cell responses and arthritis development in mice by producing reactive oxygen species. *J. Clin. Investig.* **2007**, *117*, 3020–3028. [CrossRef]
59. Kraaij, M.D.; Savage, N.D.L.; van der Kooij, S.W.; Koekkoek, K.; Wang, J.; van den Berg, J.M.; Ottenhoff, T.H.M.; Kuijpers, T.W.; Holmdahl, R.; van Kooten, C.; et al. Induction of regulatory T cells by macrophages is dependent on production of reactive oxygen species. *Proc. Natl. Acad. Sci. USA* **2010**, *107*, 17686–17691. [CrossRef]
60. Ivashkiv, L.B.; Donlin, L.T. Regulation of type I interferon responses. *Nat. Rev. Immunol.* **2014**, *14*, 36–49. [CrossRef]

61. Rossi, M.; Castiglioni, P.; Hartley, M.-A.; Eren, R.O.; Prével, F.; Desponds, C.; Utzschneider, D.T.; Zehn, D.; Cusi, M.G.; Kuhlmann, F.M.; et al. Type I interferons induced by endogenous or exogenous viral infections promote metastasis and relapse of leishmaniasis. *Proc. Natl. Acad. Sci. USA* **2017**, *114*, 4987–4992. [CrossRef]
62. Crow, M.K.; Olferiev, M.; Kirou, K.A. Type I Interferons in Autoimmune Disease. *Annu. Rev. Pathol.* **2019**, *14*, 369–393. [CrossRef]
63. Psarras, A.; Emery, P.; Vital, E.M. Type I interferon-mediated autoimmune diseases: Pathogenesis, diagnosis and targeted therapy. *Rheumatology* **2017**, *56*, 1662–1675. [CrossRef]
64. A Bengtsson, A.; Sturfelt, G.; Truedsson, L.; Blomberg, J.; Alm, G.; Vallin, H.; Rönnblom, L. Activation of type I interferon system in systemic lupus erythematosus correlates with disease activity but not with antiretroviral antibodies. *Lupus* **2000**, *9*, 664–671. [CrossRef]
65. Lee, M.H.; Chakhtoura, M.; Sriram, U.; Caricchio, R.; Gallucci, S. Conventional DCs from Male and Female Lupus-Prone B6.NZM Sle1/Sle2/Sle3 Mice Express an IFN Signature and Have a Higher Immunometabolism That Are Enhanced by Estrogen. *J. Immunol. Res.* **2018**, *2018*, 1601079. [CrossRef]
66. Braun, D.; Geraldes, P.; Demengeot, J. Type I Interferon controls the onset and severity of autoimmune manifestations in lpr mice. *J. Autoimmun.* **2003**, *20*, 15–25. [CrossRef]
67. Fairhurst, A.M.; Mathian, A.; Connolly, J.E.; Wang, A.; Gray, H.F.; George, T.A.; Boudreaux, C.D.; Zhou, X.J.; Li, Q.Z.; Koutouzov, S.; et al. Systemic IFN-alpha drives kidney nephritis in B6.Sle123 mice. *Eur. J. Immunol.* **2008**, *38*, 1948–1960. [CrossRef]
68. Hjelmervik, T.O.R.; Petersen, K.; Jonassen, I.; Jonsson, R.; Bolstad, A.I. Gene expression profiling of minor salivary glands clearly distinguishes primary Sjögren's syndrome patients from healthy control subjects. *Arthritis Rheum.* **2005**, *52*, 1534–1544. [CrossRef]
69. Kimoto, O.; Sawada, J.; Shimoyama, K.; Suzuki, D.; Nakamura, S.; Hayashi, H.; Ogawa, N. Activation of the Interferon Pathway in Peripheral Blood of Patients with Sjögren's Syndrome. *J. Rheumatol.* **2011**, *38*, 310–316. [CrossRef]
70. Lübbers, J.; Brink, M.; A van de Stadt, L.; Vosslamber, S.; Wesseling, J.G.; van Schaardenburg, D.; Rantapää-Dahlqvist, S.; Verweij, C.L. The type I IFN signature as a biomarker of preclinical rheumatoid arthritis. *Ann. Rheum. Dis.* **2013**, *72*, 776–780. [CrossRef]
71. Thurlings, R.M.; Boumans, M.; Tekstra, J.; Van Roon, J.A.; Vos, K.; Van Westing, D.M.; Van Baarsen, L.G.; Bos, C.; Kirou, K.A.; Gerlag, D.M.; et al. Relationship between the type I interferon signature and the response to rituximab in rheumatoid arthritis patients. *Arthritis Rheum.* **2010**, *62*, 3607–3614. [CrossRef]
72. Jego, G.; Palucka, A.; Blanck, J.-P.; Chalouni, C.; Pascual, V.; Banchereau, J. Plasmacytoid Dendritic Cells Induce Plasma Cell Differentiation through Type I Interferon and Interleukin 6. *Immunity* **2003**, *19*, 225–234. [CrossRef]
73. Le Bon, A.; Thompson, C.; Kamphuis, E.; Durand, V.; Rossmann, C.; Kalinke, U.; Tough, D.F. Cutting Edge: Enhancement of Antibody Responses Through Direct Stimulation of B and T Cells by Type I IFN. *J. Immunol.* **2006**, *176*, 2074–2078. [CrossRef]
74. Longhi, P.; Trumpfheller, C.; Idoyaga, J.; Caskey, M.; Matos, I.; Kluger, C.; Salazar, A.M.; Colonna, M.; Steinman, R.M. Dendritic cells require a systemic type I interferon response to mature and induce CD4+ Th1 immunity with poly IC as adjuvant. *J. Exp. Med.* **2009**, *206*, 1589–1602. [CrossRef]
75. Galibert, L.; Burdin, N.; De Saint-Vis, B.; Garrone, P.; van Kooten, C.; Banchereau, J.; Rousset, F. CD40 and B cell antigen receptor dual triggering of resting B lymphocytes turns on a partial germinal center phenotype. *J. Exp. Med.* **1996**, *183*, 77–85. [CrossRef]
76. Zupo, S.; Rugari, E.; Dono, M.; Taborelli, G.; Malavasi, F.; Ferrarini, M. CD38 signaling by agonistic monoclonal antibody prevents apoptosis of human germinal center B cells. *Eur. J. Immunol.* **1994**, *24*, 1218–1222. [CrossRef]
77. Qi, Y.; Zhou, X.; Zhang, H. Autophagy and immunological aberrations in systemic lupus erythematosus. *Eur. J. Immunol.* **2019**, *49*, 523–533. [CrossRef]
78. Xu, Y.; Shen, J.; Ran, Z. Emerging views of mitophagy in immunity and autoimmune diseases. *Autophagy* **2020**, *16*, 3–17. [CrossRef]
79. Murdaca, G.; Gulli, R.; Spanò, F.; Lantieri, F.; Burlando, M.; Parodi, A.; Mandich, P.; Puppo, F. TNF-α gene polymorphisms: Association with disease susceptibility and response to anti-TNF-α treatment in psoriatic arthritis. *J. Investig. Dermatol.* **2014**, *134*, 2503–2509. [CrossRef]

Communication

Significance of Serum Oxidative and Antioxidative Status in Congenital Central Hypoventilation Syndrome (CCHS) Patients

Elisabetta Bigagli [1], Maura Lodovici [1], Marzia Vasarri [2], Marta Peruzzi [3], Niccolò Nassi [3] and Donatella Degl'Innocenti [2,*]

[1] Department of Neuroscience, Psychology, Drug Research and Child Health (NEUROFARBA), Section of Pharmacology and Toxicology, University of Florence, Viale Pieraccini 6, 50139 Florence, Italy; elisabetta.bigagli@unifi.it (E.B.); maura.lodovici@unifi.it (M.L.)

[2] Department of Experimental and Clinical Biomedical Sciences "Mario Serio", University of Florence, Viale Morgagni 50, 50134 Florence, Italy; marzia.vasarri@unifi.it

[3] Sleep Breathing Disorders and SIDS Centre, A. Meyer Children's Hospital, Viale Pieraccini 24, 50139 Florence, Italy; marta.peruzzi@meyer.it (M.P.); niccolo.nassi@meyer.it (N.N.)

* Correspondence: donatella.deglinnocenti@unifi.it

Abstract: Congenital central hypoventilation syndrome (CCHS) is a rare neurological genetic disorder that affects sleep-related respiratory control. Currently, no drug therapy is available. In light of this, there is a need for lifelong ventilation support, at least during sleep, for these patients. The pathogenesis of several chronic diseases is influenced by oxidative stress. Thus, determining oxidative stress in CCHS may indicate further disorders in the course of this rare genetic disease. Liquid biopsies are widely used to assess circulating biomarkers of oxidative stress. In this study, ferric reducing ability of plasma, thiobarbituric acid-reactive substances, advanced oxidation protein products (AOPPs), and advanced glycation end-products were measured in the serum of CCHS patients to investigate the relationship between oxidative stress and CCHS and the significance of this balance in CCHS. Here, AOPPs were found to be the most relevant serum biomarker to monitor oxidative stress in CCHS patients. According to this communication, CCHS patients may suffer from other chronic pathophysiological processes because of the persistent levels of AOPPs.

Keywords: CCHS; rare disease; oxidative stress; circulating biomarkers; TBARS; FRAP; AOPP

Citation: Bigagli, E.; Lodovici, M.; Vasarri, M.; Peruzzi, M.; Nassi, N.; Degl'Innocenti, D. Significance of Serum Oxidative and Antioxidative Status in Congenital Central Hypoventilation Syndrome (CCHS) Patients. *Antioxidants* **2022**, *11*, 1497. https://doi.org/10.3390/antiox11081497

Academic Editor: Stanley Omaye

Received: 14 July 2022
Accepted: 28 July 2022
Published: 30 July 2022

Publisher's Note: MDPI stays neutral with regard to jurisdictional claims in published maps and institutional affiliations.

1. Introduction

The development and the progression of several chronic diseases are often associated with oxidative stress (OS), a phenomenon defined as an imbalance between the formation of reactive oxygen species (ROS) and antioxidant defense systems [1,2]. ROS are biomarkers of OS; however, being highly reactive and with a short half-life, the measurement of the products of oxidative damage caused by ROS to lipids, proteins, and nucleic acids is the preferred approach to assess oxidative stress in clinical samples [3].

Advanced oxidation protein products (AOPPs), derived from the action of hypochlorous acid and chloramine produced by neutrophil myeloperoxidase, are regarded as both oxidative stress and inflammatory biomarkers in clinical samples [4]. Lipid peroxidation is commonly used as an indicator of ROS-induced damage to polyunsaturated fatty acids in the cell membrane [5]. Malondialdehyde (MDA) is one of the best-studied end-products of lipid peroxidation and can be measured using thiobarbituric acid-reactive substances (TBARS) [6].

Advanced glycation end-products (AGEs) are heterogeneous compounds generated from the nonenzymatic glycation of macromolecules (proteins, nucleic acids, and lipids) through the Maillard reaction; despite being often associated with hyperglycemia, increased AGEs have also been documented in normoglycemic subjects suffering from other conditions characterized by increased OS such as inflammatory diseases and obstructive sleep apnea [4,7].

Physiological homeostasis is maintained by sophisticated enzymatic and nonenzymatic antioxidant defenses that counteract excessive ROS levels [8]. The FRAP assay measures the ferric reducing ability of plasma, provides estimates of the total nonenzymatic antioxidants in plasma, and represents an indication of the intrinsic ability of plasma to prevent and limit the negative effects of oxidative stress [9]. FRAP values have been found to correlate with the clinical status in several metabolic, cardiovascular, and respiratory diseases [10–13].

Congenital central hypoventilation syndrome (CCHS) is a rare autosomal dominant genetic disorder (ORPHA:661) that affects respiratory control (with an incidence of about 1/200,000 live births [14,15]. Heterozygous polyalanine repeat expansion (PolyALA) in the paired-like homeobox 2b (*PHOX2B*) gene is present in 90% of CCHS patients [16]. *PHOX2B* is a key gene for the neuronal formation and differentiation, especially in the autonomic nervous system. Indeed, CCHS patients manifest episodes of apnea/hypopnea during sleep and during routine activities of daily living. This pathological condition results from abnormal control of breathing with reduced sensitivity to hypercapnia and hypoxemia [17]. CCHS has no resolving drug therapy [16,18], but lifelong ventilation support during sleep and, in severe cases, also during wakefulness supplants the "forgotten breathing" in CCHS patients [14,15], allowing CCHS patients to live a dignified life.

In 2018, we demonstrated for the first time that erythrocytes and leukocyte subpopulations of CCHS patients had increased levels of ROS [19].

Blood cells reflect the condition of the whole organism and are an outstanding model for evaluating oxidative status, but ROS determination in blood cells requires complex and time-consuming techniques such as cytofluorimetric analysis. Therefore, there is much emphasis on circulating biomarkers of oxidative stress because they can be easily examined in liquid biopsies (e.g., serum and urine), whereas determination of ROS levels requires very rapid sample processing (within 2 h from collection). In light of these considerations, we recently performed an HPLC–MS/MS analysis to detect the three major oxidative biomarkers, i.e., o,o'-dityrosine, malondialdehyde, and 8-hydroxy-2′-deoxyguanosine, in the CCHS urine. This study showed higher levels of urinary MDA in young adults with CCHS than in control subjects of the same age, but not in minors [20].

According to the assumption that OS contributes to the onset of chronic-degenerative diseases [21,22], the determination of OS in CCHS may indicate the risk of suffering from additional disorders during the course of the CCHS rare genetic disease [23].

Therefore, in this study, circulating antioxidant/oxidant status was evaluated by measuring FRAP, TBARS, AOPP, and AGEs in CCHS patients to evaluate the relationship between OS and CCHS, and to investigate the significance of this balance in this rare genetic disease.

2. Materials and Methods

2.1. Subject Enrollment and Sample Collection

For this study, four minor healthy controls, five adult healthy controls, and nine CCHS patients (five minors and four adults) with confirmed *PHOX2B* gene mutation were enrolled. Enrolled subjects received detailed explanations of the study, and all subjects signed a written, informed consent form; for minors, a parental written informed consent form was signed. The Institutional Review Board of Meyer Children's Hospital (Florence, Italy) approved the study protocol (Ethics Committee registration number 99/2017).

All subjects were Italian and of the same ethnicity; they had no differences in height, weight, body mass index (BMI), and lipid profiles. Healthy controls were age-matched subjects without diseases, not taking drugs or alcohol, and nonsmokers. All CCHS patients had no additional phenotypic features of CCHS (i.e., neural crest tumor or Hirschsprung's disease). Table 1 reports the demographic and clinical features of CCHS patients.

Table 1. Demographic and clinical characteristics of enrolled CCHS patients.

ID	Age	Sex	*PHOX2B* Gene Mutation	Type of Ventilation
3	30	F	PolyALA 20/25	Pacemaker/noninvasive ventilation
8	26	M	PolyALA 20/26	Pacemaker/noninvasive ventilation
7	20	F	PolyALA 20/33	Pacemaker/noninvasive ventilation
2	27	F	PolyALA 20/25	Noninvasive ventilation
5	8	M	PolyALA 20/25	Noninvasive ventilation
1	11	F	PolyALA 20/26	Tracheostomy
6	11	M	PolyALA 20/26	Tracheostomy
4	1	F	Frameshift	Tracheostomy
9	11	M	PolyALA 20/25	Noninvasive ventilation

ROS production is influenced by daily activities or individual-dependent factors [24]. CCHS patients were admitted to the hospital the day before sample collection to exclude ROS production due to acute exposure to other factors or daily activities. The respiratory parameters, i.e., apnea–hypopnea index, oxygen desaturation index, and carbon dioxide levels, of all CCHS patients were within normal limits after they underwent full nocturnal polysomnographic examinations and capnography to exclude any possible interference from recent hypoxic episodes.

All blood samples were collected fasting in the morning; for CCHS patients, blood was sampled after the polysomnography analysis. Blood was collected in nonadditive serum preparation tubes (red top). The serum tubes were kept at room temperature for 60 min before centrifugation ($1000 \times g$, 15 min). Serum samples were stored at $-20\ °C$ within 2 h of collection.

2.2. Thiobarbituric Acid-Reactive Substances (TBARS) Determination

TBARS are markers of lipid peroxidation, and they were determined as reported in our previous work [25]. Briefly, the serum sample (100 μL) was deproteinized by adding 100 μL of trichloroacetic acid (Merck KGaA, Darmstadt, DA, Germany) (1:1 v/v), and 160 μL of the resulting supernatant was added to 32 μL of thiobarbituric acid (0.12 M) (Merck KGaA, Darmstadt, DA, Germany) in 0.26 M Tris, before heating at $100\ °C$ for 15 min. A 10 min ice bath and centrifugation ($1600 \times g$, $4\ °C$, 10 min) stopped the reaction, and the absorbance of the supernatant was measured at 532 nm (Perkin Elmer Wallac 1420 Victor3 Multilabel Counter).

The molar absorption coefficient ($1.56 \times 10^{-5}\ \text{M·cm}^{-1}$) was used to calculate TBARS content. TBARS were expressed as pmol/mg protein.

2.3. Advanced Oxidation Protein Product (AOPP) Determination

For AOPP measurement, the serum sample (20 μL) was diluted in 1 mL of PBS, and then mixed with 50 μL of KI (1.16 M) and 100 μL of acetic acid. The absorbance value was immediately read at 340 nm. Chloramine-T (Merck KGaA, Darmstadt, DA, Germany), as the standard for the calibration curve, was used to quantify AOPPs. AOPPs were expressed as nmol of chloramine per mg protein [26].

2.4. Advanced Glycated End-Products (AGEs)

According to Cournot and Burillo [27], AGEs were determined by spectrofluorometric detection of fluorescent products. A 96-well black plate was loaded with 100 μL of serum, and fluorescence was measured at 460 nm after excitation at 355 nm. As a result, the results were expressed in arbitrary units (AU)/mg protein.

2.5. Ferric Reducing Ability of Plasma (FRAP) Determination

Assays based on Benzie and Strain's method were used to determine FRAP [28]. As previously described [25], serum samples (10 μL) were added to 90 μL of distilled water. A FRAP solution (300 mM acetate buffer (pH 3.6), 10 mM 2,4,6-tripyridyl-S-triazine (TPTZ) in

40 mM HCl, and 20 mM $FeCl_3 \cdot 6H_2O$ (10:1:1)) was freshly prepared incubated for 30 min at 37 °C. The ferric–tripyridyl–triazine complex is reduced into the ferrous form, developing an intense blue color. The absorbance value was measured at 595 nm using the Wallac 1420 Victor 3 Multilabel Counter (Perkin Elmer, Waltham, MA, United States). $FeSO_4 \cdot 7H_2O$ was used for the standard curve, FRAP was expressed as nmol/mg protein. Merck KGaA (Darmstadt, DA, Germany) supplied all reagents.

2.6. Statistical Analysis

Statistical analyses were performed using Graph-Pad Prism 7.00. A *p*-value < 0.05 was considered statistically significant. Normality was verified with the Kolmogorov–Smirnov test. Comparison of continuous variables between the two groups was performed using the Student's *t*-test or Mann–Whitney test as applicable according to data distribution.

3. Results and Discussion

Oxidative Damage and Antioxidant Capacity in the Serum of CCHS Patients

Biomarkers of systemic OS and total antioxidant capacity are of great value in assessing a physiologically altered or increased disease state [29].

As shown in Table 2, the mean age of CCHS patients was similar to that of the controls (15.61 ± 3.3 vs. 17.89 ± 3.62), and the two groups were also matched for sex. AOPP levels were significantly higher in the CCHS serum when compared to age- and sex-matched healthy subjects (12.18 ± 1.81 vs. 2.76 ± 0.72, *p* < 0.001). This finding is consistent with the elevated ROS levels we previously detected in blood cells of CCHS patients [19]. On the contrary, no significant differences were observed when comparing TBARS and AGE levels in the two groups; similarly, the total antioxidant power of CCHS patients was similar to that of the healthy controls.

Table 2. Demographic features, and serum biomarkers in CCHS patients and healthy controls.

Demographic Characteristics and Oxidative Markers	CCHS Patients	Healthy Controls
n	9	9
Sex (males, %)	44%	33%
Age (years)	15.61 ± 3.3	17.89 ± 3.62
TBARS (pmol/mg)	4.46 ± 1.21	2.52 ± 0.38
AOPP (nmol/mg)	12.18 ± 1.81 ***	2.76 ± 0.72
FRAP (nmol Fe^{2+}/mg)	4.40 ± 0.77	5.42 ± 1.01
AGEs (AU/mg)	1.63 ± 0.12	1.84 ± 0.22

Data are reported as the mean ± SE. *** *p* < 0.001 vs. healthy controls according to Student's *t*-test. TBARS: thiobarbituric acid-reactive substances; AOPPs: advanced oxidation protein products; FRAP: ferric reducing ability of plasma; AGEs: advanced glycation end-products; AU: arbitrary units.

Since we recently demonstrated increased urinary MDA levels in young adult CCHS compared with age-matched control subjects [20], in this study, we evaluated serum OS biomarkers and antioxidant status according to two age groups: minors (<18 years) and adults (>18 years).

As depicted in Figure 1A, AOPP levels were significantly higher in both minor CCHS patients (mean age 7.9 ± 1.73 years) compared to age-matched controls (mean age 7 ± 2.16 years) (13.35 ± 3.48 nmol/mg vs. 3.99 ± 1.24 nmol/mg; *p* < 0.05) and adult CCHS patients (mean age 25.25 ± 2.13 years) compared to adult healthy controls (mean age 26.60 ± 1.28 years) (11.01 ± 1.50 nmol/mg vs. 1.78 ± 0.63 nmol/mg; *p* < 0.05).

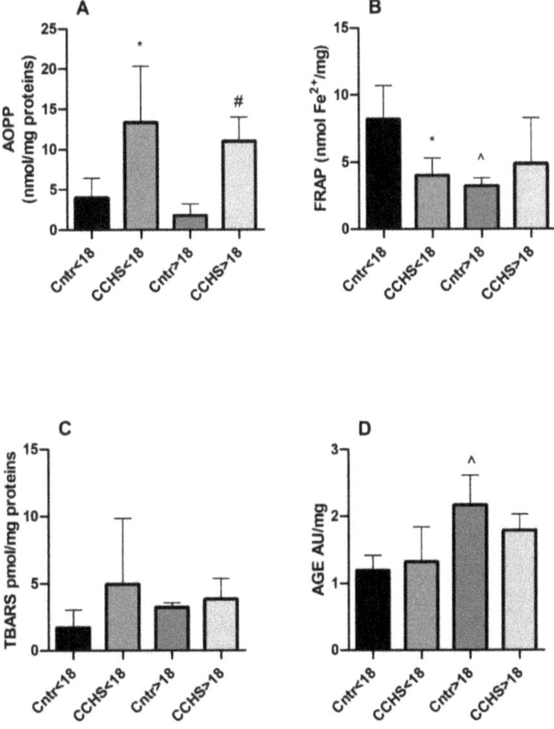

Figure 1. Levels of serum (**A**) AOPP, (**B**) FRAP, (**C**) TBARS, and (**D**) AGEs in healthy controls (Cntr) and CCHS patients aged <18 years and >18 years. Data are expressed as the mean $\pm$ SEM. * $p < 0.05$ vs. healthy minors; # $p < 0.05$ vs. healthy adults; ^ $p < 0.05$ vs. healthy minors.

AOPPs resulting from myeloperoxidase activation and hypochlorous acid production during inflammation are capable of binding to pattern recognition receptors (PRRs) whose downstream effects result in the activation of OS and inflammatory pathways including nicotinamide adenine dinucleotide phosphate (NADPH) oxidase and nuclear factor-κB (NF-κB) activation [30]. Among the PPRs activated by AOPPs, the receptor of advanced glycation end-products (RAGE) is implicated in diverse chronic inflammatory states and has been linked to vascular lesions in diabetes, renal diseases, and atherosclerosis [31,32]; AOPPs have been also found increased in patients with inflammatory conditions such as Crohn's disease [4], rheumatoid arthritis [33], and obstructive sleep apnea syndrome [34].

In addition, FRAP values were significantly lower in CCHS aged below 18 years compared to healthy minors (4.01 ± 0.57 nmol Fe^{2+}/mg vs. 8.17 ± 1.25 nmol Fe^{2+}/mg, $p < 0.05$) whereas no difference was observed when comparing FRAP levels from adult CCHS to adult controls (Figure 1B). Comparing FRAP values between healthy subjects, adults showed significantly lower levels compared to minors (3.22 ± 0.26 nmol Fe^{2+}/mg vs. 8.17 ± 1.25 nmol Fe^{2+}/mg, $p < 0.05$). Reports on FRAP values in healthy children and young adults are lacking, and data on antioxidants plasma levels are contradictory. The concentration of total thiols in the plasma of females, but not of males, progressively decreased in the group aged 20–39 compared to groups aged 6–13 and 14–19 [35]; on the contrary, a slight increase in total antioxidant capacity, measured by TAC assay, was observed in the plasma of healthy children (aged 2–14) compared to adults (aged 25 45) [36]. FRAP levels are indicative of antioxidants such as ascorbic acid, α-tocopherol, uric acid, and bilirubin, while thiols (including GSH and albumin) react slowly in this assay [9]. The relative contribution of each single antioxidant that may react in the FRAP assay deserves to be further evaluated to understand which specific antioxidant may account for the decreased FRAP levels observed in minor CCHS patients. This further analysis could also shed light on the physiological processes responsible for the reduced FRAP values in young healthy adults compared to healthy minors.

However, our data suggest that, in CCHS minors, nonenzymatic antioxidant sources were impaired, which could have contributed to the increased AOPP levels detected in these patients. We can hypothesize that while, in minors, the depletion of antioxidants could be a major contributor to AOPP formation, in adults, this seems more linked to a direct activation of inflammatory processes.

In our previous study, increased urinary MDA levels, detected by means of an HPLC–MS/MS analysis, were demonstrated in young CCHS adults compared with controls [20]. In this study, it was found that serum TBARS were not affected by the clinical status in minors or in adults (Figure 1C).

This could be explained by the different biological fluids analyzed (serum vs. urine), but most probably by the detection method; in fact, spectrophotometric TBARS determination is nonspecific for MDA since the thiobarbituric acid test measures reactive substances other than MDA, such as 2-alkenals and 2,4-alkadienals [37]. It was also demonstrated that the lack of specificity of the spectrophotometric TBARS determination limited the detection of differences in the level of lipid peroxidation in biological samples observed by HPLC detection [38].

Furthermore, this study showed that AGEs levels were not altered between patients with CCHS and healthy controls but were increased in healthy adults compared with healthy minors (Figure 1D). This event is in line with the knowledge of AGEs increase with advancing age, including the transition from pediatric to adult age, contributing to the aging process [39]. Another interesting aspect, which deserves further evaluation, is the lack of significant differences in all the measured parameters between minor and adult CCHS patients. This result suggests that OS, particularly the AOPP/RAGE axis, which operates at the crossroad between OS and inflammation, might be intrinsically linked to the CCHS pathogenesis and the role of *PHOX2B* mutations in this disease.

Overall, our results suggest increased OS in CCHS patients and that AOPPs are the most relevant serum biomarker for monitoring OS in these patients. It is noteworthy that AOPPs can cause cellular dysfunction, cell death, and tissue damage in different organs, thus representing promising biomarkers of the redox imbalance associated with CCHS and its complications.

In this study, the enrolled patients did not have hypoventilation during wakefulness; the apnea index and dyspnea index, oxygen desaturation index, and carbon dioxide levels in these patients were in the normal range during night ventilation. Therefore, we can rule out an association between the results obtained in this study and the possible ineffectiveness of artificial ventilation and sleep quality. However, to date, very little is known about the downstream cellular and molecular consequences of *PHOX2B* mutations, although some studies have suggested that transcriptional dysregulation and protein misfolding may underlie pathogenesis [18,40]

4. Conclusions

The biochemical mechanisms underlying rare diseases are generally poorly understood. In most cases, the chronic pathophysiological conditions associated with the disease, in addition to the patient's clinical manifestations, are generally asymptomatic. However, they could be responsible for the onset of other chronic collateral disorders causing patients' clinical worsening over the lifetime.

Discoveries about molecules or signaling pathways associated with these latent pathophysiological processes could provide fundamental knowledge to intervene early against worsening clinical status.

This communication adds new insights about the systemic oxidative state in CCHS patients and demonstrates for the first time that AOPPs are one of the key serum biomarkers of oxidation in CCHS.

Because AOPPs are known to be stimulating factors of inflammatory responses, this study suggests a possible involvement of the inflammatory process in CCHS. In light of these considerations, further evaluation on the inflammatory state in these patients could contribute to the understanding of the pathophysiology of this rare genetic disease.

Overall, the results obtained so far on CCHS generate hypotheses for future clinical trials aimed at improving the CCHS outcomes by hindering increased oxidative stress through the promotion of a healthy lifestyle, adequate diet, and the possible therapeutic use of antioxidant agents or pharmacological strategies targeting the AOPP/RAGE axis and activation of its downstream inflammatory pathways.

Author Contributions: Conceptualization, M.L. and D.D.; methodology, M.L. and E.B.; formal analysis, M.L., E.B. and M.V.; investigation, M.L. and E.B.; sample collection, M.P. and N.N.; data curation, M.L., E.B. and M.V.; writing—original draft preparation, M.L., E.B., M.V. and D.D.; writing—review and editing, M.L., E.B., M.V. and D.D.; supervision, M.L. and D.D.; funding acquisition, M.L. and D.D. All authors have read and agreed to the published version of the manuscript.

Funding: This research was funded by Fondi di Ateneo 2022 to M.L. and D.D.I.

Institutional Review Board Statement: The study was conducted in accordance with the Declaration of Helsinki and approved by the Institutional Review Board of Meyer Children's Hospital (Florence, Italy) (Ethics Committee registration number 99/2017).

Informed Consent Statement: Informed consent was obtained from all subjects involved in the study.

Data Availability Statement: Data is contained within the article.

Acknowledgments: The authors wish to thank A.I.S.I.C.C. and all the enrolled subjects, particularly the CCHS patients and their families, for allowing the enrolment of their children in this work.

Conflicts of Interest: The authors declare no conflict of interest.

References

1. Pizzino, G.; Irrera, N.; Cucinotta, M.; Pallio, G.; Mannino, F.; Arcoraci, V.; Squadrito, F.; Altavilla, D.; Bitto, A. Oxidative Stress: Harms and Benefits for Human Health. *Oxid. Med. Cell. Longev.* **2017**, *2017*, 8416763. [CrossRef] [PubMed]
2. Sharifi-Rad, M.; Anil Kumar, N.V.; Zucca, P.; Varoni, E.M.; Dini, L.; Panzarini, E.; Rajkovic, J.; Tsouh Fokou, P.V.; Azzini, E.; Peluso, I.; et al. Lifestyle, Oxidative Stress, and Antioxidants: Back and Forth in the Pathophysiology of Chronic Diseases. *Front. Physiol.* **2020**, *11*, 694. [CrossRef] [PubMed]
3. Dalle-Donne, I.; Rossi, R.; Colombo, R.; Giustarini, D.; Milzani, A. Biomarkers of oxidative damage in human disease. *Clin. Chem.* **2006**, *52*, 601–623. [CrossRef] [PubMed]
4. Luceri, C.; Bigagli, E.; Agostiniani, S.; Giudici, F.; Zambonin, D.; Scaringi, S.; Ficari, F.; Lodovici, M.; Malentacchi, C. Analysis of Oxidative Stress-Related Markers in Crohn's Disease Patients at Surgery and Correlations with Clinical Findings. *Antioxidants* **2019**, *8*, 378. [CrossRef]
5. Ayala, A.; Muñoz, M.F.; Argüelles, S. Lipid peroxidation: Production, metabolism, and signaling mechanisms of malondialdehyde and 4-hydroxy-2-nonenal. *Oxid. Med. Cell Longev.* **2014**, *2014*, 360438. [CrossRef]
6. Aguilar Diaz De Leon, J.; Borges, C.R. Evaluation of Oxidative Stress in Biological Samples Using the Thiobarbituric Acid Reactive Substances Assay. *J. Vis. Exp.* **2020**, *159*, e61122. [CrossRef]
7. Tan, K.C.; Chow, W.S.; Lam, J.C.; Lam, B.; Bucala, R.; Betteridge, J.; Ip, M.S. Advanced glycation endproducts in nondiabetic patients with obstructive sleep apnea. *Sleep* **2006**, *29*, 329–333. [CrossRef]
8. Finkel, T.; Holbrook, N.J. Oxidants, oxidative stress and the biology of ageing. *Nature* **2000**, *408*, 239–247. [CrossRef]
9. Benzie, I.F.; Devaki, M. The ferric reducing/antioxidant power (FRAP) assay for non-enzymatic antioxidant capacity: Concepts, procedures, limitations and applications. In *Measurement of Antioxidant Activity & Capacity*; Apak, R., Capanoglu, E., Shahidi, F., Eds.; John Wiley & Sons: Hoboken, NJ, USA, 2018. [CrossRef]
10. Causer, A.J.; Shute, J.K.; Cummings, M.H.; Shepherd, A.I.; Gruet, M.; Costello, J.T.; Bailey, S.; Lindley, M.; Pearson, C.; Connett, G.; et al. Circulating biomarkers of antioxidant status and oxidative stress in people with cystic fibrosis: A systematic review and meta-analysis. *Redox Biol.* **2020**, *32*, 101436. [CrossRef] [PubMed]
11. Lodovici, M.; Bigagli, E.; Luceri, C.; Mannucci, E.; Rotella, C.M.; Raimondi, L. Gender-related drug effect on several markers of oxidation stress in diabetes patients with and without complications. *Eur. J. Pharmacol.* **2015**, *766*, 86–90. [CrossRef] [PubMed]

12. Mancuso, M.; Bonanni, E.; LoGerfo, A.; Orsucci, D.; Maestri, M.; Chico, L.; DiCoscio, E.; Fabbrini, M.; Siciliano, G.; Murri, L. Oxidative stress biomarkers in patients with untreated obstructive sleep apnea syndrome. *Sleep Med.* **2012**, *13*, 632–636. [CrossRef]
13. Rafiei, A.; Ahmadi, R.; Kazemian, S.; Rahimzadeh-Fallah, T.; Mohammad-Rezaei, M.; Azadegan-Dehkordi, F.; Sanami, S.; Mirzaei, Y.; Aghaei, F.; Bagheri, N. Serum levels of IL-37 and correlation with inflammatory cytokines and clinical outcomes in patients with coronary artery disease. *J. Investig. Med.* **2022**, 002134. [CrossRef] [PubMed]
14. Dweik, R.A.; Boggs, P.B.; Erzurum, S.C.; Irvin, C.G.; Leigh, M.W.; Lundberg, J.O.; Olin, A.C.; Plummer, A.L.; Taylor, D.R.; American Thoracic Society Committee on Interpretation of Exhaled Nitric Oxide Levels (FENO) for Clinical Applications. An official ATS clinical practice guideline: Interpretation of exhaled nitric oxide levels (FENO) for clinical applications. *Am. J. Respir. Crit. Care Med.* **2011**, *184*, 602–615. [CrossRef]
15. Weese-Mayer, D.E.; Rand, C.M.; Zhou, A.; Carroll, M.S.; Hunt, C.E. Congenital central hypoventilation syndrome: A bedside-to-bench success story for advancing early diagnosis and treatment and improved survival and quality of life. *Pediatr. Res.* **2017**, *81*, 192–201. [CrossRef]
16. Amiel, J.; Laudier, B.; Attié-Bitach, T.; Trang, H.; de Pontual, L.; Gener, B.; Trochet, D.; Etchevers, H.; Ray, P.; Simonneau, M.; et al. Polyalanine expansion and frameshift mutations of the paired-like homeobox gene PHOX2B in congenital central hypoventilation syndrome. *Nat. Genet.* **2003**, *33*, 459–461. [CrossRef]
17. Trang, H.; Samuels, M.; Ceccherini, I.; Frerick, M.; Garcia-Teresa, M.A.; Peters, J.; Schoeber, J.; Migdal, M.; Markstrom, A.; Ottonello, G.; et al. Guidelines for diagnosis and management of congenital central hypoventilation syndrome. *Orphanet J. Rare Dis.* **2020**, *15*, 252. [CrossRef]
18. Di Lascio, S.; Benfante, R.; Cardani, S.; Fornasari, D. Research Advances on Therapeutic Approaches to Congenital Central Hypoventilation Syndrome (CCHS). *Front. Neurosci.* **2021**, *14*, 615666. [CrossRef]
19. Degl'Innocenti, D.; Becatti, M.; Peruzzi, M.; Fiorillo, C.; Ramazzotti, M.; Nassi, N.; Arzilli, C.; Piumelli, R. Systemic oxidative stress in congenital central hypoventilation syndrome. *Eur. Respir. J.* **2018**, *52*, 1801497. [CrossRef]
20. Peruzzi, M.; Ramazzotti, M.; Damiano, R.; Vasarri, M.; la Marca, G.; Arzilli, C.; Piumelli, R.; Nassi, N.; Degl'Innocenti, D. Urinary Biomarkers as a Proxy for Congenital Central Hypoventilation Syndrome Patient Follow-Up. *Antioxidants* **2022**, *11*, 929. [CrossRef]
21. Slifer, K.J.; Tucker, C.L.; Dahlquist, L.M. Helping Children and Caregivers Cope with Repeated Invasive Procedures: How Are We Doing? *J. Clin. Psychol. Med.* **2002**, *9*, 131–152. [CrossRef]
22. Il'yasova, D.; Scarbrough, P.; Spasojevic, I. Urinary biomarkers of oxidative status. *Clin. Chim. Acta* **2012**, *413*, 1446–1453. [CrossRef] [PubMed]
23. Argüelles, S.; García, S.; Maldonado, M.; Machado, A.; Ayala, A. Do the serum oxidative stress biomarkers provide a reasonable index of the general oxidative stress status? *Biochim. Biophys. Acta* **2004**, *1674*, 251–259. [CrossRef]
24. Martinez-Moral, M.P.; Kannan, K. How stable is oxidative stress level? An observational study of intra- and inter-individual variability in urinary oxidative stress biomarkers of DNA, proteins, and lipids in healthy individuals. *Environ. Int.* **2019**, *123*, 382–389. [CrossRef]
25. Bigagli, E.; Luceri, C.; Dicembrini, I.; Tatti, L.; Scavone, F.; Giovannelli, L.; Mannucci, E.; Lodovici, M. Effect of Dipeptidyl-Peptidase 4 Inhibitors on Circulating Oxidative Stress Biomarkers in Patients with Type 2 Diabetes Mellitus. *Antioxidants* **2020**, *9*, 233. [CrossRef] [PubMed]
26. Witko-Sarsat, V.; Friedlander, M.; Capeillère-Blandin, C.; Nguyen-Khoa, T.; Nguyen, A.T.; Zingraff, J.; Jungers, P.; Descamps-Latscha, B. Advanced oxidation protein products as a novel marker of oxidative stress in uremia. *Kidney Int.* **1996**, *49*, 1304–1313. [CrossRef]
27. Cournot, M.; Burillo, E.; Saulnier, P.J.; Planesse, C.; Gand, E.; Rehman, M.; Ragot, S.; Rondeau, P.; Catan, A.; Gonthier, M.P.; et al. Circulating Concentrations of Redox Biomarkers Do Not Improve the Prediction of Adverse Cardiovascular Events in Patients With Type 2 Diabetes Mellitus. *J. Am. Heart Assoc.* **2018**, *7*, e007397. [CrossRef]
28. Benzie, I.; Strain, J. The ferric reducing ability of plasma as a measure of antioxidant. *Anal. Biochem.* **1996**, *239*, 70–76. [CrossRef]
29. Marrocco, I.; Altieri, F.; Peluso, I. Measurement and Clinical Significance of Biomarkers of Oxidative Stress in Humans. *Oxid. Med. Cell Longev.* **2017**, *2017*, 6501046. [CrossRef]
30. Zhou, L.; Chen, X.; Lu, M.; Wu, Q.; Yuan, Q.; Hu, C.; Miao, J.; Zhang, Y.; Li, H.; Hou, F.F.; et al. Wnt/β-catenin links oxidative stress to podocyte injury and proteinuria. *Kidney Int.* **2019**, *95*, 830–845. [CrossRef] [PubMed]
31. Guo, Z.J.; Niu, H.X.; Hou, F.F.; Zhang, L.; Fu, N.; Nagai, R.; Lu, X.; Chen, B.H.; Shan, Y.X.; Tian, J.W.; et al. Advanced oxidation protein products activate vascular endothelial cells via a RAGE-mediated signaling pathway. *Antioxid. Redox Signal* **2008**, *10*, 1699–1712. [CrossRef] [PubMed]
32. Valente, A.J.; Yoshida, T.; Clark, R.A.; Delafontaine, P.; Siebenlist, U.; Chandrasekar, B. Advanced oxidation protein products induce cardiomyocyte death via Nox2/Rac1/superoxide-dependent TRAF3IP2/JNK signaling. *Free Radic. Biol. Med.* **2013**, *60*, 125–135. [CrossRef] [PubMed]
33. Lou, A.; Wang, L.; Lai, W.; Zhu, D.; Wu, W.; Wang, Z.; Cai, Z.; Yang, M. Advanced oxidation protein products induce inflammatory responses and invasive behaviour in fibroblast-like synoviocytes via the RAGE-NF-κB pathway. *Bone Joint Res.* **2021**, *10*, 259–268. [CrossRef] [PubMed]
34. Yağmur, A.R.; Çetin, M.A.; Karakurt, S.E.; Turhan, T.; Dere, H.H. The levels of advanced oxidation protein products in patients with obstructive sleep apnea syndrome. *Ir. J. Med. Sci.* **2020**, *189*, 1403–1409. [CrossRef] [PubMed]

35. Maciejczyk, M.; Nesterowicz, M.; Szulimowska, J.; Zalewska, A. Oxidation, Glycation, and Carbamylation of Salivary Biomolecules in Healthy Children, Adults, and the Elderly: Can Saliva Be Used in the Assessment of Aging? *J. Inflamm. Res.* **2022**, *15*, 2051–2073. [CrossRef] [PubMed]

36. Maciejczyk, M.; Zalewska, A.; Ładny, J.R. Salivary Antioxidant Barrier, Redox Status, and Oxidative Damage to Proteins and Lipids in Healthy Children, Adults, and the Elderly. *Oxid. Med. Cell Longev.* **2019**, *2019*, 4393460. [CrossRef] [PubMed]

37. Dahle, L.K.; Hill, E.G.; Holman, R.T. Reprint of: The Thiobarbituric Acid Reaction and the Autoxidations of Polyunsaturated Fatty Acid Methyl Esters. *Arch. Biochem. Biophys.* **2022**, *726*, 109250. [CrossRef] [PubMed]

38. Moselhy, H.F.; Reid, R.G.; Yousef, S.; Boyle, S.P. A specific, accurate, and sensitive measure of total plasma malondialdehyde by HPLC. *J. Lipid Res.* **2013**, *54*, 852–858. [CrossRef]

39. Semba, R.D.; Nicklett, E.J.; Ferrucci, L. Does accumulation of advanced glycation end products contribute to the aging phenotype? *J. Gerontol. A Biol. Sci. Med. Sci.* **2010**, *65*, 963–975. [CrossRef] [PubMed]

40. Bachetti, T.; Bagnasco, S.; Piumelli, R.; Palmieri, A.; Ceccherini, I. A Common 3′UTR Variant of the PHOX2B Gene Is Associated With Infant Life-Threatening and Sudden Death Events in the Italian Population. *Front. Neurol.* **2021**, *12*, 642735. [CrossRef]

Article

Salivary Oxidative Stress Markers' Relation to Oral Diseases in Children and Adolescents

Bahareh Nazemi Salman [1], Shayan Darvish [2,*], Ancuta Goriuc [3,*], Saeideh Mazloomzadeh [4],
Maryam Hossein Poor Tehrani [5] and Ionut Luchian [6]

[1]   Department of Pediatric Dentistry, School of Dentistry, Zanjan University of Medical Sciences, Zanjan 4513956184, Iran; drnazemi@zums.ac.ir

[2]   Pardis Health Center, Shahid Beheshti University of Medical Sciences, Tehran 1985717443, Iran

[3]   Department of Biochemistry, Faculty of Dental Medicine, "Grigore T. Popa" University of Medicine and Pharmacy, 700115 Iasi, Romania

[4]   Department of Epidemiology, Zanjan University of Medical Sciences, Zanjan 4513956184, Iran; smazloomzadeh@zums.ac.ir

[5]   School of Pharmacy, Tehran University of Medical Sciences, Tehran 1417614411, Iran; m250_tehrani@yahoo.com

[6]   Department of Periodontology, Faculty of Dental Medicine, "Grigore T. Popa" University of Medicine and Pharmacy, 700115 Iasi, Romania; ionut.luchian@umfiasi.ro

*   Correspondence: shayandarvish@sbmu.ac.ir (S.D.); ancuta.goriuc@umfiasi.ro (A.G.)

Citation: Salman, B.N.; Darvish, S.; Goriuc, A.; Mazloomzadeh, S.; Hossein Poor Tehrani, M.; Luchian, I. Salivary Oxidative Stress Markers' Relation to Oral Diseases in Children and Adolescents. *Antioxidants* **2021**, *10*, 1540. https://doi.org/10.3390/antiox10101540

Academic Editors: Soliman Khatib and Dana Atrahimovich Blatt

Received: 28 August 2021
Accepted: 24 September 2021
Published: 28 September 2021

Abstract: Current evidence suggests that salivary markers of oxidative stress are indicative of clinical disease indices such as the papillary bleeding index (PBI) and the caries index (CI). The aim of this study was to assess the relation of oxidative stress markers with oral dental caries and periodontal problems in a pediatric population. In our case-control study, unstimulated whole saliva was collected from individuals aged 3–18 years ($n = 177$); 14 individuals were excluded. Study subjects were divided into those with caries (CI = 2, $n = 78$) and those who were caries-free ($n = 85$). These groups were then divided into another subset consisting of children (mean age 7.3 years, $n = 121$) and adolescents (mean age 16.1 years, $n = 42$). The PBI was determined in all groups. We then assessed salivary levels of oxidative stress markers. Our results showed that, the total antioxidant capacity (TAC) level increased in patients with more gingival bleeding ($p < 0.05$) in the study group aged 3–18 years. In addition, TAC showed a significant decrease in samples with caries when compared to the caries-free group in adolescents ($p = 0.008$). In conclusion, TAC levels may be a marker of both gingival bleeding and dental caries in young adult populations. We hope that in the near future, prophylaxis, control, follow up and even possible therapeutic use of oxidative stress markers in a chairside way will become possible as antioxidants have been shown to be effective against oral diseases.

Keywords: oxidative stress; dental caries; saliva; periodontal diseases; child dentistry

1. Introduction

Dental caries is the most common oral disease found in children and adults. Although preventable, it is still the most common chronic disease in the world and results in pain and tooth loss [1]. Dental caries is five times more prevalent than asthma, which is the second common disease after dental caries in children [2]. Periodontal disease is another frequent oral problem—11% of the global population are suffering from it [3]. Bacteria can cause dental caries and periodontal problems; however, both diseases have a multifactorial etiology, with different biomarkers, that affect saliva [1,4,5]. From these biomarkers, oxidative agents such as reactive oxygen/nitrogen substances (ROS/RNS) take part [6]. They interact and are neutralized by antioxidants such as glutathione, uric acid alpha lipoic acid or vitamin C. [7] Remnants of these free radicals which are not neutralized can cause damage to biochemical molecules such as DNA, lipids, and proteins.

This produces different biochemical markers as a byproduct of the damaging process [8]. The imbalance between free radicals and antioxidants is called oxidative stress [9].

Previous studies confirm the role of oxidative stress in diabetes, cardiovascular disease, chronic inflammatory diseases, and oral diseases such as dental caries and periodontal problems [10–13]; however, it is not yet known whether oxidative stress is a cause or a result of the inflammatory response, specifically in oral diseases. Three main possible sources have been introduced for oxidative stress so far: neutrophils, plasma, and bacteria. Among different oxidative stress markers, tiobarbituric acid reactive substances (TBARS) are related to periodontal diseases as a marker for lipid oxidation in saliva [14,15], advanced oxidation protein products (AOPP) are related to dental caries as a marker for protein oxidation [16,17], and total antioxidant capacity (TAC) is a marker for antioxidant agents as it relates to both dental caries and periodontal diseases. Multiple antioxidant agents play a role throughout the body but measuring all of them in saliva is impractical. TAC is a good source to measure overall antioxidant activity [8,18,19], (Figure 1). There is still not enough evidence, however, to use these markers as a way of risk assessment for dental caries or periodontal problems [8].

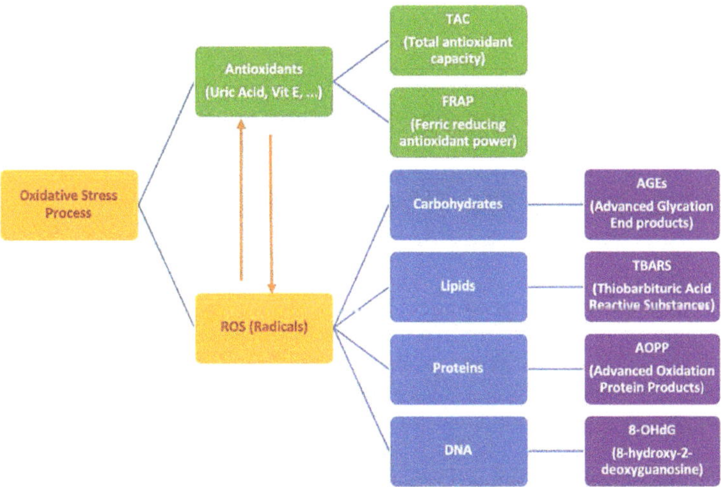

Figure 1. Oxidative stress process and its end product markers.

Lack of research on oxidative stress markers, specifically in a pediatric population, makes it necessary to study them in relation to dental caries and periodontal problems in the valuable fluid of saliva, specifically in children, as early care must be taken to prevent these diseases; thus, this study was designed to assess the possible relation between oxidative stress markers and clinical indices such as PBI and CI in a novel population of children and adolescents with the mean age of 7.3 and 16.1 years, respectively.

2. Materials and Methods

2.1. Study Group

In this case-control study, 177 healthy children and teenagers between the ages of 3 and 18 were included who had at least one dental cavity infecting the dentin or were caries-free after dental examination (Table 1). The number of cases with at least one tooth with a cavity infecting the dentin that could be identified during routine oral examination was 78 (caries index (CI) = 2). Within the control group, the number of caries-free cases was 85 (CI = 0). All of the borderline cases with possible proximal caries were excluded from the study. Examination and collection of saliva was performed in kindergartens, schools and high schools in Zanjan, Iran. All dental examinations were performed using a single-use sterile oral examination kit and a WHO-type periodontal probe (KerrHawe®️ SA, Bioggio, Switzerland) by one dental student in their final year of studies under the observation of a periodontist. All kids/teenagers were prohibited from food consumption 2 h prior to saliva collection. Parents were informed with written consent prior to sampling. This procedure was confirmed by the ethical committee of Zanjan University of Medical Sciences (ZUMS.REC.1395.252).

Table 1. Characteristic information of the study.

	Males	**Females**	*p*-**Value**
CI = 0	41 (48.8%)	43 (51.2%)	0.75
CI = 2	40 (51.3%)	38 (48.7%)	
PBI = 0	28 (49.1%)	29 (50.9%)	
PBI = 1	34 (51.5%)	32 (48.5%)	0.32
PBI = 2	11 (35.5%)	20 (64.5%)	

CI = Caries Index, PBI = Papillary Bleeding Index.

Five milliliters of whole, unstimulated saliva were collected from all children between 9 a.m. and 12 p.m. Saliva samples were stored in Eppendorf tubes and transferred to Tehran University of Medical Sciences, where they were stored at −20 degrees Celsius until the lab test day. PBI examination was performed after saliva sampling to avoid blood contamination using standardized protocols. We assigned PBI = 0 for no bleeding on probing, PBI = 1 for mild bleeding and PBI = 2 for moderate to severe bleeding. The highest score achieved during in the examination period was assigned to each person. Afterwards, the plaque index (PI) was measured on all teeth, using Fuchsin pills and was reported in a percentage.

2.2. Lab Procedures

All lab procedures were collected and followed to use the same path to compare the results with the work of Celec et al. and use a standardized method [16]. TBARS, as the marker for lipid oxidation, AOPP, as the marker for protein oxidation and TAC, as the marker for antioxidants, were measured using spectrophotometry and spectrofluorometric methods in this study. A Synergy 4 BioTek multi-mode reader (Winooski, VT, USA) was used to measure the absorbance and fluorescence. All the lab procedures were conducted in Tehran University of Medical Sciences, School of Pharmacy. Briefly, TBARS was measured with n-butanol, along with the heating procedure at excitation: 515 nm and emission at 553 nm, AOPP with glacial acetic acid at 340 nm, and TAC with TROLOX at 660 nm. Lab procedures are shown in Figure 2.

2.3. Statistical Analysis

Kolmogorov–Smirnov tests were used first to determine the normal distribution of the data. TAC, TBARS and PI had a normal distribution whereas age and AOPP did not. Independent t-tests and Mann–Whitney U tests were performed for CI, one-way ANOVA and Kruskal–Wallis tests were performed for PBI, chi-squared tests were performed for gender, and post-hoc Tukey's tests were performed for two-by-two comparisons. SPSS software version 20.0 was used to analyze the data. A *p* value of less than 0.05 was considered a significant relation. In addition to statistical analysis for all samples, samples taken from individuals aged 3–12 and 13–18 was assessed separately as well due to data that suggested different immune system functioning between these ages [20].

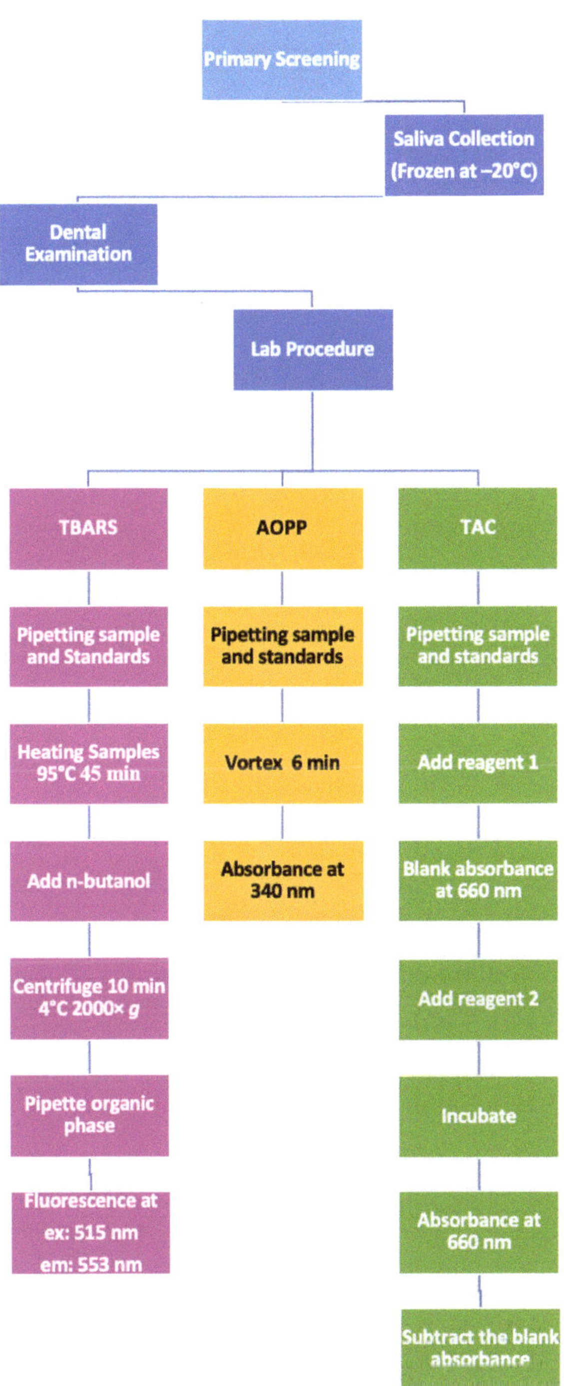

Figure 2. Brief chart for lab procedures measuring TBARS, AOPP, and TAC.

3. Results

Fourteen samples were excluded from the study due to the exclusion criteria: 1—consumption of any drugs including antibiotics in the past two weeks, 2—food consumption in the two hours prior to sampling, 3—history of any kind of medical or oral diseases except dental caries or periodontal problems. We assessed the relation between CI and oxidative stress markers' levels in saliva. AOPP, TBARS and TAC level changes were not significant between the caries-free (CI = 0) and caries-infected dentin (extreme caries) (CI = 2) group in overall samples aged between 3 and 18. This made us curious to investigate further; we decided to separate the adolescents (13 to 18, $n = 42$) from children (3 to 12, $n = 121$) due to different immune system functioning between these two age groups. After separating the groups, TAC levels, which represent the total antioxidant capacity of saliva, expressed a significant decrease in adolescents with a caries index of 2 ($p = 0.008$) when compared to the caries-free samples of the same age.

This significant TAC decrease was not expressed in the children's group aged 3 to 12. AOPP and TBARS did not have significant changes. Plaque index increased significantly in overall samples with caries infecting dentin ($p = 0.04$). Individual results and comparisons are available in Table 2.

Table 2. Individual relation of CI with oxidative stress markers, age and plaque index.

	CI = 0 Average ± Standard Deviation	CI = 2 Average ± Standard Deviation	*p*-Value
Age	4.4 ± 9.8	9.4 ± 4.5	0.39
Age (3–12)	7.6 ± 2.6	7.1 ± 2.4	0.34
Age (13–18)	16 ± 1.1	16.3 ± 1.2	0.33
Plaque index	24.3 ± 11.3	30.1 ± 16.5	0.04 *
Plaque index (3–12)	23.7 ± 11.4	27.4 ± 15.5	0.27
Plaque index (13–18)	25.5 ± 11.4	35.2 ± 17.5	0.05
TBARS (μmol/L)	3 ± 1.4	3.1 ± 1.4	0.75
TBARS (3–12) (μmol/L)	2.9 ± 1.4	3.2 ± 1.3	0.28
TBARS (13–18) (μmol/L)	3.4 ± 1.1	2.9 ± 1.5	0.21
AOPP (μmol/L)	75.9 ± 65.3	71.8 ± 78	0.48
AOPP (3–12) (μmol/L)	74.6 ± 67.6	74.6 ± 85.1	0.99
AOPP (13–18) (μmol/L)	79.8 ± 59.6	63.4 ± 53.3	0.36
TAC (μmol/L)	541.4 ± 228	522.5 ± 204.1	0.57
TAC (3–12) (μmol/L)	494.1 ± 229.7	520 ± 216	0.52
TAC (13–18) (μmol/L)	675 ± 163.9	530.1 ± 166.8	0.008 *

* Indicates *p*-value < 0.05.

Comparing oxidative stress markers and PBI, TAC showed a significant positive correlation with PBI using the ANOVA test ($p = 0.02$). Plaque index also showed a positive significant correlation with PBI ($p = 0.03$). Age of samples had a significant relation with PBI ($p = 0.0001$). No tendency was visible between CI/PBI and gender. Individual results and comparisons for PBI are shown in Table 3. Figure 3 shows the decrease and increase in TAC levels in relation to CI and PBI, respectively.

Table 3. Individual PBI relation to oxidative stress markers, age and plaque index.

	PBI = 2 Average ± Standard Deviation	PBI = 1 Average ± Standard Deviation	PBI = 0 Average ± Standard Deviation	*p*-Value
Age	14.1 ± 2.9	10.8 ± 3.2	5.25 ± 1.9	0.0001 *
Plaque index	32.1 ± 16.7	25.4 ± 12.7	19.1 ± 7.8	0.03 *
TBARS (μmol/L)	3.1 ± 1.1	3.1 ± 1.5	3 ± 1.2	0.97
AOPP (μmol/L)	84.5 ± 90.9	68.4 ± 75.3	73.7 ± 54.8	0.31
TAC (μmol/L)	623.4 ± 176.3	507.3 ± 210.4	509.2 ± 233.2	0.02 *

* Indicates *p*-value < 0.05.

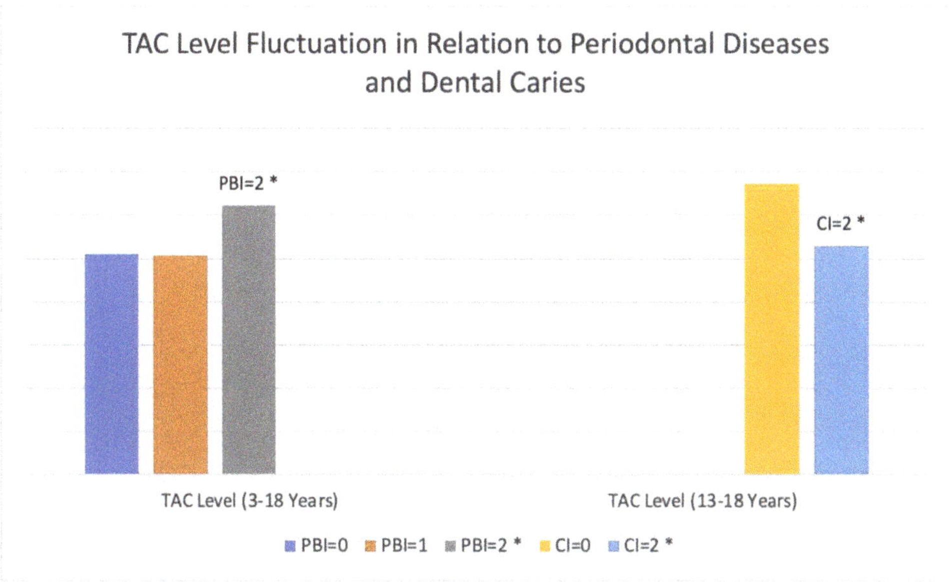

Figure 3. TAC level (μmol/L) fluctuation in relation to periodontal diseases and dental caries. * Indicates *p*-value < 0.05.

4. Discussion

We found that AOPP as a marker for protein oxidation level was less in the CI = 2 group when compared with caries-free children aged between 3 and 18, though this reduction was not significant. These findings are in opposition to the hypothesis that dental caries can release the marker of protein oxidation in saliva due to the demineralization of dentin proteins. Celecova et al. also reported a significant reduction in AOPP levels. They measured the AOPP level in adults aged between 19 and 83, whereas we measured it in children [21]. A possible explanation for this contrast could be the existence of different mechanisms of inflammation between children and adults. Vahabi et al. showed that during the inflammation process in children, bone/dental degrading materials such as prostaglandins and cytokines will not be secreted; therefore, most of the periodontal diseases will remain at only a gingivitis level in children and will not progress further [20]. Since the inflammation process will not develop completely in children, we did not observe a significant change in the AOPP level. Tothova et al. also did not find any significant relation between CI and AOPP when studying on children between 4 and 18. In their study, the only significant relation between AOPP and CI was observed in adults [16]. Whether significant or not significant, AOPP levels only showed a decrease in these studies, and this makes the hypothesis of AOPP release into saliva during the dental caries process negative. Selmeci et al. explained that AOPP might not even be a proper marker for protein oxidation and may be one of the non-enzymatic oxidative markers' system [22]. This finding can be a possible explanation for the reduction in AOPP levels in recent studies. More studies are needed, however, to confirm the role of AOPP in oral diseases.

One of the advantages of this study was to divide children and adolescents. As mentioned above, studies show different results when studying children or adults [21,23]. Pyati et al. and Ahmadi-Motamayel et al. showed in their studies that TAC levels increased in the caries active group when studying children and adolescents [24,25]. This made us separate children aged 3 to 12 years and adolescents aged 13 to 18. Again, after separate assessment in children aged 3 to 12 years old, no significant relation was expressed between oxidative stress markers and CI. This might be due to an immature immune system, lack of plasma cell production and an incomplete inflammation process. However, we found that adolescents aged 13 to 18 have a significant decrease in TAC (antioxidant) levels in CI = 2 group, which is counter to the results of the observed studies. A possible explanation for the reduction in TAC levels in teenagers with more severe dental caries could be due to the excess production of free radicals and consumption of antioxidants simultaneously. This is in line with Ahmadi-Motamayel et al.'s work, as they also reported that during the dental caries process, antioxidant levels can decrease [26]. The key point is that during the dental caries procedure, only hard tissue destruction is taking place and the immune system cannot overcome the consumption of antioxidants during the caries process due to lower interaction and involvement. However, periodontal diseases frequently involve the immune system as soft tissue destruction takes place.

We found a significant relation between the TAC marker and the PBI as well. In our study, samples taken from individuals aged 3 to 18 years old with more bleeding upon probing expressed more antioxidant markers (TACs) in their saliva. This is against the finding of multiple studies that expressed lower antioxidant (TAC) levels in the saliva of periodontal patients with more bleeding. Su et al. also showed an increase in antioxidant levels when studying 58 adult periodontal patients and 234 healthy adults. They observed more TAC in the saliva of periodontal patients. First, they attributed this to the possible consumption of antioxidant foods by the periodontal patients, but when they performed the correlation test to remove this effect, they still observed significantly higher TAC levels in periodontal patients [27]. The only explanation that can be given to the TAC increment is the overproduction of antioxidants by the immune system to protect gingival tissues against free radicals. In periodontal diseases, soft and hard tissue destruction takes place simultaneously and the immune system, unlike in dental caries, becomes more involved in the inflammation process, thus increasing the antioxidant agents.

Zhang et al. had a different result. In their study, TAC levels decreased in the saliva of periodontal patients. They attributed this to the consumption of antioxidants already available in saliva by the immune system to neutralize the free radicals during the inflammation process [28]. More studies are needed to answer the question of whether TAC levels will increase or decrease in severe periodontal conditions.

Antioxidants have been shown to be a possible effective agent in treating and evaluating oral diseases in recent studies [16]. De Caro et al. showed, in their study, that antioxidants can be used as a thin film to improve the condition of oral diseases in the simple use of cosmetics and nutraceuticals [29]. Zukowski et al. suggested the use of antioxidants in their study as a supplement for people who are exposed to more ROS in their oral cavity [30]. In our study, TAC levels showed both a significant increase and decrease in relation to PBI and CI, respectively. This is in line with the use of antioxidants in oral diseases as suggested by the above studies; thus, these markers can be a proper modifying and controlling factor in determining the oral health status of diseased children.

Celecova et al. did not find any significant relation between CI and TBARS as well but expressed a significant positive relation between TBARS and PBI. In their study, as more bleeding occurs, more periodontal destruction is taking place, and this will release more TBARS in saliva [21]. Celec et al. had the same results. Our results did not emphasize any significant relation between PBI and TBARS in comparison with other studies [26].

It is known that different factors affect oxidative stress markers, free radicals, and antioxidants and future studies should focus on elucidating the bidirectional relationship between them and oral diseases to find the actual source of these markers. The possible sources suggested for production of the free radicals are: 1—the immune system, 2—the actual bacteria present in the inflammation site causing destruction via the free radicals and 3—the exchange between plasma and saliva [8,31]. Alakin et al. studied these markers in chronic periodontitis patients. They compared oxidative stress markers' levels in saliva, serum and gingival crevicular fluid (GCF). GCF had the highest level of oxidative stress markers in their study. Therefore, the source for the interaction between antioxidants and free radicals can be from localized onsite inflammation [32]. All these factors need to be assessed further to better understand oxidative stress markers.

Plaque index in our study showed a significant increase in more severe bleeding (PBI = 2) and caries-infected dentin (CI = 2) groups. As these plaques are sources of bacteria, it may have affected the production of oxidative stress markers in saliva. Nonetheless, these bacteria and plaque are a part of the inflammation process which contributes to periodontal disease and dental caries originally, and thus can affect the production of oxidative stress markers, but to what extent is still unknown. Interestingly, in adolescents (13–18) the plaque index increased in the CI = 2 group almost significantly ($p = 0.05$) and in the children's (3–12) group, it had an insignificant increase. Further studies are needed in order to be able to determine the level of effect that plaques can have on this matter.

A possible explanation for different results we found in our study could be external factors that can affect oxidative stress markers and might be a reason for discrepancies. Kamodyova et al. showed that day rhythm can affect oxidative stress markers' levels. As an example, the AOPP level reached its maximum at 2 p.m. in their study. Another contributing factor in their study was tooth brushing; tooth brushing will increase TAC level as it will wash away the gingival crevicular fluid (GCF) and replace it with fresh saliva full of antioxidants [33]. In our study, we took saliva between 9 a.m. and 12 p.m. to prevent day rhythm difference. We prohibited children to use any food or brush two hours prior to saliva collection; thus, these factors could not have affected our results, but external factors are not limited to these two.

This is one of the first studies that assessed oxidative stress markers and their relation to dental caries and periodontal diseases that differs children and adolescents by considering the effect of puberty and dental system change from deciduous to permanent teeth. All the lab methods were selected from the Celec et al. team who have the most published work on oxidative stress markers regarding oral diseases in order to standardize the methods. Celec et al. mentioned in their recent review article that even the sources of these free radicals are not yet determined and mentions that more interventional studies are needed in the field of oxidative stress [34]. We recommend that antioxidants and their markers can play a key role in relation to oral diseases. Kumar et al. suggested the use of future drugs which rely on understanding the pathway of inflammation and oxidative stress more clearly [35]. We recommend using more interventional studies in relation to oxidative stress markers in all three possible fluids: plasma, saliva and GCF for future studies.

5. Conclusions

To conclude, we found a significant positive relation between antioxidant marker (TAC) and PBI, but a negative significant relation was expressed in adolescents aged between 13 and 18 when assessed with dental caries index in this study. TAC level increases in the PBI = 2 group may be associated with the overproduction of antioxidants by the immune system, whereas the TAC level reduction in the adolescent group with severe dental caries (CI = 2) can be attributed to consumption of antioxidant capacities. The incapacity of the immune system to regulate those variations can be influenced by the fluctuations induced by puberty and a change in the dental system.

Our study opens promising perspectives regarding the important role of a potential therapeutic use of oxidative stress markers in a chairside way, as antioxidants have shown to be effective against oral diseases.

Author Contributions: Conceptualization B.N.S.; methodology, B.N.S. and S.D.; software, S.M.; validation, S.M.; formal analysis, S.M.; investigation, S.D.; data curation, S.D. and M.H.P.T.; writing—original draft preparation, S.D.; writing—review and editing, I.L. and A.G. All authors have read and agreed to the published version of the manuscript.

Funding: This research was funded by Zanjan University of Medical Sciences, Iran.

Institutional Review Board Statement. The study was conducted according to the guidelines of the Declaration of Helsinki, and approved by the Ethics Committee of Zanjan University of Medical Sciences-Iran, protocol number ZUMS.REC.1395.252.

Informed Consent Statement: Informed consent for was obtained from all participants involved in the study and from their legal tutors.

Data Availability Statement: The data used to support the findings are available from the corresponding authors upon reasonable request.

Acknowledgments: Authors are grateful to Peter Celec for sharing the lab methods.

Conflicts of Interest: The authors declare no conflict of interest.

References

1. Selwitz, R.H.; Ismail, A.I.; Pitts, N.B. Dental caries. *Lancet* **2007**, *369*, 51–59. [CrossRef]
2. Guo, L.; Shi, W. Salivary biomarkers for caries risk assessment. *J. Calif. Dent. Assoc.* **2013**, *41*, 107. [PubMed]
3. Marcenes, W.; Kassebaum, N.J.; Bernabe, E.; Flaxman, A.; Naghavi, M.; Lopez, A.; Murray, C.J. Global burden of oral conditions in 1990–2010: A systematic analysis. *J. Dent. Res.* **2013**, *92*, 592–597. [CrossRef] [PubMed]
4. Luchian, I.; Moscalu, M.; Goriuc, A.; Nucci, L.; Tatarciuc, M.; Martu, I.; Covasa, M. Using Salivary MMP-9 to Successfully Quantify Periodontal Inflammation during Orthodontic Treatment. *J. Clin. Med.* **2021**, *10*, 379. [CrossRef] [PubMed]
5. Nazemisalman, B.; Sajedinejad, N.; Darvish, S.; Vahabi, S.; Gudarzi, H. Evaluation of inductive effects of different concentrations of cyclosporine A on MMP-1, MMP-2, MMP-3, TIMP-1, and TIMP-2 in fetal and adult human gingival fibroblasts. *J. Basic Clin. Physiol. Pharmacol.* **2019**, *30*. [CrossRef]
6. Valko, M.; Leibfritz, D.; Moncol, J.; Cronin, M.T.; Mazur, M.; Telser, J. Free radicals and antioxidants in normal physiological functions and human disease. *Int. J. Biochem. Cell Biol.* **2007**, *39*, 44–84. [CrossRef]
7. Aruoma, O.I. Free radicals, oxidative stress, and antioxidants in human health and disease. *J. Amer. Oil. Chem. Soc.* **1998**, *75*, 199–212. [CrossRef]
8. Tothova, L.; Kamodyova, N.; Cervenka, T.; Celec, P. Salivary markers of oxidative stress in oral diseases. *Front. Cell. Infect. Microbiol.* **2015**, *5*, 73. [CrossRef]
9. Sies, H. Oxidative stress: Oxidants and antioxidants. *Exp. Physiol.* **1997**, *82*, 291–295. [CrossRef]
10. Iannitti, T.; Rottigni, V.; Palmieri, B. Role of free radicals and antioxidant defences in oral cavity-related pathologies. *J. Oral Pathol. Med.* **2012**, *41*, 649–661. [CrossRef]
11. Reuter, S.; Gupta, S.C.; Chaturvedi, M.M.; Aggarwal, B.B. Oxidative stress, inflammation, and cancer: How are they linked? *Free Radic. Biol. Med.* **2010**, *49*, 1603–1616. [CrossRef] [PubMed]
12. Stephens, J.W.; Khanolkar, M.P.; Bain, S.C. The biological relevance and measurement of plasma markers of oxidative stress in diabetes and cardiovascular disease. *Atherosclerosis* **2009**, *202*, 321–329. [CrossRef]
13. Vlkova, B.; Stanko, P.; Minarik, G.; Tothova, L.; Szemes, T.; Banasova, L.; Novotnakova, D.; Hodosy, J.; Celec, P. Salivary markers of oxidative stress in patients with oral premalignant lesions. *Arch. Oral Biol.* **2012**, *57*, 1651–1656. [CrossRef]
14. Chapple, I.L.; Matthews, J.B. The role of reactive oxygen and antioxidant species in periodontal tissue destruction. *Periodontol. 2000* **2007**, *43*, 160–232. [CrossRef] [PubMed]
15. Tsai, C.C.; Chen, H.S.; Chen, S.L.; Ho, Y.P.; Ho, K.Y.; Wu, Y.M.; Hung, C.C. Lipid peroxidation: A possible role in the induction and progression of chronic periodontitis. *J. Periodontal. Res.* **2005**, *40*, 378–384. [CrossRef] [PubMed]
16. Tothova, L.; Celecova, V.; Celec, P. Salivary markers of oxidative stress and their relation to periodontal and dental status in children. *Dis. Markers* **2013**, *34*, 9–15. [CrossRef] [PubMed]
17. Witko-Sarsat, V.; Friedlander, M.; Capeillere-Blandin, C.; Nguyen-Khoa, T.; Nguyen, A.T.; Zingraff, J.; Jungers, P.; Descamps-Latscha, B. Advanced oxidation protein products as a novel marker of oxidative stress in uremia. *Kidney Int.* **1996**, *49*, 1304–1313. [CrossRef]
18. Erel, O. A novel automated direct measurement method for total antioxidant capacity using a new generation, more stable ABTS radical cation. *Clin. Biochem.* **2004**, *37*, 277–285. [CrossRef]
19. Wang, J.; Schipper, H.M.; Velly, A.M.; Mohit, S.; Gornitsky, M. Salivary biomarkers of oxidative stress: A critical review. *Free Radic. Biol. Med.* **2015**, *85*, 95–104. [CrossRef]
20. Vahabi, S.; Moslemi, M.; Nazemisalman, B.; Yadegari, Z. Phenytoin Effects on Proliferation and Induction of IL1β and PGE2 in Pediatric and Adults' Gingival Fibroblasts. *Open J. Stom.* **2014**, *4*, 452. [CrossRef]
21. Celecova, V.; Kamodyova, N.; Tothova, L.; Kudela, M.; Celec, P. Salivary markers of oxidative stress are related to age and oral health in adult non-smokers. *J. Oral Pathol. Med.* **2013**, *42*, 263–266. [CrossRef]
22. Selmeci, L. Advanced oxidation protein products (AOPP): Novel uremic toxins, or components of the non-enzymatic antioxidant system of the plasma proteome? *Free Radic. Res.* **2011**, *45*, 1115–1123. [CrossRef] [PubMed]
23. Behuliak, M.; Palffy, R.; Gardlik, R.; Hodosy, J.; Halcak, L.; Celec, P. Variability of thiobarbituric acid reacting substances in saliva. *Dis. Markers* **2009**, *26*, 49–53. [CrossRef] [PubMed]
24. Ahmadi-Motamayel, F.; Goodarzi, M.T.; Hendi, S.S.; Kasraei, S.; Moghimbeigi, A. Total antioxidant capacity of saliva and dental caries. *Med. Oral Patol. Oral Cir. Bucal* **2013**, *18*, e553. [CrossRef]

25. Pyati, S.A.; Naveen Kumar, R.; Kumar, V.; Praveen Kumar, N.; Parveen Reddy, K. Salivary flow rate, pH, buffering capacity, total protein, oxidative stress and antioxidant capacity in children with and without dental caries. *J. Clin. Ped. Dent.* **2018**, *42*, 445–449. [CrossRef]
26. Ahmadi-Motamayel, F.; Goodarzi, M.T.; Mahdavinezhad, A.; Jamshidi, Z.; Darvishi, M. Salivary and serum antioxidant and oxidative stress markers in dental caries. *Caries Res.* **2018**, *52*, 565–569. [CrossRef]
27. Su, H.; Gornitsky, M.; Velly, A.M.; Yu, H.; Benarroch, M.; Schipper, H.M. Salivary DNA, lipid, and protein oxidation in nonsmokers with periodontal disease. *Free Radic. Biol. Med.* **2009**, *46*, 914–921. [CrossRef]
28. Zhang, T.; Andrukhov, O.; Haririan, H.; Muller-Kern, M.; Liu, S.; Liu, Z.; Rausch-Fan, X. Total Antioxidant Capacity and Total Oxidant Status in Saliva of Periodontitis Patients in Relation to Bacterial Load. *Front. Cell. Infect. Microbiol.* **2015**, *5*, 97. [CrossRef]
29. De Caro, V.; Murgia, D.; Seidita, F.; Bologna, E.; Alotta, G.; Zingales, M.; Campisi, G. Enhanced in situ availability of aphanizomenon flos-aquae constituents entrapped in buccal films for the treatment of oxidative stress-related oral diseases: Biomechanical characterization and in vitro/ex vivo evaluation. *Pharmaceutics* **2019**, *11*, 35. [CrossRef]
30. Żukowski, P.; Maciejczyk, M.; Waszkiel, D. Sources of free radicals and oxidative stress in the oral cavity. *Arch. Oral Biol.* **2018**, *92*, 8–17. [CrossRef] [PubMed]
31. Vlkova, B.; Celec, P. Does Enterococcus faecalis contribute to salivary thiobarbituric acid-reacting substances? *In Vivo* **2009**, *23*, 343–345.
32. Akalın, F.A.; Baltacıoğlu, E.; Alver, A.; Karabulut, E. Lipid peroxidation levels and total oxidant status in serum, saliva and gingival crevicular fluid in patients with chronic periodontitis. *J. Clin. Periodontol.* **2007**, *34*, 558–565. [CrossRef] [PubMed]
33. Kamodyova, N.; Tothova, L.; Celec, P. Salivary markers of oxidative stress and antioxidant status: Influence of external factors. *Dis. Markers.* **2013**, *34*, 313–321. [CrossRef] [PubMed]
34. Celec, P. Oxidative stress and antioxidants in the diagnosis and therapy of periodontitis. *Front. Physiol.* **2017**, *8*, 1055.
35. Kumar, J.; Teoh, S.L.; Das, S.; Mahakknaukrauh, P. Oxidative stress in oral diseases: Understanding its relation with other systemic diseases. *Front. Physiol.* **2017**, *8*, 693. [CrossRef] [PubMed]

 *antioxidants*

Article

Human Milk Antioxidative Modifications in Mastitis: Further Beneficial Effects of Cranberry Supplementation

Victoria Valls-Bellés [1], Cristina Abad [1], María Teresa Hernández-Aguilar [1], Amalia Nacher [1], Carlos Guerrero [1], Pablo Baliño [1], Francisco J. Romero [2,*] and María Muriach [1,*]

[1] Unitat Predepartamental de Medicina, Facultat de Ciencies de la Salud, Universitat Jaume I, 12071 Castellon de la Plana, Spain; vallsv@uji.es (V.V.-B.); cristina.abad@cofcastellon.org (C.A.); lactancia_peset@gva.es (M.T.H.-A.); nacher_ama@gva.es (A.N.); cguerrer@uji.es (C.G.); balino@uji.es (P.B.)

[2] Hospital General de Requena, Conselleria de Sanitat, Generalitat Valenciana, 46340 Requena, Spain

* Correspondence: romero_fragom@gva.es (F.J.R.); muriach@uji.es (M.M.)

Abstract: Mastitis is the inflammation of one or several mammal lobes which can be accompanied by a mammary gland infection, and is the leading cause of undesired early weaning in humans. However, little information exists regarding the changes that this disease may induce in the biochemical composition of human milk, especially in terms of oxidative status. Given that newborns are subject to a significant increase in total ROS burden in their transition to neonatal life and that their antioxidant defense system is not completely developed, the aim of this study was to evaluate antioxidant defense (glutathione peroxidase (GPx), reduced glutathione (GSH), total polyphenol content (TPP), and total antioxidant capacity (TAC)) in milk samples from mothers suffering from mastitis and controls. We also measured the oxidative damage to lipids (malondyaldehyde (MDA)) and proteins (carbonyl group content (CGC)) in these samples. Finally, we tested whether dietary supplementation with cranberries (a product rich in antioxidants) in these breastfeeding mothers during 21 days could improve the oxidative status of milk. GPx activity, TPP, and TAC were increased in milk samples from mastitis-affected women, providing a protective mechanism to the newborn drinking mastitis milk. MDA concentrations were diminished in the mastitis group, confirming this proposal. Some oxidative damage might occur in the mammary gland since the CGC was increased in mastitis milk. Cranberries supplementation seems to strengthen the antioxidant system, further improving the antioxidative state of milk.

Keywords: mastitis; human milk; oxidative status; cranberry

Citation: Valls-Bellés, V.; Abad, C.; Hernández-Aguilar, M.T.; Nacher, A.; Guerrero, C.; Baliño, P.; Romero, F.J.; Muriach, M. Human Milk Antioxidative Modifications in Mastitis: Further Beneficial Effects of Cranberry Supplementation. *Antioxidants* **2022**, *11*, 51. https://doi.org/10.3390/antiox11010051

Academic Editors: Soliman Khatib and Dana Atrahimovich Blatt

Received: 1 December 2021
Accepted: 16 December 2021
Published: 27 December 2021

Publisher's Note: MDPI stays neutral with regard to jurisdictional claims in published maps and institutional affiliations.

1. Introduction

Human milk is characterized by a huge variability in its composition, as it includes nutrients as well as bioactive compounds and a vast array of microbes known as the human milk microbiota. As the neonatal immune system develops, this milk variability and composition provides the infant with well-balanced nutrition and protection against potential infectious pathogens [1,2]. The World Health Organization (WHO) recommends that infants be exclusively breastfed for the first six months of life for optimal growth, development, and health, and that breastfeeding continues to be an important part of the diet until the infant is at least two years old [3].

With regard to bioactive compounds, the antioxidant content of breast milk has been the subject of a number of studies, confirming the presence of different components that have been reported to modulate the effects of oxidative stress [4–10]. Birth represents a significant oxidative challenge because it involves the transition from the relatively hypoxic intrauterine conditions to the oxygen-rich extrauterine environment. Thus, newborns are subject to a significant increase in total ROS burden in their transition to neonatal life [11,12].

This situation is especially relevant in premature newborns due to their undeveloped antioxidant defense system [13,14]. Moreover, increasing evidence also indicates that some pathological conditions appearing in adulthood (such as type 2 diabetes, hypertension, obesity, and others) associated with oxidative stress may have their origin in the fetal or neonatal period of life [15–17].

Mastitis is the inflammation of one or several mammal lobes which can be accompanied by a mammary gland infection, and is the leading cause of unwelcome early human weaning. The primary origin of mastitis is milk stasis [18], and its incidence varies from 2% to 33% according to different authors, being more frequent during the early postpartum weeks. Infective mastitis is usually associated with the growth of certain species of staphylococci, streptococci, and/or corynebacteriae in the milk [19,20]. Treatment traditionally consists of antibiotic therapy, analgesic medication, and proper milk removal [21]. However, the WHO has highlighted significant concerns regarding the adverse side effects of antibiotics due to the emergence of antibiotic-resistant strains of microorganisms. In this context, more recently specific probiotic bacteria have been postulated to prevent mastitis due to their anti-inflammatory and immunomodulatory properties [22].

In fact, almost all of our knowledge regarding the biology of human mastitis is extrapolated from bovine or goat studies [23]. The group of Gelasakis and colleagues demonstrated lower fat and lactose content in milk from goats with subclinical mastitis, together with a mild increase in protein content [24]. Similar results were obtained by Sun et al. in bovine mastitis [25]. Furthermore, it has been reported that subclinical mastitis in goats upregulates nitric oxide-derived oxidative stress and reduces milk antioxidant properties [26]. However, it is noteworthy that two recent studies also described the effect of lactational mastitis on the macronutrient content of human breast milk [27,28] in accordance with the above-mentioned results reported in bovine and goats. In addition, Perez et al. reported that mastitis modifies the biogenic amine profile of human milk [29]. Interestingly, it has been described that human mastitis milk has the same anti-inflammatory components and characteristics of normal milk, with elevations in selected components/activities (not including antioxidant activity in terms of spontaneous cytochrome c reducing activity) that may help protect the nursing infant from developing clinical illness due to feeding on mastitis milk [30]. This work aims to study oxidative status in milk from mothers with mastitis. Moreover, we extended the present investigation to study whether dietary supplementation with cranberries (*Vaccinium* sp.) a product known to be rich in antioxidants, could improve oxidative status in human milk [31,32].

2. Materials and Methods

2.1. Experimental Design

In total, 60 non-smoking lactating mothers who had delivered to term and breastfed their babies starting from the first day postpartum were included in the study. The study was approved by the Research Ethics Committees from both, Hospital General Universitario from Castellón (4/2015) and Hospital Universitario Dr. Peset from Valencia (23/16). All the participants were fully informed of the procedure and gave their written consent to participate. They were also allowed to withdraw from the study at will. Of these mothers, 30 were healthy lactating mothers and the other 30 referred to symptoms of mastitis which included breast swelling and redness together with pain or a burning sensation continuously or while breast-feeding. No fever or abscess were present in any case. The microbiological culture confirmed the bacterial infection in all cases, and therefore all subjects in the mastitis group received antibiotic treatment during the study. A total of 15 subjects from each group accepted dietary supplementation with cranberries (20 g/day) for 21 days. Thus, subjects were finally assigned to 4 groups ($n = 15$): control (C), control + cranberries (C + C), mastitis (M), and mastitis + cranberries (M + C).

Vaccinium berry fruits are widely known for their health benefits. These particular, berry species present high concentrations of antioxidants, including phenolic compounds, and the presence of specific, particularly potent polyphenolic compounds [31,32].

Two samples of mature milk were obtained from each mother, with one taken at the beginning of the study and another after 21 days. Prior to sampling, all donors gave informed consent regarding participation in the study and completed a questionnaire on nutritional habits, with the purpose of confirming the homogeneity of the population. Furthermore, the demographic characteristics of the subjects were recorded, and dietary intake was assessed using 24 h recall. This dietary assessment was performed over 3 days (1 holiday and 2 working days). The food ingested on the different days was processed by the "Alimentador" software of the Sociedad Española de Dietética y Ciencias de la Alimentación (SEDCA).

2.2. Collection of Samples

Approximately 10 mL of mature milk was collected prior to feeding of the nursing infant using an electric breast pump fitted with a vacuum regulator (MEDELA) connected to polypropylene containers in which the samples were collected directly. After milk extraction, the samples were immediately transported to liquid N_2 storage in dark conditions to avoid oxidative processes. The samples were subsequently divided into 4 aliquots and frozen at $-80\,^{\circ}$C.

2.3. Methods

The oxidative status of human milk samples was assessed in this study. For this purpose, on one hand the total antioxidant capacity (TAC), glutathione peroxidase (GPx) activity, reduced glutathione (GSH) concentration, and total polyphenol content (TPP) values were measured. On the other hand, oxidative damage to macromolecules was assessed by measuring malondialdehyde (MDA) as a lipid peroxidation product and carbonyl group content (CGC) as a marker of oxidative damage to proteins.

2.4. TAC Determination

The ABTS assay to determine TAC was modeled after the method proposed by Miller and Rice Evans [33] and modified by Re et al. [34]. Briefly, the assay is based on the oxidation of ABTS by potassium persulphate to form the radical monocation ABTS$^{\bullet+}$, which is reduced in the presence of hydrogen-donating antioxidants. The reagent ABTS$^{\bullet+}$ was generated by exposing a 7 mM solution of ABTS to a solution of 2.45 mM potassium persulphate at a 1:1 ratio. The assay included 2.970 mL of ABTS$^{\bullet+}$ and 30 µL of plasma. Antioxidant activity was determined by measuring the decolorization of the ABTS$^{\bullet+}$ (reduction of the radical cation) through the absorbance at 734 nm.

2.5. Assay of GPx Activity

Glutathione peroxidase activity, which catalyzes the oxidation by H_2O_2 of GSH to its disulfide (GSSG), was assayed spectrophotometrically as reported by Lawrence et al. [35] by monitoring the oxidation of NADPH at 340 nm. The reaction mixture consisted of 240 mU/mL of GSH disulfide reductase, 1 mM GSH, and 0.15 mM NADPH in 0.1 M potassium phosphate buffer at pH 7.0 containing 1 mM EDTA and 1 mM sodium azide. A 50 µL sample was added to this mixture and allowed to equilibrate at 37 $^{\circ}$C for 3 min. The reaction was started by the addition of hydrogen peroxide to adjust the final volume of the assay mixture to 1 mL.

2.6. Assay of GSH Concentration

GSH concentration was quantified following the method of Reed et al. [36] based on the reaction of iodoacetic acid with the thiol groups followed by a chromophore derivatization of the amino groups with Sanger reactant (1-fluoro-2,4-dinitrobencene), giving rise to derivatives which are quickly separated by means of high-performance liquid chromatography (HPLC).

2.7. TPP Determination

To measure the total content of polyphenols in breast milk, the method of Folin Ciocalteu was used [37]. This method is based on the oxidation of polyphenols with the Folin reagent. When milk is mixed with the reagent, the formation of a blue complex is achieved, for which absorbance is measured at 750 nm. Briefly, milk samples were centrifuged at 12,000× *g* for 5 min. Then, 500 μL of supernatant was taken and 100 μL of 1.32 M metaphosphoric acid was added for protein precipitation. The mixture was centrifuged at 2700 rpm for 3 min, and the supernatant was collected. Then, 20 μL of the supernatant, 80 μL of distilled water, and 500 μL of reagent were mixed. Subsequently, 400 μL of 7.5% sodium carbonate was added and measured at 750 nm, using gallic acid as standard.

2.8. Assay of MDA Concentration

MDA concentration was measured by liquid chromatography according to a modification [38] of the original method of Richard et al. [39]. Briefly, 0.1 mL of sample (or standard solutions prepared daily from 1,1,3,3-tetramethoxypropane) and 0.75 mL of working solution (thiobarbituric acid 0.37% and perchloric acid 6.4%; 2:1, *v/v*) were mixed and heated to 95 °C for 1 h. After cooling for 10 min in an ice water bath, the flocculent precipitate was removed by centrifugation at 3200× *g* for 10 min. The supernatant was neutralized and filtered (0.22 μm) prior to injection on an ODS 5 μm column (250 × 4.6 mm). The mobile phase consisted of 50 mM phosphate buffer (pH 6.0): methanol (58:42, *v/v*). Isocratic separation was performed with 1.0 mL/min flow and detection at 532 nm.

2.9. Quantification of CGC

Carbonyl groups were determined to evaluate protein oxidation in milk samples. The CGC released during incubation with 2,4-dinitrophenylhydrazine was measured using the method reported by Levine et al. [40]. Briefly, the samples were centrifuged at 13,000× *g* for 10 min. Then, 20 mL of brain homogenate was placed in a 1.5 mL Eppendorf tube, and 400 mL of 10 mM 2,4 dinitrophenylhydrazine/2.5 M hydrochloric acid (HCl) and 400 mL of 2.5 M HCl were added. This mixture was incubated for 1 h at room temperature. Protein precipitation was performed using 1 mL of 100% TCA, and the mixture was washed twice with ethanol/ethyl acetate (1/1, *v/v*) and centrifuged at 12,600× *g* for 3 min. Finally, 1.5 mL of 6 N guanidine (pH 2.3) was added and the samples were incubated in a 37 °C water bath for 30 min and centrifuged at 12,600× *g* for 3 min. The carbonyl content was calculated from peak absorption (373 nm) using an absorption coefficient of 22,000 M^{-1} cm^{-1} and was expressed as nmol/mg protein.

2.10. Protein Concentration Measurement

Protein content was determined using Bradford method.

2.11. Statistical Analysis

The statistical analyses were performed using SPSS 24 (IBM SPSS, Chicago, IL, USA). The Kolmogorov test was used to check the normal distribution of each population ($p > 0.100$) to analyze the data based on parametric analysis of variance (ANOVA).

Generally, multifactorial ANOVA is applied to analyze the differences between data groups (in this case for example according to the influence of the cranberry supplement and considering time as a second factor). However, given that the "time" factor includes only two measurements, we finally decided to use 1-way ANOVA using the factor of "treatment", with the studied variable being the difference in the results for each parameter measured at the beginning and the end of the study. Statistical significance was considered at $p < 0.05$. Significant differences in each sampling group were assessed using the least significant difference (LSD) procedure.

3. Results

The obtained anthropometric measurements and dietetic assessment results are listed in Table 1. There were no differences in age, weight, height, and BMI among the different groups in the study. Significant differences were not found in terms of macronutrient and micronutrient intakes among the four groups.

Table 1. Anthropometric measurements and dietetic assessment (daily average).

	C	C + C	M	M + C
Age	36.5 ± 4.1	33.8 ± 3.7	33.7 ± 3.9	36.4 ± 3.3
Weight (kg)	60.6 ± 7.2	64.6 ± 8.8	66.2 ± 10.5	63.9 ± 10.5
Height (cm)	164.6 ± 4.7	165.5 ± 6.2	163.2 ± 4.8	164.6 ± 6.7
BMI	22.3 ± 2.5	23.6 ± 2.6	24.8 ± 3.9	23.6 ± 4
Kcal	1595 ± 394	1510 ± 265	1659 ± 248	1631 ± 428
Carbohydrates (g)	166.2 ± 54.5	167.8 ± 48	191.2 ± 44.7	188.6 ± 40
Proteins (g)	73.8 ± 15.9	71.6 ± 16	82.7 ± 12.5	80.8 ± 19.6
Fat (g)	77.6 ± 23	67 ± 11.2	68.9 ± 16.7	68.2 ± 30.1
Cholesterol (mg)	227.5 ± 76.3	238.7 ± 109.2	249.7 ± 90.6	231.8 ± 86.1
Fiber (g)	12.4 ± 6	9.5 ± 3.5	14.6 ± 7.5	11.5 ± 5.6

C: Control group; C + C: Control group supplemented with Cranberries; M: Mastitis group; M + C: Mastitis group supplemented with Cranberries. BMI: Body Mass Index. The results are expressed as mean ± standard deviation.

Regarding the oxidative status of human milk, TAC was higher at the beginning of the study in milk from the mastitis group than in control milk. Twenty-one days later the values for this parameter were similar to those of the control. When the TAC evolution in every group throughout the study was analyzed, a significant decrease in TAC in milk samples from the mastitis group was observed when compared to the control group (Figure 1b). Furthermore, Figure 1b shows that cranberry intake significantly increased TAC in control and mastitis milk samples.

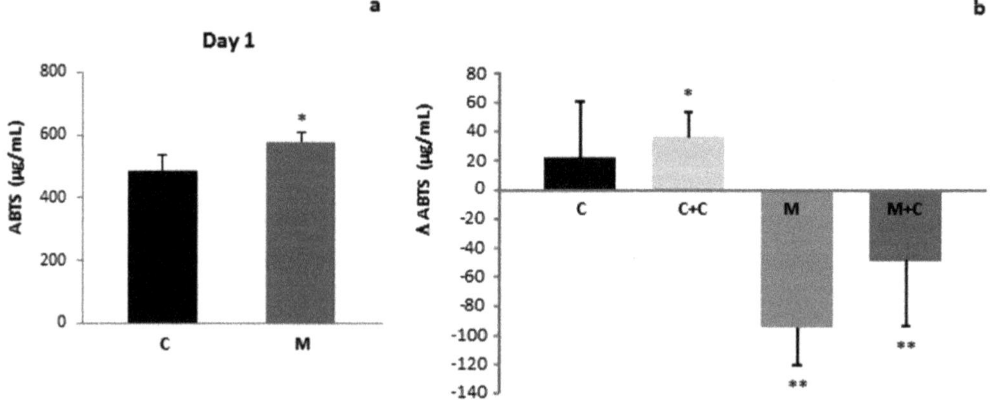

Figure 1. (**a**) Total antioxidant capacity of human milk of Control and Mastitis groups ($p < 0.05$ vs. C). (**b**) Variation of total antioxidant capacity of human milk along the study of Control, Control + Cranberries, Mastitis and Mastitis + Cranberries groups. Control (C), Control + Cranberries (C + C), Mastitis (M) and Mastitis + Cranberries (M + C). * $p < 0.05$ vs. M and M + C, ** $p < 0.05$ vs. rest of groups.

An increase in TPP content in mastitis samples at the beginning of the study was observed (Figure 2a). Similar to what occurs to TAC, TPP content normalized after 21 days. Cranberry intake increased TPP content in milk samples during the study, although this was significant only in the mastitis group (Figure 2b).

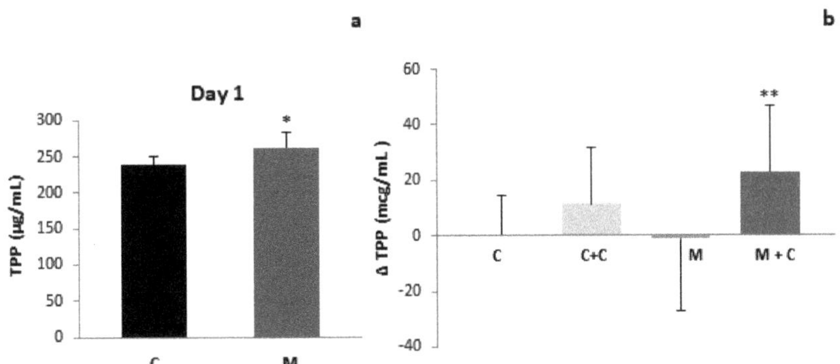

Figure 2. (**a**) Total polyphenol content of human milk of Control and Mastitis groups ($p < 0.05$ vs. C). (**b**) Variation of total polyphenol content of human milk along the study of Control, Control + Cranberries, Mastitis and Mastitis + Cranberries groups. Control (C), Control + Cranberries (C + C), Mastitis (M) and Mastitis + Cranberries (M + C). * $p < 0.05$ vs. C, ** $p < 0.05$ vs. C and M.

The GSH concentration of milk was not significantly affected by mastitis (Figure 3a). In contrast, GPx activity was increased in the milk of the mastitis group as compared to the control at the beginning of the study (Figure 3c), but again normalized 21 days later. No significant changes were observed in the evolution of this parameter in any group over the 21 days of the study. Cranberry intake had no effect on GSH concentrations nor GPx activity (Figure 3b,d, respectively).

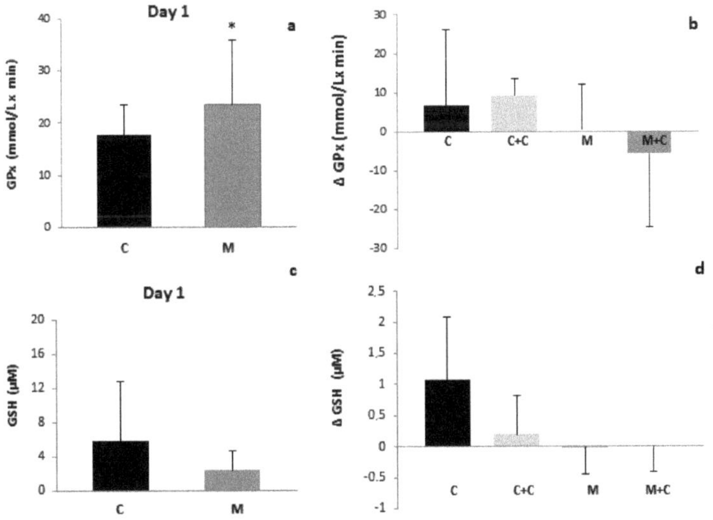

Figure 3. (**a**). GPx activity of human milk of Control and Mastitis groups ($p < 0.05$ vs. C). (**b**). Variation of GPx activity of human milk along the study of Control, Control + Cranberries, Mastitis and Mastitis + Cranberries groups. * $p < 0.05$ vs. C. (**c**). GSH concentration of human milk of Control and Mastitis groups ($p < 0.05$ vs. C). (**d**). Variation of GSH concentration of human milk along the study of Control, Control + Cranberries, Mastitis and Mastitis + Cranberries groups. Control (C), Control + Cranberries (C + C), Mastitis (M) and Mastitis + Cranberries (M + C). * $p < 0.05$ vs. C.

Greater oxidative damage to proteins was observed in mastitis milk samples as compared to the control at the beginning of the study (Figure 4a). However, this difference was not statistically significant in the samples obtained after 21 days. Nor were significant differences observed in the evolution of this parameter within groups throughout the study (Figure 4b). Contrary to the findings of CGC, MDA concentration in milk was significantly lower in the mastitis group at the beginning of the study as compared to the control (Figure 4c). Similar to CGC, after 21 days MDA concentration was normalized. Figure 4d shows that MDA concentrations were not affected by cranberry intake.

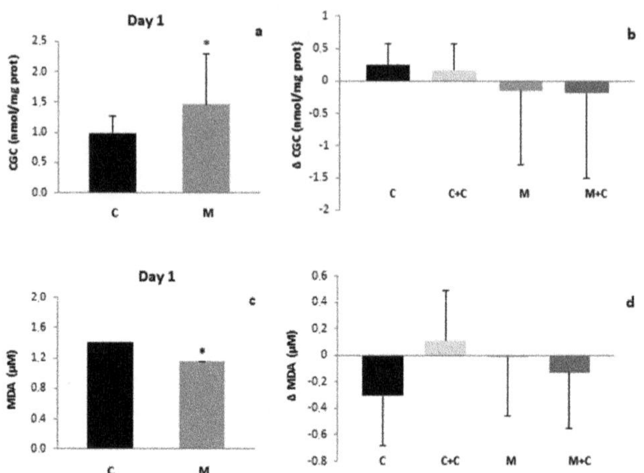

Figure 4. (**a**) Carbonyl groups content of human milk of Control and Mastitis groups ($p < 0.05$ vs. C). (**b**) Variation of Carbonyl groups content of human milk along the study of Control, Control + Cranberries, Mastitis and Mastitis + Cranberries groups. (**c**) MDA concentration of human milk of Control and Mastitis groups ($p < 0.05$ vs. C). (**d**) Variation of MDA concentration of human milk along the study of Control, Control + Cranberries, Mastitis and Mastitis + Cranberries groups. Control (C), Control + Cranberries (C + C), Mastitis (M) and Mastitis + Cranberries (M + C). * $p < 0.05$ vs. C.

4. Discussion

In recent decades there have been significant lifestyle changes in the human population, with a tendency towards an increased consumption of processed foods and a consequent rearrangement of dietary patterns [41,42]. The results from the present study demonstrate insufficient nutrient intake in the study population with respect to WHO recommendations [3], i.e., daily fruit and vegetables, as well as carbohydrate intakes were below those recommended by WHO, whereas fat and protein consumption was above WHO recommendations [3].

It has been widely reported that breast milk has a powerful and essential antioxidant composition, which is related to the combination of both exogenous and endogenous molecules including, among others, enzymes, vitamins, protein components and derivatives, oligoelements, carotenoids, and polypohenols [7,17]. Mastitis is associated with inflammatory processes and innate immune cell recruitment and activation, which in turn results in the release of proinflammatory cytokines [43]. Although leukocyte recruitment can also increase the local free radical production to unbalanced levels, thus compromising the oxidative status of milk, as has been previously described in cows, goats, and sheep [44–46], it is not clear whether the anti-inflammatory content of human milk in mastitis affected women, provides an extra protection against these changes [30]; these authors proposed that the increased contents of selected components in womens' mastitis milk (e.g., TNFa or IL-1RA) might help protect the nursing infant from clinical illness due to feeding on mastitis milk [30].

To the best of our knowledge our study shows for the first time that total antioxidant capacity (TAC) was increased (Figure 1), along with GPx activity and TPP content in milk samples from women suffering from mastitis when compared to the control group (Figures 2 and 3a,b). These results may be explained as a compensatory adaptation to oxidative stress that has previously been described in other pathological conditions, in which the antioxidative response is stimulated to reduce the oxidant content and reestablish the redox balance [47,48], or a breast-specific mechanism that increases permeability of the mammary gland for certain components. Regarding the GPx increase, it has been previously reported that lactoferrin (LF) concentration is increased in human breast milk during mastitis [27,49]. This protein is known to upregulate the expression of GPx [50] and to enhance its activity [51]. On the other hand, recent studies have demonstrated that mastitis is associated with an increase in milk selenium (Se) concentrations [27]; selenocysteine is present in the active site of GPx [52]. Therefore, the increase in milk Se concentration could certainly contribute to the enhanced GPx activity in milk from the mastitis group as compared to the control.

Polyphenols are present in human milk, of which flavonoids represent the largest subgroup [53]. Li et al. compared polyphenol content in human breast milk with that in formula [54]. However, as mentioned above, at present there are no studies addressing the effect of mastitis on human milk TPP content.

In contrast to the findings on GPx activity and TPP content, no increase in GSH concentration was observed in milk from the mastitis group (Figure 3c,d), in agreement with former results in bovine milk [44]; although these authors reported a decrease of TAC in bovine milk. Furthermore, Dimri et al. also described a decrease in milk TAC together with lower GPx activity in buffaloes [55]. These discordances may be due to interspecies differences.

Interestingly, 21 days after mastitis onset, values for all the parameters studied returned to normal, probably due to the effect of antibiotics. Moreover, 21 days cranberry supplementation increased TTP content in milk samples from both the control and mastitis group (Figure 2), since cranberries are fruits that are especially rich in polyphenols [56]. Other authors have previously demonstrated that dietary supplementation with foods rich in these compounds are able to increase the quantities of flavonoids in human milk [57,58]. Similar results were obtained for TAC in milk after supplementation with cranberries (Figure 1). Other studies in animal models of mastitis have also focused on the role of natural antioxidants such as vitamin E or ginsenoside Rg1 in protecting the oxidative status of milk [55,59], although no studies are available on human milk.

With regard to oxidative damage to macromolecules, there are no reports at present showing the effects of mastitis in terms of oxidative damage to proteins. We have demonstrated that mastitis induces oxidative damage to milk proteins (Figure 4). This result is consistent with the increase in mastitis milk protein content reported by Samuel and cols. [27]. In contrast to the observed oxidative damage to proteins, lipid peroxidation (MDA concentration) was diminished in milk from the mastitis group when compared to the control (Figure 4a,b). This is a surprising result, given that lipids are more sensitive to oxidative damage than proteins. In this respect, it is noteworthy that Say et al. recently evidenced that lactational mastitis is associated with lower breast milk fat [28], and therefore it is possible that the decrease in MDA concentration in the mastitis group was due, at least in part, to the diminished fat content. Furthermore, as mentioned above, recent studies have reported that mastitis increases LF concentrations [27,50]. LF sequesters Fe^{3+}, preventing the formation of hydroxyl radicals via the Fenton reaction, and in turn prevents lipid peroxidation [51,60]. Therefore, the increase in LF content in milk could also contribute to the decrease in MDA concentration observed in the mastitis group (Figure 4a,b). Again, these consequences of mastitis could be specific to lactating human females, since lipid peroxidation products are increased in milk samples from rats or cows with mastitis [61,62].

Both CGC and MDA concentrations normalized at the end of the study, after antimicrobial therapy (Figure 4), which probably reduces the macronutrient content impairment described previously [27,28]. It is remarkable that cranberry supplementation had no effect on these oxidative damage parameters (Figure 4), probably because cranberries are able to attenuate the oxidative insult by increasing overall antioxidant defense. Many studies have assayed therapies with antioxidants in different animal models to prevent the effects of mastitis on the oxidative status of milk [55,59,63]. The results have been mainly successful, but some controversial results have also been reported regarding the use of several antioxidants such as vitamin E [64,65]. However, at present there are no studies reporting on the antioxidative status of milk from women with mastitis or the effects of antioxidant supplementation on this status in human milk.

In conclusion, during mastitis there is a compensatory mechanism in human milk in which antioxidant defenses are increased, providing the nursing infant with further protection while feeding on mastitis milk. In addition, the use of cranberries as supplement for antibiotic therapy seems to strengthen the antioxidant system of milk.

Author Contributions: Conceptualization, M.M., V.V.-B., M.T.H.-A. and A.N.; methodology, C.A., V.V.-B. and M.T.H.-A.; formal analysis, C.G. and P.B.; investigation, C.G., C.A. and A.N.; resources, M.M., V.V.-B. and P.B.; data curation, M.M. and P.B.; writing—original draft preparation, C.A., M.M. and V.V.-B.; writing—review and editing, M.M., F.J.R. and P.B. All authors have read and agreed to the published version of the manuscript.

Funding: This research was funded by Universitat Jaume I (UJI), Spain; grant number UJI-A2016-03 and UJI-B2019-38.

Institutional Review Board Statement: The study was conducted according to the guidelines of the Declaration of Helsinki, and approved by the Institutional Review Board and Ethics Committees of Hospital Universitario Peset from Valencia (23/16) and Hospital General Universitario from Castellón (4/2015).

Informed Consent Statement: Informed consent was obtained from all subjects involved in the study.

Data Availability Statement: Data is contained within the article.

Acknowledgments: Not applicable.

Conflicts of Interest: The authors declare no conflict of interest.

References

1. Fanaro, S. The biological specificity and superiority of human milk. Scientific basis, guarantees and safety controls. *Minerva Pediatr.* **2002**, *54*, 113–129.
2. Ojo-Okunola, A.; Nicol, M.; du Toit, E. Human breast milk bacteriome in health and disease. *Nutrients* **2018**, *10*, 1643. [CrossRef]
3. WHO. *Healthy Diet. Fact Sheet, n° 394*; WHO: Geneva, Switzerland, August 2018.
4. Abbe, M.R.; Friel, J.K.J. Superoxide dismutase and glutathione peroxidase content of human milk from mothers of premature and full-term infants during the first 3 months of lactation. *Gastroenterol. Nutr.* **2000**, *31*, 174–270.
5. Turoli, D.; Testolin, G.; Zanini, R.; Bellù, R. Determination of oxidative status in breast and formula milk. *Acta Paediatr.* **2004**, *93*, 1569–1574. [CrossRef]
6. Hanson, C.; Lyden, E.; Furtado, J.; Van Ormer, M.; Anderson-Berry, A. A comparison of nutritional antioxidant content in breast milk, donor milk, and infant formulas. *Nutrients* **2016**, *8*, 681. [CrossRef]
7. Tsopmo, A. Phytochemicals in human milk and their potential antioxidative protection. *Antioxidants* **2018**, *7*, 32. [CrossRef]
8. Silvestre, D.; Miranda, M.; Muriach, M.; Almansa, I.; Jareño, E.; Romero, F.J. Frozen breast milk at −20 °C and −80 °C: A longitudinal study of glutathione peroxidase activity and malondialdehyde concentration. *J. Hum. Lactation* **2010**, *26*, 35–41. [CrossRef]
9. Silvestre, D.; Miranda, M.; Muriach, M.; Almansa, I.; Jareno, E.; Romero, F.J. Antioxidant capacity of human milk: Effect of thermal conditions for the pasteurization. *Acta Paediatr. Int. J. Paediatr.* **2008**, *97*, 1070–1074. [CrossRef]
10. Miranda, M.; Muriach, M.; Almansa, I.; Jareño, E.; Bosch-Morell, F.; Romero, F.J.; Silvestre, D. Oxidative status of human milk and its variations during cold storage. *BioFactors* **2004**, *20*, 129–137. [CrossRef]
11. Codoñer-Franch, P.; Hernández-Aguilar, M.T.; Navarro-Ruiz, A.; López-Jaén, A.B.; Borja-Herrero, C.; Valls-Bellés, V. Diet supplementation during early lactation with non-alcoholic beer increases the antioxidant properties of breastmilk and decreases the oxidative damage in breastfeeding mothers. *Breastfeed. Med.* **2013**, *8*, 164–169. [CrossRef]
12. Mutinati, M.; Pantaleo, M.; Roncetti, M.; Piccinno, M.; Rizzo, A.; Sciorsci, R.L. Oxidative stress in neonatology: A review. *Reprod. Domest. Anim.* **2014**, *49*, 7–16. [CrossRef]
13. Ledo, A.; Arduini, A.; Asensi, M.A.; Sastre, J.; Escrig, R.; Brugada, M.; Aguar, M.; Saenz, P.; Vento, M. Human milk enhances antioxidant defenses against hydroxyl radical aggression in preterm infants. *Am. J. Clin. Nutr.* **2009**, *89*, 210–215. [CrossRef]

14. Lee, J.W.; Davis, J.M. Future applications of antioxidants in premature infants. *Curr. Opin. Pediatr.* **2011**, *23*, 161–166. [CrossRef] [PubMed]

15. Parkinson, J.R.C.; Hyde, M.J.; Gale, C.; Santhakumaran, S.; Modi, N. Preterm birth and the metabolic syndrome in adult life: A systematic review and meta-analysis. *Pediatrics* **2013**, *131*, e1240–e1263. [CrossRef]

16. Lapillonne, A.; Griffin, I.J. Feeding preterm infants today for later metabolic and cardiovascular outcomes. *J. Pediatr.* **2013**, *162*, S7. [CrossRef] [PubMed]

17. Gila-Diaz, A.; Arribas, S.M.; Algara, A.; Martín-Cabrejas, M.A.; de Pablo, Á.L.L.; de Pipaón, M.S.; Ramiro-Cortijo, D. A review of bioactive factors in human breastmilk: A focus on prematurity. *Nutrients* **2019**, *11*, 1307. [CrossRef] [PubMed]

18. Le, H. Benign breast disorders: The clinician's view. *Cancer Detect. Prev.* **1992**, *16*, 1–5.

19. Contreras, G.A.; Rodríguez, J.M. Mastitis: Comparative etiology and epidemiology. *J. Mammary Gland Biol. Neoplasia* **2011**, *16*, 339–356. [CrossRef]

20. Patel, S.H.; Vaidya, Y.H.; Patel, R.J.; Pandit, R.J.; Joshi, C.G.; Kunjadiya, A.P. Culture independent assessment of human milk microbial community in lactational mastitis. *Sci. Rep.* **2017**, *7*, 7804. [CrossRef]

21. Jahanfar, S.; Ng, C.J.; Teng, C.L. Antibiotics for mastitis in breastfeeding women. *Sao Paulo Med. J.* **2016**, *134*, 273. [CrossRef]

22. Bond, D.M.; Morris, J.M.; Nassar, N. Study protocol: Evaluation of the probiotic Lactobacillus Fermentum CECT5716 for the prevention of mastitis in breastfeeding women: A randomised controlled trial. *BMC Pregnancy Childbirth* **2017**, *17*, 148. [CrossRef]

23. Johnson, H.M.; Mitchell, K.B. Lactational phlegmon: A distinct clinical entity affecting breastfeeding women within the mastitis-abscess spectrum. *Breast J.* **2020**, *26*, 149–154. [CrossRef] [PubMed]

24. Gelasakis, A.I.; Angelidis, A.S.; Giannakou, R.; Filioussis, G.; Kalamaki, M.S.; Arsenos, G. Bacterial subclinical mastitis and its effect on milk yield in low-input dairy goat herds. *J. Dairy Sci.* **2016**, *99*, 3698–3708. [CrossRef] [PubMed]

25. Sun, M.; Gao, J.; Ali, T.; Yu, D.; Zhang, S.; Khan, S.U.; Fanning, S.; Han, B. Characteristics of Aerococcus viridans isolated from bovine subclinical mastitis and its effect on milk SCC, yield, and composition. *Trop. Anim. Health Prod.* **2017**, *49*, 843–849. [CrossRef]

26. Silanikove, N.; Merin, U.; Shapiro, F.; Leitner, G. Subclinical mastitis in goats is associated with upregulation of nitric oxide-derived oxidative stress that causes reduction of milk antioxidative properties and impairment of its quality. *J. Dairy Sci.* **2014**, *97*, 3449–3455. [CrossRef]

27. Samuel, T.M.; De Castro, C.A.; Dubascoux, S.; Affolter, M.; Giuffrida, F.; Billeaud, C.; Picaud, J.-C.; Agosti, M.; Al-Jashi, I.; Barroso Pereira, A.; et al. Subclinical Mastitis in a European Mul-ticenter Cohort: Prevalence, Impact on Human Milk (HM) Composition, and Association with Infant HM Intake and Growth. *Nutrients* **2020**, *12*, 105. [CrossRef] [PubMed]

28. Say, B.; Dizdar, E.A.; Degirmencioglu, H.; Uras, N.; Sari, F.N.; Oguz, S.; Canpolat, F.E. The effect of lactational mastitis on the macronutrient content of breast milk. *Early Hum. Dev.* **2016**, *98*, 7–9. [CrossRef]

29. Perez, M.; Ladero, V.; Redruello, B.; Del Rio, B.; Fernandez, L.; Rodriguez, J.M.; Martín, M.C.; Fernandez, M.; Alvarez, M.A. Mastitis modifies the biogenic amines profile in human milk, with significant changes in the presence of histamine, putrescine and spermine. *PLoS ONE* **2016**, *11*, e0162426. [CrossRef]

30. Buescher, E.S.; Hair, P.S. Human milk anti-inflammatory component contents during acute mastitis. *Cell. Immunol.* **2001**, *210*, 87–95. [CrossRef]

31. Nardi, G.M.; De Farias Januário, A.G.; Freire, C.G.; Megiolaro, F.; Schneider, K.; Perazzoli, M.R.A.; Do Nascimento, S.R.; Gon, A.C.; Mariano, L.N.B.; Wagner, G.; et al. Anti-inflammatory activity of berry fruits in mice model of inflammation is based on oxidative stress modulation. *Pharmacogn. Res.* **2016**, *8*, S42–S49. [CrossRef]

32. Grace, M.H.; Esposito, D.; Dunlap, K.L.; Lila, M.A. Comparative analysis of phenolic content and profile, antioxidant capacity, and anti-inflammatory bioactivity in wild alaskan and commercial vaccinium berries. *J. Agric. Food Chem.* **2014**, *62*, 4007–4017. [CrossRef] [PubMed]

33. Miller, N.J.; Rice-Evans, C.A. Factors influencing the antioxidant activity determined by the ABTS$^{\bullet+}$ radical cation assay. *Free Radic. Res.* **1997**, *26*, 195–199. [CrossRef]

34. Re, R.; Pellegrini, N.; Proteggente, A.; Pannala, A.; Yang, M.; Rice-Evans, C. Antioxidant activity applying an improved ABTS radical cation decolorization assay. *Free Radic. Biol. Med.* **1999**, *26*, 1231–1237. [CrossRef]

35. Lawrence, R.A.; Parkhill, L.K.; Burk, R.F. Hepatic cytosolic non selenium-dependent glutathione peroxidase activity: Its nature and the effect of selenium deficiency. *J. Nutr.* **1978**, *108*, 981–987. [CrossRef] [PubMed]

36. Reed, D.J.; Babson, J.R.; Beatty, P.W.; Brodie, A.E.; Ellis, W.W.; Potter, D.W. High-performance liquid chromatography analysis of nanomole levels of glutathione, glutathione disulfide, and related thiols and disulfides. *Anal. Biochem.* **1980**, *106*, 55–62. [CrossRef]

37. Singleton, V.L.; Rossi, J.A.J. Colorimetry to total phenolics with phosphomolybdic acid reagents. *Am. J. Enol. Vinic.* **1965**, *16*, 144–158.

38. Romero, M.J.; Bosch-Morell, F.; Romero, B.; Rodrigo, J.M.; Serra, M.A.; Romero, F.J. Serum malondialdehyde: Possible use for the clinical management of chronic hepatitis C patients. *Free Radic. Biol. Med.* **1998**, *25*, 993–997. [CrossRef]

39. Richard, M.J.; Guiraud, P.; Meo, J.; Favier, A. High-performance liquid chromatographic separation of malondialdehyde-thiobarbituric acid adduct in biological materials (plasma and human cells) using a commercially available reagent. *J. Chromatogr. B Biomed. Sci. Appl.* **1992**, *577*, 9–18. [CrossRef]

40. Levine, R.L.; Garland, D.; Oliver, C.N.; Amici, A.; Climent, I.; Lenz, A.G.; Ahn, B.W.; Shaltiel, S.; Stadtman, E.R. Determination of Carbonyl Content in Oxidatively Modified Proteins. *Methods Enzymol.* **1990**, *186*, 464–478. [CrossRef]
41. Calder, P.C. Very long-chain n-3 fatty acids and human health: Fact, fiction and the future. *Proc. Nutr. Soc.* **2018**, *77*, 52–72. [CrossRef]
42. Steffl, M.; Kinkorova, I.; Kokstejn, J.; Petr, M. Macronutrient intake in soccer players—A meta-analysis. *Nutrients* **2019**, *11*, 1305. [CrossRef]
43. Espinosa-Martos, I.; Jiménez, E.; De Andrés, J.; Rodríguez-Alcalá, L.M.; Tavárez, S.; Manzano, S.; Fernández, L.; Alonso, E.; Fontecha, J.; Rodríguez, J.M. Milk and blood biomarkers associated to the clinical efficacy of a probiotic for the treatment of infectious mastitis. *Benef. Microbes* **2016**, *7*, 305–318. [CrossRef]
44. Sadek, K.; Saleh, E.; Ayoub, M. Selective, reliable blood and milk bio-markers for diagnosing clinical and subclinical bovine mastitis. *Trop. Anim. Health Prod.* **2017**, *49*, 431–437. [CrossRef]
45. Alba, D.F.; da Rosa, G.; Hanauer, D.; Saldanha, T.F.; Souza, C.F.; Baldissera, M.D.; da Silva dos Santos, D.; Piovezan, A.P.; Girardini, L.K.; Schafer Da Silva, A. Subclinical mastitis in Lacaune sheep: Causative agents, impacts on milk production, milk quality, oxidative profiles and treatment efficacy of ceftiofur. *Microb. Pathog.* **2019**, *137*, 103732. [CrossRef]
46. Wang, Y.; Zhang, Y.; Chi, X.; Ma, X.; Xu, W.; Shi, F.; Hu, S. Anti-inflammatory mechanism of ginsenoside Rg1: Proteomic analysis of milk from goats with mastitis induced with lipopolysaccharide. *Int. Immunopharmacol.* **2019**, *71*, 382–391. [CrossRef]
47. Baliño, P.; Romero-Cano, R.; Sánchez-Andrés, J.V.; Valls, V.; Aragón, C.G.; Muriach, M. Effects of Acute Ethanol Administration on Brain Oxidative Status: The Role of Acetaldehyde. *Alcohol. Clin. Exp. Res.* **2019**, *43*, 1672–1681. [CrossRef]
48. Fromenty, B.; Vadrot, N.; Massart, J.; Turlin, B.; Barri-Ova, N.; Lettéron, P.; Fautrel, A.; Robin, M.A. Chronic ethanol consumption lessens the gain of body weight, liver triglycerides, and diabetes in obese ob/ob mice. *J. Pharmacol. Exp. Ther.* **2009**, *331*, 23–34. [CrossRef]
49. Fetherston, C.M.; Lai, C.T.; Hartmann, P.E. Relationships between symptoms and changes in breast physiology during lactation mastitis. *Breastfeed. Med. Off. J. Acad. Breastfeed. Med.* **2006**, *1*, 136–145. [CrossRef]
50. Kruzel, M.L.; Zimecki, M.; Actor, J.K. Lactoferrin in a context of inflammation-induced pathology. *Front. Immunol.* **2017**, *8*, 1438. [CrossRef]
51. Han, N.; Li, H.; Li, G.; Shen, Y.; Fei, M.; Nan, Y. Effect of bovine lactoferrin as a novel therapeutic agent in a rat model of sepsis-induced acute lung injury. *AMB Express* **2019**, *9*, 177. [CrossRef]
52. Rotruck, J.T.; Pope, A.L.; Ganther, H.E.; Swanson, A.B.; Hafeman, D.G.; Hoekstra, W.G. Selenium: Biochemical role as a component of glatathione peroxidase. *Science* **1973**, *179*, 588–590. [CrossRef]
53. Song, B.J.; Jouni, Z.E.; Ferruzzi, M.G. Assessment of phytochemical content in human milk during different stages of lactation. *Nutrition* **2013**, *29*, 195–202. [CrossRef]
54. Li, W.; Hosseinian, F.S.; Tsopmo, A.; Friel, J.K.; Beta, T. Evaluation of antioxidant capacity and aroma quality of breast milk. *Nutrition* **2009**, *25*, 105–114. [CrossRef]
55. Dimri, U.; Sharma, M.C.; Singh, S.K.; Kumar, P.; Jhambh, R.; Singh, B.; Bandhyopadhyay, S.; Verma, M.R. Amelioration of altered oxidant/antioxidant balance of Indian water buffaloes with subclinical mastitis by vitamins A, D3, E, and H supplementation. *Trop. Anim. Health Prod.* **2013**, *45*, 971–978. [CrossRef]
56. Li, C.; Feng, J.; Huang, W.Y.; An, X.T. Composition of polyphenols and antioxidant activity of rabbiteye blueberry (Vaccinium ashei) in Nanjing. *J. Agric. Food Chem.* **2013**, *61*, 523–531. [CrossRef]
57. Franke, A.A.; Halm, B.M.; Custer, L.J.; Tatsumura, Y.; Hebshi, S. Isoflavones in breastfed infants after mothers consume soy. *Am. J. Clin. Nutr.* **2006**, *84*, 406–413. [CrossRef]
58. Jochum, F.; Alteheld, B.; Meinardus, P.; Dahlinger, N.; Nomayo, A.; Stehle, P. Mothers' Consumption of Soy Drink but Not Black Tea Increases the Flavonoid Content of Term Breast Milk: A Pilot Randomized, Controlled Intervention Study. *Ann. Nutr. Metab.* **2017**, *70*, 147–153. [CrossRef]
59. Dai, H.; Coleman, D.N.; Hu, L.; Martinez-Cortés, I.; Wang, M.; Parys, C.; Shen, X.; Loor, J.J. Methionine and arginine supplementation alter inflammatory and oxidative stress responses during lipopolysaccharide challenge in bovine mammary epithelial cells in vitro. *J. Dairy Sci.* **2020**, *103*, 676–689. [CrossRef]
60. Alhalwani, A.Y.; Davey, R.L.; Kaul, N.; Barbee, S.A.; Alex Huffman, J. Modification of lactoferrin by peroxynitrite reduces its antibacterial activity and changes protein structure. *Proteins Struct. Funct. Bioinform.* **2020**, *88*, 166–174. [CrossRef]
61. Eslami, H.; Batavani, R.A.; Asr I-Rezaei, S.; Hobbenaghi, R. Changes of stress oxidative enzymes in rat mammary tissue, blood and milk after experimental mastitis induced by E. coli lipopolysaccharide. *Vet. Res. Forum Int. Q. J.* **2015**, *6*, 131–136.
62. Nedić, S.; Vakanjac, S.; Samardžija, M.; Borozan, S. Paraoxonase 1 in bovine milk and blood as marker of subclinical mastitis caused by Staphylococcus aureus. *Res. Vet. Sci.* **2019**, *125*, 323–332. [CrossRef]
63. Abouzed, T.K.; Sadek, K.M.; Ayoub, M.M.; Saleh, E.A.; Nasr, S.M.; El-Sayed, Y.S.; Shoukry, M. Papaya extract upregulates the immune and antioxidants-related genes, and proteins expression in milk somatic cells of Friesian dairy cows. *J. Anim. Physiol. Anim. Nutr.* **2019**, *103*, 407–415. [CrossRef]

64. Mukherjee, R. Selenium and vitamin E increases polymorphonuclear cell phagocytosis and antioxidant levels during acute mastitis in riverine buffaloes. *Vet. Res. Commun.* **2008**, *32*, 305–313. [CrossRef]
65. Politis, I.; Theodorou, G.; Lampidonis, A.D.; Kominakis, A.; Baldi, A. Short communication: Oxidative status and incidence of mastitis relative to blood α-tocopherol concentrations in the postpartum period in dairy cows. *J. Dairy Sci.* **2012**, *95*, 7331–7335. [CrossRef]

Article

Platinum Nanoparticles: The Potential Antioxidant in the Human Lung Cancer Cells

Noor Akmal Shareela Ismail *, Jun Xin Lee and Fatimah Yusof

Biochemistry Department, Faculty of Medicine, Universiti Kebangsaan Malaysia, Jalan Yaacob Latif, Bandar Tun Razak, Kuala Lumpur 56000, Malaysia; a161478@siswa.ukm.edu.my (J.X.L.); p74488@siswa.ukm.edu.my (F.Y.)
* Correspondence: nasismail@ukm.edu.my

Abstract: Oxidative stress-related conditions associated with lung cells, specifically lung cancer, often lead to a poor prognosis. We hypothesized that platinum nanoparticles (PtNPs) can play a role in reversing oxidative stress in human lung adenocarcinoma A549 epithelial lung cell lines. Hydrogen peroxide (H_2O_2) was used to induce oxidative stress in cells, and the ability of PtNPs to lower the oxidative stress in the H_2O_2 treated epithelial lung cell line was determined. The differential capacity of PtNPs to remove H_2O_2 was studied through cell viability, nanoparticle uptake, DNA damage, ROS production, and antioxidant enzymes (superoxide dismutase, glutathione peroxidase, and catalase). Results indicated that a higher concentration of PtNPs exhibited a higher antioxidant capacity and was able to reduce DNA damage and quench ROS production in the presence of 350 μM H_2O_2. All antioxidant enzymes' activities also increased in the PtNPs treatment. Our data suggested that PtNPs could be a promising antioxidant in the treatment of lung cancer.

Keywords: platinum nanoparticles; lung cancer; A549 epithelial lung cell; antioxidant; oxidative stress; reactive oxygen species

Citation: Ismail, N.A.S.; Lee, J.X.; Yusof, F. Platinum Nanoparticles: The Potential Antioxidant in the Human Lung Cancer Cells. *Antioxidants* **2022**, *11*, 986. https://doi.org/10.3390/antiox11050986

Academic Editor: Stanley Omaye

Received: 9 April 2022
Accepted: 16 May 2022
Published: 18 May 2022

Publisher's Note: MDPI stays neutral with regard to jurisdictional claims in published maps and institutional affiliations.

1. Introduction

The balance between oxidant and antioxidant defense mechanisms is important to maintain the metabolic state of normal cells. Oxidative stress conditions or diseases are the endpoints of the imbalance between these mechanisms [1]. This is due to changes in the structure of proteins and nucleic acids and an increased permeability of the membrane to injury and lipid peroxidation [1]. The accumulation of reactive oxygen species (ROS) takes place and subsequently induces cellular oxidative damage. Reactive oxygen species (ROS) are highly reactive molecules generated from the oxidation process when molecular oxygen (O_2) is reduced to produce superoxide ($O_2^{\bullet-}$), which is the precursor to most other reactive oxygen species [2]. Cells that undergo oxidative stress are unable to function normally and ROS will attack various cell components including lipids, enzymes, and DNA [3]. Some organs will prevent oxidative damage through enzymatic and non-enzymatic antioxidant defense mechanisms [4]. Enzymatic antioxidant defense mechanisms consist of superoxide dismutase (SOD), glutathione peroxidase (GPx), and catalase (CAT), while non-enzymatic antioxidant defense mechanisms originate from various sources such as vitamin C, vitamin E, and uric acid [5].

Lung cancer remains one of the most common types of cancer and the leading cause of cancer-related deaths worldwide despite considerable advancements in diagnostics and management [6]. Oxidative stress in lung cells can be a deleterious process that leads to lung cancer that carries a poor prognosis. It plays a crucial role in carcinogenesis by promoting tumor growth and inducing neoplastic transformation [7]. While exogenous antioxidants through daily diet consumption and derivatives of medicinal plants have been intensively explored as potential preventive agents [8], the role of endogenous antioxidants in lung cancer is ambiguous as various antioxidant enzymes have been found to promote neoplastic cell viability [9]. Hence, ROS regulation is one of the essential fields of study, especially in the presence of antioxidant agents as a pharmacological manipulation in lung inflammation and injury.

The use of cisplatin and platinum-based chemotherapy has been the fundamental treatment of lung cancer, but it has also been shown to cause organ toxicities, especially in the kidney, gastrointestinal, and neuron cells [10]. These adverse events can interfere with the treatment of lung cancer. Therefore, the elucidation of its mechanism to treat lung cells is important to explore the potential of platinum nanoparticles (PtNPs) in reducing cell toxicity. Metal nanoparticles have raised considerable interest in research due to their material properties, availability, capabilities, specific targeting, and sustainable release [11]. PtNPs specifically are chosen as an antioxidant candidate in this study as PtNPs can reverse antioxidant activity in several induced oxidative conditions [12–15] and have quenched the production of free radicals, thus reducing the impact of oxidative damage [16,17]. PtNPs have garnered attention in medicine due to their reactivity to cells and are renowned for their therapeutic application in different cell lines. PtNPs have successfully induced cell death in cervical cancer cells [18], breast cancer cells [19], and human colon carcinoma cell lines HT29 [20,21], notified by reduced DNA damage and antioxidant response. In addition, PtNPs do not show a cytotoxicity effect in several established cultured cells (TIG-1, WI-38, MRC-5, HeLa, and HepG2) whilst the cellular uptake of PtNPs does occur in a time- and dose-dependent manner [22]. The exposure of PtNPs through in vivo experiments has been found to only affect lung cells but not the brain, the kidney, the heart, and liver cells [13]. It was first discovered as an antioxidant through the ability to protect the inflamed lung cells from further oxidation-induced inflammation by scavenging superoxide anions ($O_2^{\bullet-}$) and hydroxyl radicals ($OH^{\bullet}$) from aqueous solutions [13].

Thus, this study aimed to determine the antioxidant properties of PtNPs in balancing oxidative status in human lung adenocarcinoma A549 epithelial lung cell line as an in vitro model; an area in which the available evidence is scarce. As such, the tests that were being carried out included the Ferric Reducing Antioxidant Power (FRAP) assay, the MTS assay, the comet assay, and antioxidant enzymes (superoxide dismutase, glutathione peroxidase, catalase) activity to further elucidate the antioxidative role of PtNPs.

2. Materials and Methods

2.1. Platinum Nanoparticles and Hydrogen Peroxide

Platinum nanoparticles (PtNPs) colloids were purchased from Sigma-Aldrich, St. Louis, MI, USA. The PtNPs were 3 nm in size, without any prior modification or purification done. This is according to a study made by the previous literature that suggested a similar size of diameter of each PAA-Pt nanoparticle (2.0 ± 0.4 nm) as these PtNPs exhibit the chemical composition, biological reactivity, and antioxidant properties of the nanoparticles [13]. Hydrogen peroxide (H_2O_2) was purchased from Merck, Germany. This was to induce an oxidative condition in the cell line.

2.2. Treatment Procedure

A549 cells were cultured between passages P17 to P25 that were at the growth phase. Six treatment groups were used during this study that were the untreated group (control), cells with PtNPs and H_2O_2, respectively, and H_2O_2/PtNPs treatment. There were two durations, 3 h representing the acute treatment, and 24 h for the chronic treatment. The number of cells used per well is 0.5×10^6 for a petri dish and 5×10^3 for each of the 96-well.

2.3. Cell Culture

Human lung adenocarcinoma epithelial cell line A549 was purchased from American Type Culture Collection (ATCC), item number CCL-185. A549 cells were cultured in Dulbecco's Modified Eagle's Medium-F12 (DMEM-F12), with additional L-glutamine supplemented with 10% Fetal Bovine Serum (FBS), 1% penicillin/streptomycin, and 1%

amphotericin B. Cells were incubated in a humidified atmosphere at 37 °C in 5% CO_2. All chemicals for cell culture were obtained from Biowest, USA.

2.4. Ferric Reducing Antioxidant Power (FRAP) Assay

FRAP assay was carried out to study the antioxidant capacity, based on the ability of the sample to reduce Fe^{3+} to Fe^{2+}. FRAP reagent was freshly prepared by mixing acetate buffer, $FeCl_3 \cdot 6H_2O$ solution, and 2,4,6-Tris(2-pyridyl)-s-triazine (TPTZ) in a 10:1:1 ratio [23]. The concentration of PtNPs stock used was 1000 µg/mL and the concentration of ascorbic acid (Sigma-Aldrich) stock was 5000 µg/mL. Ascorbic acid acts as a positive control for antioxidant properties. Absorbance readings were taken at 593 nm using a microplate reader (PerkinElmer, Waltham, MA, USA). The antioxidant capacity of the sample was determined by plotting a standard curve of $FeSO_4 \cdot 7H_2O$.

2.5. MTS Assay

This colorimetric assay was used to determine the number of living cells through the how much NAD(P)H-dependent cellular oxidoreductase enzymes convert 3-(4,5-dimethylthiazol-2-yl)-5-(3-carboxymethoxyphenyl)-2-(4-sulfophenyl)-2H-tetrazolium (MTS) (Invitrogen, Paisley, UK) to formazan and is soluble in aqueous media cell culture [24]. Thus, the number of living cells can be measured directly by measuring the quantity of formazan produced at an absorbance of 490 nm. After treatment duration, cells were washed with phosphate-buffered saline (PBS). The mixture of 20 µL MTS solution and 100 µL media was added to each well for 2 h. Cell viability percentage was calculated based on the ratio of mean optical density from each treatment group to the mean OD control group.

2.6. Nanoparticles Uptake

This protocol is adapted from the processing tissue and cells for transmission electron microscopy protocol [25]. A549 cells were treated with 100 µg/mL PtNPs for 3 and 24 h. After the treatment duration, cells were washed with PBS three times. Next, the cells were trypsinized and washed with PBS again to get rid of unbound PtNPs. Cells were fixed with 3% glutaraldehyde for 30 min and underwent the second fixation with 1% osmium tetroxide for 20 min. Cells were washed with PBS before being blocked by 0.5–3.0% uranyl acetate for an hour. Serial ethanol percentage was carried out before infiltration with resin. Cells were embedded in a 100% resin capsule and hardened at 70 °C. Subsequently, cells were cut ultrathin and stained with uranyl acetate. Cells in grid form were analyzed using a transmission electron microscope (Philips, Hillsboro, OR, USA).

2.7. Comet Assay

DNA damage in A549 cells was measured by the comet assay based on the protocol by Olive and Banáth, 2006 [26]. Cells were embedded in agarose and lysed by lysis solution in a Coplin jar. Electrophoresis was done in an alkaline solution to separate DNA in the nucleus. DNA damage formed a comet-like tail due to DNA fragments that move faster than the nucleus. After undergoing the neutralization process, cells were stained with Ethidium Bromide (EtBr). Cells with intact nucleus appeared as intact orange dots under a fluorescence microscope (Carl Zeiss, Jena, Germany). A total of 500 cells that did not overlap were selected randomly from each of the slides. Grading was done on each cell according to a scale of 0 to 4 based on DNA damage level. The total cell count for each grade was multiplied by the number of each grade and summed together for total DNA damage.

2.8. Reactive Oxygen Species (ROS) Production

2′,7′-dichlorodihydrofluorescein diacetate (H_2DCF-DA) and dihydroethidium (DHE) were used to detect ROS production. DHE produced red fluorescence when oxidized by superoxide radicals while H_2DCFDA emitted green fluorescence when oxidized by H_2O_2.

After treatment, cells were washed with 50 mM PBS before staining with either 5 mM DHE or 50 mM H_2DCFDA. Plates were incubated for 30 min. Then, cells were washed again to remove the excess dye in the cell before the media was added to each well for 30 min. Next, cells were washed before being trypsinized using 500 µL trypsin. The cell suspension was added to 800 µL of 50 mM PBS before being transferred into 1.5 mL centrifuge tubes and spun at $3000\times g$ for 5 min. The supernatant was removed, and the pellet was dissolved in 200 µL of 50 mM PBS. Changes in the fluorescence intensity of the cell population to the right side of the X-axis than in the control group showed ROS production increased. A total of 10,000 cells were used for each sample and results were obtained using a flow cytometer (BD Facsverse). This protocol was adapted from Wojtala et al. [27].

2.9. Antioxidant Enzymes

Superoxide dismutase (SOD) and glutathione peroxidase (GPx) activity were measured using the commercially available kit (Cayman Chemicals, USA) following the protocol provided by the manufacturer. Catalase (CAT) enzyme activity was determined by using 50 mM PBS buffer, pH 7.0, based on a protocol by Aebi et al. [28]. The protein concentration of each sample was determined in advance using the Bradford reagent. During the experiment, 1.5 ug/mL of extract enzymes were added to each well. Control wells were added to with 50 mM PBS buffer meanwhile 0.01 M H_2O_2 was inserted into the sample wells. Absorbance readings were recorded at 240 nm at 0 and 30 s.

2.10. Statistical Analysis

All data were analyzed as mean ± standard deviation for the value obtained at least in three independent experiments. The data were analyzed using the one-way analysis of variance (ANOVA) by SPSS version 22. p-values of less than 0.05 ($p < 0.05$) were accepted as significant results and denoted as a comparison between two groups in an alphabet.

3. Results

3.1. Antioxidant Capacity of PtNPs was Lower Than Ascorbic Acid

An antioxidant capacity of PtNPs was first initiated to determine the optimal dosage prior to subsequent experiments. The half-maximal inhibitory concentration (IC50) has been determined through the half percentage of the cell viability in the oxidation of H_2O_2 at 350 µM (Figure 1A). The percentage of cell viability reduces when the duration of different PtNPs concentration treatments is increased. PtNPs treatment groups that are pre-treated with H_2O_2 also show a lower percentage of cell viability in prolonged treatment (Figure 1B).

Next, the antioxidant capacity of the PtNPs that was measured with the FRAP assay was significantly lower than ascorbic acid (Figure 1B). PtNPs were able to convert ferric ions to ferrous ions lower than in ascorbic acid. The scavenging of free radical activity of PtNPs was still durable even though the reading of the FRAP value in PtNPs was much lower than ascorbic acid. The amount of PtNPs in the FRAP assay can only be measured up until 1000 µg/mL as that was the highest concentration in the initial stock. Ascorbic acid (AA) shows higher antioxidant capacity at all concentrations than platinum nanoparticles (PtNPs).

3.2. Cell Viability Was Better in Acute PtNPs

The half-maximal inhibitory concentration (IC50) had shown the half percentage of the cell viability through the oxidation of H_2O_2 was at 350 µM (Figure 1A). Based on this data, this dosage was used in all experiments. We had chosen two different durations, acute (3 h) and chronic (24 h) exposure of PtNPs. A longer incubation of PtNPs revealed less cell viability (Figure 2B) as compared to 3 h (Figure 2A). In the acute treatment (3 h), 100 ug/mL showed the highest cell viability where the 50% cell viability is noted at 300 ug/mL (Figure 2C). In the prolonged treatment (24 h), all ranges of concentration used in this experiment have shown low cell viability (Figure 2D). When PtNPs were introduced in an oxidized condition (pre-treated with H_2O_2), PtNPs can reverse the effect caused by the oxidant in 100 µg/mL (Figure 2C).

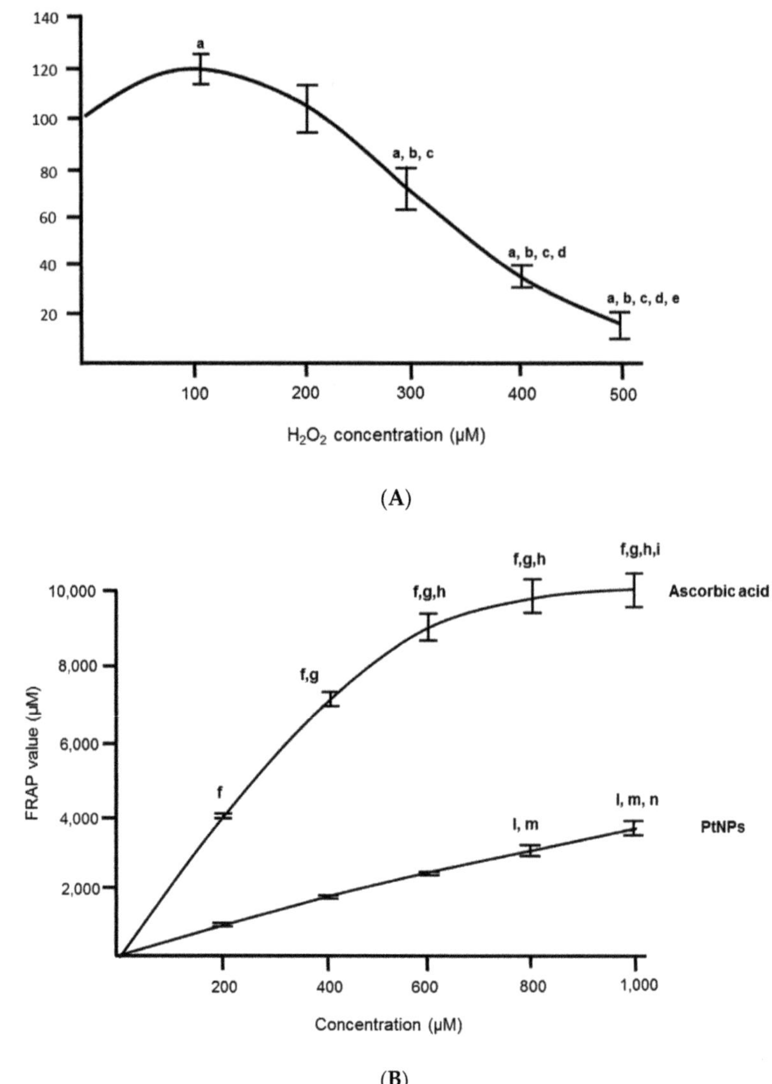

(A)

(B)

Figure 1. (A) Cell viability percentage of H_2O_2 concentration (n = 3). A statistically significant difference ($p < 0.05$) denoted in alphabet a–e represents a comparison between a group to, a: untreated H_2O_2 (0 µM), b: 100 µM, c: 200 µM, d: 300 µM, and e: 400 µM. **(B)** antioxidant capacity of PtNPs and ascorbic acid using FRAP assay (n = 3). A statistically significant difference ($p < 0.05$) denoted in alphabet f–i (ascorbic acid) and l–n(PtNPs) that represents a comparison between a group to, f and l: untreated cells (0 µg/mL), g and m: 200 µg/mL, h and n: 400 µg/mL, i: 600 µg/mL.

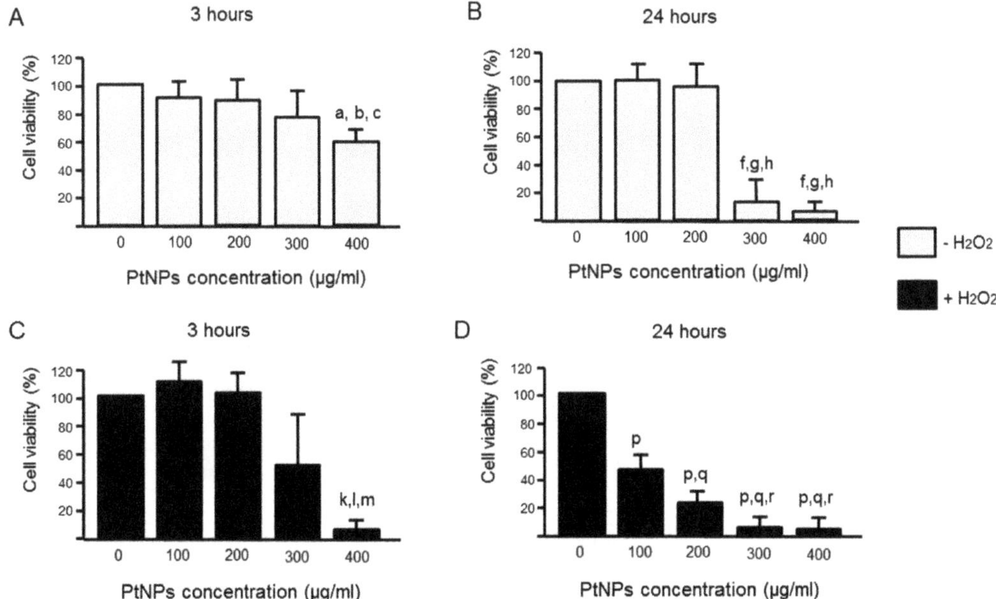

Figure 2. Cell viability in different PtNPs concentrations in two different time points (n = 3), (**A**) 3 h and (**B**) 24 h are measured by determining antioxidant capacity using FRAP assay. The 50% cell viability is determined at 400 ug/mL for 3 h and in 24 h, 300 and 400 ug/mL have shown minimal cell viability as compared to other dosages. Subsequently, several concentrations of PtNPs were introduced after cells were pretreated with H_2O_2 in two different durations, (**C**) 3 h and (**D**) 24 h. A statistically significant difference ($p < 0.05$) denoted in alphabet a–c (**A**), f–h (**B**), k–m (**C**), and p–r (**D**) that represents a comparison between a group to, a/f/k/p: untreated H_2O_2 (0 µg/mL), b/g/l/q: 100 µg/mL, c/h/m/r: 200 µg/mL.

3.3. The Level of Antioxidant Enzymes was Elevated in the Presence of PtNPs

For the analysis of the level of antioxidant enzymes, the PtNPs treatment had significantly increased the antioxidant activity of SOD, GPx, and CAT in pre-treated H_2O_2 cells when compared to the H_2O_2 treatment group (Figure 3A–C).

3.4. PtNPs Uptake was Seen in the Nucleus

This study was conducted using 3 nm PtNPs without any surface protectant during PtNPs synthesis. We aimed to show the direct effect of PtNPs on the cells. Indeed, we found that PtNPs can penetrate the epithelial cell in as early as 3 h of treatment and can enter the nucleus membrane in both treatments (Figure 4A). However, 24 h of treatment allowed more time for PtNPs to accumulate in the cytoplasm and nucleus (Figure 4B).

3.5. PtNPs Can Reduce Oxidative Condition in H_2O_2 Treated Cells

DNA damage can be seen clearly in the H_2O_2 treatment group, but this damage is reduced significantly both qualitatively (Figure 5A) and quantitatively (Figure 5B) in the presence of PtNPs. The PtNPs treatment group showed significantly lower DNA damage as compared to the control group. The production of ROS was significantly reduced by the H_2O_2/PtNPs treatment group as compared to the H_2O_2 treatment group. In parallel with DNA damage data, the PtNPs treatment group also showed a significant reduction in H_2DCFDA (Figure 5C) and DHE (Figure 5D) compared to the control group.

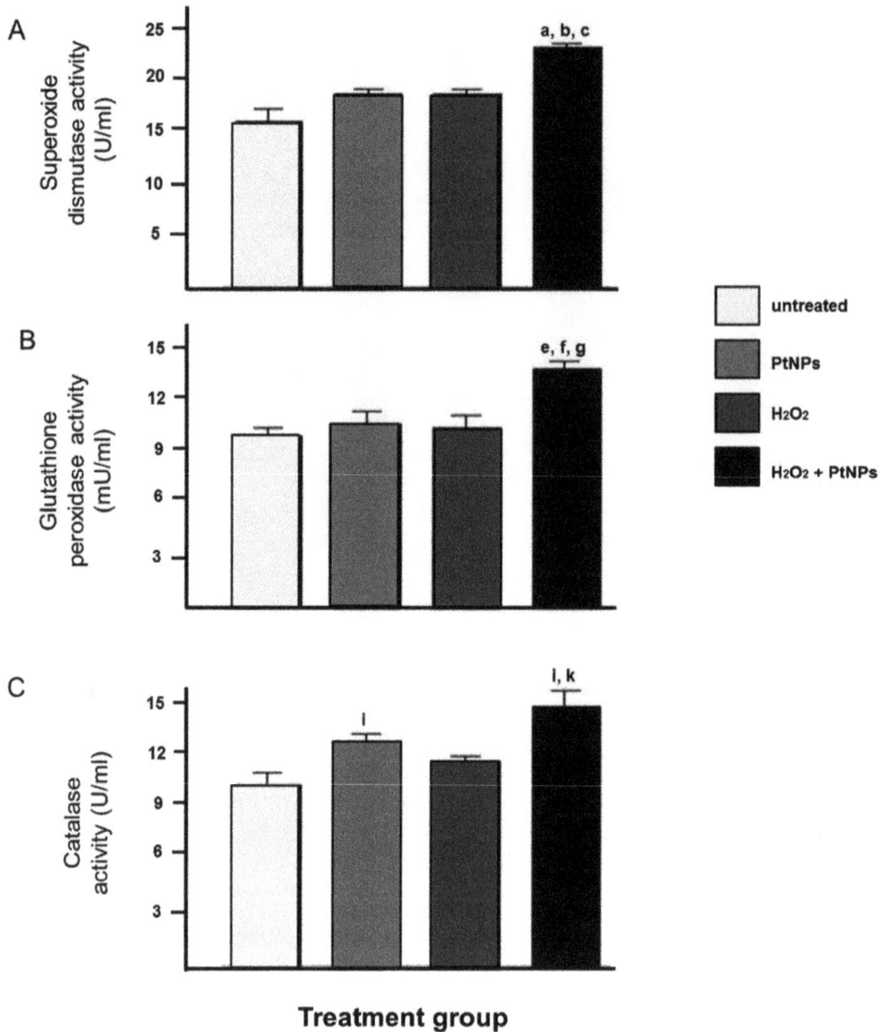

Figure 3. Cells were treated with H_2O_2 and PtNPs for 3 h and subjected to the level of antioxidants activity. PtNPs were shown to increase the activity of antioxidants enzymes of (**A**) superoxide dismutase, (**B**) glutathione peroxidase, and (**C**) catalase activities. A statistically significant difference ($p < 0.05$) denoted in alphabet a–c (**A**), e–g (**B**), and i, k (**C**) that represents a comparison between a group to, a/e/i: untreated cells, b/f: PtNPs, and c/g/k: H_2O_2.

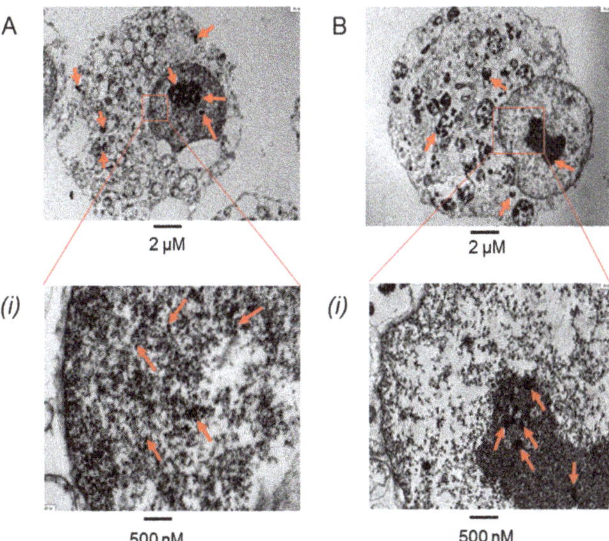

Figure 4. TEM image of PtNPs uptake by A549 cells (**A**) 3 h and (**B**) 24 h treatment (Scale bar: whole cell, 2 nm). The scale was enlarged to focus on the nucleus (500 μm). Red arrows indicate the accumulation of PtNPs. PtNPs can penetrate the cells as early as 3 h however the cells keep accumulating in the cytoplasm and nucleus in prolonged treatment (24 h).

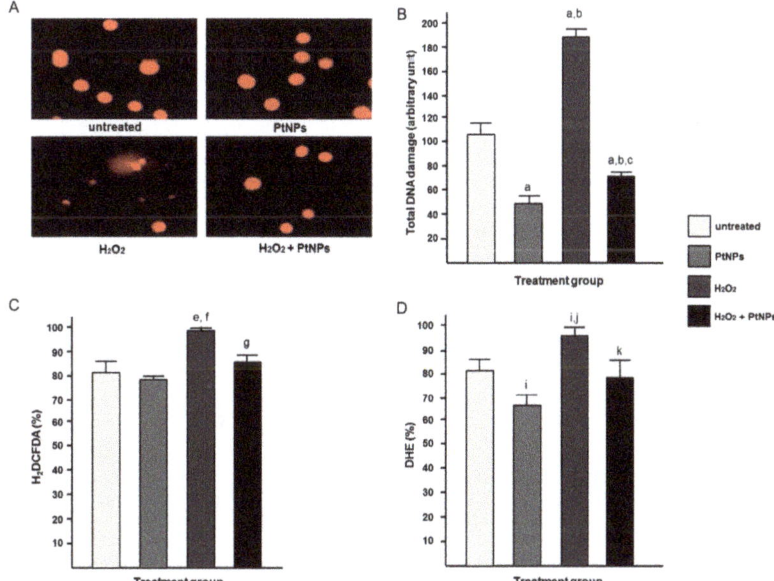

Figure 5. (**A**) Oxidative status through DNA damage represented with comet assay qualitatively. Comet tail can be seen in DNA damaged cells and quantitatively measured in (**B**) total DNA damage and ROS production represented (**C**) H_2DCFDA and (**D**) DHE, respectively. A statistically significant difference ($p < 0.05$) denotes in alphabet a–c (**A**), e–g(**B**), i–k (**C**) that represents a comparison between a group to a/c/i: untreated cells, b/f/j: PtNPs, and c/g/k: H_2O_2.

4. Discussion

Platinum-based compounds, mainly cisplatin and carboplatin, have been the most commonly used chemotherapeutic agents in the practice of lung cancer treatment. Nevertheless, there are several profound complications, especially neuro- and nephrotoxicity [10], whilst the response and survival rate among lung cancer patients are still poor. Hence, our study studied the prospective role of PtNPs against lung cancer through an in vitro model to see its antioxidative effect in an oxidant condition. Our study revealed that PtNPs have exhibited antioxidant properties by scavenging free radicals induced by hydrogen peroxide. Whilst its antioxidant capacity is relatively lower than vitamin C, results from SOD, GPx, and CAT were significantly increased when compared to the untreated control group. This is parallel to the first finding suggesting that PtNPs exhibit antioxidant properties in reducing oxidative damage in mouse lung models induced by inflammation [13]. This data is also in parallel to the previous study where they found a similar reduction in antioxidant status but with a different lung cell line model [29,30]. Besides, our study also revealed a significant reduction in ROS production with PtNPs treatment that was similar to other studies worldwide [16,17]. Among all metal nanoparticles, PtNPs were seen to be the strongest quencher of reactive oxygen species, namely hydrogen peroxide and superoxide anion, and able to act as SOD mimetics in a dose-dependent manner [16] whilst another study also demonstrated the capability of PtNPs in inhibiting hyperthermia-induced apoptosis in lymphoma cells [31]. The superoxide dismutase and catalase activities exhibited by PtNPs have also been shown in studies to further potentiate its interventional effect in aging-related skin [12], skeletal muscle [32], vascular endothelium [33], and monocytic leukemia [34] in vitro models. This is evident by our study that also revealed that PtNPs had significantly increased the antioxidant activity of SOD, GPx, and CAT in the A549 cell line in supporting the capability of PtNPs as antioxidants, as established in other studies [35–39]. In comparison to different metal nanoparticles that exhibit antioxidant activities, platinum nanoparticles are among one the best antioxidants to increase all CAT, GPx, and SOD [39]. PtNPs were also proposed to be a catalase (CAT) mimicker and have been used to enhance radiation efficacy in the treatment of cancer [38,39]. PtNPs are a suitable candidate as the smaller size of PtNPs produces a lower oxidation of H_2DCFDA compared with the bigger size due to the larger surface area to enhance the catalytic activity [29]. Looking at numerous data that support PtNPs as a potential antioxidant, it is worth to further scrutinize the mechanisms involved in between exerting the antioxidant properties and inducing apoptosis. This could be due to the blockage in different stages of the cell cycle, changes to the cell membrane, disruption and alteration in the cell homeostasis, and enzyme activity. Hence, it is worth to subsequently perform cell cycle analysis and establish the mechanism of cell death in various concentrations of PtNPs and exposures.

The unique features of PtNPs, such as surface functionalities; size; size distribution; shape; porosity; surface area; composition; crystalline nature; agglomeration; and electro, catalytic, thermal, and plasmonic properties, making their application desirable in various fields [40,41]. The capping of nanoparticles with stabilizing agents aims to prevent aggregation while in liquid suspension. However, the porous organic capping agents have open spaces that allow reactant and product molecules to reach the metal catalysts, causing a reduction in catalytic activity as certain elements in the capping agents can block the reactive sites on nanoparticles [42]. Hence, the uncapping of PtNPs was used throughout this study to ensure better penetration to the nucleus and to observe its antioxidant capacity, which fits the main aim of our study. A study was also conducted on pure PtNPs in vitro, which showed that there was no release of metal ions into the media [43]. This suggests that unmodified PtNPs do not contribute to cytotoxicity based on metal ion release. Even though different sizes of PtNPs (5–10 nm) can penetrate A549 and HaCaT cells within 24 h of treatment, these nanoparticles were not found to be present in the nucleus in acute treatments [43], which coincides with our data.

The efficacy of PtNPs treatment as an antioxidant is strongly dependent on the type of cell, size, concentration, and duration of care settings. The size of PtNPs used in this study is 3 nM, which was relatively small, therefore, cytotoxicity could be caused by prolonged exposure as it can penetrate to the cell membrane of epithelial lung cancer cells over time. In fact, an in vitro study has shown that 1 nm PtNPs are capable of reducing superoxide anion radicals better than 5 nm PtNPs [22]. Furthermore, our data supported that the cells did not survive well with chronic exposure to PtNPs by keeping the PtNPs longer than 3 h. There were similar findings in other studies; notably, smaller sizes (<100 nm) of PtNPs can penetrate directly into the cell when on the other hand, larger sizes (>100 nm) must be taken in through conventional endocytosis [43], which could lead to a higher cytotoxicity toward the cells.

With the chronic exposure in a longer duration, PtNPs were seen to have accumulated in both the cytoplasm and the nucleus, which could lead to less antioxidative effects. The coagulation and sedimentation of the nanoparticles may limit the antioxidant activity due to the instability of the nanoparticles in the medium. Nanoparticles that tend to clump may disrupt the function of organelles within the cell. The addition of FBS and human serum albumin (HSA) can inhibit the clotting process thus improving the stability of the nanoparticles [44], and this can be a suggestion to further ameliorate the effect of PtNPs as an antioxidative agent in in vitro models. Studies have revealed that higher dosages of PtNPs, and prolonged (48 h) exposure in the A549 lung cell line can cause less cell viability [45], and a shorter exposure duration was able to decrease nanotoxicity to the cells [39]. This data could serve as a great starting point for several treatments in quenching ROS, especially in the chemodynamic therapy, synergistic therapy, and the controlled drug release of nanomaterials [46].

This study has its own limitation in the route to decipher the antioxidative role of PtNPs. Since the sizes of PtNPs are small, chronic exposure could lead to DNA damage and surpass the beneficial role of PtNPs. Hence, a study needs to be conducted to further compare different agents in the capping of PtNPs using various amphiphilic molecules comprised of a polar head group and a non-polar hydrocarbon tail [47,48] to further stabilize the nanoparticle and specifically target the cancer cells [49,50]. This might include polyethylene glycol (PEG) [51], polyvinylpyrrolidone (PVP) [40], polycaprolactone (PCL) [52], chitosan [53], and plant extracts [15,54–56]. All of these capping agents are capable of further alleviating cellular toxicity [57] through minimal agglomeration of the nanoparticles [58], especially in the nucleus.

5. Conclusions

The present study has shown the antioxidant property of PtNPs in reversing oxidative stress conditions in epithelial lung cancer in an in vitro model. Despite the low antioxidative effect shown by the PtNPs, the nanoparticles are capable of reversing the oxidative stress status at a lower concentration. The small size of PtNPs exhibited better penetration of the nucleus and was able to reduce DNA damage and ROS production when cells were under oxidative stress. Hence, we recommend the potential of PtNPs as an antioxidant in reducing oxidative stress, subjected to further testing on different cell lines and in vivo models. This antioxidative potential should be observed in a low concentration of PtNPs to ensure a non-toxic environment for the cells. Modification of PtNPs, especially through encapsulation, can reduce the reaction time and energy required and offer ambient conditions of fabrication that would be another promising way forward in the treatment of lung cancer.

Author Contributions: Conceptualization, N.A.S.I.; methodology, F.Y.; validation, N.A.S.I. and J.X.L.; formal analysis, N.A.S.I. and F.Y.; resources, N.A.S.I.; data curation, N.A.S.I., J.X.L. and F.Y.; writing—original draft preparation, N.A.S.I., J.X.L. and F.Y.; writing—review and editing, N.A.S.I. and J.X.L.; visualization, N.A.S.I.; supervision, N.A.S.I.; project administration, N.A.S.I.; funding acquisition, N.A.S.I. All authors have read and agreed to the published version of the manuscript.

Funding: This research was funded by a grant from the Ministry of Education, Malaysia (FRGS/1/2013/SKK01/UKM/03/1).

Institutional Review Board Statement: Not applicable.

Informed Consent Statement: Not applicable.

Data Availability Statement: Data is contained within the article.

Conflicts of Interest: The authors declare no conflict of interest.

References

1. Humphries, K.M.; Szweda, P.A.; Szweda, L.I. Aging: A shift from redox regulation to oxidative damage. *Free Radic. Res.* **2006**, *40*, 1239–1243. [CrossRef] [PubMed]
2. Wu, D.; Yotnda, P. Production and detection of reactive oxygen species (ROS) in cancers. *J. Vis. Exp.* **2011**, *57*, e3357. [CrossRef]
3. Juan, C.A.; Pérez de la Lastra, J.M.; Plou, F.J.; Pérez-Lebeña, E. The Chemistry of Reactive Oxygen Species (ROS) Revisited: Outlining Their Role in Biological Macromolecules (DNA, Lipids and Proteins) and Induced Pathologies. *Int. J. Mol. Sci.* **2021**, *22*, 4642. [CrossRef] [PubMed]
4. York-Duran, M.J.; Godoy-Gallardo, M.; Jansman, M.M.T.; Hosta-Rigau, L. A dual-component carrier with both non-enzymatic and enzymatic antioxidant activity towards ROS depletion. *Biomater. Sci.* **2019**, *7*, 4813–4826. [CrossRef] [PubMed]
5. Ighodaro, O.M.; Akinloye, O.A. First line defence antioxidants-superoxide dismutase (SOD), catalase (CAT) and glutathione peroxidase (GPX): Their fundamental role in the entire antioxidant defence grid. *Alex. J. Med.* **2018**, *54*, 287–293. [CrossRef]
6. Sung, H.; Ferlay, J.; Siegel, R.L.; Laversanne, M.; Soerjomataram, I.; Jemal, A.; Bray, F. Global cancer statistics 2020: GLOBOCAN estimates of incidence and mortality worldwide for 36 cancers in 185 countries. *CA A Cancer J. Clin.* **2021**, *71*, 209–249. [CrossRef] [PubMed]
7. Reuter, S.; Gupta, S.C.; Chaturvedi, M.M.; Aggarwal, B.B. Oxidative stress, inflammation, and cancer: How are they linked? *Free Radic. Biol. Med.* **2010**, *49*, 1603–1616. [CrossRef]
8. Xu, D.-P.; Li, Y.; Meng, X.; Zhou, T.; Zhou, Y.; Zheng, J.; Zhang, J.-J.; Li, H.-B. Natural Antioxidants in Foods and Medicinal Plants: Extraction, Assessment and Resources. *Int. J. Mol. Sci.* **2017**, *18*, 96. [CrossRef]
9. Harris, I.S.; DeNicola, G.M. The Complex Interplay between Antioxidants and ROS in Cancer. *Trends Cell Biol.* **2020**, *30*, 440–451. [CrossRef]
10. El-Sheikh, A.; Khired, Z. Interactions of Analgesics with Cisplatin: Modulation of Anticancer Efficacy and Potential Organ Toxicity. *Medicina* **2022**, *58*, 46. [CrossRef]
11. Khan, I.; Saeed, K.; Khan, I. Nanoparticles: Properties, applications and toxicities. *Arab. J. Chem.* **2019**, *12*, 908–931. [CrossRef]
12. Shibuya, S.; Ozawa, Y.; Watanabe, K.; Izuo, N.; Toda, T.; Yokote, K.; Shimizu, T. Palladium and platinum nanoparticles attenuate aging-like skin atrophy via antioxidant activity in mice. *PLoS ONE* **2014**, *9*, e109288. [CrossRef] [PubMed]
13. Onizawa, S.; Aoshiba, K.; Kajita, M.; Miyamoto, Y.; Nagai, A. Platinum nanoparticle antioxidants inhibit pulmonary inflammation in mice exposed to cigarette smoke. *Pulm. Pharmacol. Ther.* **2009**, *22*, 340–349. [CrossRef] [PubMed]
14. Schmid, M.; Zimmermann, S.; Krug, H.F.; Sures, B. Influence of platinum, palladium and rhodium as compared with cadmium, nickel and chromium on cell viability and oxidative stress in human bronchial epithelial cells. *Environ. Int.* **2007**, *33*, 385–390. [CrossRef]
15. Shakibaie, M.; Torabi-Shamsabad, R.; Forootanfar, H.; Amiri-Moghadam, P.; Amirheidari, B.; Adeli-Sardou, M.; Ameri, A. Rapid microwave-assisted biosynthesis of platinum nanoparticles and evaluation of their antioxidant properties and cytotoxic effects against MCF-7 and A549 cell lines. *3 Biotech* **2021**, *11*, 511. [CrossRef]
16. Kajita, M.; Hikosaka, K.; Iitsuka, M.; Kanayama, A.; Toshima, N.; Miyamoto, Y. Platinum nanoparticle is a useful scavenger of superoxide anion and hydrogen peroxide. *Free Radic. Res.* **2007**, *41*, 615–626. [CrossRef]
17. Hikosaka, K.; Kim, J.; Kajita, M.; Kanayama, A.; Miyamoto, Y. Platinum nanoparticles have an activity similar to mitochondrial NADH:ubiquinone oxidoreductase. *Colloids Surf. B Biointerfaces* **2008**, *66*, 195–200. [CrossRef]
18. Alshatwi, A.A.; Athinarayanan, J.; Vaiyapuri Subbarayan, P. Green synthesis of platinum nanoparticles that induce cell death and G2/M-phase cell cycle arrest in human cervical cancer cells. *J. Mater. Sci. Mater. Med.* **2015**, *26*, 5330. [CrossRef]
19. Hullo, M.; Grall, R.; Perrot, Y.; Mathé, C.; Ménard, V.; Yang, X.; Lacombe, S.; Porcel, E.; Villagrasa, C.; Chevillard, S.; et al. Radiation Enhancer Effect of Platinum Nanoparticles in Breast Cancer Cell Lines: In Vitro and In Silico Analyses. *Int. J. Mol. Sci.* **2021**, *22*, 4436. [CrossRef]
20. Pelka, J.; Gehrke, H.; Esselen, M.; Türk, M.; Crone, M.; Bräse, S.; Muller, T.; Blank, H.; Send, W.; Zibat, V.; et al. Cellular uptake of platinum nanoparticles in human colon carcinoma cells and their impact on cellular redox systems and DNA integrity. *Chem. Res. Toxicol.* **2009**, *22*, 649–659. [CrossRef]
21. Gehrke, H.; Pelka, J.; Hartinger, C.G.; Blank, H.; Bleimund, F.; Schneider, R.; Gerthsen, D.; Bräse, S.; Crone, M.; Türk, M.; et al. Platinum nanoparticles and their cellular uptake and DNA platination at non-cytotoxic concentrations. *Arch. Toxicol.* **2011**, *85*, 799–812. [CrossRef] [PubMed]
22. Hamasaki, T.; Kashiwagi, T.; Imada, T.; Nakamichi, N.; Aramaki, S.; Toh, K.; Morisawa, S.; Shimakoshi, H.; Hisaeda, Y.; Shirahata, S. Kinetic analysis of superoxide anion radical-scavenging and hydroxyl radical-scavenging activities of platinum nanoparticles. *Langmuir* **2008**, *24*, 7354–7364. [CrossRef] [PubMed]
23. Benzie, I.F.F.; Strain, J.J. The Ferric Reducing Ability of Plasma (FRAP) as a Measure of "Antioxidant Power": The FRAP Assay. *Anal. Biochem.* **1996**, *239*, 70–76. [CrossRef] [PubMed]

24. Riss, T.L.; Moravec, R.A.; Niles, A.L.; Duellman, S.; Benink, H.A.; Worzella, T.J.; Minor, L. Cell viability assays. In *Assay Guidance Manual*; Markossian, S., Grossman, A., Brimacombe, K., Arkin, M., Auld, D., Austin, C.P., Baell, J., Chung, T.D.Y., Coussens, N.P., Dahlin, J.L., et al., Eds.; Eli Lilly & Company and the National Center for Advancing Translational Sciences: Bethesda, MD, USA, 2016.
25. Graham, L.; Orenstein, J.M. Processing tissue and cells for transmission electron microscopy in diagnostic pathology and research. *Nat. Protoc.* **2007**, *2*, 2439–2450. [CrossRef] [PubMed]
26. Olive, P.L.; Banáth, J.P. The comet assay: A method to measure DNA damage in individual cells. *Nat. Protoc.* **2006**, *1*, 23–29. [CrossRef]
27. Wojtala, A.; Bonora, M.; Malinska, D.; Pinton, P.; Duszynski, J.; Wieckowski, M.R. Chapter Thirteen—Methods to Monitor ROS Production by Fluorescence Microscopy and Fluorometry. In *Methods in Enzymology*; Galluzzi, L., Kroemer, G., Eds.; Academic Press: Cambridge, MA, USA, 2014; Volume 542, pp. 243–262.
28. Aebi, H. Catalase in vitro. In *Methods in Enzymology*; Academic Press: Cambridge, MA, USA, 1984; Volume 105, pp. 121–126.
29. Elder, A.; Yang, H.; Gwiazda, R.; Teng, X.; Thurston, S.; He, H.; Oberdörster, G. Testing nanomaterials of unknown toxicity: An example based on platinum nanoparticles of different shapes. *Adv. Mater.* **2007**, *19*, 3124–3129. [CrossRef]
30. Lee, J.H.; Kim, H.; Lee, Y.S.; Jung, D.Y. Enhanced catalytic activity of platinum nanoparticles by exfoliated metal hydroxide nanosheets. *ChemCatChem* **2014**, *6*, 113–118. [CrossRef]
31. Yoshihisa, Y.; Zhao, Q.L.; Hassan, M.A.; Wei, Z.L.; Furuichi, M.; Miyamoto, Y.; Kondo, T.; Shimizu, T. SOD/catalase mimetic platinum nanoparticles inhibit heat-induced apoptosis in human lymphoma U937 and HH cells. *Free Radic. Res.* **2011**, *45*, 326–335. [CrossRef]
32. Nakanishi, H.; Hamasaki, T.; Kinjo, T.; Yan, H.; Nakamichi, N.; Kabayama, S.; Teruya, K.; Shirahata, S. Low Concentration Platinum Nanoparticles Effectively Scavenge Reactive Oxygen Species in Rat Skeletal L6 Cells. *Nano Biomed. Eng.* **2013**, *5*, 76–85. [CrossRef]
33. Wen, T.; Yang, A.; Piao, L.; Hao, S.; Du, L.; Meng, J.; Liu, J.; Xu, H. Comparative study of in vitro effects of different nanoparticles at non-cytotoxic concentration on the adherens junction of human vascular endothelial cells. *Int. J. Nanomed.* **2019**, *14*, 4475. [CrossRef]
34. Gurunathan, S.; Jeyaraj, M.; La, H.; Yoo, H.; Choi, Y.; Do, J.T.; Park, C.; Kim, J.-H.; Hong, K. Anisotropic Platinum Nanoparticle-Induced Cytotoxicity, Apoptosis, Inflammatory Response, and Transcriptomic and Molecular Pathways in Human Acute Monocytic Leukemia Cells. *Int. J. Mol. Sci.* **2020**, *21*, 440. [CrossRef] [PubMed]
35. Tian, R.; Xu, J.; Luo, Q.; Hou, C.; Liu, J. Rational Design and Biological Application of Antioxidant Nanozymes. *Front. Chem.* **2020**, *8*, 831. [CrossRef] [PubMed]
36. Ivanova, P.; Dzięgielewski, K.; Drozd, M.; Skorupska, S.; Grabowska-Jadach, I.; Pietrzak, M. Nanoparticles of chosen noble metals as reactive oxygen species scavengers. *Nanotechnology* **2021**, *32*, 055704. [CrossRef]
37. Zheng, W.; Jiang, B.; Hao, Y.; Zhao, Y.; Zhang, W.; Jiang, X. Screening reactive oxygen species scavenging properties of platinum nanoparticles on a microfluidic chip. *Biofabrication* **2014**, *6*, 045004. [CrossRef] [PubMed]
38. Kinoshita, A.; Lima, I.; Guidelli, É.J.; Baffa Filho, O. Antioxidative activity of gold and platinum nanoparticles assessed through electron spin resonance. *Eclética Química* **2021**, *46*, 68–74. [CrossRef]
39. Bloch, K.; Pardesi, K.; Satriano, C.; Ghosh, S. Bacteriogenic Platinum Nanoparticles for Application in Nanomedicine. *Front. Chem.* **2021**, *9*, 624344. [CrossRef]
40. Jan, H.; Gul, R.; Andleeb, A.; Ullah, S.; Shah, M.; Khanum, M.; Ullah, I.; Hano, C.; Abbasi, B.H. A detailed review on biosynthesis of platinum nanoparticles (PtNPs), their potential antimicrobial and biomedical applications. *J. Saudi Chem. Soc.* **2021**, *25*, 101297. [CrossRef]
41. Jeyaraj, M.; Gurunathan, S.; Qasim, M.; Kang, M.-H.; Kim, J.-H. A Comprehensive Review on the Synthesis, Characterization, and Biomedical Application of Platinum Nanoparticles. *Nanomaterials* **2019**, *9*, 1719. [CrossRef]
42. Abed, A.; Derakhshan, M.; Karimi, M.; Shirazinia, M.; Mahjoubin-Tehran, M.; Homayonfal, M.; Hamblin, M.R.; Mirzaei, S.A.; Soleimanpour, H.; Dehghani, S.; et al. Platinum Nanoparticles in Biomedicine: Preparation, Anti-Cancer Activity, and Drug Delivery Vehicles. *Front. Pharmacol.* **2022**, *13*, 797804. [CrossRef]
43. Horie, M.; Kato, H.; Endoh, S.; Fujita, K.; Nishio, K.; Komaba, L.K.; Fukui, H.; Nakamura, A.; Miyauchi, A.; Nakazato, T.; et al. Evaluation of cellular influences of platinum nanoparticles by stable medium dispersion. *Metallomics* **2011**, *3*, 1244–1252. [CrossRef]
44. Torrano, A.A.; Herrmann, R.; Strobel, C.; Rennhak, M.; Engelke, H.; Reller, A.; Hilger, I.; Wixforth, A.; Bräuchle, C. Cell membrane penetration and mitochondrial targeting by platinum-decorated ceria nanoparticles. *Nanoscale* **2016**, *8*, 13352–13367. [CrossRef] [PubMed]
45. Allouni, Z.E.; Cimpan, M.R.; Høl, P.J.; Skodvin, T.; Gjerdet, N.R. Agglomeration and sedimentation of TiO$_2$ nanoparticles in cell culture medium. *Colloids Surf. B Biointerfaces* **2009**, *68*, 83–87. [CrossRef] [PubMed]
46. Bendale, Y.; Bendale, V.; Natu, R.; Paul, S. Biosynthesized Platinum Nanoparticles Inhibit the Proliferation of Human Lung-Cancer Cells in vitro and Delay the Growth of a Human Lung-Tumor Xenograft in vivo: -In vitro and in vivo Anticancer Activity of bio-Pt NPs. *J. Pharmacopunct.* **2016**, *19*, 114–121. [CrossRef]
47. Javed, R.; Zia, M.; Naz, S.; Aisida, S.O.; Ain, N.u.; Ao, Q. Role of capping agents in the application of nanoparticles in biomedicine and environmental remediation: Recent trends and future prospects. *J. Nanobiotechnol.* **2020**, *18*, 172. [CrossRef]

48. Ullah, S.; Ahmad, A.; Wang, A.; Raza, M.; Jan, A.U.; Tahir, K.; Rahman, A.U.; Qipeng, Y. Bio-fabrication of catalytic platinum nanoparticles and their in vitro efficac.cy against lungs cancer cells line (A549). *J. Photochem. Photobiol. B* **2017**, *173*, 368–375. [CrossRef]

49. Zhao, N.; Xin, H.; Zhang, L. Advanced Biomedical Applications of Reactive Oxygen Species-Based Nanomaterials in Lung Cancer. *Front. Chem.* **2021**, *9*, 649772. [CrossRef]

50. Pawar, A.A.; Sahoo, J.; Verma, A.; Lodh, A.; Lakkakula, J. Usage of Platinum Nanoparticles for Anticancer Therapy over Last Decade: A Review. *Part. Part. Syst. Charact.* **2021**, *38*, 2100115. [CrossRef]

51. Akbari, E.; Mousazadeh, H.; Hanifehpour, Y.; Mostafavi, E.; Gorabi, A.M.; Nejati, K.; Keyhanvar, P.; Pazoki-Toroudi, H.; Mohammadhosseini, M.; Akbarzadeh, A. Co-Loading of Cisplatin and Methotrexate in Nanoparticle-Based PCL-PEG System Enhances Lung Cancer Chemotherapy Effects. *J. Clust. Sci.* **2019**, *4*, 209–219. [CrossRef]

52. Esim, O.; Bakirhan, N.K.; Yildirim, N.; Sarper, M.; Savaser, A.; Ozkan, S.A.; Ozkan, Y. Development, optimization and in vitro evaluation of oxaliplatin loaded nanoparticles in non-small cell lung cancer. *Daru* **2020**, *28*, 673–684. [CrossRef]

53. Guo, X.; Zhuang, Q.; Ji, T.; Zhang, Y.; Li, C.; Wang, Y.; Li, H.; Jia, H.; Liu, Y.; Du, L. Multi-functionalized chitosan nanoparticles for enhanced chemotherapy in lung cancer. *Carbohydr. Polym.* **2018**, *195*, 311–320. [CrossRef]

54. Fahmy, S.A.; Preis, E.; Bakowsky, U.; Azzazy, H.M.E.-S. Platinum Nanoparticles: Green Synthesis and Biomedical Applications. *Molecules* **2020**, *25*, 4981. [CrossRef] [PubMed]

55. Hosny, M.; Fawzy, M.; El-Fakharany, E.M.; Omer, A.M.; El-Monaem, E.M.A.; Khalifa, R.E.; Eltaweil, A.S. Biogenic synthesis, characterization, antimicrobial, antioxidant, antidiabetic, and catalytic applications of platinum nanoparticles synthesized from Polygonum salicifolium leaves. *J. Environ. Chem. Eng.* **2022**, *10*, 106806. [CrossRef]

56. Eltaweil, A.S.; Fawzy, M.; Hosny, M.; Abd El-Monaem, E.M.; Tamer, T.M.; Omer, A.M. Green synthesis of platinum nanoparticles using Atriplex halimus leaves for potential antimicrobial, antioxidant, and catalytic applications. *Arab. J. Chem.* **2022**, *15*, 103517. [CrossRef]

57. Rajendran, K.; Sen, S. Effect of capping agent on antimicrobial activity of nanoparticles. *Der Pharm. Lett.* **2015**, *7*, 37–42.

58. Javed, R.; Usman, M.; Tabassum, S.; Zia, M. Effect of capping agents: Structural, optical and biological properties of ZnO nanoparticles. *Appl. Surf. Sci.* **2016**, *386*, 319–326. [CrossRef]

antioxidants

MDPI

Article

Antioxidant Ascorbic Acid Modulates NLRP3 Inflammasome in LPS-G Treated Oral Stem Cells through NFκB/Caspase-1/IL-1β Pathway

Jacopo Pizzicannella [1], Luigia Fonticoli [2], Simone Guarnieri [3], Guya D. Marconi [4], Thangavelu Soundara Rajan [5], Oriana Trubiani [2] and Francesca Diomede [2,*]

[1] "Ss. Annunziata" Hospital, ASL 02 Lanciano-Vasto-Chieti, 66100 Chieti, Italy; jacopo.pizzicannella@unich.it
[2] Department of Innovative Technologies in Medicine & Dentistry, University "G. d'Annunzio" Chieti-Pescara, via dei Vestini, 31, 66100 Chieti, Italy; luigia.fonticoli@unich.it (L.F.); oriana.trubiani@unich.it (O.T.)
[3] Department of Neuroscience, Imaging and Clinical Sciences, Center for Advanced Studies and Technology (CAST), University "G. d'Annunzio" Chieti-Pescara, via dei Vestini, 31, 66100 Chieti, Italy; simone.guarnieri@unich.it
[4] Department of Medical, Oral and Biotechnological Sciences, University "G. d'Annunzio" Chieti-Pescara, via dei Vestini, 31, 66100 Chieti, Italy; guya.marconi@unich.it
[5] Department of Biotechnology, Karpagam Academy of Higher Education, Coimbatore 641 021, India; tsrajanpillai@gmail.com
* Correspondence: francesca.diomede@unich.it; Tel.: +39-08713554080

Citation: Pizzicannella, J.; Fonticoli, L.; Guarnieri, S.; Marconi, G.D.; Rajan, T.S.; Trubiani, O.; Diomede, F. Antioxidant Ascorbic Acid Modulates NLRP3 Inflammasome in LPS-G Treated Oral Stem Cells through NFκB/Caspase-1/IL-1β Pathway. *Antioxidants* 2021, 10, 797. https://doi.org/10.3390/antiox10050797

Academic Editors: Soliman Khatib and Dana Atrahimovich Blatt

Received: 1 April 2021
Accepted: 12 May 2021
Published: 18 May 2021

Publisher's Note: MDPI stays neutral with regard to jurisdictional claims in published maps and institutional affiliations.

Abstract: Human gingival mesenchymal stem cells (hGMSCs) and endothelial committed hGMSCs (e-hGMSCs) have considerable potential to serve as an in vitro model to replicate the inflammation sustained by *Porphyromonas gingivalis* in periodontal and cardiovascular diseases. The present study aimed to investigate the effect of ascorbic acid (AA) on the inflammatory reverting action of lipopolysaccharide (LPS-G) on the cell metabolic activity, inflammation pathway and reactive oxygen species (ROS) generation in hGMSCs and e-hGMSCs. Cells were treated with LPS-G (5 μg mL^{-1}) or AA (50 μg mL^{-1}) and analyzed by 3-(4,5-Dimethylthiazol-2-yl)-2,5-Diphenyltetrazolium Bromide (MTT) assay, immunofluorescence and Western blot methods. The rate of cell metabolic activity was decreased significantly in LPS-G-treated groups, while groups co-treated with LPS-G and AA showed a logarithmic cell metabolic activity rate similar to untreated cells. AA treatment attenuated the inflammatory effect of LPS-G by reducing the expression of TLR4/MyD88/NFκB/NLRP3/Caspase-1/IL-1β, as demonstrated by Western blot analysis and immunofluorescence acquisition. LPS-G-induced cells displayed an increase in ROS production, while AA co-treated cells showed a protective effect. In summary, our work suggests that AA attenuated LPS-G-mediated inflammation and ROS generation in hGMSCs and e-hGMSCs via suppressing the NFκB/Caspase-1/IL-1β pathway. These findings indicate that AA may be considered as a potential factor involved in the modulation of the inflammatory pathway triggered by LPS-G in an vitro cellular model.

Keywords: periodontal disease; NLRP3 inflammasome; ascorbic acid; gingival mesenchymal stem cells; *Porphyromonas gingivalis*

1. Introduction

Ascorbic acid (AA) is well known to perform a key role in the maintenance of tissue integrity, scavenging free radicals, and has demonstrated immunomodulatory properties during chronic inflammatory diseases. In periodontitis, AA showed a deceleration in the progression of tissue loss and induced regeneration by stimulating the progenitor cells of the periodontal ligament commitment [1]. Periodontal disease (PD) can be defined as a complex oral disease and its treatment represents a challenging condition for clinicians [2]. Periodontal disease involves chronic inflammation sustained by various types of bacteria that accumulate in dental plaque and cause localized inflammation by producing various pro-inflammatory factors, which include C-reactive protein (CRP), interleukin (IL)-1β, IL-6, tumor necrosis factor (TNF)-α and matrix metalloproteinases (MMP) [3]. The progression of periodontitis leads to the destruction of deep periodontal pockets and causes systemic disease processes upon the increased expression of pro-inflammatory factors. Indeed, periodontitis represents a possible risk factor for several systemic diseases, including cardiovascular disease (CVD) [4].

There are a wide range of disorders that affect the heart and blood vessels with complex pathogenic mechanisms. The pathogenesis is based on the presence of a high level of low-density lipoprotein cholesterol in the blood, which affects the cellular permeability and modulates the integrity of arterial walls. A strong link has been found between periodontitis and cardiovascular diseases as both diseases show common inflammatory pathways, share similar risk factors and start with tissue damage [5–7].

The presence of Gram-negative anaerobic pathogens in the periodontal pockets is critical for the invasion of the deeper tissues, reaching the blood circulation and inducing a systemic immune response away from the original niche [8].

Several studies reported that periodontal pathogens are associated with chronic inflammation and lead to epithelial barrier dysfunction, causing a loss of epithelial sheet integrity and producing microulceration [9–11]. *Porphyromonas gingivalis* (*P. gingivalis*) is one of the bacteria involved in the biofilm development of bacterial plaque and plays a vital part in the advancement of periodontal illness. Lipopolysaccharide (LPS-G) is the virulence factor of *P. gingivalisis*, which induces the host immune response via activating the innate and acquired immunity [12].

To study the in vitro response to LPS-G and inflammatory pathway modulation, different mesenchymal stem cell (MSC) populations have been used [13,14]. Recently, human gingival mesenchymal stem cells (hGMSCs) have attracted the attention of many researchers. They met the minimal criteria proposed by Dominici et al. [15] to characterize MSCs. hGMSCs showed self-renewal capacity, mesengenic differentiation ability under induction conditions and the expression of MSC markers [16].

The inflammasome NLRP3 is activated during periodontal disease, which is most often associated with TLR4 activation. NLRP3 initiates the inflammatory cascade by activating the release of interleukin (IL)-1β that regulates the degree and progression of the inflammation.

Ascorbic acid is a water-soluble molecule required for human health. The human body is unable to synthesize vitamin C endogenously and thus requires external sources of vitamin C from dietary intake and supplementation [17]. AA is an effective antioxidant that relieves oxidative stress and participates in a variety of biochemical reactions. Moreover, AA is able to decrease the responses of some inflammatory biomarkers and several pro-inflammatory cytokines [18–20].

The role of AA in the suppression of the activation of inflammasome cascade TLR4/NFκB signaling that leads to the release of IL-1β in periodontitis is yet to be determined. Accordingly, the aim of our research was to explore the role of AA in LPS-G-stimulated hGMSCs and endothelial differentiated hGMSCs (e-hGMSCs) to evaluate the regulation of the inflammatory cascade.

2. Materials and Methods

2.1. Ethic Statement

The present research project was approved by the Medical Ethics Committee at the Medical School, "G. d'Annunzio" University, Chieti, Italy (n°266/14). All enrolled subjects signed an informative consent form.

2.2. Cell Culture

Cells were isolated from human gingival tissue biopsies from two different donors as previously described [21]. Biopsies were washed with phosphate-buffered saline (PBS) (Lonza, Basel, Switzerland), cut into small pieces and then placed in a TheraPEAK™MSCGM-

CD™ Bullet Kit serum free, chemically defined medium for the growth of human mesenchymal stem cells (MSCGM-CD, Lonza, Basel, Switzerland) at 37 °C in a controlled atmosphere (5% CO_2). The medium was refreshed twice a week. After 2 weeks of culture, cells were spontaneously migrated from the tissue explants. After reaching around 80% of confluence, cells were trypsinized using trypsin-EDTA (Lonza, Basel, Switzerland) and subcultured until passage 2 (P2). All the following experiments were conducted in triplicate and were repeated 3 times.

2.3. Cell Characterization

Human Gingival Mesenchymal Stem Cells (GMSC)s were characterized by cytofluorimetric and morphological analysis and the capacity to differentiate into osteogenic and adipogenic lineage. Fluorescence-Activated Cell Sorting (FACS) analysis showed the expression of CD14, CD34, CD45, CD73, CD90 and CD105 as previously reported [22]. Briefly, cells were stained for CD45, CD73 and CD90 with fluorescein isothiocyanate-conjugated anti-human antibodies and for CD14, CD34 and CD105 with phycoerythrin-conjugated antibodies. After staining procedures, a FACStar-plus flow-cytometry system running Cell-Quest software (version 5.2.1, Becton-Dickinson, Mountain View, CA, USA) was used. All reagents used for flow cytometry were purchased from Becton Dickinson (Milan, Italy). To study the morphological features, hGMSCs were fixed with paraformaldehyde, stained with toluidine blue and observed with a Leica microsystem microscope. To evaluate the capacity to differentiate into osteogenic and adipogenic lineages, hGMSCs were incubated with MSCGM-CD (Lonza, Basel, Switzerland) medium with osteogenic supplements and in adipogenesis induction/maintenance medium (Lonza, Basel, Switzerland), respectively. Osteogenic differentiation was evaluated using alizarin red S staining (Sigma-Aldrich, Milan, Italy) and the adipogenic commitment was stained with oil red O solution (Lonza Basel, Switzerland). To validate the colorimetric detection, a real-time polymerase chain reaction (RT-PCR) was performed, evaluating the expression of Runt-related transcription factor-2 (RUNX-2), alkaline phosphatase (ALP), fatty acid-binding protein 4 (FABP4) and peroxisome proliferator-activated receptor γ (PPARγ) after 28 days of differentiation. Commercially available TaqMan Gene Expression Assays (RUNX-2Hs00231692_m1; ALP Hs01029144_m1; FABP4 Hs01086177_m1;PPARγ Hs01115513_m1) and the TaqManUniversal PCR Master Mix (Applied Biosystems, Foster City, CA, USA) were used according to standard protocols. Beta-2 microglobulin (B2MHs99999907_m1) (Applied Biosystems, Foster City, CA, USA) was used for template normalization [23].

2.4. Endothelial Differentiation

Endothelial differentiation procedures to obtain the e-hGMSCs were started with 50–60% of cell confluency. Endothelial Growth Medium (EGM-2, Lonza, Basel, Switzerland), composed of EGM-2 Bullet Kit (Lonza, Basel, Switzerland) growth supplements containing hydrocortisone, human Fibroblast Growth Factor (hFGF-b), R3-Insulin-like Growth Factor-1 (R3-IGF-1), ascorbic acid, human Epithelial Growth Factor (hEGF), GA-1000, heparin, 5% FBS and 50 ng/mL of Vascular Endothelial Growth Factor-165 (VEGF-165), was used. Cells were maintained at 37 °C with 5% CO_2 for 14 days [24].

2.5. Tube Formation Test

In this study, 12-well culture plates pretreated with Cultrex® Basement Membrane Extract (Trevigen Inc., Gaithersburg, MD, USA) (300 µL/well) were used in the tube formation assay. Human GMSCs and e-hGMSCs were seeded at a density of 2×10^5 cells per well after matrix solidification (at 37 °C for 30 min). Capillary-like tube structures have been observed using an inverted light microscope at phase contrast after 4 h of culture to 24 h [24].

2.6. Study Design

All experiments were performed in triplicate with hGMSCs at P2.

The study was organized with the following groups:

- Untreated hGMSCs, used as negative control (CTRL);
- hGMSCs treated for 24 h with 50 μg mL^{-1} AA (AA);
- hGMSCs treated for 24 h with 5 μg mL^{-1} ultrapure LPS-G from *P. gingivalis* (tlrl-ppglps, InvivoGen, San Diego, CA, USA) (LPS-G);
- hGMSCs co-treated for 24 h with 50 μg mL^{-1} AA and 5 μgmL^{-1} LPS-G (AA + LPS-G);
- Untreated e-GMSCs, used as negative control (e-CTRL);
- e-hGMSCs treated for 24 h with 50 μg mL^{-1} AA (e-AA);
- e-hGMSCs treated for 24 h with 5 μg mL^{-1} ultrapure LPS-G (tlrl-ppglps, InvivoGen, San Diego, CA, USA; e-LPS-G);
- e-hGMSCs co-treated for 24 h with 50 μgmL^{-1} AA and 5 μg mL^{-1} LPS-G (AA + e-LPS-G).

Based on our previously obtained data (data not shown), we found, after an LPS-G titration tested on oral stem cells, that 5 μg mL^{-1} LPS-G was the optimal concentration to be utilized in the current study [25].

2.7. Cell Metabolic Activity

3-(4,5-Dimethylthiazol-2-yl)-2,5-Diphenyltetrazolium Bromide (MTT) colorimetric assay was used for all experimental groups to determine the cell viability as previously reported [26]. Human GMSCs were seeded with a density of 2×10^3 cells/well in a 96-well plate. At 24, 48 and 72 h of culture, 20 μL of MTT (CellTiter 96 AQueous One Solution reagent, Promega, Milan, Italy) solution was added to each well. Plates were maintained in the incubator for 3 h and then were read at 490 nm wavelength using a microplate reader (Synergy HT, BioTek Instruments, Winooski, VT, USA).

2.8. Molecular Pathway

Samples were fixed with 4% paraformaldehyde in 0.1 M of PBS (Lonza, Basel, Switzerland). Then, cells were permeabilized with 0.5% Triton X-100 in PBS (Lonza, Basel, Switzerland) for 10 min and blocked with 5% skimmed milk in PBS for 30 min [27]. The following primary antibodies were used in the study: anti-TLR4 (1:200; Santa Cruz Biotechnology, Dallas, TX, USA), anti-MyD88 (1:200; Santa Cruz Biotechnology, Dallas, TX, USA), anti-NFκB (1:200; Santa Cruz Biotechnology, Dallas, TX, USA), anti-NLRP3(1:500; Novus, Centennial, CO, USA), anti-Caspase-1 (1:200; Santa Cruz Biotechnology, Dallas, TX, USA) and anti-IL-1β (5 μg/mL; ThermoFisher, Waltham, MA, USA). Cells were incubated with primary antibody for 2 h at room temperature. Then, samples were incubated with Alexa Fluor 568 red fluorescence conjugated goat anti-rabbit secondary antibody (1:200; Molecular Probes, Invitrogen, Eugene, OR, USA) for 1 h at 37 °C. To stain the cytoskeleton actin, cells were treated with Alexa Fluor 488 phalloidin green fluorescent conjugate (1:400, Molecular Probes, Invitrogen, Eugene, OR, USA) for 1 h, and to stain the nuclei, cells were stained with TOPRO (1:200; Molecular Probes, Invitrogen, Eugene, OR, USA) for 1 h. The Zeiss LSM800 confocal system (Zeiss, Jena, Germany) was used to acquire microphotographs.

2.9. Western Blot Analysis

Proteins (50 μg) from all samples were processed as previously described [28]. Membranes were incubated for 12 h at 4 °C with primary antibodies to anti-TLR4 (1:500; Santa Cruz Biotechnology, Dallas, TX, USA), anti-MyD88 (1:500; Santa Cruz Biotechnology, Dallas, TX, USA), anti-NFκB (1:500; Santa Cruz Biotechnology, Dallas, TX, USA), anti-NLRP3 (3 μg/mL; Novus, Dallas, TX, USA), anti-Caspase-1(1:500; Santa Cruz Biotechnology, Dallas, TX, USA), anti-IL-1β (1 μg/mL; Thermo Fisher, Waltham, MA, USA) and β-actin (1:1000; Santa Cruz Biotechnology, Dallas, TX, USA). Samples were maintained at room temperature for 30 min with peroxidase-conjugated secondary antibody diluted 1:1000 in 1 × Tris Buffered Saline (TBS), 5% milk and 0.05% Tween-20 (Signa-Aldrich, Milan, Italy) [29]. To visualize protein bands, the Electrochemiluminescence (ECL) method was used, and the protein levels were measured by means of the Bio-Rad Protein Assay (Bio-Rad Laboratories, Hercules, CA, USA).

2.10. Reactive Oxygen Species (ROS) Evaluation

Human and endothelial differentiated GMSCs were seeded in a 35 mm imaging dish (μ-Dish, ibidi GmbH, Gräfelfing, Germany) and treated for 24 h in culture medium containing 5 μg mL^{-1} LPS-G (hGMSCs + LPS-G or e-hGMSCs + LPS-G) or 5 μg mL^{-1} LPS-G plus 50μg/mL ascorbic acid (hGMSCs + AA/LPS-G or e-hGMSCs + AA/LPS-G) or 50μg/mL ascorbic acid (hGMSCs + AA or e-hGMSCs+ AA) or culture medium alone (hGMSCs, or e-hGMSCs). At the end of the expected time, incubation medium was removed and the cells were washed with Normal External Solution (NES) containing (in mM) 125 NaCl, 5 KCl, 1 MgSO4, 1 KH$_2$PO$_4$, 5.5 glucose, 1 CaCl$_2$, 20 4-(2-hydroxyethyl)-1-piperazineethanesulfonic acid (HEPES), pH 7.4 and incubated with 10 μM of 2',7'-dichlorodihydrofluorescein diacetate (H2DCFDA, Thermo Fisher, Waltham, MA, USA) at 37 °C in a humidified incubator (for 30 min) maintaining for all procedures the respective culture media treatments. At the end of dye incubation, the cells were washed with NES and observed in NES alone (hGMSCs or e-hGMSCs) or maintained in NES plus LPS-G, LPS-G and AA or AA alone as expected. For each condition, confocal images were randomly acquired by means of motorized table SMC 2009 and multiple single position acquisition function (Tiles-Advanced setup, carrier 35 mm petri dish) of Zen Blue software (Zen 3.0 SR, Carl Zeiss, Jena, Germany) using a Zeiss LSM800 microscope (Carl Zeiss, Jena, Germany), equipped with an inverted microscope Axio-obserber D1 (Carl Zeiss, Jena, Germany) and an objective W-Plan-Apo 40 X/1.3 DIC (Carl Zeiss, Jena, Germany). Excitation was fixed at 488 nm and emission collected with the filter set over 505–530 nm. The acquisition settings were kept constant between specimens. Offline image analyses were performed using Fiji distribution of ImageJ (version 1.53c, National Institutes of Health, Bethesda, MD, USA) measuring for each acquired cell the mean of fluorescence intensity (arbitrary units, F) and the area of the measured cells (μm^2). Quantitative data of ROS production are expressed as ratio F/μm^2.

2.11. Statistical Analysis

Statistical evaluation was performed using GraphPad Prism 4.0 software (Graph-Pad, San Diego, CA, USA) using t-test and ordinary one-way ANOVA followed by post hoc Bonferroni's multiple comparisons tests. Values of $p < 0.01$ were considered statistically significant.

3. Results

3.1. Immunophenotype and In Vitro Differentiation Ability of Isolated hGMSCs

Characterization analysis showed that hGMSCs were positive for CD73, CD90 and CD105, respectively. Confirming their mesenchymal phenotype, cells were negative for hematopoietic markers (Figure 1A). Data presented are mean $\pm$ SD ($n = 3$). To assay differentiation capacity to adipocytes and osteoblasts, hGMSCs were induced with specific differentiation kit media (Lonza, Basel, Switzerland). Cells treated with adipogenic medium produced numerous vacuoles detected with adipo oil red solution (Figure 1B, central panel). Furthermore, after treating hGMSCs with the osteogenic differentiation medium, the formation of calcium deposits revealed by alizarin red staining was observable (Figure 1B, right panel). These results indicate that hGMSCs can differentiate into both lineages. To validate the differentiation ability, RT-PCR was performed to evaluate the expression for adipogenic- and osteogenic-specific markers. In differentiated cells, FABP4, PPARγ, RUNX2 and ALP were upregulated when compared to the control cells (Figure 1B, down panel).

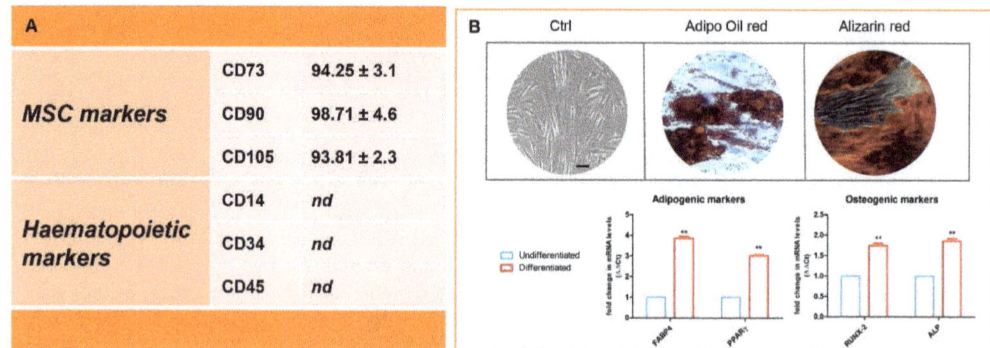

Figure 1. Human Gingival Mesenchymal Stem Cells (hGMSCs) characterization. (**A**) Antigen expression quantification by flow cytometry in human Gingival Mesenchymal Stem Cells (hGMSCs). Cells do not express hematopoietic antigens CD14, CD34 and CD45 and co-express CD73, CD90 and CD105. (**B**) Differentiation assay: cultured hGMSCs were induced to differentiate into adipocytes and osteoblast. Adipogenic differentiation was demonstrated by lipid vacuole detection with the lipid adipo oil red staining; and calcium deposits in the osteoblast differentiation were detected with alizarin red staining. Graph bars showed the expression of specific markers in cells undergoing adipogenic and osteogenic differentiation. Results represent three independent experiments performed in triplicate ($n = 3$). MSC: mesenchymal stem cell; nd: not detectable; FABP4: fatty acid-binding protein 4; PPARγ: peroxisome proliferator-activated receptor γ; RUNX-2: Runt-related transcription factor-2; ALP: alkaline phosphatase. Scale bar: 20 µm. ** $p < 0.01$.

3.2. Endothelial Differentiation and Tube Formation of hGMSCs

An evaluation of the impact of the differentiation medium kit exposure on hGMSCs to induce the endothelial transition (e-hGMSCs) was performed by examining the typical endothelial markers using CD31 immunofluorescence via staining. As reported, differentiated cells showed the expression of CD31 observed under confocal laser scanning microscopy (Figure 2A1–A4). Results of the tube formation assay showed that a number of capillary-like structures were observed after 24 h of differentiation induction (Figure 2B–D).

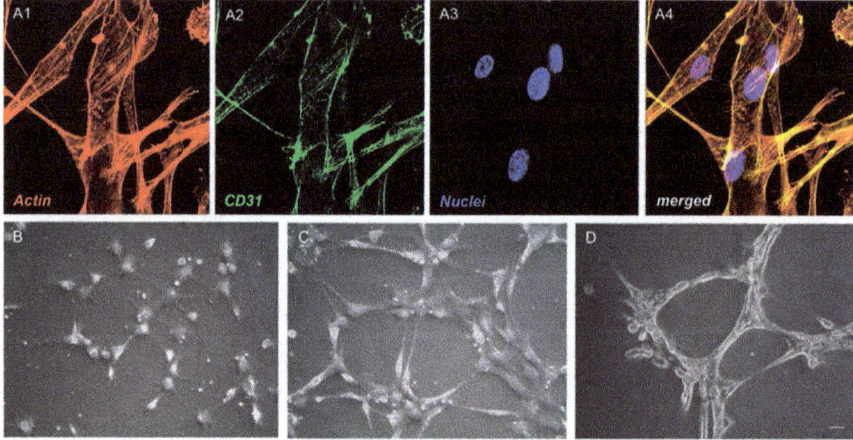

Figure 2. Endothelial differentiation and tube formation of hGMSCs. (**A1–A4**) After 14 days of exposure to endothelial differentiation, endothelial differentiated- human Gingival Mesenchymal Stem Cells (e-hGMSCs) showed positivity for endothelial marker, CD31. (**B–D**) After culture under endothelial differentiation conditions for 24 h, hGMSCs were seeded onto Cultrex to allow the formation of capillary-like structures. Cells were observed by means light microscopy after 2 (h) (**B**), 4 h (**C**) and 6 h (**D**) after seeding on dishes pre-treated with Cultrex. (**A1–A4**): magnification, $40\times$; (**B–D**): scale bar, 20 µm.

3.3. LPS-G Affect the Cell Metabolic Activity of hGMSCs

At defined time points, 24, 48 and 72 h after cultivation in all considered culture conditions hGMSCs, the cell metabolic activity was evaluated. Human GMSC and e-hGMSC cell metabolic activity was reduced when treated with LPS-G compared to the untreated cells. Conversely, constant cell numbers were recorded in the untreated and AA-treated groups. In addition, cells co-treated with AA and LPS-G showed a better trend in cell metabolic activity at all considered end points (Figure 3).

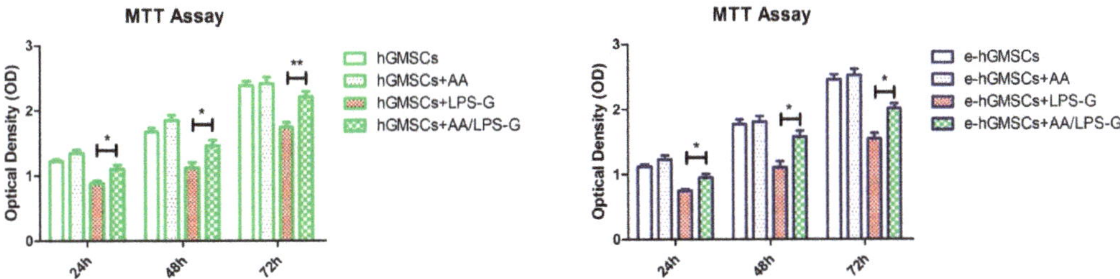

Figure 3. Effects of Lipopolysaccharide from *Porphyromonas gingivalis* (LPS-G) and ascorbic acid (AA) on the cell metabolic activity of hGMSCs and e-hGMSCs. Cell growth curve of hGMSCs and e-hGMSCs detected by absorbance at 490 nm showed the different cell metabolic activity of cells under treatment with AA, LPS-G and LPS-G + AA. * $p > 0.05$; ** $p < 0.01$.

3.4. The TLR4/MyD88/NFκB/NLRP3/Caspase-1/IL-1β Signaling Pathway Was Involved in AA Anti-inflammatory Effects on LPS-G-Stimulated Cells

Stimulation of hGMSCs and e-hGMSCs with LPS-G induced the expression of the inflammatory pathway. Cells showed an overexpression of TLR4, MyD88, NF-κB, NLRP3, Caspase-1 and IL-1β as observed under confocal microscopy (Figure 4C1–C3, Figure 5C1–C3, Figure 6C1–C3, Figure 7C1–C3). Untreated hGMSCs and e-hGMSCs (CTRL) and AA-treated cells showed lower positivity for TLR4, MyD88, NFκB, NLRP3, Caspase-1 and IL-1β compared to LPS-G-stimulated cells (Figure 4A1–B3, Figure 5A1–B3, Figure 6A1–B3, Figure 7A1–B3). The co-treatment of AA and LPS-G in all experimental groups (hGMSCS and e-hGMSCs) showed a reduction in the expression of studied proteins, indicating a possible protective effect exerted by AA (Figure 4D1–D3, Figure 5D1–D3, Figure 6D1–D3, Figure 7D1–D3). This finding was further confirmed by Western blot analyses, which showed that the expression of TLR4, MyD88, NFκB, NLRP3, Caspase-1 and IL-1β was increased in LPS-G-treated samples. The cells co-treated with AA and LPS-G showed a reduction in the expression of proteins involved in the inflammatory pathway (Figures 4E, 5E, 6E and 7E).

3.5. ROS Production

ROS production induced by LPS-G has been studied in hGMSCs and e-hGMSCs loaded with the cell-permeant H2DCFDA. Once penetrated inside the cell, the cleavage of the acetate groups by intracellular esterases makes the molecule active and able to undergo oxidation. In this way, the nonfluorescent H2DCFDA is converted to the highly fluorescent 2′,7′-dichlorofluorescein (DCF) by ROS. Images were acquired in live cells by means of confocal microscopy as reported in Figure 8A and the single cells' fluorescence recorded was analyzed offline. Quantitative results (Figure 8B) showed a significative increase in ROS production both in 5 µg mL^{-1} LPS-G-treated hGMSCs and e-hGMSCs vs. the control condition (hGMSCs and e-hGMSCs, respectively). Interestingly, the co-incubation of LPS-G together with AA counteracted the LPS-G increase in ROS production. Of note, comparing the results obtained in the hGMSCs and e-hGMSCs LPS-G-treated cells, we observed that the latter were more prone to produce ROS.

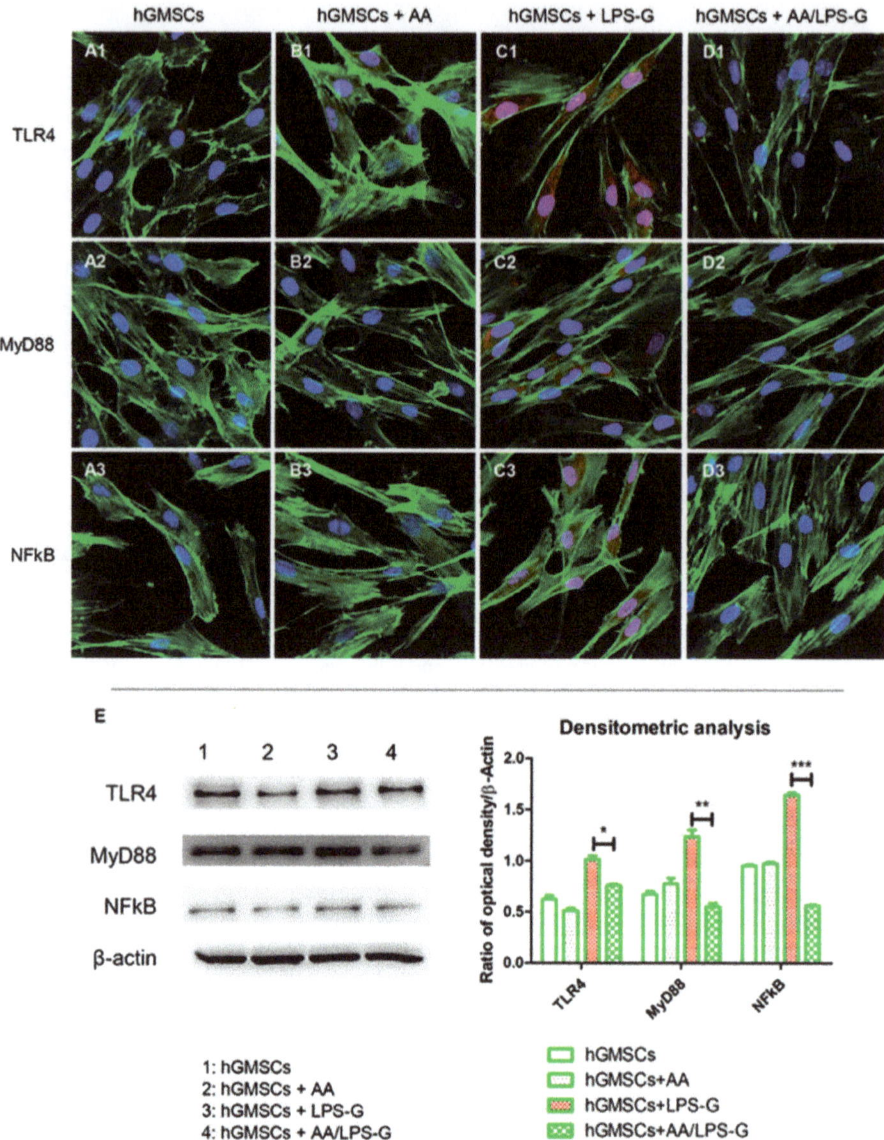

Figure 4. TLR4/MyD88/NFκB signaling pathway and differences in protein levels in hGMSCs. Expression of TLR4, MYD88 and NFκB, examined by confocal microscopy and Western blot experiments. (**A1–D1**) TLR4 expression in hGMSCs (CTRL), hGMSCs + AA, hGMSCs + LPS-G, hGMSCs + AA/LPS-G. (**A2–D2**) MyD88 expression in hGMSCs (CTRL), hGMSCs + AA, hGMSCs + LPS-G, hGMSCs + AA/LPS-G. (**A3–D3**) NFκB expression in hGMSCs (CTRL), hGMSCs + AA, hGMSCs + LPS-G, hGMSCs + AA/LPS-G. (**E**) TLR4, MyD88 and NFκB protein levels. β-actin was used as a loading control. Green fluorescence: cytoskeleton actin. Red fluorescence: TLR4, MYD88, NFκB. Blue fluorescence: cell nuclei. Scale bar: 20 μm. * $p < 0.05$; ** $p < 0.01$; *** $p < 0.001$.

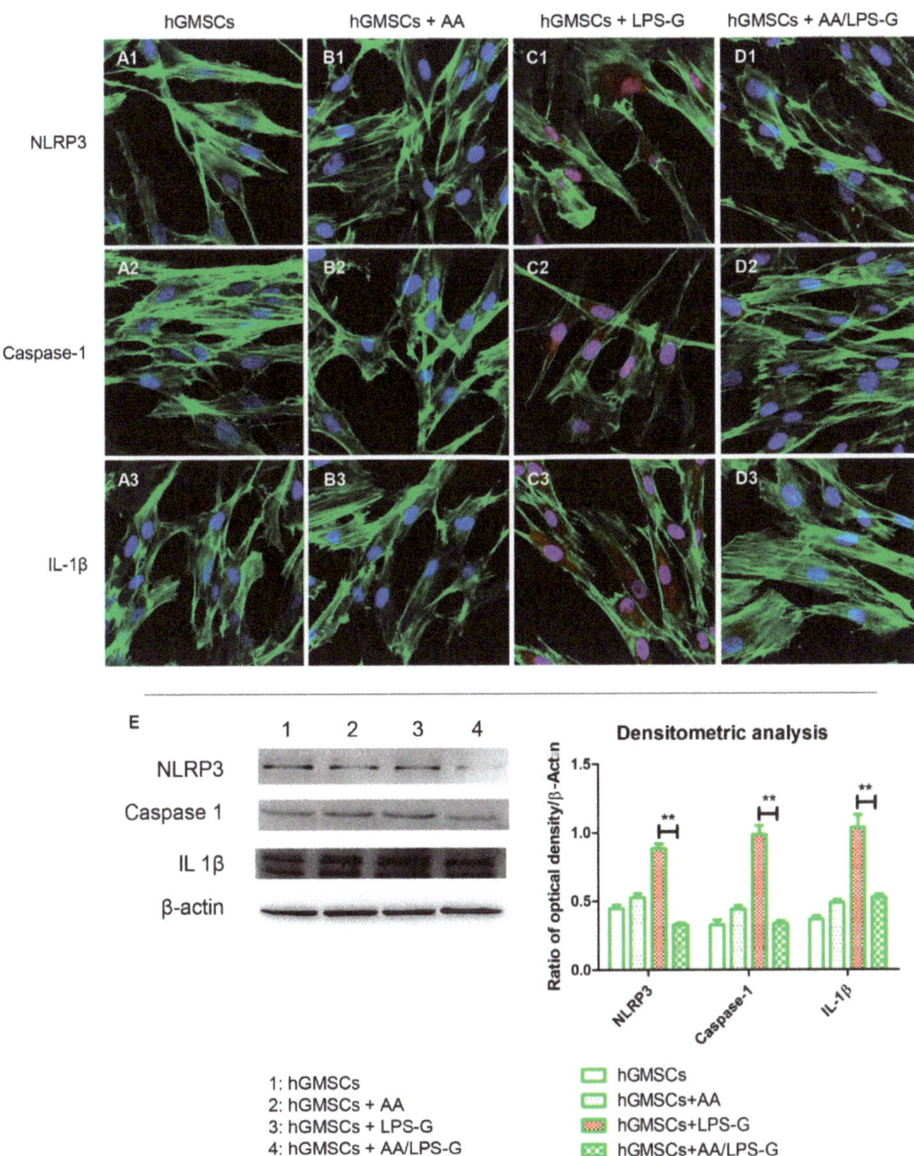

Figure 5. NLRP3/Caspase-1/IL-1β signaling pathway and differences in protein levels in hGMSCs. Expression of NLRP3, Caspase-1 and IL-1β, examined by confocal microscopy and Western blot experiments. (**A1–D1**) NLRP3 expression in hGM-SCs (CTRL), hGMSCs + AA, hGMSCs + LPS-G, hGMSCs + AA/LPS-G. (**A2–D2**) Caspase-1 expression in hGMSCs (CTRL), hGMSCs + AA, hGMSCs + LPS-G, hGMSCs + AA/LPS-G. (**A3–D3**) IL-1β expression in hGMSCs (CTRL), hGMSCs + AA, hGMSCs + LPS-G, hGMSCs + AA/LPS-G. (**E**) NLRP3, Caspase -1 and IL-1β protein levels. β-actin was used as a loading control. Green fluorescence: cytoskeleton actin. Red fluorescence: NLRP3, Caspase-1, IL-1β. Blue fluorescence: cell nuclei. Scale bar: 20 μm. ** $p < 0.01$.

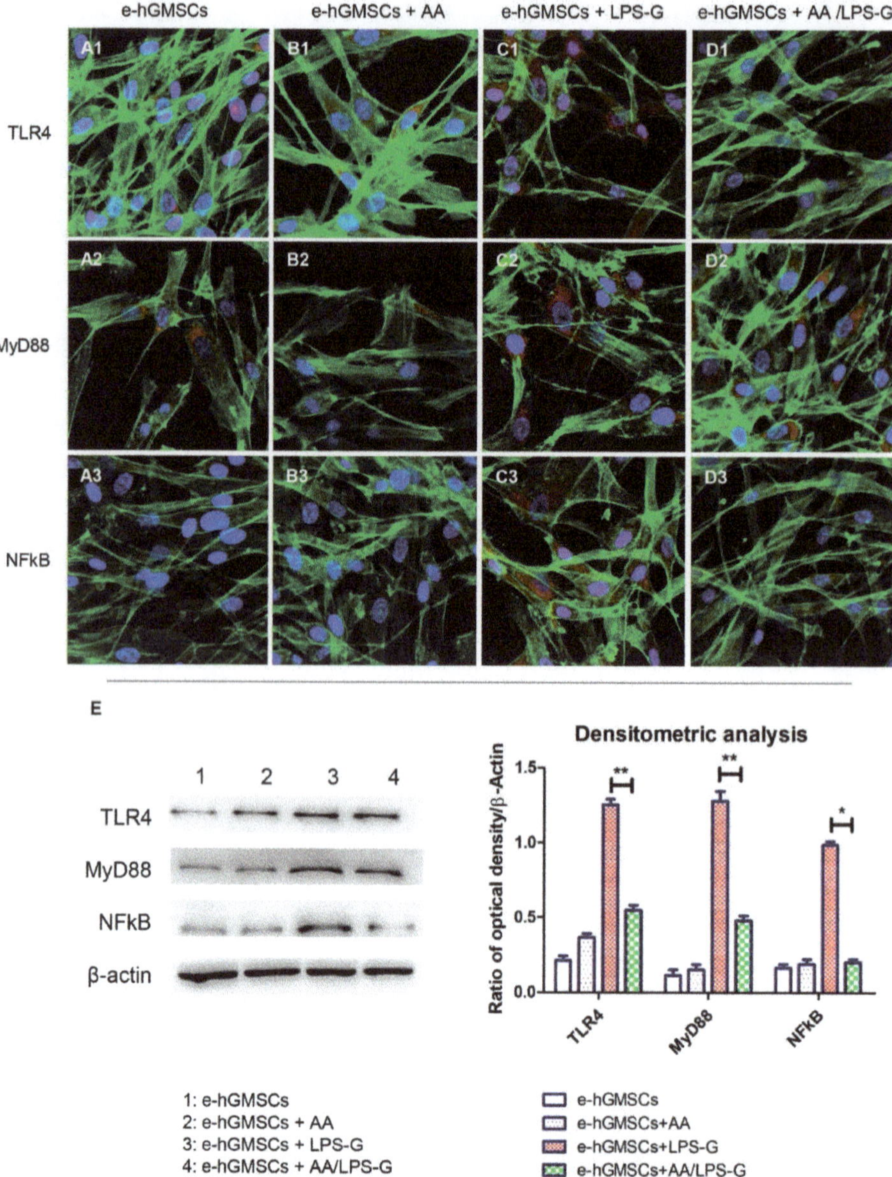

Figure 6. TLR4/MyD88/NFκB signaling pathway and differences in protein levels in e-hGMSCs. Expression of TLR4, MYD88 and NFκB, examined by confocal microscopy and Western blot experiments. (**A1–D1**) TLR4 expression in e-hGMSCs (CTRL), e-hGMSCs + AA, e-hGMSCs + LPS-G, e-hGMSCs + AA/LPS-G. (**A2–D2**) MyD88 expression in e-hGMSCs (CTRL), e-hGMSCs + AA, e-hGMSCs + LPS-G, e-hGMSCs + AA/LPS-G. (**A3–D3**) NFκB expression in e-hGMSCs (CTRL), e-hGMSCs + AA, e-hGMSCs + LPS-G, e-hGMSCs + AA/LPS-G. (**E**) TLR4, MyD88 and NFκB protein levels. β-actin was used as a loading control. Green fluorescence: cytoskeleton actin. Red fluorescence: TLR4, MYD88, NFκB. Blue fluorescence: cell nuclei. Scale bar: 20 μm. * $p < 0.05$; ** $p < 0.01$.

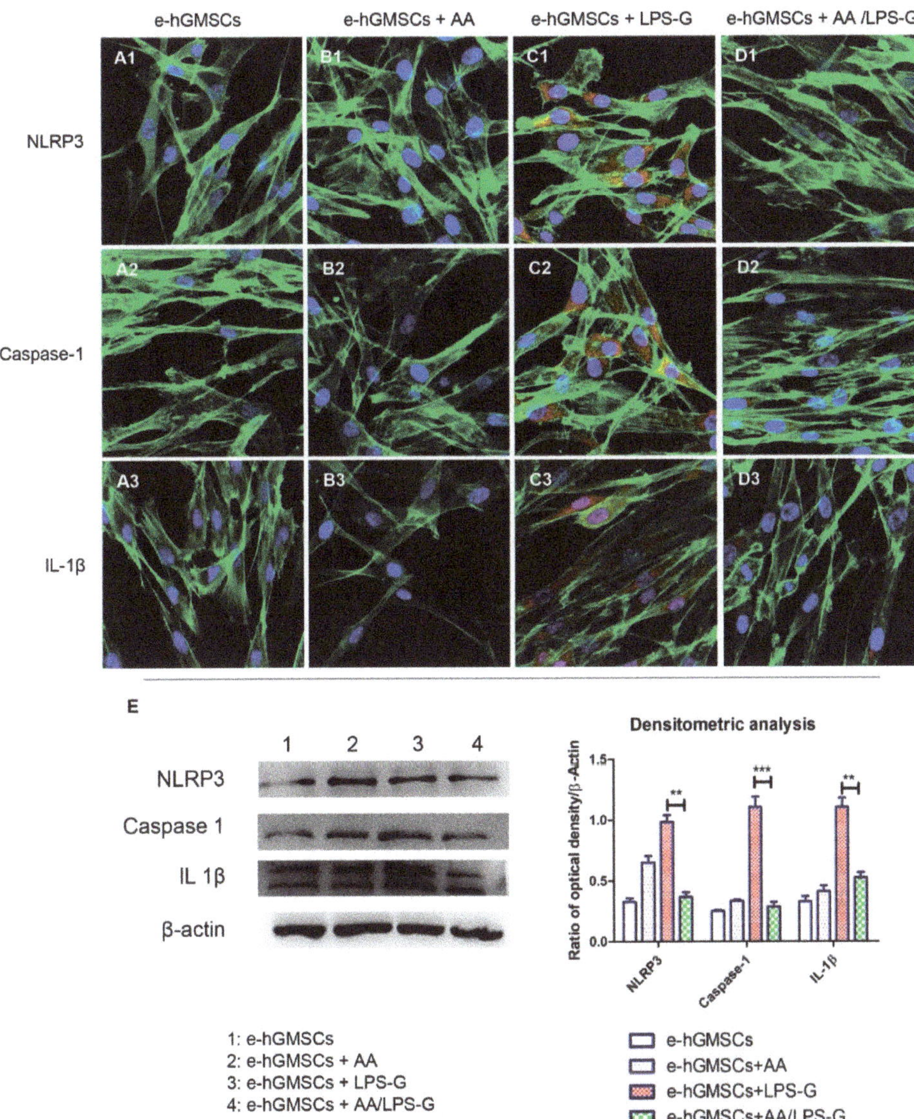

Figure 7. NLRP3/Caspase-1/IL-1β signaling pathway and differences in protein levels in e-hGMSCs. Expression of NLRP3, Caspase-1 and IL-1β, examined by confocal microscopy and Western blot experiments. (**A1–D1**) NLRP3 expression in e-hGMSCs (CTRL), e-hGMSCs + AA, e-hGMSCs + LPS-G, e-hGMSCs + AA/LPS-G. (**A2–D2**) Caspase-1 expression in e-hGMSCs (CTRL), e-hGMSCs + AA, e-hGMSCs + LPS-G, e-hGMSCs + AA/LPS-G. (**A3–D3**) IL-1β expression in e-hGMSCs (CTRL), e-hGMSCs + AA, e-hGMSCs + LPS-G, e-hGMSCs + AA/LPS-G. (**E**) NLRP3, Caspase-1 and IL-1β protein levels. β-actin was used as a loading control. Green fluorescence: cytoskeleton actin. Red fluorescence: NLRP3, Caspase-1, IL-1β. Blue fluorescence: cell nuclei. Scale bar: 20 μm. ** $p < 0.01$; *** $p < 0.001$.

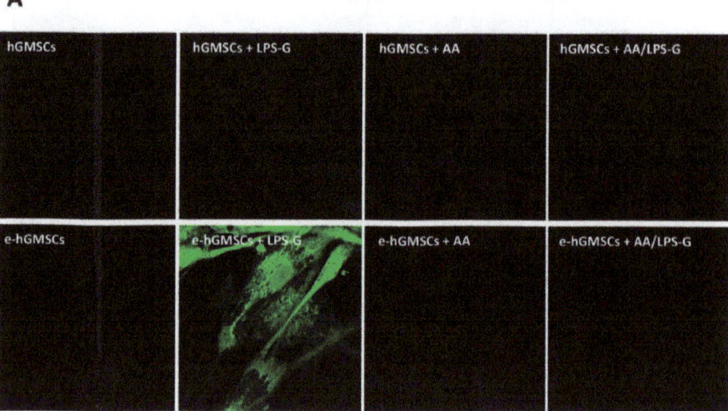

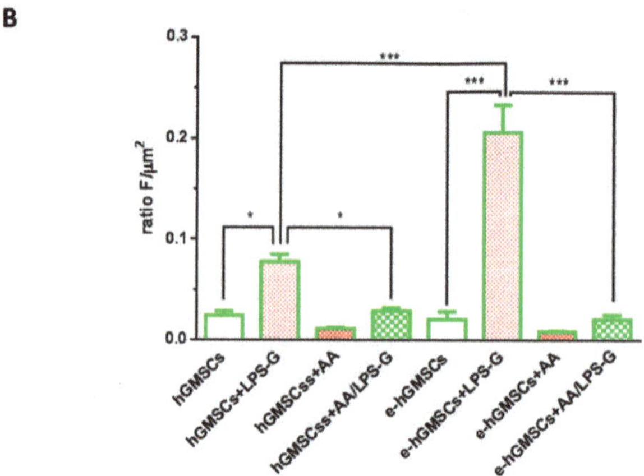

Figure 8. ROS measurements. (**A**) Representative images of H2DCFDA-loaded cells. (**B**) Analysis of ROS production calculated as arbitrary unit of fluorescence per cell surface unit (F/μm2). Data are expressed as mean± S.E.M (hGMSCs n = 166, hGMSCs + LPS-G n = 148, hGMSCs + AA n = 100; hGMSCs + AA/LPS-G n = 176; e-hGMSCs n = 85, e-hGMSCs + LPS-G n = 195, e-hGMSCs + AA n = 159; e-hGMSCs + AA/LPS-G n = 125, for each experimental condition n = 3). * $p < 0.05$; *** $p < 0.001$. Statistical analysis was performed by one-way ANOVA and post hoc Bonferroni. Scale bar: 20 μm.

4. Discussion

Periodontitis is a form of chronic inflammatory disease that invades teeth-supporting tissues, sustained mostly by Gram-negative microbes such as *Porphyromonas gingivalis* through lipopolysaccharide (LPS-G) action, which represents one of the major virulence factors in periodontitis progression [30]. Outer membrane endotoxins can trigger multiple signaling pathways that are involved several intracellular events, which eventually stimulate the production of many inflammatory factors and cause the destruction of the periodontal soft tissue, the resorption of alveolar bone and, finally, tooth loss [31].

The inflammasome activation is strictly related to the development and progression of periodontal disease. The establishment of an in vitro model of periodontal inflammation is necessary to study periodontitis' pathogenesis and the potential novel treatments for this complex disease [32].

The in vitro model established by Pizzicannella et al. [24] attempts to reproduce the microenvironment of periodontal disease using hGMSCs and LPS-G stimulation to obtain more meaningful data.

Human GMSCs are resident in the gingival connective tissues, showing the ability to maintain, repair, and remodel the extracellular matrix for tissue homeostasis [33]. Several studies reported that hGMSCs showed the features and functions of mesenchymal stem cells, such as proliferation, migration and cell differentiation capacity. As demonstrated by our obtained results, the hGMSCs can be considered as MSCs as they follow Dominici's criteria: they showed the capacity to adhere to a plastic substrate, the fibroblast-like morphology, the capacity to differentiate into adipogenic and osteogenic lineages, positivity to CD73, CD90 and CD105 and negativity for CD14, CD34 and CD45.

The effect of LPS-G stimulation may induce an inflammatory response in hGMSCs that could reproduce similarly the periodontitis microenvironment [34]. The most studied complex involved in intracellular inflammation is the NLRP3 inflammasome, which can be activated when cells are exposed to a wide range of bacterial ligands, including LPS-G [35]. The NLRP3 inflammasome contains a sensor molecule, which can initiate the inflammatory responses [36]. Both the NLRP3 inflammasome and TLR4 signaling play a key role in tissue injury. TLR4 acts as a fundamental factor that promotes the inflammatory response chain in the body [37]. The inhibition of or decrease in the NLRP3/TLR4 signaling pathway is a promising potential target in several chronic diseases [38,39]. Indeed, TLR4 activation triggers the myeloid differentiation factor 88 (MyD88) signaling pathway, which initiates the rapid activation of nuclear factor-kB (NFκB), which leads to an increase in IL-18, IL-6, IL-1β, tumor necrosis factor-α (TNF-α) and monocyte chemotactic protein-1 (MCP-1) [40,41]. Caspase-1 is considered another inflammasome component with a specific role in periodontal disease. Moreover, Caspase-1 was expressed in clinical samples collected from patients affected by periodontal disease.

During periodontal disease progression, IL-1β plays a pivotal role in the induction and maintenance of the host immune response, with a negative effect on periodontal soft tissues. IL-1β expression is regulated by the activation of the NLRP3 inflammasome complexes and subsequent modulation of this intracellular pathway may cause a reduction in IL-1β release [42,43]

The current experiments were designed to evaluate the potential positive effect of AA in response to the inflammasome activation of LPS-G stimulated cells. In the present study, an in vitro model was established where hGMSCs were treated with LPS-G to mimic the periodontal microenvironment. The use of endothelial committed hGMSCs could represent a starting point, as an in vitro model, to evaluate the intracellular signaling pathway that may link the periodontal and cardiovascular diseases. The endothelial commitment of hGMSCs was demonstrated through the CD31 expression evaluated by confocal microscopy observation; cells showed positivity for CD31 at the cytoplasmic level. Moreover, the e-hGMSCs showed the ability to form a capillary-like tube network in a plate coated with Cultrex, indicating that they were committed towards an endothelial phenotype [44]. In fact, tubule formation is one of the hallmarks of angiogenesis, along with cell proliferation and migration [45].

AA (or vitamin C) is a hydrophilic vitamin with high activity as an antioxidant [46]. AA showed the capacity to downregulate ROS production and exert a positive regulatory effect to maintain periodontal health [47]. Moreover, previous findings evidenced that the risk of CVD is correlated inversely with vitamin C diet supplementation [48,49]. The molecular mechanisms involved in the antioxidant property of AA to prevent and to treat the cardiovascular disorders remain elusive [50,51]. Earlier studies have investigated the signaling pathways involved in the effects of AA on CVD related to periodontal disease [52].

In our in vitro model, we determined that stimulation with LPS-G showed a reduction in cell metabolic activity and upregulation of TLR4, MyD88, NFκB, NLRP3, Caspase-1 and IL-1β in hGMSCs and in e-hGMSCs compared to the untreated cells, as demonstrated by immunofluorescence detection and protein level quantization. Indeed, the co-treatment of

LPS-G and AA showed a downregulation of TLR4, MyD88, NFκB, NLRP3, Caspase-1 and IL-1β, as demonstrated by qualitative and quantitative analyses, when compared to the cells stimulated with LPS-G alone. Accordingly, the modulation of TLR4 signaling and the NLRP3 inflammasome may provide a novel and valid approach to investigate, in vitro, the progression of the inflammatory cascade in cells exposed to LPS-G stimulation.

In agreement with previous studies [53–55], the data obtained in the present work demonstrated that treatment with AA showed a reduction in the expression levels of the NLP3 inflammasome that led to a reduction in the production of IL-1β through the TLR4/MyD88/NFκB signaling pathway in the hGMSCs and e-hGMSCs. Moreover, our findings evidenced a protective role played by AA through the modulation of TLR4/MyD88/NFκB/NLRP3/Caspase-1/IL-1β, reestablishing the microenvironment before LPS-G stimulation.

Furthermore, in the same in vitro model, we investigated ROS production in the same experimental groups. Several studies have demonstrated that ROS production and NLRP3 inflammasome activation are stimulated by various pathogens and metabolic stresses [56]. A high level of ROS induced the activation of the NLRP3 inflammasome in the livers of fructose-fed rats and also enhanced the secretion of IL-1β and IL-18 in rat hepatocytes [56,57]. This study showed that in cells treated with ascorbic acid, high production of ROS is present in hGMSCS and e-hGMSCs stimulated with LPS-G, while the ROS levels are reduced in cells co-treated with AA and LPS-G. Untreated and AA-treated cells showed no differences in ROS production. In fact, as reported by Jung-Yoon Choe et al., AA showed a protective effect against ROS formation and consequent activation of the NLRP3 inflammasome in human macrophages [58].

Although the present study possessed some limitations due to the use of an in vitro model considering one inflammation factor, it may provide a starting point for the evaluation of one of the intracellular mechanisms that occurs during periodontitis. Further studies are required to verify and expand our knowledge on the role of AA in the regulation of the protein and genetic factors that are involved in periodontal disease.

5. Limitations of the Study

Although the current in vitro study has provided important information on basic intracellular mechanisms that occur during periodontal disease, it is still far from addressing specific experimental questions presented in the in vivo system. Moreover, to better evaluate the specific role of AA in terms of the neutralization of LPS-G effects or the combination with cell epitopes blocking access to TLR4, further studies are needed to better clarify the molecular mechanisms that are the basis of the use of multiple products. Furthermore, this study was conducted in vitro, and the findings need to be evaluated also through in vivo assays.

6. Conclusions

In summary, the present study demonstrated the protective effects of AA through the modulation of TLR4/MyD88/NFκB/NLRP3/Caspase-1/IL-1β in an in vitro model of LPS-G-stimulated cells. These findings suggest the use of AA as a promising potential factor for the prevention of the triggering of the inflammatory cascade that may lead to periodontal and vascular damage. Further studies are necessary to better investigate the molecular mechanisms of these pathways and their role in the progression of periodontal disease.

Author Contributions: Conceptualization, J.P. and F.D.; methodology, L.F., S.G. and G.D.M.; software, G.D.M.; validation, S.G. and T.S.R.; formal analysis, G.D.M., L.F. and S.G.; investigation, J.P., O.T. and F.D.; data curation, L.F. and S.G.; writing—original draft preparation, J.P., S.G. and F.D.; writing—review and editing, T.S.R. and O.T.; visualization, S.G. and T.S.R.; supervision, J.P., O.T. and F.D.; project administration, J.P., O.T. and F.D.; funding acquisition, G.D.M., O.T. and F.D. All authors have read and agreed to the published version of the manuscript.

Antioxidants **2021**, 10, 797

Funding: This research was funded by OT, FD and GDM Research Funds of University "G. d'Annunzio" Chieti-Pescara, grant number: OT60%-2019, FD60%-2020 and GDM60%-2019.

Institutional Review Board Statement: The study was conducted according to the guidelines of the Declaration of Helsinki and approved by the Institutional Ethics Committee of University "G. d'Annunzio" Chieti-Pescara (n°266/2017).

Informed Consent Statement: Informed consent was obtained from all subjects involved in the study.

Data Availability Statement: Not applicable.

Acknowledgments: T.S.R. would like to thank DST-FIST India for support of the Department of Biotechnology, Karpagam Academy of Higher Education (SR/FST/LS-1/2018/187; Dated 26 December 2018).

Conflicts of Interest: The authors declare no conflict of interest.

References

1. Venkataiah, V.S.; Handa, K.; Njuguna, M.M.; Hasegawa, T.; Maruyama, K.; Nemoto, E.; Yamada, S.; Sugawara, S.; Lu, L.; Takedachi, M.; et al. Periodontal Regeneration by Allogeneic Transplantation of Adipose Tissue Derived Multi-Lineage Progenitor Stem Cells in vivo. *Sci. Rep.* **2019**, *9*, 921. [CrossRef] [PubMed]
2. Loos, B.G.; Van Dyke, T.E. The role of inflammation and genetics in periodontal disease. *Periodontol. 2000* **2020**, *83*, 26–39. [CrossRef] [PubMed]
3. Atilla, G.; Sorsa, T.; Ronka, H.; Emingil, G. Matrix metalloproteinases (MMP-8 and -9) and neutrophil elastase in gingival crevicular fluid of cyclosporin-treated patients. *J. Periodontol.* **2001**, *72*, 354–360. [CrossRef] [PubMed]
4. Carrion, J.; Scisci, E.; Miles, B.; Sabino, G.J.; Zeituni, A.E.; Gu, Y.; Bear, A.; Genco, C.A.; Brown, D.L.; Cutler, C.W. Microbial carriage state of peripheral blood dendritic cells (DCs) in chronic periodontitis influences DC differentiation, atherogenic potential. *J. Immunol.* **2012**, *189*, 3178–3187. [CrossRef] [PubMed]
5. Mustapha, I.Z.; Debrey, S.; Oladubu, M.; Ugarte, R. Markers of systemic bacterial exposure in periodontal disease and cardiovascular disease risk: A systematic review and meta-analysis. *J. Periodontol.* **2007**, *78*, 2289–2302. [CrossRef]
6. Orlandi, M.; Graziani, F.; D'Aiuto, F. Periodontal therapy and cardiovascular risk. *Periodontol. 2000* **2020**, *83*, 107–124. [CrossRef]
7. Tonetti, M.S.; D'Aiuto, F.; Nibali, L.; Donald, A.; Storry, C.; Parkar, M.; Suvan, J.; Hingorani, A.D.; Vallance, P.; Deanfield, J. Treatment of periodontitis and endothelial function. *N. Engl. J. Med.* **2007**, *356*, 911–920. [CrossRef]
8. Velsko, I.M.; Chukkapalli, S.S.; Rivera, M.F.; Lee, J.Y.; Chen, H.; Zheng, D.; Bhattacharyya, I.; Gangula, P.R.; Lucas, A.R.; Kesavalu, L. Active invasion of oral and aortic tissues by Porphyromonas gingivalis in mice causally links periodontitis and atherosclerosis. *PLoS ONE* **2014**, *9*, e97811. [CrossRef]
9. Lee, J.; Roberts, J.S.; Atanasova, K.R.; Chowdhury, N.; Han, K.; Yilmaz, O. Human Primary Epithelial Cells Acquire an Epithelial-Mesenchymal-Transition Phenotype during Long-Term Infection by the Oral Opportunistic Pathogen, Porphyromonas gingivalis. *Front. Cell. Infect. Microbiol.* **2017**, *7*, 493. [CrossRef]
10. Abdulkareem, A.A.; Shelton, R.M.; Landini, G.; Cooper, P.R.; Milward, M.R. Potential role of periodontal pathogens in compromising epithelial barrier function by inducing epithelial-mesenchymal transition. *J. Periodontal Res.* **2018**, *53*, 565–574. [CrossRef]
11. Yamada, M.; Takahashi, N.; Matsuda, Y.; Sato, K.; Yokoji, M.; Sulijaya, B.; Maekawa, T.; Ushiki, T.; Mikami, Y.; Hayatsu, M.; et al. A bacterial metabolite ameliorates periodontal pathogen-induced gingival epithelial barrier disruption via GPR40 signaling. *Sci. Rep.* **2018**, *8*, 9008. [CrossRef] [PubMed]
12. Gao, A.C.; Wang, X.C.; Yu, H.Y.; Li, N.; Hou, Y.B.; Yu, W.X. Effect of Porphyromonas gingivalis lipopolysaccharide (Pg-LPS) on the expression of EphA2 in osteoblasts and osteoclasts. *In Vitro Cell. Dev. Biol. Anim.* **2016**, *52*, 228–234. [CrossRef] [PubMed]
13. Li, C.H.; Li, B.; Dong, Z.W.; Gao, L.; He, X.N.; Liao, L.; Hu, C.H.; Wang, Q.T.; Jin, Y. Lipopolysaccharide differentially affects the osteogenic differentiation of periodontal ligament stem cells and bone marrow mesenchymal stem cells through Toll-like receptor 4 mediated nuclear factor kappa B pathway. *Stem Cell Res. Ther.* **2014**, *5*. [CrossRef]
14. Diomede, F.; Thangavelu, S.R.; Merciaro, I.; D'Orazio, M.; Bramanti, P.; Mazzon, E.; Trubiani, O. Porphyromonas gingivalis lipopolysaccharide stimulation in human periodontal ligament stem cells: Role of epigenetic modifications to the inflammation. *Eur. J. Histochem.* **2017**, *61*, 2826. [CrossRef] [PubMed]
15. Dominici, M.; Le Blanc, K.; Mueller, I.; Slaper-Cortenbach, I.; Marini, F.; Krause, D.; Deans, R.; Keating, A.; Prockop, D.; Horwitz, E. Minimal criteria for defining multipotent mesenchymal stromal cells. The International Society for Cellular Therapy position statement. *Cytotherapy* **2006**, *8*, 315–317. [CrossRef]
16. Trubiani, O.; Marconi, G.D.; Pierdomenico, S.D.; Piattelli, A.; Diomede, F.; Pizzicannella, J. Human Oral Stem Cells, Biomaterials and Extracellular Vesicles: A Promising Tool in Bone Tissue Repair. *Int. J. Mol. Sci.* **2019**, *20*, 4987. [CrossRef] [PubMed]
17. Mikirova, N.; Casciari, J.; Rogers, A.; Taylor, P. Effect of high-dose intravenous vitamin C on inflammation in cancer patients. *J. Transl. Med.* **2012**, *10*, 189. [CrossRef] [PubMed]

18. Liang, X.P.; Li, Y.; Hou, Y.M.; Qiu, H.; Zhou, Q.C. Effect of dietary vitamin C on the growth performance, antioxidant ability and innate immunity of juvenile yellow catfish (Pelteobagrus fulvidraco Richardson). *Aquac. Res.* **2017**, *48*, 149–160. [CrossRef]

19. Kurutas, E.B. The importance of antioxidants which play the role in cellular response against oxidative/nitrosative stress: Current state. *Nutr. J.* **2016**, *15*. [CrossRef]

20. Li, X.; Tang, L.; Lin, Y.F.; Xie, G.F. Role of vitamin C in wound healing after dental implant surgery in patients treated with bone grafts and patients with chronic periodontitis. *Clin. Implant Dent. Relat. Res.* **2018**, *20*, 793–798. [CrossRef] [PubMed]

21. Mammana, S.; Gugliandolo, A.; Cavalli, E.; Diomede, F.; Iori, R.; Zappacosta, R.; Bramanti, P.; Conti, P.; Fontana, A.; Pizzicannella, J.; et al. Human gingival mesenchymal stem cells pretreated with vesicular moringin nanostructures as a new therapeutic approach in a mouse model of spinal cord injury. *J. Tissue Eng. Regener. Med.* **2019**, *13*, 1109–1121. [CrossRef] [PubMed]

22. Libro, R.; Diomede, F.; Scionti, D.; Piattelli, A.; Grassi, G.; Pollastro, F.; Bramanti, P.; Mazzon, E.; Trubiani, O. Cannabidiol Modulates the Expression of Alzheimer's Disease-Related Genes in Mesenchymal Stem Cells. *Int. J. Mol. Sci.* **2016**, *18*, 26. [CrossRef]

23. Pizzicannella, J.; Cavalcanti, M.; Trubiani, O.; Diomede, F. MicroRNA 210 Mediates VEGF Upregulation in Human Periodontal Ligament Stem Cells Cultured on 3DHydroxyapatite Ceramic Scaffold. *Int. J. Mol. Sci.* **2018**, *19*, 3916. [CrossRef] [PubMed]

24. Pizzicannella, J.; Diomede, F.; Merciaro, I.; Caputi, S.; Tartaro, A.; Guarnieri, S.; Trubiani, O. Endothelial committed oral stem cells as modelling in the relationship between periodontal and cardiovascular disease. *J. Cell. Physiol.* **2018**, *233*, 6734–6747. [CrossRef] [PubMed]

25. Ballerini, P.; Diomede, F.; Petragnani, N.; Cicchitti, S.; Merciaro, I.; Cavalcanti, M.; Trubiani, O. Conditioned medium from relapsing-remitting multiple sclerosis patients reduces the expression and release of inflammatory cytokines induced by LPS-gingivalis in THP-1 and MO3.13 cell lines. *Cytokine* **2017**, *96*, 261–272. [CrossRef]

26. Sinjari, B.; Pizzicannella, J.; D'Aurora, M.; Zappacosta, R.; Gatta, V.; Fontana, A.; Trubiani, O.; Diomede, F. Curcumin/Liposome Nanotechnology as Delivery Platform for Anti-inflammatory Activities via NFkB/ERK/pERK Pathway in Human Dental Pulp Treated With 2-HydroxyEthyl MethAcrylate (HEMA). *Front. Physiol.* **2019**, *10*, 633. [CrossRef] [PubMed]

27. Pizzicannella, J.; Diomede, F.; Gugliandolo, A.; Chiricosta, L.; Bramanti, P.; Merciaro, I.; Orsini, T.; Mazzon, E.; Trubiani, O. 3D Printing PLA/Gingival Stem Cells/EVs Upregulate miR-2861 and -210 during Osteoangiogenesis Commitment. *Int. J. Mol. Sci.* **2019**, *20*, 3256. [CrossRef]

28. Diomede, F.; D'Aurora, M.; Gugliandolo, A.; Merciaro, I.; Orsini, T.; Gatta, V.; Piattelli, A.; Trubiani, O.; Mazzon, E. Biofunctionalized Scaffold in Bone Tissue Repair. *Int. J. Mol. Sci.* **2018**, *19*, 1022. [CrossRef] [PubMed]

29. Orciani, M.; Trubiani, O.; Guarnieri, S.; Ferrero, E.; Di Primio, R. CD38 Is Constitutively Expressed in the Nucleus of Human Hematopoietic Cells. *J. Cell. Biochem.* **2008**, *105*, 905–912. [CrossRef]

30. Moutsopoulos, N.M.; Chalmers, N.I.; Barb, J.J.; Abusleme, L.; Greenwell-Wild, T.; Dutzan, N.; Paster, B.J.; Munson, P.J.; Fine, D.H.; Uzel, G.; et al. Subgingival Microbial Communities in Leukocyte Adhesion Deficiency and Their Relationship with Local Immunopathology. *PLoS Pathog.* **2015**, *11*. [CrossRef]

31. Fitzsimmons, T.R.; Ge, S.; Bartold, P.M. Compromised inflammatory cytokine response to *P. gingivalis* LPS by fibroblasts from inflamed human gingiva. *Clin. Oral Investig.* **2018**, *22*, 919–927. [CrossRef] [PubMed]

32. Chen, C.; Ma, X.Q.; Yang, C.J.; Nie, W.; Zhang, J.; Li, H.D.; Rong, P.F.; Yi, S.N.; Wang, W. Hypoxia potentiates LPS-induced inflammatory response and increases cell death by promoting NLRP3 inflammasome activation in pancreatic beta cells. *Biochem. Biophys. Res. Commun.* **2018**, *495*, 2512–2518. [CrossRef] [PubMed]

33. McCulloch, C.A. Origins and functions of cells essential for periodontal repair: The role of fibroblasts in tissue homeostasis. *Oral Dis.* **1995**, *1*, 271–278. [CrossRef] [PubMed]

34. Taskan, M.M.; Karatas, O.; Balci Yuce, H.; Isiker Kara, G.; Gevrek, F.; Ucan Yarkac, F. Hypoxia and collagen crosslinking in the healthy and affected sites of periodontitis patients. *Acta Odontol. Scand.* **2019**, *77*, 600–607. [CrossRef] [PubMed]

35. Huck, O.; Elkaim, R.; Davideau, J.L.; Tenenbaum, H. Porphyromonas gingivalis-impaired innate immune response via NLRP3 proteolysis in endothelial cells. *Innate Immun.* **2015**, *21*, 65–72. [CrossRef] [PubMed]

36. He, C.L.; Zhao, Y.; Jiang, X.L.; Liang, X.X.; Yin, L.Z.; Yin, Z.Q.; Geng, Y.; Zhong, Z.J.; Song, X.; Zou, Y.F.; et al. Protective effect of Ketone musk on LPS/ATP-induced pyroptosis in J774A.1 cells through suppressing NLRP3/GSDMD pathway. *Int. Immunopharmacol.* **2019**, *71*, 328–335. [CrossRef] [PubMed]

37. Roger, T.; David, J.; Glauser, M.P.; Calandra, T. MIF regulates innate immune responses through modulation of Toll-like receptor 4. *Nature* **2001**, *414*, 920–924. [CrossRef] [PubMed]

38. De Nardo, D.; Latz, E. NLRP3 inflammasomes link inflammation and metabolic disease. *Trends Immunol.* **2011**, *32*, 373–379. [CrossRef] [PubMed]

39. Osborn, O.; Olefsky, J.M. The cellular and signaling networks linking the immune system and metabolism in disease. *Nat. Med.* **2012**, *18*, 363–374. [CrossRef]

40. Wang, M.X.; Zhao, X.J.; Chen, T.Y.; Liu, Y.L.; Jiao, R.Q.; Zhang, J.H.; Ma, C.H.; Liu, J.H.; Pan, Y.; Kong, L.D. Nuciferine Alleviates Renal Injury by Inhibiting Inflammatory Responses in Fructose-Fed Rats. *J. Agric. Food Chem.* **2016**, *64*, 7899–7910. [CrossRef]

41. Akira, S.; Takeda, K.; Kaisho, T. Toll-like receptors: Critical proteins linking innate and acquired immunity. *Nat. Immunol.* **2001**, *2*, 675–680. [CrossRef] [PubMed]

42. Aral, K.; Berdeli, E.; Cooper, P.R.; Milward, M.R.; Kapila, Y.; Unal, B.K.; Aral, C.A.; Berdeli, A. Differential expression of inflammasome regulatory transcripts in periodontal disease. *J. Periodontol.* **2020**, *91*, 606–616. [CrossRef]
43. Aral, K.; Milward, M.R.; Cooper, P.R. Inflammasome dysregulation in human gingival fibroblasts in response to periodontal pathogens. *Oral Dis.* **2020**. [CrossRef]
44. Weber, B.; Kehl, D.; Bleul, U.; Behr, L.; Sammut, S.; Frese, L.; Ksiazek, A.; Achermann, J.; Stranzinger, G.; Robert, J.; et al. In vitro fabrication of autologous living tissue-engineered vascular grafts based on prenatally harvested ovine amniotic fluid-derived stem cells. *J. Tissue Eng. Regener. Med.* **2016**, *10*, 52–70. [CrossRef]
45. Barachini, S.; Danti, S.; Pacini, S.; D'Alessandro, D.; Carnicelli, V.; Trombi, L.; Moscato, S.; Mannari, C.; Cei, S.; Petrini, M. Plasticity of human dental pulp stromal cells with bioengineering platforms: A versatile tool for regenerative medicine. *Micron* **2014**, *67*, 155–168. [CrossRef]
46. May, J.M.; Harrison, F.E. Role of Vitamin C in the Function of the Vascular Endothelium. *Antioxid. Redox Signal.* **2013**, *19*, 2068–2083. [CrossRef] [PubMed]
47. Pizzino, G.; Irrera, N.; Cucinotta, M.; Pallio, G.; Mannino, F.; Arcoraci, V.; Squadrito, F.; Altavilla, D.; Bitto, A. Oxidative Stress: Harms and Benefits for Human Health. *Oxid. Med. Cell. Longev.* **2017**, *2017*, 8416763. [CrossRef]
48. Salonen, R.M.; Nyyssonen, K.; Kaikkonen, J.; Porkkala-Sarataho, E.; Voutilainen, S.; Rissanen, T.H.; Tuomainen, T.P.; Valkonen, V.P.; Ristonmaa, U.; Lakka, H.M.; et al. Six-year effect of combined vitamin C and E supplementation on atherosclerotic progression—The Antioxidant Supplementation in Atherosclerosis Prevention (ASAP) study. *Circulation* **2003**, *107*, 947–953. [CrossRef]
49. Wang, Y.; Chun, O.K.; Song, W.O. Plasma and Dietary Antioxidant Status as Cardiovascular Disease Risk Factors: A Review of Human Studies. *Nutrients* **2013**, *5*, 2969–3004. [CrossRef]
50. Ingles, D.P.; Rodriguez, J.C.B.; Garcia, H. Supplemental Vitamins and Minerals for Cardiovascular Disease Prevention and Treatment. *Curr. Cardiol. Rep.* **2020**, *22*. [CrossRef]
51. Zhu, N.; Huang, B.; Jiang, W. Targets of Vitamin C With Therapeutic Potential for Cardiovascular Disease and Underlying Mechanisms: A Study of Network Pharmacology. *Front. Pharmacol.* **2020**, *11*, 591337. [CrossRef] [PubMed]
52. Teles, R.; Wang, C.Y. Mechanisms involved in the association between peridontal diseases and cardiovascular disease. *Oral Dis.* **2011**, *17*, 450–461. [CrossRef]
53. Li, T.; Li, F.; Liu, X.Y.; Liu, J.H.; Li, D.P. Synergistic anti-inflammatory effects of quercetin and catechin via inhibiting activation of TLR4-MyD88-mediated NF-kappa B and MAPK signaling pathways. *Phytother. Res.* **2019**, *33*, 756–767. [CrossRef] [PubMed]
54. Marconi, G.D.; Fonticoli, L.; Guarnieri, S.; Cavalcanti, M.; Franchi, S.; Gatta, V.; Trubiani, O.; Pizzicannella, J.; Diomede, F. Ascorbic Acid: A New Player of Epigenetic Regulation in LPS-gingivalis Treated Human Periodontal Ligament Stem Cells. *Oxid. Med. Cell. Longev.* **2021**, *2021*, 6679708. [CrossRef] [PubMed]
55. Diomede, F.; Marconi, G.D.; Guarnieri, S.; D'Attilio, M.; Cavalcanti, M.; Mariggio, M.A.; Pizzicannella, J.; Trubiani, O. A Novel Role of Ascorbic Acid in Anti-Inflammatory Pathway and ROS Generation in HEMA Treated Dental Pulp Stem Cells. *Materials* **2019**, *13*, 130. [CrossRef]
56. Schroder, K.; Tschopp, J. The Inflammasomes. *Cell* **2010**, *140*, 821–832. [CrossRef]
57. Zhang, X.; Zhang, J.H.; Chen, X.Y.; Hu, Q.H.; Wang, M.X.; Jin, R.; Zhang, Q.Y.; Wang, W.; Wang, R.; Kang, L.L.; et al. Reactive Oxygen Species-Induced TXNIP Drives Fructose-Mediated Hepatic Inflammation and Lipid Accumulation Through NLRP3 Inflammasome Activation. *Antioxid. Redox Signal.* **2015**, *22*, 848–870. [CrossRef]
58. Choe, J.Y.; Kim, S.K. Quercetin and Ascorbic Acid Suppress Fructose-Induced NLRP3 Inflammasome Activation by Blocking Intracellular Shuttling of TXNIP in Human Macrophage Cell Lines. *Inflammation* **2017**, *40*, 980–994. [CrossRef]

Article

Verbascoside Protects Gingival Cells against High Glucose-Induced Oxidative Stress via PKC/HMGB1/RAGE/NFκB Pathway

Pei-Fang Hsieh [1,2,†]**, Cheng-Chia Yu** [3,4,5,†]**, Pei-Ming Chu** [6] **and Pei-Ling Hsieh** [6,*]

[1] Department of Urology, E-Da Hospital, Kaohsiung 82445, Taiwan; n52022@gmail.com
[2] Department of Medical Laboratory Science and Biotechnology, Chung-Hwa University of Medical Technology, Tainan 71703, Taiwan
[3] Institute of Oral Sciences, Chung Shan Medical University, Taichung 40201, Taiwan; ccyu@csmu.edu.tw
[4] Department of Dentistry, Chung Shan Medical University Hospital, Taichung 40201, Taiwan
[5] School of Dentistry, Chung Shan Medical University, Taichung 40201, Taiwan
[6] Department of Anatomy, School of Medicine, China Medical University, Taichung 404333, Taiwan; pmchu@mail.cmu.edu.tw
* Correspondence: plhsieh@mail.cmu.edu.tw
† These authors contributed equally to this work.

Abstract: Impaired wound healing often occurs in patients with diabetes and causes great inconvenience to them. Aside from the presence of prolonged inflammation, the accumulation of oxidative stress is also implicated in the delayed wound healing. In the present study, we tested the effect of verbascoside, a caffeoyl phenylethanoid glycoside, on the improvement of cell viability and wound healing capacity of gingival epithelial cells under high glucose condition. We showed that verbascoside attenuated the high glucose-induced cytotoxicity and impaired healing, which may be associated with the downregulation of oxidative stress. Our results demonstrated that verbascoside increased the activity of the antioxidant enzyme SOD and reduced the oxidative stress indicator, 8-OHdG, as well as apoptosis. Moreover, verbascoside upregulated the PGC1-α and NRF1 expression and promoted mitochondrial biogenesis, which was mediated by suppression of PKC/HMGB1/RAGE/NFκB signaling. Likewise, we showed the inhibitory effect of verbascoside on oxidative stress was via repression of PKC/HMGB1/RAGE/NFκB activation. Also, our data suggested that the PKC-mediated oxidative stress may lead to the elevated production of inflammatory cytokines, IL-6 and IL-1β. Collectively, we demonstrated that verbascoside may be beneficial to ameliorate impaired oral wound healing for diabetic patients.

Keywords: verbascoside; PKC; HMGB1; impaired wound healing; antioxidant activity

Citation: Hsieh, P.-F.; Yu, C.-C.; Chu, P.-M.; Hsieh, P.-L. Verbascoside Protects Gingival Cells against High Glucose-Induced Oxidative Stress via PKC/HMGB1/RAGE/NFκB Pathway. *Antioxidants* **2021**, *10*, 1445. https://doi.org/10.3390/antiox10091445

Academic Editors: Soliman Khatib and Dana Atrahimovich Blatt

Received: 24 August 2021
Accepted: 9 September 2021
Published: 12 September 2021

Publisher's Note: MDPI stays neutral with regard to jurisdictional claims in published maps and institutional affiliations.

1. Introduction

Diabetes mellitus is a multifaceted metabolic disorder that represents a major concern globally. According to the International Diabetes Federation (IDF), it was estimated that there were approximately 415 million people with diabetes worldwide, and by 2040, the predicted number may rise to 642 million [1]. A significant amount of health expenditure due to diabetes becomes a disease burden to be reckoned with. Furthermore, diabetes and its complications, such as macrovascular and microvascular diseases, constitute major causes of morbidity and mortality [2]. This chronic inflammatory disease not only affects the cardiovascular system, but it also causes dysregulation of wound healing, including in the oral tissues [3]. Diabetes has been known to be a risk factor of periodontitis, which is characterized by the destruction of the supporting structures of the teeth [4]. It has been indicated that diminished cell proliferation and migration, the elevation of pro-inflammatory cytokines, and reduced formation of new connective tissue and bone were observed in diabetic oral wound healing [3]. Nevertheless, the molecular mechanisms underlying this impaired healing process remain poorly understood.

During gingival healing, epithelial cells migrate to the injured site to cover denuded surfaces, and basal epithelial cells adjacent to migrating cells start to proliferate for wound closure [5,6]. Defective gingival healing of diabetic patients may cause invasion of microbes or other agents into tissues, which leads to chronic infection and aggravate the inflammatory condition. Apart from inflammation, mounting evidence has shown that oxidative stress is another contributing factor for prolonged diabetic wound healing [7]. As a key mediator of oxidative stress and inflammation, the role of the high mobility group box-1 (HMGB-1), a representative damage-associated molecular pattern (DAMP), in diabetes is gaining increasing attention [8]. It has been shown that hyperglycemia-induced reactive oxygen species (ROS) elevates the expression of HMGB1 and the receptor for advanced glycation end products (RAGE) in human aortic endothelial cells [9]. In addition, HMGB-1 signaling via RAGE has been thought to promote the production of pro-inflammatory cytokines through the activation of nuclear factor-κB (NF-κB) [10]. Whether this HMGB1/RAGE/NF-κB cascade involves in the disturbed oral wound healing in diabetes remains to be determined.

Verbascoside, also known as acteoside ($C_{29}H_{36}O_{15}$), is a phenylethanoid glycoside isolated from various herbal plants, such as *Verbascum phlomoides* [11] or *Buddleja globose* [12]. It has been known to possess numerous bioactivities, such as anti-tumor, anti-viral, anti-inflammatory, and anti-oxidant properties [13]. Besides, verbascoside has been shown to serve as a protein kinase C (PKC) inhibitor [14]. It has been revealed that elevation of PKC in fibroblasts derived from diabetic patients contributed to impaired skin wound healing [15]. Since HMGB1 can be phosphorylated by PKC for secretion [16], we aimed to investigate the effect of verbascosides on the expression of PKC/HMGB1/RAGE/NF-κB signaling and if verbascosides treatment improved the compromised wound repair of gingival cells under high glucose condition through suppression of oxidative stress and inflammation.

2. Materials and Methods

2.1. Cells and Chemicals

The Smulow-Glickman (S-G) gingival epithelial cell line used in this study was originally derived from human attached gingiva [17], and obtained from Kasten et al., East Tennessee State University, Quillen College of Medicine, Johnson City, TN, USA [18]. Cells were maintained in Dulbecco's modified Eagle's medium (DMEM; 11965092, Thermo Fisher Scientific, Waltham, MA, USA) supplemented with 10% fetal bovine serum (Gibco, Grand Island, NY, USA) at 37 °C in the presence of 5% CO_2 and medium were changed every 2–3 days. For high glucose groups, cells were exposed to 70 mmol/L concentration of D-glucose for 24 h followed by 25, 50, 100 µM verbascoside (00820580, HWI Analytik Gmbh, Rulzheim, Germany) or 5 mM N-Acetyl-L-cysteine (NAC; A9165, Sigma–Aldrich, St. Louis, MO, USA) for another 24 h. Bisindolylmalcimide I (Bis I; 5 µM) was purchased from Calbiochem (La Jolla, CA, USA) and used as a PKC inhibitor [19]. Glycyrrhizic acid (200 µM), FPS-ZM1 (1 µM), and dehydroxymethylepoxyquinomicin (DHMEQ) (10 µg/mL) were all obtained from Sigma-Aldrich and used as a HMGB1 inhibitor [20], a RAGE antagonist [21], and a NF-κB inhibitor [22], respectively. The concentrations of Bis I, glycyrrhizic acid, FPS-ZM1 and DHMEQ were referred from published studies [20,23–25].

2.2. Prestoblue, MTT, and LDH Assays

PrestoBlue™ Cell Viability Reagent (A13262, Invitrogen, Carlsbad, CA, USA) was utilized for cell viability and this assay was carried out in a 96-well plate according to the manufacturer's instructions. Briefly, each well contained the cells to be tested with cultured medium or verbascoside-containing medium and was incubated at 37 °C for 2 h. After incubation of 100 µL 10X PrestoBlue reagent, the absorbance was measured at 570/600 nm using iMark™ Microplate Absorbance Reader (Bio-Rad, Hercules, CA, USA).

MTT assay was conducted to measure cell proliferation. Cells were seeded into a 24-well plate for 24 h and then treated with various concentrations of verbascoside for another 24 h. Afterward, the cells were incubated with 3-(4,5-dimethylthiazol-2-y1)-2,5-diphenyltetrazolium bromide (MTT) in a humidified atmosphere containing 5% of CO_2 at 37 °C. The blue formazan crystals of viable cells were dissolved using isopropanol and then evaluated spectrophotometrically at 595 nm.

Lactate dehydrogenase (LDH) Cytotoxicity Detection Kit (630117, Clontech Laboratories, Mountain View, CA, USA) was used to assess the level of LDH released from the plasma membrane of damaged cells according to the manufacturer's instructions. Briefly, cells were added to a 96-well plate and cultured at 37 °C in 5% CO_2 overnight. The supernatant was then collected and transferred to another clear 96-well plate, and 100 µL of the reaction mixture from the LDH assay kit was added to each well for 30 min. Cells treated with 1% Triton X-100 were used as a positive control group for maximum LDH release. Absorbance was then measured at 490 nm (reference wavelength greater than 600 nm).

2.3. Wound Healing Assay

Cell suspension was added into two chambers of the silicone Culture-Insert 2 Well in µ-Dish 35 mm (ibidi GmbH, Martinsried, Germany), and cells were allowed to adhere and spread on the substrate (cells in two chambers were separated by 500 µm). After removing the culture-insert, cultured medium or verbascoside-containing medium were provided. Each well was photographed at 0 and 48 h to capture the two different fields of each group at each time point.

2.4. Examination of Antioxidant Capacity and Oxidative Stress

Superoxide Dismutase (SOD) Activity Colorimetric Assay Kit (K335-100, BioVision, Milpitas, CA, USA) was used for the examination of antioxidant capacity by assessing the water-soluble formazan dye upon reduction with superoxide anion to detect the inhibition activity of SOD. Briefly, the WST-1 working solution and the SOD enzyme working solution were added to the wells and mixed thoroughly. After incubation for 20 min, the inhibition activity of SOD was estimated using a microplate reader at 450 nm. The SOD activity (inhibition rate %) was calculated according to the manufacturer's instructions.

8-hydroxy 2 deoxyguanosine (8-OHdG) ELISA Kit (ab201734, Abcam, Cambridge, UK) was utilized to measure the oxidative damage of DNA by oxidative stress according to the manufacturer's protocol. Briefly, standards and samples were added into each well as instructed followed by incubation of HRP conjugated 8-OHdG antibody preparation for an hour. The enzymatic color reaction was developed after the TMB substrate solution was added and then the absorbance was measured at 450 nm.

2.5. Western Blot

Cells were lysed by lysis buffer (10 mM Tris, 1 m MEDTA,1% Triton X-100, 1 mM Na_3VO_4, 20 µg/mL aprotinin, 20 µg/mL leupeptin, 1 mM dithiothreitol, and 50 µg/mL phenylmethylsulfonyl fluoride). Cell lysates were subjected to 10% SDS-PAGE and proteins were transferred to the PVDF membrane by an electro-transfer unit. Membranes were probed with primary antibodies overnight at 4 °C. The primary antibodies included anti-Bcl-2 (ab692, Abcam; 1:2000), anti-caspase-3 (ab32351; 1:2000), anti-Bax (#89477, cell signaling technology, Beverly, MA, USA; 1:2000), anti-PGC1-α (ab54481; 1:2000), anti-NRF1 (ab34682; 1:2000), anti-PKC (ab181558; 1:2000), anti-HMGB1 (ab18256; 1:2000), anti-RAGE (ab216329; 1:2000), and anti-NFκB (ab16502; 1:2000). Following incubation of primary antibodies, the membranes were rinsed three times and incubated with corresponding secondary antibodies. The immunoreactive bands were developed using a Western-Ready™ ECL Substrate Plus Kit (426316, BioLegend, San Diego, CA, USA) and detected by MultiGel-21® image system. β-actin was used as an internal control.

2.6. Assessment of Mitochondrial Biogenesis

Mitochondrial biogenesis was analyzed using MitoBiogenesis™ InCell ELISA Kit (ab110217) according to the manufacturer's protocol. Briefly, cells were seeded in a 96-well plate and fixed with 4% paraformaldehyde after adhering to the bottom of the plate. 0.5% acetic acid was added for 5 min to block endogenous alkaline phosphatase activity. Afterward, 0.1% Triton X-100 was used to permeabilize cells followed by 2X blocking solution addition. These cells were then incubated with primary antibodies against mtDNA encoded COX-I, and nuclear-DNA encoded SDHA proteins. After washing, the reaction was sequentially developed first with AP reagent and then HRP development solution. A 15-min kinetic reaction with a 1 min interval was recorded using a microplate reader at 405 nm (for AP detection of SDH-A) and 600 nm (for HRP detection of COX-I). The ratio of COX-I/SDHA was calculated to determine mitochondrial biogenesis.

2.7. Quantitative Real-Time PCR (qRT-PCR)

Total RNA of each sample was extracted using an Absolutely PureLink™ RNA Mini Kit (Invitrogen Life Technologies, Carlsbad, CA, USA) and subjected to RT by SuperScript™ First-Strand Synthesis System (Invitrogen) for RT-PCR. All cDNA samples were diluted as a working template in TaqMan™ Fast Advanced Master Mix (Applied Biosystems, Waltham, MA, USA), which was carried out on an ABI 7500 Fast Real-Time PCR System (Applied Biosystems). PCR amplification was conducted by using the TaqMan primers and probes (Assay ID: Hs00176973_m1 for PKC α; Assay ID: Hs01923466 for HMGB1; Assay ID: Hs00179504 for RAGE; Assay ID: Hs00765730 for NF-κB) using the TaqMan Universal PCR Master Mix Protocol (Applied Biosystems) according to the manufactures instructions.

2.8. Measurement of Inflammatory Cytokine Production

The secretion of IL-6 (BSKH1007) and IL-1β (BSKH1001) were determined using an enzyme-linked immunosorbent assay (ELISA) kit (Bioss, Woburn, MA, USA) and quantified at 450 nm according to the manufacturer's instructions. Cells were cultured in 6-well plates with verbascoside, NAC, or PKC inhibitor for 24 h. Cell supernatants were collected and centrifuged to remove dead cells. Each sample was analyzed in triplicate.

2.9. Statistical Analysis

The results were expressed as mean $\pm$ SEM. Statistical analyses were performed using one-way ANOVA followed by an LSD post hoc test as appropriate. A p-value < 0.05 was considered statistically significant.

3. Results

3.1. Treatment of Verbascoside Mitigates the Suppressed Cell Proliferation and Wound Healing Capacity of Gingival Epithelial Cells under High Glucose Condition

To assess the cytoprotective effect of verbascoside, PrestoBlue and MTT assays were employed to examine cell viability and proliferation of high glucose-treated S-G gingival epithelial cells. As shown in Figure 1A,B, both PrestoBlue and MTT assays showed that verbascoside dose-dependently reversed the repressed cell proliferation in high glucose-cultured cells. By using PrestoBlue assay, we observed that the cell death caused by high glucose was attenuated following the treatment of a commonly used antioxidant N-acetyl-l-cysteine (NAC), suggesting that oxidative stress may mediate the downregulation of cell viability in response to high glucose. Likewise, lactate dehydrogenase (LDH) cytotoxicity assay demonstrated that incubation of high glucose-treated gingival epithelial cells with verbascoside markedly suppressed LDH release and improved cell survival in a dose-dependent manner, and NAC exhibited a similar effect (Figure 1C). Besides, epithelial cells migrate and proliferate at the injured site during the healing process, and we showed that these capacities of epithelial cells were inhibited under the high glucose condition (Figure 1D). Nevertheless, these impairments were improved in high glucose-cultured groups by adding verbascoside or NAC (Figure 1D). These results indicated the beneficial effect of verbascoside against high glucose-induced cell death and disturbed wound healing ability.

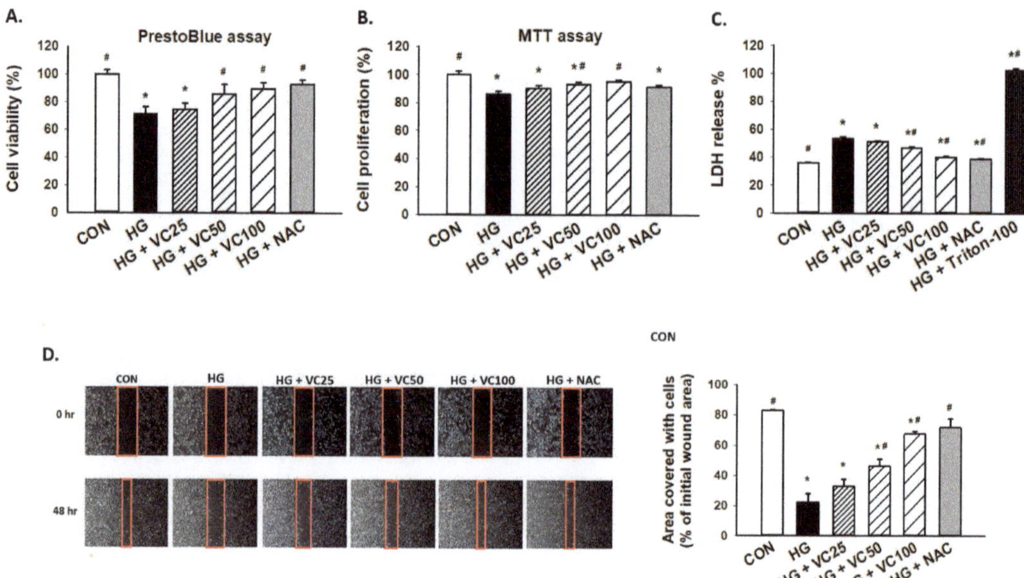

Figure 1. Verbascoside prevents the high glucose-inhibited cell proliferation and wound healing capacity of gingival epithelial cells. (**A**) PrestoBlue and (**B**) MTT assays were carried out to examine the cell viability of high glucose-treated cells in response to various concentrations of verbascoside (25, 50 and 100 µM) or N-acetyl-l-cysteine (NAC; 5 mM); (**C**) LDH assay was conducted to show that verbascoside protected cells against high glucose-induced cytotoxicity. Triton X-100 served as a positive control group; (**D**) wound healing capacity was tested using a two-well culture-insert. * indicates $p < 0.05$ compared to the control group (CON); # indicates $p < 0.05$ compared to the high glucose group (HG).

3.2. Verbascoside Attenuates the Oxidative Stress and Apoptosis in High Glucose-Cultured Gingival Epithelial Cells

Superoxide dismutase (SOD) is an antioxidant enzyme that has been shown to be implicated in the protracted wound healing of diabetic rats [26]. We showed that verbascoside upregulated SOD activity of high glucose-cultured cells in a dose-dependent fashion, whereas administration of NAC alone failed to increase enzyme activity of SOD (Figure 2A), possibly due to the insufficient amount of NAC. Besides, 8-hydroxy-2-deoxyguanosine (8-OHdG), a byproduct of oxidative DNA damage, was found to be abundantly upregulated in high glucose-treated cells (Figure 2B). As expected, this indicator of oxidative stress in high glucose-cultured groups was markedly decreased in the presence of verbascoside and NAC (Figure 2B).

Furthermore, a concentration-dependent decrease in apoptosis was observed (Figure 2C). We demonstrated that administration of 100 µM verbascoside resulted in the upregulation of Bcl-2 (Figure 2C,D), and downregulation of Bax (Figure 2C,E) and caspase 3 (Figure 2C,F). Also, inhibition of oxidative stress by NAC displayed a comparable effect (Figure 2C–F). Overall, these findings supported that verbascoside may serve as a promising agent to alleviate high glucose-stimulated apoptosis in gingiva via repression of oxidative stress.

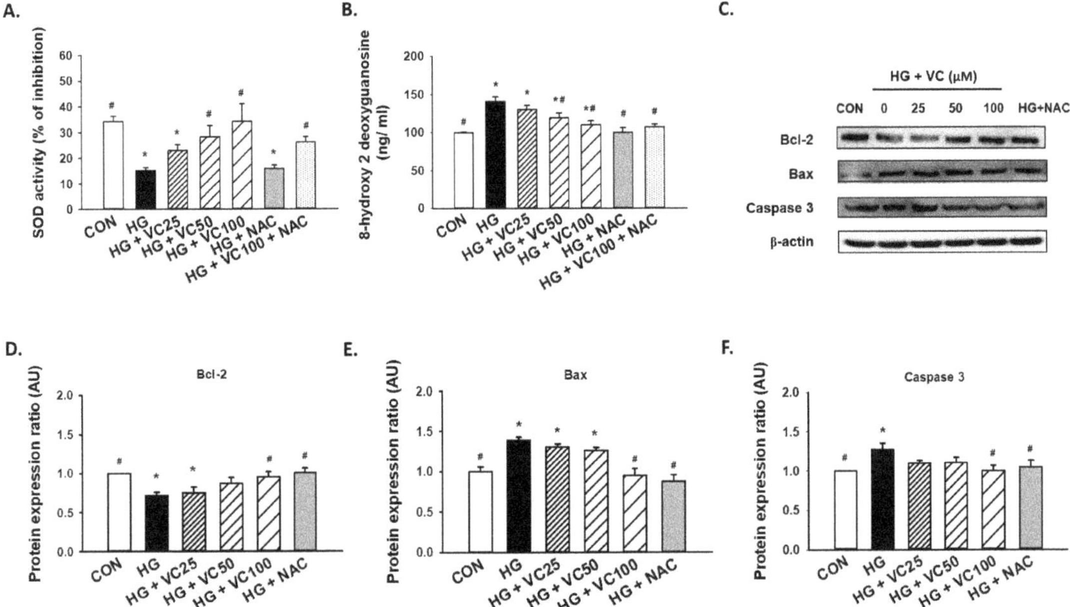

Figure 2. Verbascoside diminishes the high glucose-induced oxidative stress and apoptosis of gingival epithelial cells. (**A**) superoxide dismutase (SOD) and (**B**) 8-hydroxy-2-deoxyguanosine (8-OHdG) assays were utilized to assess the antioxidant capacity and oxidative stress of high glucose-cultured cells following treatment with various concentrations of verbascoside (25, 50 and 100 μM) or N-acetyl-l-cysteine (NAC; 5 mM); (**C**) representatives of western blots and densitometric analysis of Bcl-2 (**D**), Bax (**E**), and caspase 3 (**F**) in high glucose-treated cells incubated with various concentrations of verbascoside (25, 50 and 100 μM) or N-acetyl-l-cysteine (NAC; 5 mM) were presented. * indicates $p < 0.05$ compared to the control group (CON); # indicates $p < 0.05$ compared to the high glucose group (HG).

3.3. The High Glucose-Induced Mitochondrial Dysfunction Is Ameliorated by Verbascoside

Peroxisomal proliferator activator receptor γ coactivator (PGC1)-α and nuclear respiratory factor 1 (NRF1) are primary regulators of mitochondrial biogenesis [27]. Mitochondrial dysfunction has been known to be associated with the increased reactive oxygen species (ROS) production [28], and it has been previously reported that both PGC1-α and NRF1 are reduced in diabetic subjects [29]. Here, we showed a consistent result that high glucose decreased the expression of PGC1-α and NRF1, while treatment of verbascoside dose-dependently increased the expression levels of these two factors (Figure 3A–C). Similarly, administration of NAC reversed the downregulation of NRF1 and PGC1-α (Figure 3A–C), indicating that the high glucose-inhibited PGC1-α and NRF1 expression may be due to the accumulation of oxidative stress.

Additionally, we assessed the ratio of cytochrome c oxidase I (COX1)/succinate dehydrogenase complex, subunit A (SDH-A) to measure mitochondrial biogenesis. COX-I is encoded by mitochondrial DNA (mtDNA) and its proper synthesis is greatly relied on mtDNA integrity [30]. SDH-A is encoded by nuclear DNA and its activity is typically not affected by impaired mtDNA [31]. As shown in Figure 3D, exposure to high glucose conditions caused suppression of mitochondrial biogenesis. However, treatment of various concentrations of verbascoside increased the COX1/SDH-A ratio (Figure 3D), suggesting that it may promote mitochondrial function, at least partially, through upregulation of PGC1-α and NRF1.

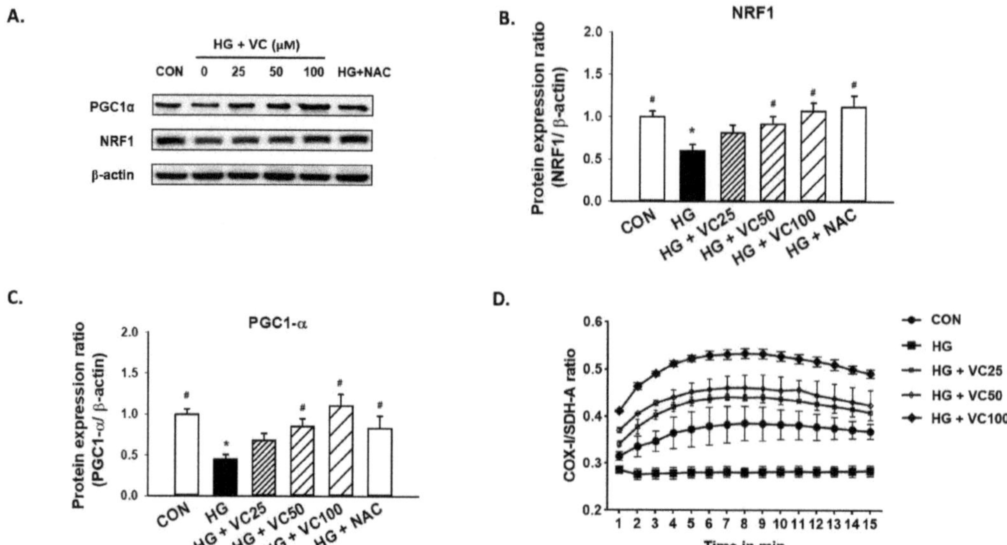

Figure 3. Verbascoside attenuates the high glucose-induced mitochondrial dysfunction of gingival epithelial cells. Representatives of Western blots (**A**), and densitometric analysis of PGC1-α (**B**) and NRF1 (**C**) in high glucose-treated cells incubated with various concentrations of verbascoside (25, 50 and 100 µM) or NAC (5 mM) were presented. (**D**) The ratio of COX1/SDH-A was used to show mitochondrial biogenesis over the course of 15 min. * indicates $p < 0.05$ compared to the control group (CON); # indicates $p < 0.05$ compared to the high glucose group (HG).

3.4. Verbascoside Inhibits the High Glucose-Elicited PKC/HMGB1/RAGE/NFκB Pathway

Results from qRT-PCR (Figure 4A–D) and western blot (Figure 4E–H) revealed that both gene and protein expression levels of PKC, HMGB1, RAGE, and NFκB were suppressed by verbascoside. First, we observed that PKC was decreased in the verbascoside-treated cells (Figure 4A,E). Moreover, we showed the expression of HMGB1 was reduced by PKC inhibitor (Figure 4B,F), and the expression of RAGE was downregulated by PKC inhibitor and HMGB1 inhibitor (Figure 4C,G). Also, NFκB was suppressed by inhibitors of PKC and HMGB1 as well as the antagonist of RAGE (Figure 4D,H). Taken together, these results revealed that verbascoside possesses the inhibitory effect on the high glucose-induced PKC/HMGB1/RAGE/NFκB activation.

3.5. Suppression of PKC/HMGB1/RAGE/NFκB Signaling Reverses the Downregulation of PGC1-α and NRF1

Subsequently, we aimed to evaluate if the activation of the PKC/HMGB1/RAGE/NFκB pathway participated in the reduced expression of PGC1-α and NRF1 under high glucose conditions since it has been reported that PKC modulates mitochondrial function after oxidant injury in renal cells [32] and NF-κB regulates mitochondrial biogenesis by increasing the expression or activity of PGC-1α and NRF 1 [33]. We showed that inhibition of PKC reverted the reduction of PGC1-α and NRF1 caused by high glucose (Figure 5A–C). Similarly, we demonstrated that administration of HMGB1 inhibitor, RAGE antagonist, and NFκB inhibitor all upregulated the expression levels of PGC1-α and NRF1 in gingival epithelial cells in response to high glucose stimulation as the verbascoside-treated group (Figure 5A–C). These findings showed that blockade of the PKC/HMGB1/RAGE/NFκB pathway (e.g., administration of verbascoside) holds the potential to enhance mitochondrial function via upregulation of PGC1-α and NRF1.

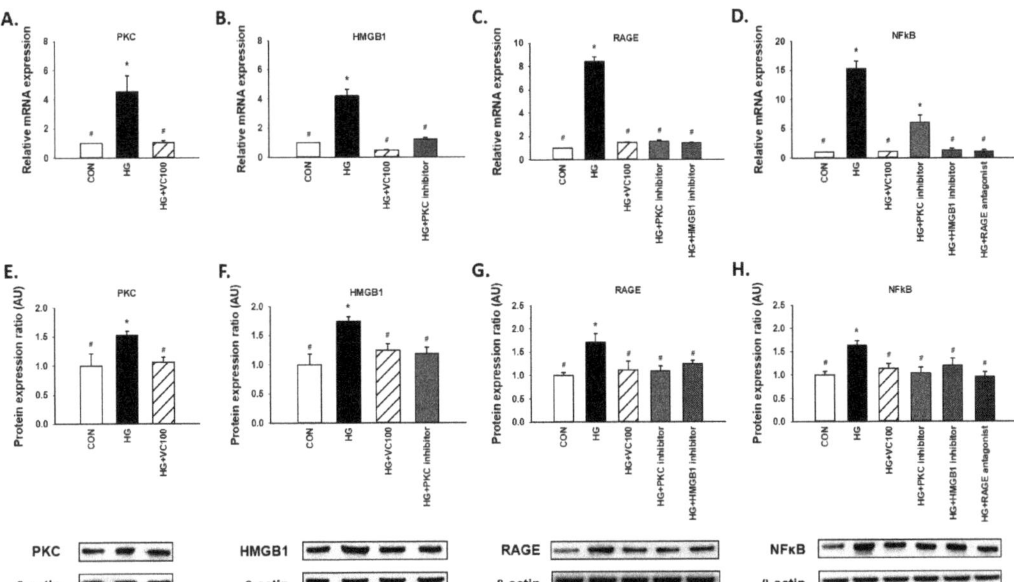

Figure 4. Verbascoside suppresses the high glucose-activated PKC/HMGB1/RAGE/NFκB pathway in gingival epithelial cells. Relative mRNA (**A–D**) and protein expression with representatives of Western blots (**E–H**) of PKC, HMGB1, RAGE, and NFκB in high glucose-stimulated cells with verbascoside treatment were presented. In some groups, coincubation with PKC inhibitor (Bis I; 5 μM), HMGB1 inhibitor (Glycyrrhizic acid; 200 μM), or RAGE antagonist (FPS-ZM1; 1 μM) was performed as suggested. * indicates $p < 0.05$ compared to the control group (CON); # indicates $p < 0.05$ compared to the high glucose group (HG).

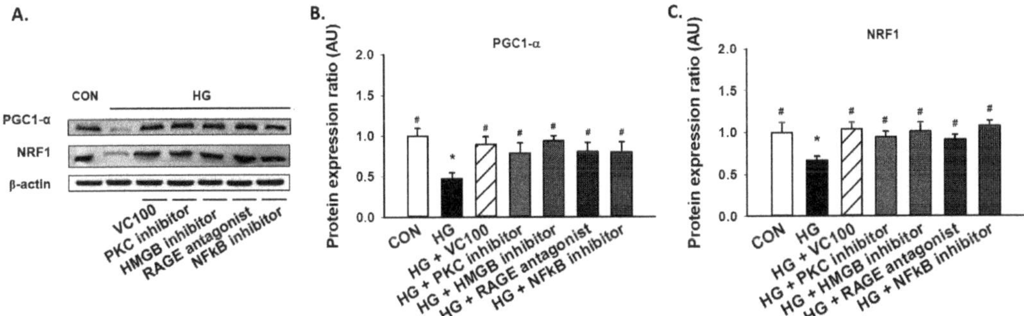

Figure 5. Downregulation of the PKC/HMGB1/RAGE/NFκB pathway promotes the high glucose-inhibited PGC1-α and NRF1 expression in gingival epithelial cells. Representatives of Western blots (**A**), and densitometric analysis of PGC1-α (**B**) and NRF1 (**C**) in high glucose-treated cells along with verbascoside, PKC inhibitor, HMGB1 inhibitor, RAGE antagonist, or NFκB inhibitor (DHMEQ; 10 μg/mL). * indicates $p < 0.05$ compared to the control group (CON); # indicates $p < 0.05$ compared to the high glucose group (HG).

3.6. Inhibition of the PKC/HMGB1/RAGE/NFκB Pathway Diminishes High Glucose-Induced Oxidative Stress and the Subsequent Inflammation

Furthermore, we tested whether PKC/HMGB1/RAGE/NFκB signaling mediated the accumulation of oxidative stress in high glucose-cultured cells. Our results showed that the SOD activity was increased in verbascoside-treated cells or cells with inhibition of this signaling pathway (Figure 6A), whereas 8-OHdG production was markedly decreased in verbascoside-treated cells or cells with repression of PKC/HMGB1/RAGE/NFκB signaling (Figure 6B). Besides, we demonstrated that downregulation of oxidative stress (NAC-treated) successfully mitigated the high glucose-elicited production of interleukin (IL)-6 (Figure 6C) and IL-1β (Figure 6D), which was also observed in the PKC inhibitor-treated group (Figure 6C,D). As such, our results suggested that the high glucose-induced elevation of IL-6 or IL-1β was attributed, at least partially, to oxidative stress which was mediated by the PKC/HMGB1/RAGE/NFκB pathway.

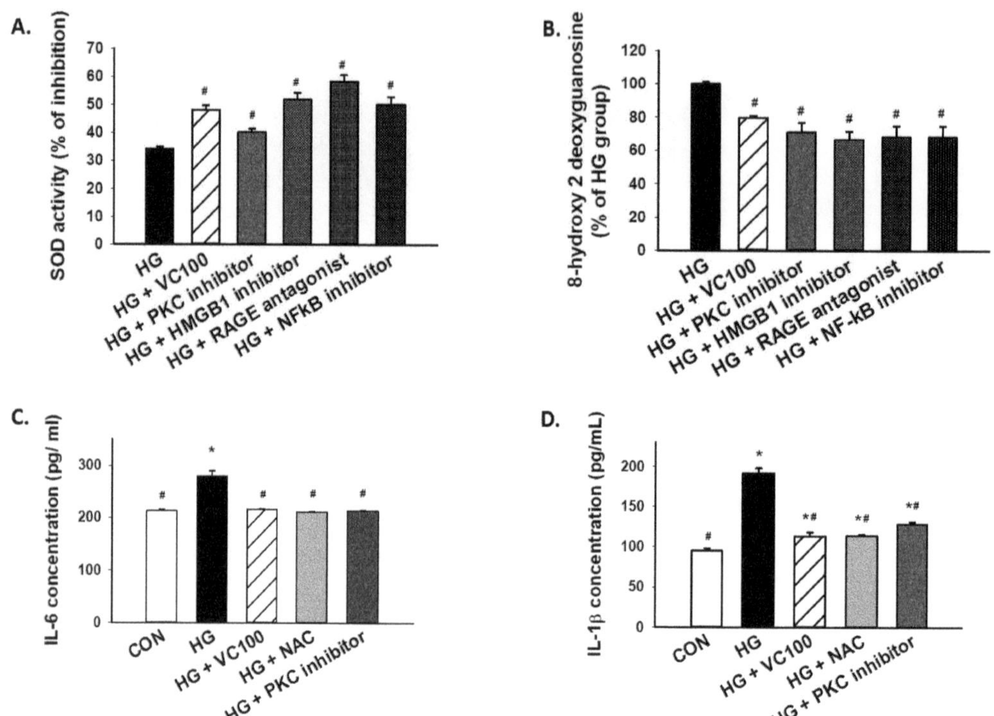

Figure 6. Inhibition of high glucose-induced oxidative stress and inflammation through suppression of PKC/HMGB1/RAGE/NFκB pathway by verbascoside. (**A**) SOD and (**B**) 8-OHdG assays were used to assess oxidative stress of high glucose-cultured cells following inhibition of PKC/HMGB1/RAGE/NFκB signaling; the high glucose-elicited production of IL-6 (**C**) and IL-1β (**D**) were downregulated in cells treated with verbascoside, NAC, or PKC inhibitor. * indicates $p < 0.05$ compared to the control group (CON); # indicates $p < 0.05$ compared to the high glucose group (HG).

4. Discussion

Verbascoside has been recognized as a widespread polyphenol with anti-oxidant and anti-inflammatory properties. For instance, it has been shown to attenuate the X-ray-induced intracellular ROS and apoptosis in human skin fibroblasts via mitogen-activated protein kinase (MAPK) signaling [34]. Another study demonstrated that verbascoside inhibited various inflammatory mediators in the primary rat chondrocytes treated with IL-1β through the MAPK pathway as well [35]. Moreover, recent studies have suggested verbascoside possessed the anti-diabetic potential since it has been reported to lower hyperglycemia and hyperlipidemia in diet/STZ-induced diabetic rats [36]. In a KK-Ay mouse model of type 2 diabetes, verbascoside was found to enhance lipolysis and fatty acid oxidation by inducing mRNA expression of adipose triglyceride lipase through the AMP-activated protein kinase (AMPK) pathway [37]. One of the more recent studies has suggested that verbascoside may be beneficial in terms of prevention and treatment of diabetes since it exerted protective effects against ER stress-associated dysfunctions in human β-cells via reduction of protein kinase RNA-like ER kinase (PERK) expression [38]. In the current study, we showed verbascoside modulated mitochondrial biogenesis and redox homeostasis via PKC/HMGB1/RAGE/NFκB pathway under high glucose conditions, leading to mitigation of inflammation and apoptosis.

It has long been known that high glucose increased diacylglycerol (DAG), which activated PKC [39]. Various studies have revealed that the DAG/PKC axis contributed to insulin resistance [40] and hyperglycemia-induced oxidative stress in diabetic rat kidneys [41]. Apart from DAG, it has been revealed that the accumulation of advanced glycation end products (AGEs) elevated the translocation of PKC and oxidative stress in rat neonatal mesangial cells as well [42]. Several lines of evidence suggested that PKC mediated the production of oxidative stress in diabetes via NADPH oxidase activation using aortic smooth muscle cells [43], retinal endothelial cells [44], and human renal mesangial cells [45]. In line with these findings, we also showed that high glucose-induced the expression of PKC in gingival epithelial cells and suppression of PKC resulted in an increase in anti-oxidant enzyme activity and a reduction of oxidative stress marker. On the other hand, it has been shown that high glucose induced NF-κB activity and inflammatory cytokine expression via toll-like receptor 2 (TLR2) and TLR4 expression in human monocytes [46]. Dasu et al. showed that the upregulation of TLR2 and TLR4 expression was through PKC-α and PKC-δ, respectively, by stimulating NADPH oxidase [46]. Another study demonstrated a similar finding showing that high glucose elevated the NF-κB expression and secretion of inflammatory cytokine through PKC-induced TLR2 pathway in gingival fibroblasts [47]. Here, we revealed that PKC can activate NF-κB by HMGB1/RAGE axis in addition to TLR2 pathway, which resulted in the subsequent aggravation of inflammation via generation of oxidative stress.

As a nuclear protein, HMGB1 has been found to trigger inflammation and oxidative stress [8,48]. It has been shown that HMGB1 can be phosphorylated by PKC and shuttled from nuclear to cytoplasmic compartments that direct it toward secretion [16,49]. The extracellular HMGB1 has been known to act as a multifunctional cytokine and induce NF-κB, leading to the secretion of various proinflammatory cytokines [8]. Emerging evidence has revealed the significance of the HMGB1/RAGE/NF-κB axis in diabetic complications. For instance, it has been shown that the retinal expression levels of HMGB1 and RAGE and the activity of NF-κB were elevated in the diabetic retina [50]. In addition, Sohn et al. demonstrated the direct binding of NF-κB p65 to the RAGE promoter using ChIP assays [50]. The HMGB1/RAGE/NF-κB signaling also contributed to neuroinflammation [51] and periodontitis [52] in diabetic mice. It has been revealed that high glucose prompted the translocation of HMGB1 from the nucleus to the cytosol through an NADPH oxidase and PKC-dependent pathway in vascular smooth muscle cells [53]. Also, they showed the upregulation of HMGB1 was crucial to high-glucose-induced vascular calcification through regulation of NFκB activation and bone morphogenetic protein-2 (BMP-2) [53]. Our results demonstrated that PKC/HMGB1/RAGE/NF-κB signaling may participate in diabetes-associated impairment of oral wound healing.

In the oral cavity, higher expression of HMGB1 was detected in gingival tissues and gingival crevicular fluid (GCF) of patients with periodontitis and peri-implantitis, which was accompanied by the elevated concentrations of pro-inflammatory cytokines, such as IL-1β, IL-6, and IL-8 [54]. In diabetic patients, the GCF level of IL-1β was increased and some reports suggested that the GCF IL-6 level was upregulated as well [55,56]. Besides, the increased mRNA expression of RAGE was found in the type 2 diabetes gingival epithelium [57]. It has been shown that gingival epithelial cells secreted HMGB1 upon stimulation of tumor necrosis factor alpha (TNF α) [58] or IL 1β [59]. Ito et al. reported that HMGB1 and RAGE were markedly expressed in gingiva and promptly released during gingival inflammation [59]. Another study showed that the HMGB1/RAGE axis involved in the intraoral palatal wound healing and knockdown of RAGE inhibited the cell migration and cell proliferation in gingival epithelial cells [60]. Our results were consistent with these findings and showed that the HMGB1/RAGE axis and pro-inflammatory cytokines were upregulated in high glucose-stimulated gingival epithelial cells.

It is becoming increasingly evident that the high glucose-induced inflammation and oxidative stress may contribute to the impaired healing processes. The elevation of pro-inflammatory cytokines, such as IL-1β, has been found to impair diabetic wound healing [61]. The high glucose-induced oxidative stress also has been shown to impede the proliferation and migration of human gingival fibroblasts [62]. In agreement with these results, we showed that suppression of the PKC/HMGB1/RAGE/NFκB-mediated oxidative stress and pro-inflammatory cytokine production may improve the high glucose-inhibited cell proliferation and wound healing. Furthermore, we demonstrated that this signaling pathway was also associated with high glucose-related mitochondrial dysfunction, which was pivotal to oxidative stress and apoptosis in various diabetic complications [63,64].

5. Conclusions

In this study, we showed that administration of verbascoside improved the cell viability and wound healing capacity of gingival epithelial cells in a dose-dependent fashion under high glucose conditions, which may be associated with the downregulation of oxidative stress. Our results demonstrated that verbascoside dose-dependently increased the anti-oxidant enzyme, SOD, and reduced the DNA damage marker, 8-OHdG, along with suppression of apoptosis in high glucose-cultured cells. We showed that the treatment of verbascoside reversed the high glucose-inhibited PGC1-α and NRF1 expression as well as promoted mitochondrial biogenesis. Furthermore, our data suggested that these effects were mediated by downregulation of the PKC/HMGB1/RAGE/NFκB pathway. Also, we demonstrated that a decrease of oxidative stress by verbascoside via PKC signaling mitigated the high glucose-induced inflammatory cytokines, both IL-6 and IL-1β (Figure 7). Taken together, this study revealed that administration of verbascoside may be beneficial to individuals with diabetes in terms of amelioration of impaired oral wound healing.

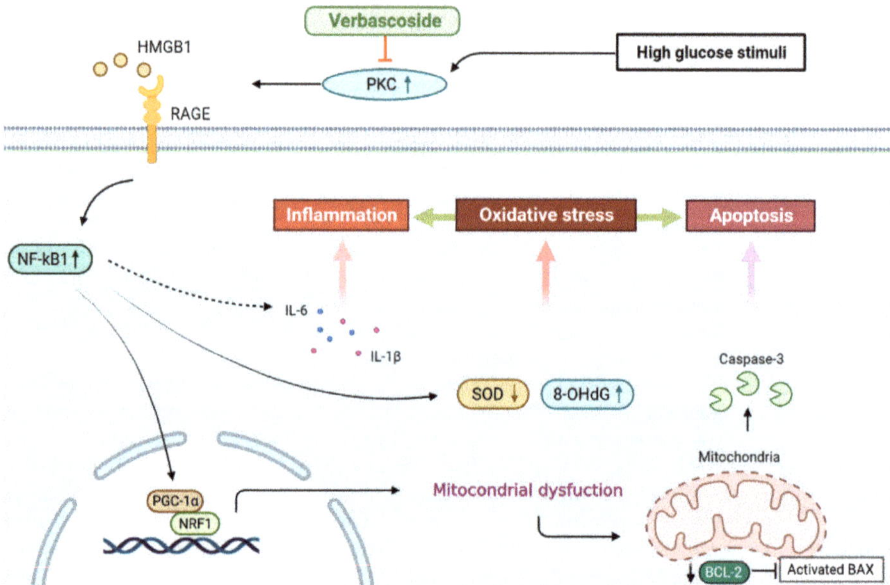

Figure 7. Schematic diagram showing the possible mechanism underlying the protective effect of verbascoside on high glucose-induced impairment of oral wound healing.

Author Contributions: Conceptualization, P.-L.H.; methodology, C.-C.Y. and P.-F.H.; formal analysis, C.-C.Y. and P.-M.C.; investigation, P.-F.H. and P.-L.H.; writing—original draft preparation, P.-L.H.; funding acquisition, P.-L.H. All authors have read and agreed to the published version of the manuscript.

Funding: This study was funded by China Medical University (CMU109-N-10).

Institutional Review Board Statement: Not applicable.

Informed Consent Statement: Not applicable.

Data Availability Statement: The data is contained within this article.

Conflicts of Interest: The authors declare no conflict of interest.

References

1. Ogurtsova, K.; da Rocha Fernandes, J.; Huang, Y.; Linnenkamp, U.; Guariguata, L.; Cho, N.H.; Cavan, D.; Shaw, J.; Makaroff, L.E. IDF Diabetes Atlas: Global estimates for the prevalence of diabetes for 2015 and 2040. *Diabetes Res. Clin. Pract.* **2017**, *128*, 40–50. [CrossRef]
2. Raghavan, S.; Vassy, J.; Ho, Y.; Song, R.J.; Gagnon, D.R.; Cho, K.; Wilson, P.W.F.; Phillips, L.S. Diabetes Mellitus–Related All-Cause and Cardiovascular Mortality in a National Cohort of Adults. *J. Am. Hear. Assoc.* **2019**, *8*, e011295. [CrossRef] [PubMed]
3. Ko, K.I.; Sculean, A.; Graves, D.T. Diabetic wound healing in soft and hard oral tissues. *Transl. Res.* **2021**, *236*, 72–86. [CrossRef] [PubMed]
4. Preshaw, P.M.; Alba, A.L.; Herrera, D.; Jepsen, S.; Konstantinidis, A.; Makrilakis, K.; Taylor, R. Periodontitis and diabetes: A two-way relationship. *Diabetologia* **2011**, *55*, 21–31. [CrossRef]
5. Smith, P.C.; Cáceres, M.; Martínez, C.; Oyarzún, A. Gingival Wound Healing: An essential response disturbed by aging? *J. Dent. Res.* **2014**, *94*, 395–402. [CrossRef]
6. Leoni, G.; Neumann, P.-A.; Sumagin, R.; Denning, T.; Nusrat, A. Wound repair: Role of immune–epithelial interactions. *Mucosal Immunol.* **2015**, *8*, 959–968. [CrossRef]
7. Deng, L.; Du, C.; Song, P.; Chen, T.; Rui, S.; Armstrong, D.G.; Deng, W. The Role of Oxidative Stress and Antioxidants in Diabetic Wound Healing. *Oxidative Med. Cell. Longev.* **2021**, *2021*, 1–11. [CrossRef]
8. Wu, H.; Chen, Z.; Xie, J.; Kang, L.-N.; Wang, L.; Xu, B. High Mobility Group Box-1: A Missing Link between Diabetes and Its Complications. *Mediat. Inflamm.* **2016**, *2016*, 1–11. [CrossRef] [PubMed]
9. Yao, D.; Brownlee, M. Hyperglycemia-Induced Reactive Oxygen Species Increase Expression of the Receptor for Advanced Glycation End Products (RAGE) and RAGE Ligands. *Diabetes* **2009**, *59*, 249–255. [CrossRef]
10. Park, J.S.; Arcaroli, J.; Yum, H.-K.; Yang, H.; Wang, H.; Yang, K.-Y.; Choe, K.-H.; Strassheim, D.; Pitts, T.M.; Tracey, K.J.; et al. Activation of gene expression in human neutrophils by high mobility group box 1 protein. *Am. J. Physiol. Physiol.* **2003**, *284*, C870–C879. [CrossRef]
11. Klimek, B. 6′-0-apiosyl-verbascoside in the flowers of mullein (Verbascum species). *Acta Pol. Pharm.-Drug Res.* **1996**, *53*, 137–140.
12. Pardo, F.; Perich, F.; Villarroel, L.; Torres, R. Isolation of verbascoside, an antimicrobial constituent of Buddleja globosa leaves. *J. Ethnopharmacol.* **1993**, *39*, 221–222. [CrossRef]
13. Tian, X.-Y.; Li, M.-X.; Lin, T.; Qiu, Y.; Zhu, Y.-T.; Li, X.-L.; Tao, W.-D.; Wang, P.; Ren, X.-X.; Chen, L.-P. A review on the structure and pharmacological activity of phenylethanoid glycosides. *Eur. J. Med. Chem.* **2021**, *209*, 112563. [CrossRef] [PubMed]
14. Herbert, J.M.; Maffrand, J.P.; Taoubi, K.; Augereau, J.M.; Fouraste, I.; Gleye, J. Verbascoside Isolated from Lantana camara, an Inhibitor of Protein Kinase C. *J. Nat. Prod.* **1991**, *54*, 1595–1600. [CrossRef] [PubMed]
15. Khamaisi, M.; Katagiri, S.; Keenan, H.; Park, K.; Maeda, Y.; Li, Q.; Qi, W.; Thomou, T.; Eschuk, D.; Tellechea, A.; et al. PKCδ inhibition normalizes the wound-healing capacity of diabetic human fibroblasts. *J. Clin. Investig.* **2016**, *126*, 837–853. [CrossRef]
16. Oh, Y.J.; Youn, J.H.; Ji, Y.; Lee, S.E.; Lim, K.J.; Choi, J.E.; Shin, J.-S. HMGB1 Is Phosphorylated by Classical Protein Kinase C and Is Secreted by a Calcium-Dependent Mechanism. *J. Immunol.* **2009**, *182*, 5800–5809. [CrossRef]
17. Smulow, J.B.; Glickman, I. An Epithelial-Like Cell Line in Continuous Culture from Normal Adult Human Gingiva. *Exp. Biol. Med.* **1966**, *121*, 1294–1296. [CrossRef] [PubMed]
18. Kasten, F.H.; Soileau, K.; Meffert, R.M. Quantitative evaluation of human gingival epithelial cell attachment to implant surfaces in vitro. *Int. J. Periodontics Restor. Dent.* **1990**, *10*, 68–79.
19. Toullec, D.; Pianetti, P.; Coste, H.; Bellevergue, P.; Grand-Perret, T.; Ajakane, M.; Baudet, V.; Boissin, P.; Boursier, E.; Loriolle, F. The bisindolylmaleimide GF 109203X is a potent and selective inhibitor of protein kinase C. *J. Biol. Chem.* **1991**, *266*, 15771–15781. [CrossRef]
20. Mollica, L.; De Marchis, F.; Spitaleri, A.; Dallacosta, C.; Pennacchini, D.; Zamai, M.; Agresti, A.; Trisciuoglio, L.; Musco, G.; Bianchi, M.E. Glycyrrhizin Binds to High-Mobility Group Box 1 Protein and Inhibits Its Cytokine Activities. *Chem. Biol.* **2007**, *14*, 431–441. [CrossRef] [PubMed]
21. Hong, Y.; Shen, C.; Yin, Q.; Sun, M.; Ma, Y.; Liu, X. Effects of RAGE-Specific Inhibitor FPS-ZM1 on Amyloid-β Metabolism and AGEs-Induced Inflammation and Oxidative Stress in Rat Hippocampus. *Neurochem. Res.* **2016**, *41*, 1192–1199. [CrossRef] [PubMed]

22. Horie, R.; Watanabe, M.; Okamura, T.; Taira, M.; Shoda, M.; Motoji, T.; Utsunomiya, A.; Higashihara, M.; Umezawa, K.; Watanabe, T. DHMEQ, a new NF-κB inhibitor, induces apoptosis and enhances fludarabine effects on chronic lymphocytic leukemia cells. *Leukemia* **2006**, *20*, 800–806. [CrossRef] [PubMed]

23. Vivo, M.; Ranieri, M.; Sansone, F.; Santoriello, C.; Calogero, R.A.; Calabro, V.; Pollice, A.; La Mantia, G. Mimicking p14ARF Phosphorylation Influences Its Ability to Restrain Cell Proliferation. *PLoS ONE* **2013**, *8*, e53631. [CrossRef]

24. Downs, C.A.; Kreiner, L.H.; Johnson, N.M.; Brown, L.A.; Helms, M.N. Receptor for Advanced Glycation End-Products Regulates Lung Fluid Balance via Protein Kinase C–gp91phox Signaling to Epithelial Sodium Channels. *Am. J. Respir. Cell Mol. Biol.* **2015**, *52*, 75–87. [CrossRef]

25. Ariga, A.; Namekawa, J.-I.; Matsumoto, N.; Inoue, J.-I.; Umezawa, K. Inhibition of Tumor Necrosis Factor-α-induced Nuclear Translocation and Activation of NF-κB by Dehydroxymethylepoxyquinomicin. *J. Biol. Chem.* **2002**, *277*, 24625–24630. [CrossRef] [PubMed]

26. Marrotte, E.J.; Chen, D.-D.; Hakim, J.; Chen, A.F. Manganese superoxide dismutase expression in endothelial progenitor cells accelerates wound healing in diabetic mice. *J. Clin. Investig.* **2010**, *120*, 4207–4219. [CrossRef]

27. Wu, Z.; Puigserver, P.; Andersson, U.; Zhang, C.; Adelmant, G.; Mootha, V.; Troy, A.; Cinti, S.; Lowell, B.; Scarpulla, R.C.; et al. Mechanisms Controlling Mitochondrial Biogenesis and Respiration through the Thermogenic Coactivator PGC-1. *Cell* **1999**, *98*, 115–124. [CrossRef]

28. Murphy, M.P. Mitochondrial Dysfunction Indirectly Elevates ROS Production by the Endoplasmic Reticulum. *Cell Metab.* **2013**, *18*, 145–146. [CrossRef]

29. Patti, M.E.; Butte, A.J.; Landaker, E.J.; Goldfine, A.B.; Mun, E.; DeFronzo, R.; Finlayson, J.; Kahn, C.R.; Mandarino, L.J.; Crunkhorn, S.; et al. Coordinated reduction of genes of oxidative metabolism in humans with insulin resistance and diabetes: Potential role ofPGC1andNRF1. *Proc. Natl. Acad. Sci. USA* **2003**, *100*, 8466–8471. [CrossRef]

30. Cottrell, D.A.; Blakely, E.L.; Borthwick, G.M.; Johnson, M.A.; Taylor, G.A.; Brierley, E.J.; Ince, P.; Turnbull, D.M. Role of Mitochondrial DNA Mutations in Disease and Aging. *Ann. N. Y. Acad. Sci.* **2006**, *908*, 199–207. [CrossRef]

31. Edgar, D.; Shabalina, I.; Camara, Y.; Wredenberg, A.; Calvaruso, M.A.; Nijtmans, L.; Nedergaard, J.; Cannon, B.; Larsson, N.-G.; Trifunovic, A. Random Point Mutations with Major Effects on Protein-Coding Genes Are the Driving Force behind Premature Aging in mtDNA Mutator Mice. *Cell Metab.* **2009**, *10*, 131–138. [CrossRef] [PubMed]

32. Nowak, G.; Bakajsova, D.; Clifton, G.L. Protein kinase C-ε modulates mitochondrial function and active Na+transport after oxidant injury in renal cells. *Am. J. Physiol. Physiol.* **2004**, *286*, F307–F316. [CrossRef]

33. Cherry, A.D.; Piantadosi, C.A. Regulation of Mitochondrial Biogenesis and Its Intersection with Inflammatory Responses. *Antioxidants Redox Signal.* **2015**, *22*, 965–976. [CrossRef] [PubMed]

34. Yang, J.; Yan, Y.; Liu, H.; Hu, J. Protective effects of acteoside against X-ray-induced damage in human skin fibroblasts. *Mol. Med. Rep.* **2015**, *12*, 2301–2306. [CrossRef]

35. Lim, H.; Kim, D.K.; Kim, T.-H.; Kang, K.-R.; Seo, J.-Y.; Cho, S.S.; Yun, Y.; Choi, Y.-Y.; Leem, J.; Kim, H.-W.; et al. Acteoside Counteracts Interleukin-1β-Induced Catabolic Processes through the Modulation of Mitogen-Activated Protein Kinases and the NFκB Cellular Signaling Pathway. *Oxidative Med. Cell. Longev.* **2021**, *2021*, 1–16. [CrossRef]

36. Zhu, K.; Meng, Z.; Tian, Y.; Gu, R.; Xu, Z.; Fang, H.; Liu, W.; Huang, W.; Ding, G.; Xiao, W. Hypoglycemic and hypolipidemic effects of total glycosides of Cistanche tubulosa in diet/streptozotocin-induced diabetic rats. *J. Ethnopharmacol.* **2021**, *276*, 113991. [CrossRef] [PubMed]

37. Zhang, Y.; Liu, M.; Chen, Q.; Wang, T.; Yu, H.; Xu, J.; Wang, T. Leaves of Lippia triphylla improve hepatic lipid metabolism via activating AMPK to regulate lipid synthesis and degradation. *J. Nat. Med.* **2019**, *73*, 707–716. [CrossRef]

38. Galli, A.; Marciani, P.; Marku, A.; Ghislanzoni, S.; Bertuzzi, F.; Rossi, R.; Di Giancamillo, A.; Castagna, M.; Perego, C. Verbascoside Protects Pancreatic β-Cells against ER-Stress. *Biomedicines* **2020**, *8*, 582. [CrossRef]

39. Ayo, S.H.; Radnik, R.; Garoni, J.A.; Troyer, D.A.; Kreisberg, J.I. High glucose increases diacylglycerol mass and activates protein kinase C in mesangial cell cultures. *Am. J. Physiol. Physiol.* **1991**, *261*, F571–F577. [CrossRef]

40. Shmueli, E.; Alberti, K.G.M.M.; Record, C.O. Diacylglycerol/protein kinase C signalling: A mechanism for insulin resistance? *J. Intern. Med.* **1993**, *234*, 397–400. [CrossRef]

41. Ha, H.; Yu, M.-R.; Choi, Y.J.; Lee, H.B. Activation of protein kinase C-δ and C-ε by oxidative stress in early diabetic rat kidney. *Am. J. Kidney Dis.* **2001**, *38*, S204–S207. [CrossRef]

42. Scivittaro, V.; Ganz, M.B.; Weiss, M.F. AGEs induce oxidative stress and activate protein kinase C-βII in neonatal mesangial cells. *Am. J. Physiol. Physiol.* **2000**, *278*, F676–F683. [CrossRef] [PubMed]

43. Inoguchi, T.; Li, P.; Umeda, F.; Yu, H.Y.; Kakimoto, M.; Imamura, M.; Aoki, T.; Etoh, T.; Hashimoto, T.; Naruse, M.; et al. High glucose level and free fatty acid stimulate reactive oxygen species production through protein kinase C–dependent activation of NAD(P)H oxidase in cultured vascular cells. *Diabetes* **2000**, *49*, 1939–1945. [CrossRef] [PubMed]

44. Giordo, R.; Nasrallah, G.K.; Posadino, A.M.; Galimi, F.; Capobianco, G.; Eid, A.H.; Pintus, G. Resveratrol-Elicited PKC Inhibition Counteracts NOX-Mediated Endothelial to Mesenchymal Transition in Human Retinal Endothelial Cells Exposed to High Glucose. *Antioxidants* **2021**, *10*, 224. [CrossRef] [PubMed]

45. Qin, J.; Peng, Z.; Yuan, Q.; Li, Q.; Peng, Y.; Wen, R.; Hu, Z.; Liu, J.; Xia, X.; Deng, H.; et al. AKF-PD alleviates diabetic nephropathy via blocking the RAGE/AGEs/NOX and PKC/NOX Pathways. *Sci. Rep.* **2019**, *9*, 1–12. [CrossRef]

46. Dasu, M.R.; Devaraj, S.; Zhao, L.; Hwang, D.H.; Jialal, I. High Glucose Induces Toll-Like Receptor Expression in Human Monocytes: Mechanism of Activation. *Diabetes* **2008**, *57*, 3090–3098. [CrossRef]
47. Jiang, S.-Y.; Wei, C.-C.; Shang, T.-T.; Lian, Q.; Wu, C.-X.; Deng, J.-Y. High glucose induces inflammatory cytokine through protein kinase C-induced toll-like receptor 2 pathway in gingival fibroblasts. *Biochem. Biophys. Res. Commun.* **2012**, *427*, 666–670. [CrossRef]
48. Tang, D.; Kang, R.; Zeh, H.J., III; Lotze, M.T. High-Mobility Group Box 1, Oxidative Stress, and Disease. *Antioxid. Redox Signal.* **2011**, *14*, 1315–1335. [CrossRef]
49. Youn, J.H.; Shin, J.-S. Nucleocytoplasmic Shuttling of HMGB1 Is Regulated by Phosphorylation That Redirects It toward Secretion. *J. Immunol.* **2006**, *177*, 7889–7897. [CrossRef]
50. Sohn, E.; Kim, J.; Kim, C.-S.; Lee, Y.M.; Kim, J.S. Extract of Polygonum cuspidatum Attenuates Diabetic Retinopathy by Inhibiting the High-Mobility Group Box-1 (HMGB1) Signaling Pathway in Streptozotocin-Induced Diabetic Rats. *Nutrients* **2016**, *8*, 140. [CrossRef]
51. Han, R.; Liu, Z.; Sun, N.; Liu, S.; Li, L.; Shen, Y.; Xiu, J.; Xu, Q. BDNF Alleviates Neuroinflammation in the Hippocampus of Type 1 Diabetic Mice via Blocking the Aberrant HMGB1/RAGE/NF-κB Pathway. *Aging Dis.* **2019**, *10*, 611–625. [CrossRef] [PubMed]
52. Akutagawa, K.; Fujita, T.; Ouhara, K.; Takemura, T.; Tari, M.; Kajiya, M.; Matsuda, S.; Kuramitsu, S.; Mizuno, N.; Shiba, H.; et al. Glycyrrhizic acid suppresses inflammation and reduces the increased glucose levels induced by the combination of Porphyromonas gulae and ligature placement in diabetic model mice. *Int. Immunopharmacol.* **2019**, *68*, 30–38. [CrossRef]
53. Wang, Y.; Shan, J.; Yang, W.; Zheng, H.; Xue, S. High Mobility Group Box 1 (HMGB1) Mediates High-Glucose-Induced Calcification in Vascular Smooth Muscle Cells of Saphenous Veins. *Inflammation* **2013**, *36*, 1592–1604. [CrossRef] [PubMed]
54. Luo, L.; Xie, P.; Gong, P.; Tang, X.-H.; Ding, Y.; Deng, L.-X. Expression of HMGB1 and HMGN2 in gingival tissues, GCF and PICF of periodontitis patients and peri-implantitis. *Arch. Oral Biol.* **2011**, *56*, 1106–1111. [CrossRef] [PubMed]
55. Atieh, M.A.; Faggion, C.M.; Seymour, G.J. Cytokines in patients with type 2 diabetes and chronic periodontitis: A systematic review and meta-analysis. *Diabetes Res. Clin. Pract.* **2014**, *104*, e38–e45. [CrossRef] [PubMed]
56. Andriankaja, O.; Barros, S.; Moss, K.; Panagakos, F.; DeVizio, W.; Beck, J.; Offenbacher, S. Levels of Serum Interleukin (IL)-6 and Gingival Crevicular Fluid of IL-1β and Prostaglandin E2 Among Non-Smoking Subjects With Gingivitis and Type 2 Diabetes. *J. Periodontol.* **2009**, *80*, 307–316. [CrossRef]
57. Katz, J.; Bhattacharyya, I.; Farkhondeh-Kish, F.; Perez, F.M.; Caudle, R.M.; Heft, M.W. Expression of the receptor of advanced glycation end products in gingival tissues of type 2 diabetes patients with chronic periodontal disease: A study utilizing immunohistochemistry and RT-PCR. *J. Clin. Periodontol.* **2005**, *32*, 40–44. [CrossRef]
58. Morimoto, Y.; Kawahara, K.-I.; Tancharoen, S.; Kikuchi, K.; Matsuyama, T.; Hashiguchi, T.; Izumi, Y.; Maruyama, I. Tumor necrosis factor-α stimulates gingival epithelial cells to release high mobility-group box 1. *J. Periodontal Res.* **2007**, *43*, 76–83. [CrossRef]
59. Ito, Y.; Bhawal, U.K.; Sasahira, T.; Toyama, T.; Sato, T.; Matsuda, D.; Nishikiori, H.; Kobayashi, M.; Sugiyama, M.; Hamada, N.; et al. Involvement of HMGB1 and RAGE in IL-1β-induced gingival inflammation. *Arch. Oral Biol.* **2012**, *57*, 73–80. [CrossRef]
60. Tancharoen, S.; Gando, S.; Binita, S.; Nagasato, T.; Kikuchi, K.; Nawa, Y.; Dararat, P.; Yamamoto, M.; Narkpinit, S.; Maruyama, I. HMGB1 Promotes Intraoral Palatal Wound Healing through RAGE-Dependent Mechanisms. *Int. J. Mol. Sci.* **2016**, *17*, 1961. [CrossRef]
61. Dai, J.; Shen, J.; Chai, Y.; Chen, H. IL-1β Impaired Diabetic Wound Healing by Regulating MMP-2 and MMP-9 through the p38 Pathway. *Mediat. Inflamm.* **2021**, *2021*, 1–10. [CrossRef] [PubMed]
62. Buranasin, P.; Mizutani, K.; Iwasaki, K.; Na Mahasarakham, C.P.; Kido, D.; Takeda, K.; Izumi, Y. High glucose-induced oxidative stress impairs proliferation and migration of human gingival fibroblasts. *PLoS ONE* **2018**, *13*, e0201855. [CrossRef] [PubMed]
63. Sifuentes-Franco, S.; Padilla-Tejeda, D.E.; Carrillo-Ibarra, S.; Miranda-Díaz, A.G. Oxidative Stress, Apoptosis, and Mitochondrial Function in Diabetic Nephropathy. *Int. J. Endocrinol.* **2018**, *2018*, 1–13. [CrossRef] [PubMed]
64. Sifuentes-Franco, S.; Pacheco Moisés, F.P.; Rodríguez-Carrizalez, A.D.; Miranda-Díaz, A.G. The Role of Oxidative Stress, Mitochondrial Function, and Autophagy in Diabetic Polyneuropathy. *J. Diabetes Res.* **2017**, *2017*, 1673081. [CrossRef]

Article

The Protective Effect of Carotenoids, Polyphenols, and Estradiol on Dermal Fibroblasts under Oxidative Stress

Aya Darawsha, Aviram Trachtenberg, Joseph Levy [†] and Yoav Sharoni [*]

Department of Clinical Biochemistry and Pharmacology, Faculty of Health Sciences, Ben-Gurion University of the Negev, Beer Sheva 8410501, Israel; ayadar@post.bgu.ac.il (A.D.); aviramtr@post.bgu.ac.il (A.T.)
* Correspondence: yoav@bgu.ac.il; Tel.: +972-52-4830883
† Deceased.

Abstract: Skin ageing is influenced by several factors including environmental exposure and hormonal changes. Reactive oxygen species (ROS), which mediate many of the effects of these factors, induce inflammatory processes in the skin and increase the production of matrix metalloproteinases (MMPs) in dermal fibroblasts, which leads to collagen degradation. Several studies have shown the protective role of estrogens and a diet rich in fruits and vegetables on skin physiology. Previous studies have shown that dietary carotenoids and polyphenols activate the cell's antioxidant defense system by increasing antioxidant response element/Nrf2 (ARE/Nrf2) transcriptional activity and reducing the inflammatory response. The aim of the current study was to examine the protective effect of such dietary-derived compounds and estradiol on dermal fibroblasts under oxidative stress induced by H_2O_2. Human dermal fibroblasts were used to study the effect of H_2O_2 on cell number and apoptosis, MMP-1, and pro-collagen secretion as markers of skin damage. Treatment of cells with H_2O_2 led to cell death, increased secretion of MMP-1, and decreased pro-collagen secretion. Pre-treatment with tomato and rosemary extracts, and with estradiol, reversed the effects of the oxidative stress. This was associated with a reduction in intracellular ROS levels, probably through the measured increased activity of ARE/Nrf2. Conclusions: This study indicates that carotenoids, polyphenols, and estradiol protect dermal fibroblasts from oxidative stress-induced damage through a reduction in ROS levels.

Keywords: lycopene; carnosic acid; apoptosis; matrix metalloproteinase (MMP); collagen; antioxidant response element/Nrf2 (ARE/Nrf2); reactive oxygen species (ROS); NHDF cells

Citation: Darawsha, A.; Trachtenberg, A.; Levy, J.; Sharoni, Y. The Protective Effect of Carotenoids, Polyphenols, and Estradiol on Dermal Fibroblasts under Oxidative Stress. *Antioxidants* **2021**, *10*, 2023. https://doi.org/10.3390/antiox10122023

Academic Editors: Soliman Khatib and Dana Atrahimovich Blatt

Received: 14 November 2021
Accepted: 18 December 2021
Published: 20 December 2021

Publisher's Note: MDPI stays neutral with regard to jurisdictional claims in published maps and institutional affiliations.

1. Introduction

The skin is affected by environmental factors such as exposure to UV radiation, air pollutants, xenobiotics, and cigarette smoke [1,2], and by endogenous changes that occur with ageing [3]. These factors generally increase oxidative stress by generating reactive oxygen species (ROS) [1,2]. Oxidative stress, defined as a disturbance in the balance between the production of ROS and the antioxidant defenses [4], leads to changes in DNA, proteins, and lipids; and activation of inflammatory processes and signaling pathways that may cause various chronic and degenerative diseases [5], such as diabetes [4], cancer [6], neurological [7], and cardiovascular diseases [8]. Oxidative stress can also cause skin damage due to induction of apoptotic cell death [9,10]; activation of inflammatory processes [1]; an increase in metalloproteinases (MMPs), which degrade collagen in the extracellular matrix [11,12]; and a reduction in collagen production [13]. All of these changes accelerate skin ageing.

Epidemiological and clinical studies indicate that a diet rich in fruits and vegetables improves skin health [14], and that plant-derived compounds (phytonutrients) reduce oxidative stress-induced skin damage [15,16]. Among these active phytonutrients are carotenoids and polyphenols [16]. It was suggested that the protective effects of phytonutrients are mediated by the modulation of signaling pathways that are involved in the adverse effects of oxidative stress. This modulation includes reduction in the activities of the nuclear factor-κB (NF-κB) transcription system, activator protein 1 (AP-1), and mitogen-activated protein kinase (MAPK) [17,18], and improving the antioxidant defense mechanism [19]. Several studies, including our own, suggest that modulation of these signaling pathways by phytonutrients can lead to a reduction in inflammatory processes [20–22] and MMP expression [17,23], and increase collagen synthesis [17].

Estradiol plays an important role in maintaining skin health [18], which is impaired in menopause due to a dramatic reduction in estradiol levels [18]. Treatment with estrogens (hormonal replacement therapy) increases collagen levels which improve skin elasticity [24]. These effects are partially associated with a reduction in oxidative stress [18] through increasing the expression of antioxidant enzymes and modulating the activity of the antioxidant response element/Nrf2 (ARE/Nrf2) transcription system [25]. Carotenoids were shown to activate the ARE/Nrf2 transcription system in several cell types [26], including keratinocytes [21], the main cell type of the skin epidermal layer. This system is also activated by polyphenols [21,27]. The activation of ARE/Nrf2 by carotenoids is probably exerted by carotenoid-oxidized derivatives which can be formed in the cells, but not by the intact molecules [28].

The aim of the current study was to investigate the effect of oxidative stress on primary human dermal fibroblasts and the ability of phytonutrients and estradiol to protect the cells from ROS-induced damage. These fibroblasts, which originate in the dermis layer of the skin, were used because they are the main source of collagen in the skin. Incubation with H_2O_2 was used to increase intracellular ROS. The effect of both H_2O_2 and the tested compounds (tomato and rosemary extracts and estradiol) was examined on cell death, MMP-1 and pro-collagen 1a1 secretions, and ROS level. The combined effect of H_2O_2 and the tested compounds on the activation of the ARE/Nrf2 transcription system was examined as a possible mechanism for the protection of dermal fibroblasts from oxidative stress.

2. Materials and Methods

2.1. Materials

Tomato extract (Lycomato^{TM}) and rosemary extract were a gift of Lycored Ltd., Beer Sheva, Israel). The tomato extract, prepared by ethyl acetate extraction of tomato pulp, contained 6% lycopene, other tomato carotenoids (phytoene and phytofluene above 1%, beta-carotene above 0.2%), and additional fat-soluble tomato components such as natural tocopherols (above 1.5%) and phytosterols (1.1–2.5%). The remainder were triacylglycerols (70–72%), monoacylglycerols (8–9%), and phospholipids (7–8%). The rosemary extract was prepared by extraction with 80% ethanol. Its composition was only partially determined to contain carnosic acid (20.2%) and carnosol (2.5%) as the main polyphenols. 17β-estradiol was purchased from Sigma-Aldrich (Rehovot, Israel). Carotenoids were dissolved in tetrahydrofuran (THF), solubilized in cell culture medium, and their final concentration was measured as described previously [28–30]. Rosemary extract and estradiol were dissolved in ethanol. H_2O_2 30% and THF, containing 0.025% butylated hydroxytoluene as an antioxidant, were purchased from Sigma-Aldrich. Dulbecco's modified Eagle's medium (DMEM), dextran-coated charcoal-treated fetal bovine serum (DCC-FBS), Hanks' solution, and 1M HEPES buffer were purchased from Biological Industries (Beth Haemek, Israel).

2.2. Cell Culture

Normal human dermal fibroblasts (NHDF) were purchased from PromoCell GmbH (Heidelberg, Germany). The cells were grown in PromoCell fibroblast growth medium 2, according to the manufacturer's instructions, in a humidified atmosphere at 37 °C in 5% CO_2. Before each experiment, cells were depleted of steroid hormones by maintaining them for 3–5 days in phenol red-free DMEM supplemented with 10% DCC-FBS (DMEM-

DCC-FBS). This medium was used throughout all experiments because it does not contain steroid hormones or any other compound with estrogenic activity, such as phenol red.

2.3. Determination of Cell Number, and Secretion of MMP-1 and Procollagen 1a1

NHDF cells were seeded in 96-well plates at a density of 10^4 cells/well in DMEM-DCC-FBS medium. Twenty-four hours later, cells were pre-incubated with phytonutrients or estradiol for 24 h. Vehicle-treated control cells were incubated with the relevant amounts of solvents used in a particular experiment, which had no effect on the measured parameters. The medium was then replaced with one containing the treatment compounds, with or without H_2O_2 and incubated for an additional 24 h. Thereafter, medium was removed and frozen for the analysis of secreted proteins, and the cell number was determined by the XTT Cell Proliferation Kit (Biological Industries, Beth Haemek, Israel), according to the manufacturer's instructions. MMP-1 and pro-collagen 1a1 protein levels in cell culture supernatants were quantified by ELISA using the Human Total MMP-1 DuoSet and Human Pro-Collagen 1a1 DuoSet, ELISA kits (R&D Systems, Inc., Minneapolis, MN, USA), according to the manufacturer's instructions. Optical density was measured using a VERSAmax tunable microplate reader (Molecular Devices, Menlo Park, CA, USA). Results of the cell number, MMP-1, and pro-collagen were calculated as a percent of the values obtained in control cells, treated with a vehicle without H_2O_2.

2.4. Assessment of Apoptosis

Cells were seeded in 6-well plates at 3×10^5 cells/well. Twenty-four hours later, they were pre-incubated with the phytonutrients or estradiol for 24 h. The medium was then replaced with one containing the treatment compounds, with or without H_2O_2, and incubated for an additional 12 h. Cells were washed in PBS and stained with Annexin V and 7-aminoactinomycin D (7-AAD), using the Annexin V-FITC/7-AAD Apoptosis Detection Kit (Biogems, Cat# 62700-50, Chai Wan, Hong Kong), according to the manufacturer's protocol. The percentages of apoptotic cells were determined by flow cytometry on a Gallios instrument (Beckman Coulter Gallios Flow Cytometer, Brea, CA, USA). For each analysis, 10,000 events were recorded, and the data were processed using Kaluza software, version #2.1, (Gallios^{TM} Kaluza). As a positive control, cells were incubated with 1.25 µM staurosporine for 18 h [31]. Annexin V positive/7AAD-negative cells were considered to be in the early apoptotic phase, and cells positive for both Annexin V and 7AAD were considered to be late apoptotic.

2.5. Detection of Intracellular ROS

Cells were seeded in 6-well plates at 3×10^5 cells/well. Twenty-four hours later, cells were pre-incubated with the phytonutrients or estradiol for 24 h. The medium was then replaced with one containing the treatment compounds with or without H_2O_2 and incubated for an additional 90 min. Intracellular ROS were detected by 2′,7′-dichlorofluorescin diacetate (DCFH-DA, Sigma-Aldrich) staining. Briefly, cells were trypsinized, and washed by Hanks' solution containing 10 mM HEPES buffer, pH = 7.4, followed by cell staining with 5 µM DCFH-DA for 30 min at 37 °C in the dark. Cells were analyzed by flow cytometry using the Gallios instrument. Cells treated with 0.5 mM H_2O_2 for 15 min were used as a positive control. Untreated and unstained cells were used as a negative control.

2.6. Transient Transfection and ARE/Nrf2 Reporter Gene Assay

NHDF cells were seeded in 24-well plates at 10^5 cells/well. Twenty-four hours later, they were transfected using jetPEI reagent (Polyplus Transfection, Illkrich, France). Briefly, cells were rinsed once with serum-free medium, followed by the addition of 0.45 mL of medium and 50 µL of a mixture containing DNA and jetPEI reagent at a charge ratio of 1:5. The total amount of DNA was 0.25 µg, containing 0.2 µg 4xARE reporter construct [32] that was kindly provided by Dr. M. Hannink (University of Missouri-Columbia, Columbia, MO, USA) and 0.05 µg Renilla luciferase (P-RL-null) expression vectors, which served as an internal transfection standard, and was purchased from Promega (Madison, WI, USA). These conditions were found to be optimal for the dermal fibroblasts. The cells were then incubated for 6 h at 37 °C. Next, the medium was replaced with DMEM-DCC-FBS plus the test compounds, and cells were incubated for an additional 16 h. Cell extracts were prepared for luciferase reporter assay (Dual Luciferase Reporter Assay System, Promega) according to the manufacturer's instructions, and luminescence was determined using a Turner Biosystems luminometer (Sunnyvale, CA, USA).

2.7. Statistical Analysis

All experiments were performed in triplicate or duplicate and repeated two to seven times as indicated in the figure legends. Statistically significant differences between two experimental groups were determined using one-way ANOVA with Dunnett's multiple comparison post-hoc analysis. Data are presented as the mean $\pm$ SEM. $p < 0.05$ was considered statistically significant. The statistical analyses were performed using GraphPad Prism 6.0 software (Graph-Pad Software, San Diego, CA, USA).

3. Results

3.1. H_2O_2 Induces Cell Death of Dermal Fibroblasts, Increases MMP-1 Secretion, and Decreases Pro-Collagen 1a1

Oxidative stress in human primary dermal fibroblasts was induced by incubating the cells with increasing concentrations of H_2O_2 (0–75 µM) for 24 h. The cell number, checked by the XTT method, decreased by H_2O_2 dose-dependently. At 50 µM, only about 20% of the cells were viable, and at higher concentrations, almost all cells died (Figure 1a). MMP-1 and pro-collagen 1a1 secretion were measured in the media using specific ELISA assays. H_2O_2 increased MMP-1 secretion and decreased pro-collagen 1a1 secretion (Figure 1b). These effects of H_2O_2 in dermal fibroblasts, cell death, and the inverse changes in MMP-1 and pro-collagen secretion may cause thinning of the dermis and a reduction in skin elasticity.

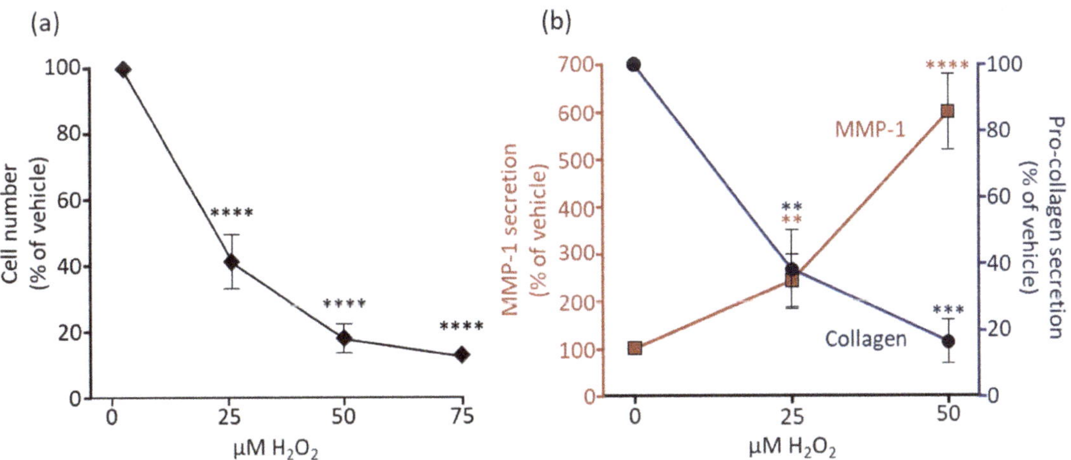

Figure 1. NHDF cells were seeded in 96-well plates (10^4 cells/well). Twenty-four hours later, the cells were incubated with the indicated concentrations of H_2O_2 for 24 h. (**a**) The cell number was determined by the XTT method as described in Section 2.3. (**b**) Secretion of MMP-1 and pro-collagen 1a1 were determined in the media using ELISA assays as described in Section 2.3. Results are presented as a percent of the values without H_2O_2. Y-axis scales: 100% of the cell number was 35,534 $\pm$ 1251 cells/well; 100% of the MMP-1 secretion was 47 + 12 pmol per 1000 cells; 100% of the pro-collagen secretion was 107 $\pm$ 60 pmol per 1000 cells. Values are the means $\pm$ SEM of 3–5 experiments, each performed in triplicate. ** $p < 0.01$, *** $p < 0.001$, **** $p < 0.0001$, significant difference between the vehicle with and without H_2O_2.

3.2. The Phytonutrients of Tomato and Rosemary Extracts and the Hormone Estradiol Protect Fibroblasts from H_2O_2-Induced Cell Damage

To investigate whether phytonutrients can protect NHDF cells from H_2O_2-induced damage, cells were pre-incubated for 24 h with increasing concentrations of tomato or rosemary extracts before exposing them to H_2O_2 for an additional 24 h in the presence of the phytonutrients. Partial protection from cell death was evident at most tested concentrations of the extracts (Figure 2a,b); however, these effects were not significant. Nonetheless, large and significant effects were obtained in the secretion of MMP-1 and pro-collagen, and both tomato and rosemary extracts reversed the effects of H_2O_2. Tomato extract reduced the H_2O_2-induced rise in MMP-1 secretion by about 50% (Figure 2c) and completely restored pro-collagen secretion (Figure 2e). Similarly, rosemary extract decreased MMP-1 secretion (Figure 2d) and increased pro-collagen secretion (Figure 2f) to their basal levels.

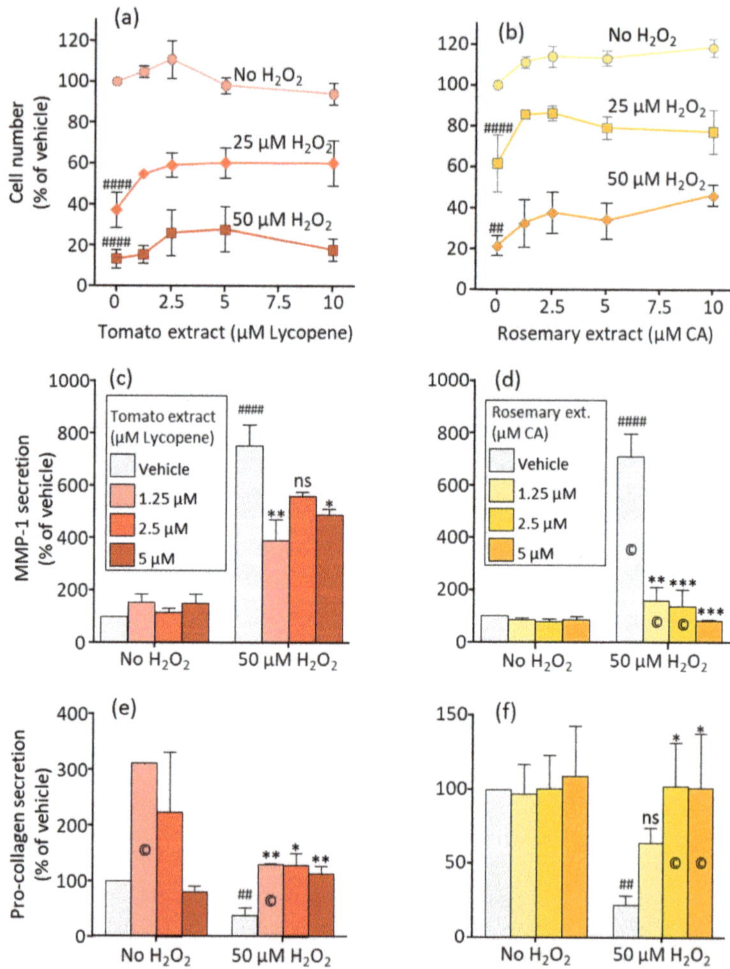

Figure 2. Tomato and rosemary extracts protect fibroblasts from H_2O_2-induced cell damage. NHDF cells were pre-incubated for 24 h with tomato extract or rosemary extract at the indicated concentrations of lycopene or carnosic acid, respectively. Then, H_2O_2 was added at the indicated concentrations for an additional 24 h. The cell number (**a,b**), and secretion of MMP-1 (**c,d**) and pro-collagen 1a1 (**e,f**),

were determined as described in Figure 1. Results are presented as a percent of the values without H_2O_2. Y-axis scales: 100% of the cell number was 36,609 $\pm$ 2029 cells/well; 100% of the MMP-1 secretion was 110 $\pm$ 37 pmol per 1000 cells; 100% of the pro-collagen secretion was 112 $\pm$ 50 pmol per 1000 cells. Values are the means $\pm$ SEM of 2–7 experiments, each performed in triplicate. The experimental groups that were repeated in only two experiments are marked with the symbol © in the center of the relevant column. ## $p < 0.01$, #### $p < 0.0001$, significant difference between the vehicle with and without H_2O_2. * $p < 0.005$, ** $p < 0.01$, *** $p < 0.001$), significant difference between the vehicle and other treatments in the presence of H_2O_2 (ns—non-significant).

Although the protection from oxidative stress-induced damage by the tomato and rosemary extracts was significant, it was examined whether more protection can be achieved by their combination. Indeed, the combination of rosemary extract with tomato extract resulted in better protection of the cells compared to each compound alone (Figure 3). The effect of the combination on the cell number was not significantly different from the single compounds (Figure 3a); however, in these experiments, a significant reduction in MMP-1 secretion was evident only when the cells were treated with the phytonutrient combination (Figure 3b). Although the effect of single phytonutrients on pro-collagen secretion was significant, treatment with their combination induced a significantly larger increase compared to each compound alone (Figure 3c). The treatment of cells under oxidative stress, with the combination of tomato and rosemary extracts, resulted in a pro-collagen level that was higher than the basal level without H_2O_2.

The protective effect of estradiol on cell survival was stronger than the effect of the phytonutrients, and in the presence of 10 nM estradiol, the cells were completely resistant to cell death induced by 25 µM H_2O_2 (Figure 4a). Similar to the effects of the phytonutrients, estradiol decreased MMP-1 secretion (Figure 4b) and increased pro-collagen secretion back to basal levels without H_2O_2 (Figure 4c).

3.3. Tomato Extract, Rosemary Extract, and Estradiol Protect Dermal Fibroblasts from H_2O_2-Induced Apoptosis

It was demonstrated above that phytonutrients and estradiol protect dermal fibroblasts from H_2O_2-induced cell death. Consequently, it was examined whether H_2O_2 induced apoptosis and whether the tested compounds protected against this route of cell death. Using flow cytometry and Annexin-V/7AAD staining, it was found that H_2O_2 causes apoptotic cell death (Figure 5). The typical flow cytometric data of one experiment with estradiol is shown in Figure 5a. Only about 1% of cells were apoptotic when the cells were treated with the vehicle, with or without estradiol (almost all the cells appear in the unstained lower left square). After 12 h with H_2O_2, 10% of the cells were in an early apoptotic state (lower right panel), and 81% were in late apoptosis (upper right square). Pre-incubation with 10 nM estradiol resulted in a reduction to 8% and 44% of early- and late-apoptotic cells, respectively. The average total apoptosis (early + late) which was negligible in the control cells, increased after H_2O_2 treatment to more than 90% (Figure 5b). Estradiol and the phytonutrients reduced total apoptosis to less than 30%.

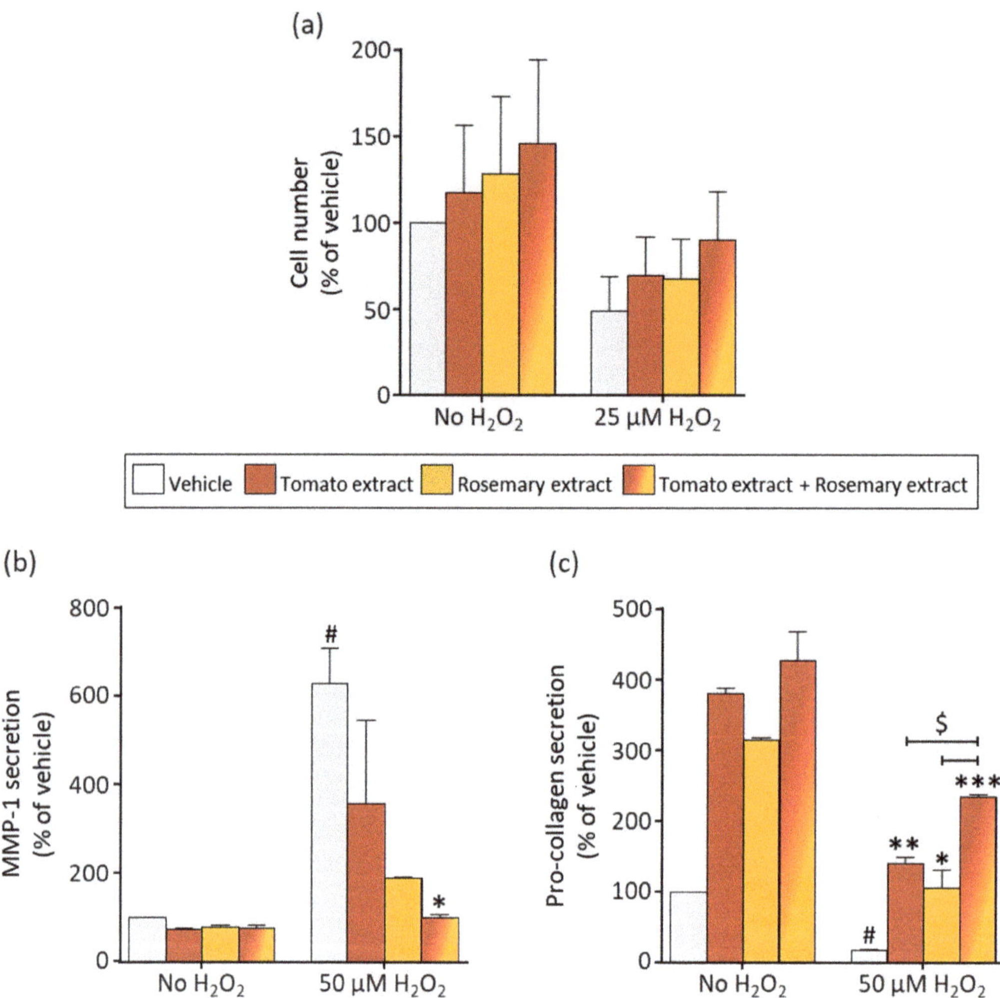

Figure 3. Protection from H_2O_2-induced cell damage was improved by a combination of tomato and rosemary extracts. NHDF cells were pre-incubated for 24 h with tomato extract (2.5 μM lycopene), rosemary extract (2.5 μM carnosic acid), or their combination. Then, H_2O_2 was added at the indicated concentrations for an additional 24 h. The cell number (**a**), and secretion of MMP-1 (**b**) and pro-collagen 1a1 (**c**), were determined as described in the legend of Figure 1. Results are presented as a percent of the values without H_2O_2. Y-axis scales: 100% of cell number was 8067 ± 6141 cells/well; 100% of the MMP-1 secretion was 84 ± 5 pmol per 1000 cells; 100% of the pro-collagen secretion was 16.5 ± 0.5, much lower than other experiments' pmol per 1000 cells. Values are the means ± SD of 2–3 experiments, each performed in triplicate. #, $p < 0.05$, significant difference between the vehicle with and without H_2O_2. * $p < 0.005$, ** $p < 0.01$, *** $p < 0.001$), significant difference between the vehicle and other treatments in the presence of H_2O_2. §, $p < 0.05$, between the combination and individual phytonutrients.

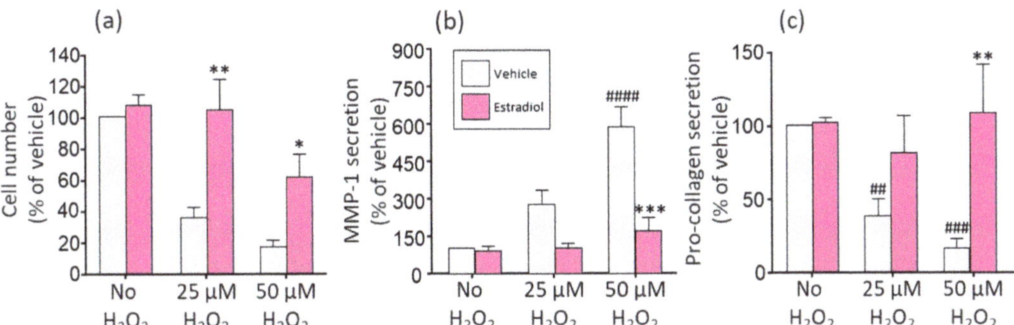

Figure 4. Estradiol protects fibroblasts from H_2O_2-induced cell damage. NHDF cells were pre-incubated for 24 h with 10 nM estradiol. Then, H_2O_2 was added at the indicated concentrations for an additional 24 h. The cell number (**a**), and secretion of MMP-1 (**b**) and pro-collagen 1a1 (**c**), were determined as described in the legend of Figure 1. Results are presented as a percent of the values without H_2O_2. Y-axis scales: 100% of the cell number was 40,040 $\pm$ 3011 cells/well; 100% of the MMP-1 secretion was 54.3 $\pm$ 7.6 pmol per 1000 cells; 100% of the pro-collagen secretion was 32.7 $\pm$ 16.7 pmol per 1000 cells. Values are the means $\pm$ SD of 2–7 experiments, each performed in triplicate. ## $p < 0.01$, ### $p < 0.001$, #### $p < 0.0001$, significant difference between the vehicle with and without H_2O_2. * $p < 0.05$, ** $p < 0.01$, *** $p < 0.001$, significant difference between the vehicle and other treatments in the presence of H_2O_2.

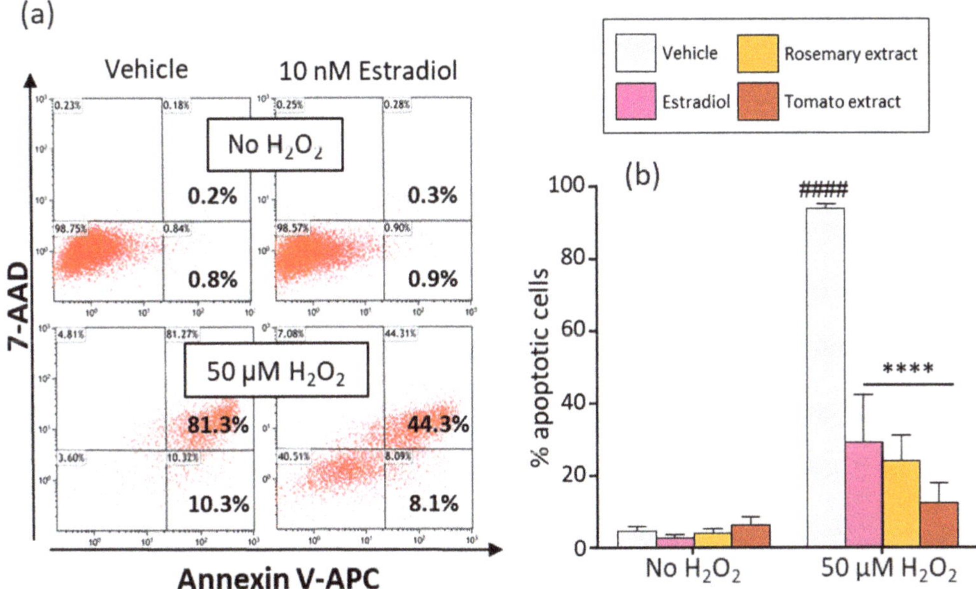

Figure 5. Estradiol and tomato and rosemary extracts protect fibroblasts against H_2O_2-induced apoptotic cell death. Cells were seeded in 6-well plates (3×10^5 cells/well) and were pre-incubated for 24 h with estradiol (10 nM), rosemary extract (5 μM carnosic acid), or tomato extract (10 μM lycopene). Then, H_2O_2 (50 μM) was added for an additional 12 h, and cell apoptosis was determined by flow cytometry as described in Section 2.4. (**a**) Typical flow cytometric data of Annexin-V and 7-AAD fluorescence, obtained in a representative experiment. (**b**) Averaged Annexin-V-stained cells (total = early + late apoptosis). Values (% apoptotic cells) are the means $\pm$ SEM of four experiments, each performed in duplicate. ####, $p < 0.0001$, significant difference between the vehicle with and without H_2O_2. ****, $p < 0.0001$, significant difference between the vehicle and other treatments in the presence of H_2O_2.

3.4. Pre-Treatment of Dermal Fibroblasts with Tomato Extract, Rosemary Extract, and Estradiol Reduces ROS Generated by H_2O_2

To verify that fibroblast cell damage was associated with oxidative stress, the cytoplasmic ROS level was determined by flow cytometry using a DCFH probe. Cells that were treated with H_2O_2 for 90 min showed a large increase in DCF fluorescence (Figure 6a, red tracing). The average geometric means of fluorescence intensities (MFI), indicating average ROS levels, increased by about six-fold (Figure 6c). Pre-incubation with estradiol (Figure 6a, black tracing) resulted in a tracing similar to the control (green tracing). ROS levels, as indicated by the MFI values, were reduced by estradiol, and the tomato and rosemary extracts back to basal level (Figure 6c). In a preliminary experiment, it was found that a reduction in ROS also occurred when the protective agents were added only during the pre-incubation, but not together with H_2O_2 (data not shown). This suggests that pre-incubation is required in order to reduce the ROS level. Indeed, when estradiol or the phytonutrients were added to cells at the same time as H_2O_2 without pre-incubation, there was no reduction in ROS levels (Figure 6d and the black tracing in Figure 6b). These results indicate that pre-incubation of the fibroblasts with the phytonutrients and estradiol is necessary to reduce these levels, and that the reduction in the ROS level did not occur by a chemical scavenging reaction between the protective agents and ROS. The increase in ROS levels by H_2O_2 and their reduction by the phytonutrients and estradiol was associated with cell damage and its attenuation, respectively. Thus, it was tested whether, without pre-incubation, cells can be rescued from ROS-induced death. However, when the phytonutrients or estradiol were added at the same time as H_2O_2, the cell number remained low and was the same as that without the protecting compounds (Figure 6e).

3.5. Oxidative Stress Increases the Induction of the ARE/Nrf2 Transcription System by the Phytonutrients and Estradiol

The results in the previous section suggest that pre-incubation with the phytonutrients and estradiol is required to decrease ROS levels and protect fibroblasts from H_2O_2-induced adverse effects. This implies that during pre-incubation, the cells increased their antioxidant defense capacity. One probable explanation for this increased capacity is the induction of the ARE/Nrf2 transcription system, which induces the transcription of antioxidant enzymes and is known to be activated by the tested compounds. The ARE/Nrf2 transcriptional activity was determined by a reporter gene assay. As expected, this transcription system was activated by rosemary extract (Figure 7a), tomato extract (Figure 7b), and estradiol (Figure 7c) from two- to four-fold in the absence of H_2O_2. H_2O_2 alone slightly increased this activity by two- to three-fold. In the presence of H_2O_2, the activation by the phytonutrients and estradiol was significantly higher than that of the compounds or H_2O_2 alone.

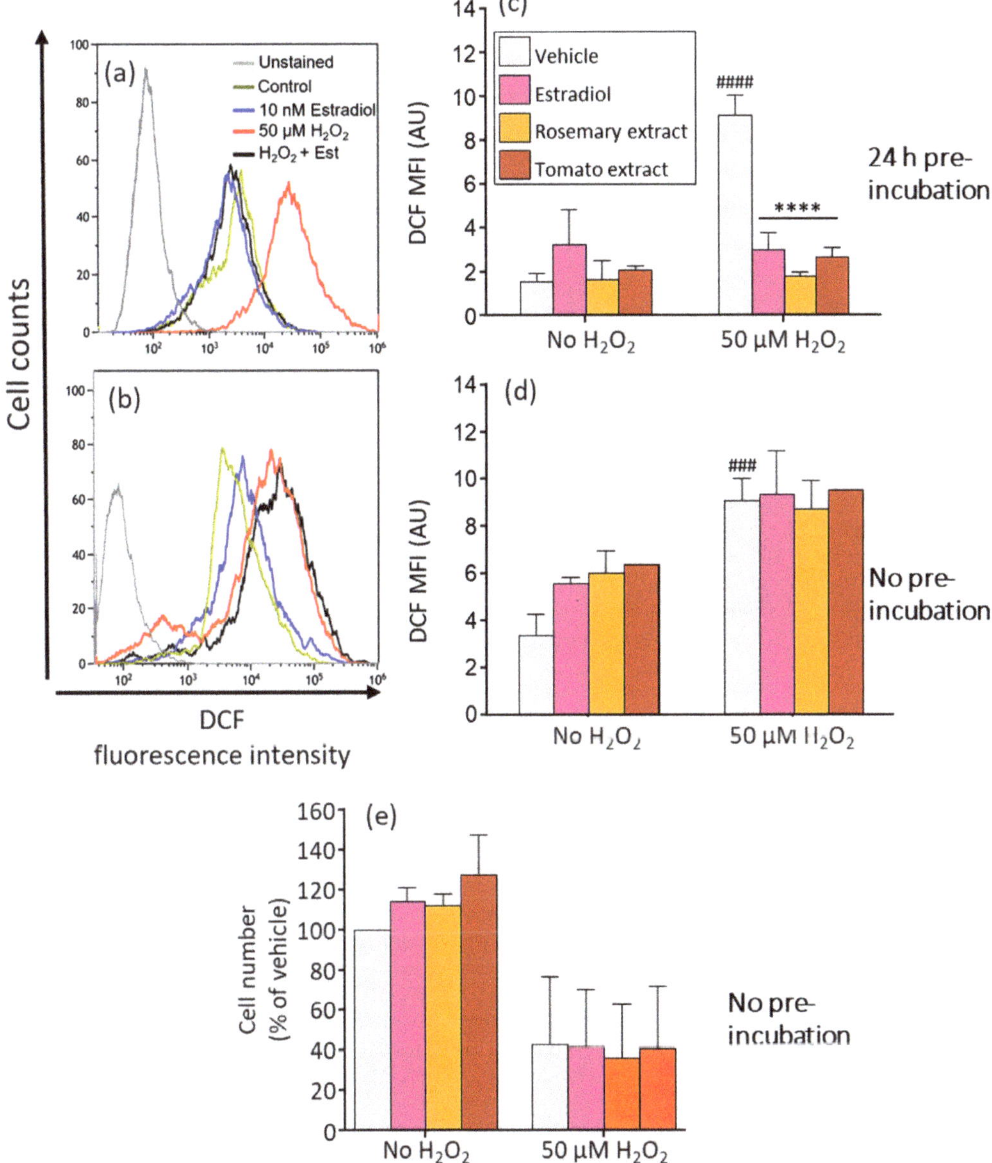

Figure 6. Pre-incubation with the phytonutrients or estradiol is required to reduce H_2O_2-generated ROS. Cells were seeded in 6-well plates (3×10^5 cells/well) and were pre-incubated for 24 h with (**a,c**) or without (**b,d,e**) estradiol (10 nM), rosemary extract (5 µM carnosic acid), or tomato extract (10 µM lycopene). Then, H_2O_2 (50 µM) was added together with the treatment compounds, and ROS levels (after 90 min, **a–d**) or cell number (after 24 h, **e**) were determined as described in Sections 2.3 and 2.5, respectively. (**a,b**) Typical flow cytometric histograms of DCF fluorescence obtained in a representative experiment with (**a**) or without (**b**) pre-incubation. (**c,d**) Averaged geometric means of DCF fluorescence intensities (MFI) with (**c**) or without (**d**) pre-incubation. (**e**) Cell numbers without pre-incubation, 100% of the cell number was 37,858 ± 676 cells/well. Values are the means ± SEM of 2–4 experiments, each performed in duplicate. ### $p < 0.001$, #### $p < 0.0001$, significant difference between the vehicle with and without H_2O_2. ****, $p < 0.0001$, significant difference between the vehicle and other treatments in the presence of H_2O_2.

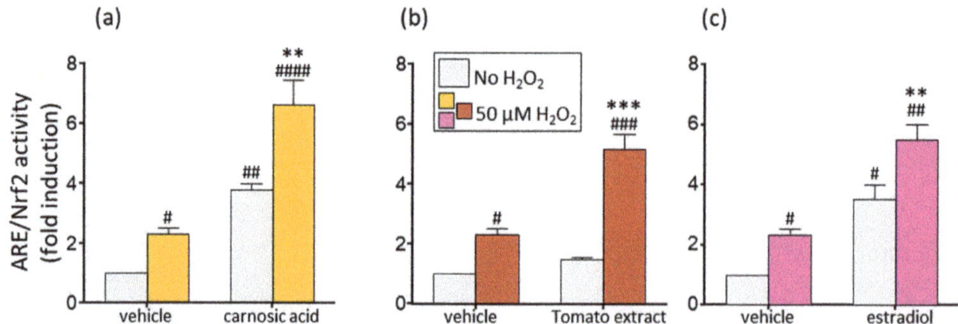

Figure 7. Activation of ARE/Nrf2 by the phytonutrients and estradiol in the presence of H_2O_2. Cells were seeded in 24-well plates (10^5 cells/well), and 24 h later were transfected with the ARE/Nrf2 reporter gene, and ARE/Nrf2 transcriptional activity was determined as described in Section 2.6. After transfection, cells were incubated for 16 h with (**a**) rosemary extract (5 µM carnosic acid), (**b**) tomato extract (10 µM lycopene), or (**c**) estradiol (10 nM), and with (dark columns) or without (light columns) H_2O_2. Values (fold induction) are the means ± SEM of 3–4 experiments, each performed in triplicate. # $p < 0.05$, ## $p < 0.01$, ### $p < 0.001$, #### $p < 0.0001$, significant difference between treatment and the vehicle without H_2O_2. ** $p < 0.01$, *** $p < 0.001$, significant difference between treatments with and without H_2O_2.

4. Discussion

The current study suggests that dietary and hormonal supplementation can prevent oxidative stress-induced damage to dermal fibroblasts. The inhibition of dermal fibroblasts cell death and the reversal of oxidative stress-induced increase in MMP-1 secretion and reduction in pro-collagen levels, that were demonstrated here, suggest that ingestion or topical application of these and similar compounds can prevent oxidative stress-induced reduction in skin depth and elasticity. The protection of dermal fibroblasts by carotenoids, polyphenols, and estradiol shown in the current study may be relevant to skin damage that is caused by several factors such as UV radiation and environmental pollutants, inducing the generation of ROS [2] (Scheme 1). In this study, the dietary carotenoids, polyphenols, and estradiol reduced oxidative stress-induced apoptosis of dermal fibroblasts, reduced MMP-1 secretion, and increased pro-collagen levels. Previous studies have shown the induction of apoptosis by oxidative stress in keratinocytes [33] and fibroblast [10] skin cells. Carotenoids, including lutein [34] and astaxanthin [35], were shown to increase the keratinocyte cell number by reducing UV-induced apoptotic cell death. An in vivo study showed that the topical application of lycopene was capable of suppressing a UVB-induced cascade of apoptosis, as indicated by the reduced expression of caspase-3 in murine skin [36]. Moreover, the protective role of polyphenols in skin cells has been shown in several studies. The survival of human dermal fibroblasts under H_2O_2-induced oxidative stress was increased by strawberry extract [37]. Various dietary polyphenols, including rosemary extract, were shown to increase keratinocyte cell survival after UVB irradiation [38–41]. Apoptosis inhibition was evident in most of these studies [38,39,41], and one compound, delphinidin, found in pomegranates, berries, and other pigmented fruits and vegetables, was also shown to inhibit apoptosis in mice skin [38].

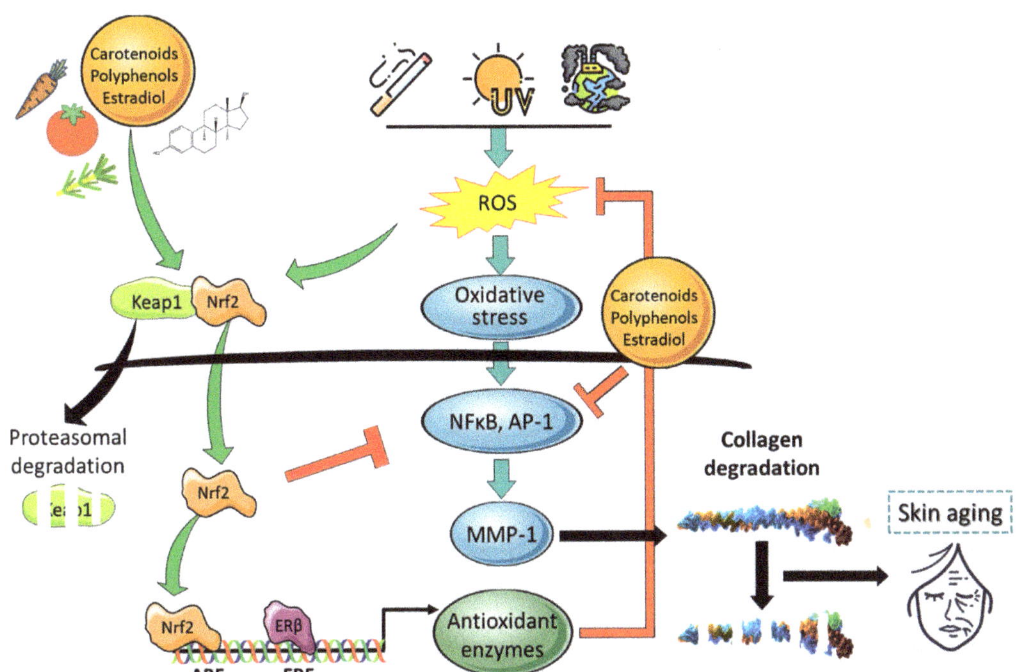

Scheme 1. A model summarizing the suggested effects of phytonutrients and estradiol on the parameters studied in dermal fibroblasts under oxidative stress. Increased cellular ROS is caused by several factors (e.g., UVB, air pollution, and smoking). Oxidative stress activates the NFκB and AP-1 transcription systems, which increases MMP-1 secretion and collagen degradation and decreases collagen secretion, leading to skin damage and aging. As a feedback mechanism, ROS activates the Nrf2 transcription system, which increases the expression of antioxidant enzymes that reduce ROS. ARE/Nrf2 is also induced by carotenoids, polyphenols, and estradiol. In addition, the expression of antioxidant enzymes can also be upregulated by estrogen receptor binding to an estrogen response element within the promoter region of the antioxidant enzyme. Carotenoids, polyphenols, and estradiol also inhibit NFκB and AP-1 activities, leading to a reversal of ROS damaging effects, delaying skin ageing.

In the skin extracellular matrix, similar to other tissues, there is a balance between the synthesis of new pro-collagen and the degradation of matrix collagen. This balance is disrupted by oxidative stress, which causes an increase in the production of MMPs [11,12,42] and a reduction in pro-collagen expression [11,13] (Scheme 1). Thus, the larger amount of pro-collagen attained by phytonutrient and estradiol treatments, and the reduced level of MMP-1, can restore this balance and, thus, improve skin elasticity. Previous studies have shown that polyphenols reduce MMP expression, both in vivo [43] and in vitro [17]. A similar reduction in MMP expression was shown by carotenoids in dermal fibroblasts [44] and in human skin [45,46]. Topical estradiol treatment reduced MMP-1 levels and increased pro-collagen expression in human skin [47], with similar effects found in murine bones [48]. Several signaling pathways are involved in the mechanism of oxidative stress-induced changes in MMP and collagen. A number of studies have shown that ROS induces the activation of MAPKs, which activate the NFκB and AP-1 transcription systems, leading to the increased expression of pro-inflammatory genes and MMPs [49–51]. Thus, inhibition of MAPKs, NFκB, and AP-1 by phytonutrients [49] may be the mechanism for reducing MMP-1 by the phytonutrients and estradiol in the current study (Scheme 1). A study from our laboratory showed that carotenoid derivatives inhibit NFκB and suggested a molecular mechanism for this inhibition [52]. β-carotene was found to inhibit NFκB in cancer cells [53], and lutein inhibited this transcription system in various cells of the choroid complex of the eye, both in vivo and in vitro [54]. The polyphenol, resveratrol, inhibited NFκB activity in mouse skin by blocking IκB kinase activity [55], and a study from our laboratory demonstrated inhibition of UVB-induced NFκB activity by several carotenoids and polyphenols in human keratinocytes [21]. Estradiol was also shown to inhibit NFκB activity by increasing the expression of IκBα, the inhibitor of NFκB [56].

The protective effects of the tested compounds in the current study were associated with a marked reduction in ROS levels only when the cells were pre-incubated with the compounds before adding H_2O_2. Thus, as discussed in Section 3.5, antioxidant defense was indeed improved during the pre-incubation, and this was possibly achieved by increased transcriptional activity of ARE/Nrf2, which can increase the expression of antioxidant enzymes, which can lead to ROS reduction (Scheme 1). Regulation of transcription and gene expression, including ARE/Nrf2, has been found to play a significant role in the effect of phytonutrients in many cellular processes [57]. As was found in the current study, carotenoids were shown to activate ARE/Nrf2 [26,28]. This transcription system can also be activated by polyphenols such as carnosic acid [58]. ARE/Nrf2 transcriptional activity is also activated by increased ROS levels, in order to lessen oxidative stress [19,59] (Scheme 1). Notably, in our study, the application of the phytonutrients and estradiol during oxidative stress increased ARE/Nrf2 activity to a level that was higher than that of the compounds or H_2O_2 alone. Similar to the activation of ARE/Nrf2 and the reduction in ROS in dermal fibroblasts described in the current study, estradiol activated ARE/Nrf2 and decreased ROS levels in neuronal cells, leading to cell survival and attenuation of homocysteine cytotoxicity [25]. In addition, estradiol can increase the expression of antioxidant enzymes by binding of estrogen receptor β to an estrogen response element within the promoter region of antioxidant enzymes [60] (Scheme 1). An additional mechanism for the antioxidant effect of estradiol was suggested by Borrás et al. [61], who demonstrated that estradiol decreases hydrogen peroxide in isolated hepatic mitochondria, probably via interaction with specific receptors present in the mitochondria. Several publications suggest that the antioxidant effects of estradiol and other estrogens is mediated by direct chemical scavenging of ROS and not by the estrogen receptors [62]. This potential mechanism is controversial because scavenging of ROS occurs only at high, non-physiological estrogen concentrations. For example, 1 µM estradiol was required to inhibit superoxide radical production by bovine heart endothelial cells, and about 50 µM was used to inhibit lipid and low-density lipoprotein peroxidation in non-cellular experimental systems [63]. Lower estradiol concentrations (100 nM) were used to reduce ROS formation in Friedreich's ataxia skin fibroblasts [62], a concentration that is ten times higher than the 10 nM estradiol used in the current study. Moreover, in the experimental protocol used in the current study, direct interaction between estradiol and ROS was possible only in the oxidative stress phase of the experiments, when H_2O_2 and estradiol were both present. However, as shown in Figure 6, the protective effect of estradiol (and the phytonutrients) depends on the pre-incubation phase, when H_2O_2 is not present. These results suggest that the interaction between H_2O_2 and estradiol probably did not occur in the current experimental protocol. Based on these considerations, it is concluded that activation of defense mechanisms during the pre-incubation phase probably explains the protective effects of estradiol and the phytonutrients in oxidative stress-challenged dermal fibroblasts. Activation of ARE/Nrf2, which was demonstrated in the current study, and inhibition of several signaling pathways such as MAPKs, NFκB, and AP-1, are possibly involved in the protection of dermal fibroblasts.

5. Conclusions

This study indicates that carotenoids, polyphenols, and estradiol protect dermal fibroblasts from ROS-induced damage and suggests that a balanced diet rich in phytonutrients may improve skin health and appearance. In addition, improvement in skin health can be achieved by topical application of estradiol and crude or purified dietary compounds. Oxidative stress can result from numerous environmental factors, and also occurs during ageing. Because this stress may negatively affect many human tissues, the implications of this study may also be relevant to various other pathologies such as inflammation, neurodegenerative and cardiovascular diseases, and cancer.

Author Contributions: Conceptualization: A.D., J.L. and Y.S.; methodology: A.D. and A.T.; formal analysis: A.D. and A.T.; investigation: A.D.; writing—original draft preparation: A.D. and Y.S.; writing—review and editing: A.D., A.T. and Y.S.; visualization: A.D., A.T. and Y.S.; supervision: J.L. and Y.S.; project administration: J.L. and Y.S.; funding acquisition: J.L. and Y.S. All authors have read and agreed to the published version of the manuscript.

Funding: This research was funded by Lycored Ltd., Beer Sheva, Israel (to Y.S. and J.L.), grant numbers 87717111 and 87784411.

Institutional Review Board Statement: Not applicable.

Informed Consent Statement: Not applicable.

Data Availability Statement: Data is contained within the article.

Acknowledgments: We thank M. Hannink (University of Missouri-Columbia, Columbia, MO, USA) for donating the 4xARE reporter construct used in this study and Robin Miller for English editing.

Conflicts of Interest: J.L. and Y.S. are consultants for Lycored Ltd., Beer Sheva, Israel. J.L. and Y.S. received research funding from Lycored. All other authors declare no conflicts of interest. Lycored is a supplier to the dietary supplement and functional food industries worldwide. Lycored Ltd. had no role in the design of the study; in the collection, analyses, or interpretation of data; in the writing of the manuscript, or in the decision to publish the results.

References

1. Bickers, D.R.; Athar, M. Oxidative Stress in the Pathogenesis of Skin Disease. *J. Investig. Dermatol.* **2006**, *126*, 2565–2575. [CrossRef]
2. Valacchi, G.; Sticozzi, C.; Pecorelli, A.; Cervellati, F.; Cervellati, C.; Maioli, E. Cutaneous responses to environmental stressors. *Ann. N. Y. Acad. Sci.* **2012**, *1271*, 75–81. [CrossRef]
3. Lephart, E.D. Skin aging and oxidative stress: Equol's anti-aging effects via biochemical and molecular mechanisms. *Ageing Res. Rev.* **2016**, *31*, 36–54. [CrossRef]
4. Betteridge, D.J. What is oxidative stress? *Metabolism* **2000**, *49*, 3–8. [CrossRef]
5. Pizzino, G.; Irrera, N.; Cucinotta, M.; Pallio, G.; Mannino, F.; Arcoraci, V.; Squadrito, F.; Altavilla, D.; Bitto, A. Oxidative Stress: Harms and Benefits for Human Health. *Oxid. Med. Cell. Longev.* **2017**, *2017*, 8416763. [CrossRef]
6. Chen, K.; Lu, P.; Beeraka, N.M.; Sukocheva, O.A.; Madhunapantula, S.V.; Liu, J.; Sinelnikov, M.Y.; Nikolenko, V.N.; Bulygin, K.V.; Mikhaleva, L.M.; et al. Mitochondrial mutations and mitoepigenetics: Focus on regulation of oxidative stress-induced responses in breast cancers. *Semin. Cancer Biol.* **2020**. [CrossRef] [PubMed]
7. Fang, C.; Gu, L.; Smerin, D.; Mao, S.; Xiong, X. The Interrelation between Reactive Oxygen Species and Autophagy in Neurological Disorders. *Oxid. Med. Cell. Longev.* **2017**, *2017*, 8495160. [CrossRef]
8. Panth, N.; Paudel, K.R.; Parajuli, K. Reactive Oxygen Species: A Key Hallmark of Cardiovascular Disease. *Adv. Med.* **2016**, *2016*, 9152732. [CrossRef]
9. Redza-Dutordoir, M.; Averill-Bates, D.A. Activation of apoptosis signalling pathways by reactive oxygen species. *Biochim. Biophys. Acta (BBA)-Mol. Cell Res.* **2016**, *1863*, 2977–2992. [CrossRef] [PubMed]
10. Zhou, B.-R.; Yin, H.-B.; Xu, Y.; Wu, D.; Zhang, Z.-H.; Yin, Z.-Q.; Permatasari, F.; Luo, D. Baicalin protects human skin fibroblasts from ultraviolet A radiation-induced oxidative damage and apoptosis. *Free Radic. Res.* **2012**, *46*, 1458–1471. [CrossRef] [PubMed]
11. Rinnerthaler, M.; Bischof, J.; Streubel, M.K.; Trost, A.; Richter, K. Oxidative Stress in Aging Human Skin. *Biomolecules* **2015**, *5*, 545–589. [CrossRef]
12. Wang, L.; Lee, W.; Jayawardena, T.U.; Cha, S.H.; Jeon, Y.J. Dieckol, an algae-derived phenolic compound, suppresses airborne particulate matter-induced skin aging by inhibiting the expressions of pro-inflammatory cytokines and matrix metalloproteinases through regulating NF-κB, AP-1, and MAPKs signaling pathways. *Food Chem. Toxicol.* **2020**, *146*, 111823. [CrossRef] [PubMed]
13. Varani, J.; Spearman, D.; Perone, P.; Fligiel, S.E.; Datta, S.C.; Wang, Z.Q.; Shao, Y.; Kang, S.; Fisher, G.J.; Voorhees, J.J. Inhibition of type I procollagen synthesis by damaged collagen in photoaged skin and by collagenase-degraded collagen in vitro. *Am. J. Pathol.* **2001**, *158*, 931–942. [CrossRef]
14. Fam, V.W.; Charoenwoodhipong, P.; Sivamani, R.K.; Holt, R.R.; Keen, C.L.; Hackman, R.M. Plant Based Foods for Skin Health: A Narrative Review. *J. Acad. Nutr. Diet.* **2021**, in press. [CrossRef] [PubMed]
15. Michalak, M.; Pierzak, M.; Krecisz, B.; Suliga, E. Bioactive Compounds for Skin Health: A Review. *Nutrients* **2021**, *13*, 203. [CrossRef]
16. Sies, H.; Stahl, W. Nutritional protection against skin damage from sunlight. *Annu. Rev. Nutr.* **2004**, *24*, 173–200. [CrossRef]
17. Heo, H.; Lee, H.; Yang, J.; Sung, J.; Kim, Y.; Jeong, H.S.; Lee, J. Protective Activity and Underlying Mechanism of Ginseng Seeds against UVB-Induced Damage in Human Fibroblasts. *Antioxidants* **2021**, *10*, 403. [CrossRef]

18. Savoia, P.; Raina, G.; Camillo, L.; Farruggio, S.; Mary, D.; Veronese, F.; Graziola, F.; Zavattaro, E.; Tiberio, R.; Grossini, E. Anti-oxidative effects of 17 β-estradiol and genistein in human skin fibroblasts and keratinocytes. *J. Dermatol. Sci.* **2018**, *92*, 62–77. [CrossRef] [PubMed]

19. Park, C.; Lee, H.; Noh, J.S.; Jin, C.-Y.; Kim, G.-Y.; Hyun, J.W.; Leem, S.-H.; Choi, Y.H. Hemistepsin A protects human keratinocytes against hydrogen peroxide-induced oxidative stress through activation of the Nrf2/HO-1 signaling pathway. *Arch. Biochem. Biophys.* **2020**, *691*, 108512. [CrossRef]

20. Abu Zaid, M.; Afaq, F.; Syed, D.N.; Dreher, M.; Mukhtar, H. Inhibition of UVB-mediated oxidative stress and markers of photoaging in immortalized HaCaT keratinocytes by pomegranate polyphenol extract POMx. *Photochem. Photobiol.* **2007**, *83*, 882–888. [CrossRef]

21. Calniquer, G.; Khanin, M.; Ovadia, H.; Linnewiel-Hermoni, K.; Stepensky, D.; Trachtenberg, A.; Sedlov, T.; Braverman, O.; Levy, J.; Sharoni, Y. Combined Effects of Carotenoids and Polyphenols in Balancing the Response of Skin Cells to UV Irradiation. *Molecules* **2021**, *26*, 1931. [CrossRef]

22. Lim, S.-R.; Kim, D.-W.; Sung, J.; Kim, T.H.; Choi, C.-H.; Lee, S.-J. Astaxanthin Inhibits Autophagic Cell Death Induced by Bisphenol A in Human Dermal Fibroblasts. *Antioxidants* **2021**, *10*, 1273. [CrossRef] [PubMed]

23. Afaq, F.; Zaid, M.A.; Khan, N.; Dreher, M.; Mukhtar, H. Protective effect of pomegranate-derived products on UVB-mediated damage in human reconstituted skin. *Exp. Dermatol.* **2009**, *18*, 553–561. [CrossRef]

24. Lephart, E.D. A review of the role of estrogen in dermal aging and facial attractiveness in women. *J. Cosmet. Dermatol.* **2018**, *17*, 282–288. [CrossRef] [PubMed]

25. Chen, C.S.; Tseng, Y.T.; Hsu, Y.Y.; Lo, Y.C. Nrf2-Keap1 Antioxidant Defense and Cell Survival Signaling Are Upregulated by 17β-Estradiol in Homocysteine-Treated Dopaminergic SH-SY5Y Cells. *Neuroendocrinology* **2013**, *97*, 232–241. [CrossRef] [PubMed]

26. Ben-Dor, A.; Steiner, M.; Gheber, L.; Danilenko, M.; Dubi, N.; Linnewiel, K.; Zick, A.; Sharoni, Y.; Levy, J. Carotenoids activate the antioxidant response element transcription system. *Mol. Cancer* **2005**, *4*, 177–186.

27. Zhou, Y.; Jiang, Z.; Lu, H.; Xu, Z.; Tong, R.; Shi, J.; Jia, G. Recent Advances of Natural Polyphenols Activators for Keap1-Nrf2 Signaling Pathway. *Chem. Biodivers.* **2019**, *16*, e1900400. [CrossRef]

28. Linnewiel, K.; Ernst, H.; Caris-Veyrat, C.; Ben-Dor, A.; Kampf, A.; Salman, H.; Danilenko, M.; Levy, J.; Sharoni, Y. Structure activity relationship of carotenoid derivatives in activation of the electrophile/antioxidant response element transcription system. *Free Radic. Biol. Med.* **2009**, *47*, 659–667. [CrossRef]

29. Cooney, R.V.; Joseph Kappock, T.; Pung, A.; Bertram, J.S. Solubilization, cellular uptake, and activity of β-carotene and other carotenoids as inhibitors of neoplastic transformation in cultured cells. In *Methods in Enzymology*; Academic Press: Cambridge, MA, USA, 1993; Volume 214, pp. 55–68.

30. Levy, J.; Bosin, E.; Feldman, B.; Giat, Y.; Miinster, A.; Danilenko, M.; Sharoni, Y. Lycopene is a more potent inhibitor of human cancer cell proliferation than either alpha-carotene or beta-carotene. *Nutr. Cancer* **1995**, *24*, 257–266. [CrossRef]

31. Pesakhov, S.; Nachliely, M.; Barvish, Z.; Aqaqe, N.; Schwartzman, B.; Voronov, E.; Sharoni, Y.; Studzinski, G.P.; Fishman, D.; Danilenko, M. Cancer-selective cytotoxic Ca^{2+} overload in acute myeloid leukemia cells and attenuation of disease progression in mice by synergistically acting polyphenols curcumin and carnosic acid. *Oncotarget* **2016**, *7*, 31847–31861. [CrossRef]

32. Cullinan, S.B.; Zhang, D.; Hannink, M.; Arvisais, E.; Kaufman, R.J.; Diehl, J.A. Nrf2 is a direct PERK substrate and effector of PERK-dependent cell survival. *Mol. Cell. Biol.* **2003**, *23*, 7198–7209. [CrossRef]

33. Takasawa, R.; Nakamura, H.; Mori, T.; Tanuma, S. Differential apoptotic pathways in human keratinocyte HaCaT cells exposed to UVB and UVC. *Apoptosis* **2005**, *10*, 1121–1130. [CrossRef] [PubMed]

34. Pongcharoen, S.; Warnnissorn, P.; Lertkajornsin, O.; Limpeanchob, N.; Sutheerawattananonda, M. Protective effect of silk lutein on ultraviolet B-irradiated human keratinocytes. *Biol. Res.* **2013**, *46*, 39–45. [CrossRef]

35. Yoshihisa, Y.; Rehman, M.u.; Shimizu, T. Astaxanthin, a xanthophyll carotenoid, inhibits ultraviolet-induced apoptosis in keratinocytes. *Exp. Dermatol.* **2014**, *23*, 178–183. [CrossRef]

36. Fazekas, Z.; Gao, D.; Saladi, R.N.; Lu, Y.; Lebwohl, M.; Wei, H. Protective effects of lycopene against ultraviolet B-induced photodamage. *Nutr. Cancer* **2003**, *47*, 181–187. [CrossRef] [PubMed]

37. Giampieri, F.; Alvarez-Suarez, J.M.; Mazzoni, L.; Forbes-Hernandez, T.Y.; Gasparrini, M.; Gonzàlez-Paramàs, A.M.; Santos-Buelga, C.; Quiles, J.L.; Bompadre, S.; Mezzetti, B.; et al. Polyphenol-Rich Strawberry Extract Protects Human Dermal Fibroblasts against Hydrogen Peroxide Oxidative Damage and Improves Mitochondrial Functionality. *Molecules* **2014**, *19*, 7798–7816. [CrossRef]

38. Afaq, F.; Syed, D.N.; Malik, A.; Hadi, N.; Sarfaraz, S.; Kweon, M.H.; Khan, N.; Zaid, M.A.; Mukhtar, H. Delphinidin, an anthocyanidin in pigmented fruits and vegetables, protects human HaCaT keratinocytes and mouse skin against UVB-mediated oxidative stress and apoptosis. *J. Invest. Dermatol.* **2007**, *127*, 222–232. [CrossRef]

39. Li, L.-H.; Wu, L.-J.; Tashiro, S.-i.; Onodera, S.; Uchiumi, F.; Ikejima, T. Silibinin Prevents UV-Induced HaCaT Cell Apoptosis Partly through Inhibition of Caspase-8 Pathway. *Biol. Pharm. Bull.* **2006**, *29*, 1096–1101. [CrossRef] [PubMed]

40. Pérez-Sánchez, A.; Barrajón-Catalán, E.; Caturla, N.; Castillo, J.; Benavente-García, O.; Alcaraz, M.; Micol, V. Protective effects of citrus and rosemary extracts on UV-induced damage in skin cell model and human volunteers. *J. Photochem. Photobiol. B* **2014**, *136*, 12–18. [CrossRef]

41. Shin, S.W.; Jung, E.; Kim, S.; Lee, K.-E.; Youm, J.-K.; Park, D. Antagonist Effects of Veratric Acid against UVB-Induced Cell Damages. *Molecules* **2013**, *18*, 5405–5419. [CrossRef]

42. Park, J.-H.; Shin, J.-M.; Yang, H.-W.; Kim, T.H.; Lee, S.H.; Lee, H.-M.; Cho, J.-G.; Park, I.-H. Cigarette Smoke Extract Stimulates MMP-2 Production in Nasal Fibroblasts via ROS/PI3K, Akt, and NF-κB Signaling Pathways. *Antioxidants* **2020**, *9*, 739. [CrossRef] [PubMed]

43. Yi, R.; Zhang, J.; Sun, P.; Qian, Y.; Zhao, X. Protective Effects of Kuding Tea (Ilex kudingcha C. J. Tseng) Polyphenols on UVB-Induced Skin Aging in SKH1 Hairless Mice. *Molecules* **2019**, *24*, 1016. [CrossRef] [PubMed]

44. Yamawaki, Y.; Mizutani, T.; Okano, Y.; Masaki, H. Xanthophyll Carotenoids Reduce the Dysfunction of Dermal Fibroblasts to Reconstruct the Dermal Matrix Damaged by Carbonylated Proteins. *J. Oleo Sci.* **2021**, *70*, 647–655. [CrossRef]

45. Grether-Beck, S.; Marini, A.; Jaenicke, T.; Stahl, W.; Krutmann, J. Molecular evidence that oral supplementation with lycopene or lutein protects human skin against ultraviolet radiation: Results from a double-blinded, placebo-controlled, crossover study. *Br. J. Dermatol.* **2017**, *176*, 1231–1240. [CrossRef]

46. Rizwan, M.; Rodriguez-Blanco, I.; Harbottle, A.; Birch-Machin, M.A.; Watson, R.E.; Rhodes, L.E. Tomato paste rich in lycopene protects against cutaneous photodamage in humans in vivo: A randomized controlled trial. *Br. J. Dermatol.* **2011**, *164*, 154–162. [CrossRef] [PubMed]

47. Son, E.D.; Lee, J.Y.; Lee, S.; Kim, M.S.; Lee, B.G.; Chang, I.S.; Chung, J.H. Topical application of 17beta-estradiol increases extracellular matrix protein synthesis by stimulating tgf-Beta signaling in aged human skin in vivo. *J. Invest. Dermatol.* **2005**, *124*, 1149–1161. [CrossRef]

48. Liu, S.; Yang, S.-D.; Huo, X.-W.; Yang, D.-L.; Ma, L.; Ding, W.-Y. 17β-Estradiol inhibits intervertebral disc degeneration by down-regulating MMP-3 and MMP-13 and up-regulating type II collagen in a rat model. *Artif. Cells Nanomed. Biotechnol.* **2018**, *46*, 182–191. [CrossRef]

49. Bosch, R.; Philips, N.; Suarez-Perez, J.A.; Juarranz, A.; Devmurari, A.; Chalensouk-Khaosaat, J.; Gonzalez, S. Mechanisms of Photoaging and Cutaneous Photocarcinogenesis, and Photoprotective Strategies with Phytochemicals. *Antioxidants* **2015**, *4*, 248–268. [CrossRef]

50. Debacq-Chainiaux, F.; Leduc, C.; Verbeke, A.; Toussaint, O. UV, stress and aging. *Dermato-Endocrinology* **2012**, *4*, 236–240. [CrossRef]

51. Karthikeyan, R.; Kanimozhi, G.; Prasad, N.R.; Agilan, B.; Ganesan, M.; Mohana, S.; Srithar, G. 7-Hydroxycoumarin prevents UVB-induced activation of NF-κB and subsequent overexpression of matrix metalloproteinases and inflammatory markers in human dermal fibroblast cells. *J. Photochem. Photobiol. B Biol.* **2016**, *161*, 170–176. [CrossRef]

52. Linnewiel-Hermoni, K.; Motro, Y.; Miller, Y.; Levy, J.; Sharoni, Y. Carotenoid derivatives inhibit nuclear factor kappa B activity in bone and cancer cells by targeting key thiol groups. *Free Radic. Biol. Med.* **2014**, *75*, 105–120. [CrossRef]

53. Guruvayoorappan, C.; Kuttan, G. β-Carotene Inhibits Tumor-Specific Angiogenesis by Altering the Cytokine Profile and Inhibits the Nuclear Translocation of Transcription Factors in B16F-10 Melanoma Cells. *Integr. Cancer Ther.* **2007**, *6*, 258–270. [CrossRef]

54. Izumi-Nagai, K.; Nagai, N.; Ohgami, K.; Satofuka, S.; Ozawa, Y.; Tsubota, K.; Umezawa, K.; Ohno, S.; Oike, Y.; Ishida, S. Macular pigment lutein is antiinflammatory in preventing choroidal neovascularization. *Arter. Thromb. Vasc. Biol.* **2007**, *27*, 2555–2562. [CrossRef] [PubMed]

55. Kundu, J.K.; Shin, Y.K.; Kim, S.H.; Surh, Y.-J. Resveratrol inhibits phorbol ester-induced expression of COX-2 and activation of NF-κB in mouse skin by blocking IκB kinase activity. *Carcinogenesis* **2006**, *27*, 1465–1474. [CrossRef] [PubMed]

56. McMurray, R.W.; Ndebele, K.; Hardy, K.J.; Jenkins, J.K. 17-β-estradiol suppresses IL-2 and IL-2 receptor. *Cytokine* **2001**, *14*, 324–333. [CrossRef]

57. Sharoni, Y.; Linnewiel-Hermoni, K.; Zango, G.; Khanin, M.; Salman, H.; Veprik, A.; Danilenko, M.; Levy, J. The role of lycopene and its derivatives in the regulation of transcription systems: Implications for cancer prevention. *Am. J. Clin. Nutr.* **2012**, *96*, 1173S–1178S. [CrossRef]

58. Kosaka, K.; Mimura, J.; Itoh, K.; Satoh, T.; Shimojo, Y.; Kitajima, C.; Maruyama, A.; Yamamoto, M.; Shirasawa, T. Role of Nrf2 and p62/ZIP in the neurite outgrowth by carnosic acid in PC12h cells. *J. Biochem.* **2009**, *147*, 73–81. [CrossRef] [PubMed]

59. Itoh, K.; Tong, K.I.; Yamamoto, M. Molecular mechanism activating Nrf2-Keap1 pathway in regulation of adaptive response to electrophiles. *Free Radic. Biol. Med.* **2004**, *36*, 1208–1213. [CrossRef]

60. Jackson, R.L.; Greiwe, J.S.; Schwen, R.J. Ageing skin: Oestrogen receptor β agonists offer an approach to change the outcome. *Exp. Dermatol.* **2011**, *20*, 879–882. [CrossRef]

61. Borrás, C.; Gambini, J.; López-Grueso, R.; Pallardó, F.V.; Viña, J. Direct antioxidant and protective effect of estradiol on isolated mitochondria. *Biochim. Biophys. Acta* **2010**, *1802*, 205–211. [CrossRef]

62. Richardson, T.E.; Yang, S.H.; Wen, Y.; Simpkins, J.W. Estrogen protection in Friedreich's ataxia skin fibroblasts. *Endocrinology* **2011**, *152*, 2742–2749. [CrossRef] [PubMed]

63. Ayres, S.; Abplanalp, W.; Liu, J.H.; Subbiah, M.T.R. Mechanisms involved in the protective effect of estradiol-17β on lipid peroxidation and DNA damage. *Am. J. Physiol.-Endocrinol. Metab.* **1998**, *274*, E1002–E1008. [CrossRef] [PubMed]

 *antioxidants*

Article

Unraveling the Effects of Carotenoids Accumulation in Human Papillary Thyroid Carcinoma

Alessandra di Masi [1,*], Rosario Luigi Sessa [1], Ylenia Cerrato [1], Gianni Pastore [2], Barbara Guantario [2], Roberto Ambra [2], Michael Di Gioacchino [1], Armida Sodo [1], Martina Verri [3], Pierfilippo Crucitti [4], Filippo Longo [4], Anda Mihaela Naciu [5], Andrea Palermo [5], Chiara Taffon [3], Filippo Acconcia [1], Fabrizio Bianchi [6], Paolo Ascenzi [1], Maria Antonietta Ricci [1] and Anna Crescenzi [3]

[1] Department of Sciences, Roma Tre University, 00146 Rome, Italy; ros.sessa1@stud.uniroma3.it (R.L.S.); yle.cerrato@stud.uniroma3.it (Y.C.); michael.digioacchino@uniroma3.it (M.D.G.); armida.sodo@uniroma3.it (A.S.); filipo.acconcia@uniroma3.it (F.A.); paolo.ascenzi@uniroma3.it (P.A.); mariaantonietta.ricci@uniroma3.it (M.A.R.)

[2] CREA (Council for Agricultural Research and Economics), Research Centre for Food and Nutrition, 00178 Rome, Italy; giovanni.pastore@crea.gov.it (G.P.); barbara.guantario@crea.gov.it (B.G.); roberto.ambra@crea.gov.it (R.A.)

[3] Pathology Unit, Fondazione Policlinico Universitario Campus Bio-Medico, 00128 Rome, Italy; m.verri@policlinicocampus.it (M.V.); c.taffon@unicampus.it (C.T.); a.crescenzi@policlinicocampus.it (A.C.)

[4] Unit of Thoracic Surgery, Fondazione Policlinico Universitario Campus Bio-Medico, 00128 Rome, Italy; p.crucitti@unicampus.it (P.C.); filippo.longo@policlinicocampus.it (F.L.)

[5] Unit of Metabolic Bone and Thyroid Disorders, Fondazione Policlinico Universitario Campus Bio-Medico, 00128 Rome, Italy; a.naciu@policlinicocampus.it (A.M.N.); a.palermo@unicampus.it (A.P.)

[6] Cancer Biomarkers Unit, Fondazione IRCCS Casa Sollievo della Sofferenza, 71013 San Giovanni Rotondo, FG, Italy; f.bianchi@operapadrepio.it

* Correspondence: alessandra.dimasi@uniroma3.it; Tel.: +39-06-57336363

Citation: di Masi, A.; Sessa, R.L.; Cerrato, Y.; Pastore, G.; Guantario, B.; Ambra, R.; Di Gioacchino, M.; Sodo, A.; Verri, M.; Crucitti, P.; et al. Unraveling the Effects of Carotenoids Accumulation in Human Papillary Thyroid Carcinoma. *Antioxidants* 2022, *11*, 1463. https://doi.org/10.3390/antiox11081463

Academic Editors: Soliman Khatib and Dana Atrahimovich Blatt

Received: 18 June 2022
Accepted: 20 July 2022
Published: 27 July 2022

Publisher's Note: MDPI stays neutral with regard to jurisdictional claims in published maps and institutional affiliations.

Abstract: Among the thyroid cancers, papillary thyroid cancer (PTC) accounts for 90% of the cases. In addition to the necessity to identify new targets for PTC treatment, early diagnosis and management are highly demanded. Previous data indicated that the multivariate statistical analysis of the Raman spectra allows the discrimination of healthy tissues from PTC ones; this is characterized by bands typical of carotenoids. Here, we dissected the molecular effects of carotenoid accumulation in PTC patients by analyzing whether they were required to provide increased retinoic acid (RA) synthesis and signaling and/or to sustain antioxidant functions. HPLC analysis revealed the lack of a significant difference in the overall content of carotenoids. For this reason, we wondered whether the carotenoid accumulation in PTC patients could be related to vitamin A derivative retinoic acid (RA) biosynthesis and, consequently, the RA-related pathway activation. The transcriptomic analysis performed using a dedicated PCR array revealed a significant downregulation of RA-related pathways in PTCs, suggesting that the carotenoid accumulation in PTC could be related to a lower metabolic conversion into RA compared to that of healthy tissues. In addition, the gene expression profile of 474 PTC cases previously published in the framework of the Cancer Genome Atlas (TGCA) project was examined by hierarchical clustering and heatmap analyses. This metanalysis study indicated that the RA-related pathways resulted in being significantly downregulated in PTCs and being associated with the follicular variant of PTC (FV-PTC). To assess whether the possible fate of the carotenoids accumulated in PTCs is associated with the oxidative stress response, the expression of enzymes involved in ROS scavenging was checked. An increased oxidative stress status and a reduced antioxidant defense response were observed in PTCs compared to matched healthy thyroids; this was possibly associated with the prooxidant effects of high levels of carotenoids. Finally, the DepMap datasets were used to profile the levels of 225 metabolites in 12 thyroid cancer cell lines. The results obtained suggested that the high carotenoid content in PTCs correlates with tryptophan metabolism. This pilot provided novel possible markers and possible therapeutic targets for PTC diagnosis and therapy. For the future, a larger study including a higher number of PTC patients will be necessary to further validate the molecular data reported here.

Keywords: antioxidant; carotenoid; papillary thyroid cancer; retinoic acid

1. Introduction

In the past 30 years, the incidence of thyroid cancer (TC) has significantly increased, making it the fastest rising malignant solid tumor worldwide [1]. Among the TCs, papillary thyroid cancer (PTC) accounts for 90% of thyroid cancer cases [2]. In addition to the even more urgent necessity to discover new biological and therapeutic targets for PTC treatment, early diagnosis and management represent key points required for the improvement of patient prognosis. Indeed, the trauma caused by surgery and the recurrence of the tumor are still intractable issues, bringing a certain life and economic burden to patients [3]. Therefore, it is of great significance to further explore the pathogenesis of PTC and to find molecular targets with the potential for early diagnosis, prevention, and treatment. In recent years, Raman spectroscopy (RS) has emerged as a powerful tool for TC diagnosis. Indeed, our group has reported that the multivariate statistical analysis of RS spectra allows the discrimination of healthy from thyroid cancerous tissues and makes it possible to distinguish between the most common TC subtypes [4,5]. In particular, RS has allowed the identification of bands typical of carotenoids in PTC cells, suggesting that the presence of these molecules could represent a specific fingerprint of this type of TC and consequently a potential candidate for differential diagnosis [4–6].

Carotenoids are isoprenoids naturally occurring in fruits and vegetables and are synthesized in plants and microorganisms [7,8]. This wide family of C40 molecules has important functions not only in carotenoid-producing organisms, but also in animals that absorb these molecules in their diets. The primary carotenoids found in human plasma are α-carotene, β-carotene, γ-cryptoxanthin, lycopene, and lutein [9,10]. All these carotenoids display an antioxidant activity, and among them, α-carotene, β-carotene, and γ-cryptoxanthin are also precursors of vitamin A in animals [10]. Besides being essential for vision, the vitamin A derivative retinoic acid (RA) represents in vertebrates a major signal, controlling a wide range of biological processes thanks to its ability to bind two classes of nuclear receptors, i.e., the retinoic acid receptor (RAR) and the retinoid X receptor (RXR) [11–13]. RAR is part of the RAR/RXR heterodimer that binds DNA regulatory sequences, its RA-induced signaling regulating the transcription of genes involved in an array of biological processes such as pattern formation during embryonic development, cell differentiation, and control of certain metabolic activities [13].

Here, we aim to determine in 11 PTC patients whether the carotenoid accumulated in this type of TC, as reported by RS analyses [4–6], is metabolized to RA or, rather, is used by cancer cells to support antioxidant functions. Considering the low availability of patients to be enrolled in this study, this was intended as a pilot study, providing a molecular fingerprint of PTC useful for the identification of novel possible markers for diagnosis and therapy.

2. Materials and Methods

2.1. Study Enrollment

We screened and enrolled subjects with thyroid nodular pathologies who were referred to the thyroid outpatient clinic of the Metabolic Bone and Thyroid Disorders Unit of Fondazione Policlinico Universitario Campus Bio-Medico (Rome, Italy) between January 2018 and January 2021. Our protocol adhered to the Declaration of Helsinki and to the International Conference on Harmonization Good Clinical Practice, receiving the approval of the local ethics committees. The participants were recruited from and were managed at Fondazione Policlinico Universitario Campus Bio-Medico (Rome, Italy), all of them granting written informed consent that allowed the use of their anonymized information for data analysis. The patient eligibility criteria were as follows: (i) $\geq$18 years old; (ii) one or more thyroid nodules with a medium-high ultrasound risk of malignancy (Thyroid

Imaging—Reporting and Data System (TI-RADS) score ≥3) [14]; (iii) at least one thyroid nodule with a fine needle aspiration (FNA) report categorized as indeterminate, suspicious, or malignant and warranting thyroid surgery according to clinical guidelines [15]; (iv) final histological diagnosis of PTC; and (v) willingness to provide material for RS and carotenoid analysis. Once thyroid histology was reported and RS analysis performed, only those subjects with material available for the transcriptomic and proteomic analyses were included in the final analysis (Table 1).

Table 1. List of all the patients investigated, along with clinical information. US Classification was conducted according to [14]. FA Oxy, follicular thyroid adenoma oncocytic variant; FC, follicular thyroid carcinoma; FC Oxy, follicular thyroid carcinoma oncocytic variant; FV-PTC, follicular variant-papillary thyroid carcinoma; n.a., not available; PTC, papillary thyroid carcinoma; PTC Oxy, papillary thyroid carcinoma oncocytic variant; TI-RADS, Thyroid Imaging—Reporting and Data System.

Sample	Gender	Age at Diagnosis	US Classification (TI-RADS)	Histological Diagnosis	Raman Cluster
TIR46	F	42	5	PTC	PTC
TIR47	M	52	4	FC Oxy	FC
TIR48	F	43	4	PTC Oxy	PTC
TIR49	F	45	4	PTC	PTC
TIR50	F	30	5	PTC	PTC
TIR54	F	36	4	PTC	n.a.
TIR68	F	24	4	PTC	PTC
TIR70	F	48	4	PTC	PTC
TIR77	M	45	4	PTC	FV-PTC
TIR94	M	50	5	PTC Hobnail	PTC
TIR99	M	74	3	FA Oxy	PTC

2.2. Patients' Evaluation

All subjects were submitted to thyroid US evaluation, using a frequency range of 10–12 MHz on a MyLab 50 (Esaote, Genova, Italy). The US scans of the thyroid gland and neck area were performed by 2 experienced endocrinologists at the Endocrinology Unit at Fondazione Policlinico Universitario Campus Bio-Medico (Rome, Italy). Nodules were classified according to the American College of Radiology (ACR) TI-RADS risk stratification criteria [14] without prior knowledge of the cytological results. Patients harboring one or more thyroid nodules of medium-high malignancy risk by US (TI-RADS score ≥ 3) and submitted to thyroid FNA represented the criteria for patient selection. Once thyroid FNA cytology was reported, only those subjects with at least one nodule categorized as indeterminate, suspicious, or malignant, with a formal indication of thyroid surgery according to the international guidelines [15], and a final histological diagnosis of PTC were enrolled in the study. Histological diagnosis was reported in agreement with the current edition of the World Health Organization (WHO) classification of endocrine tumors [16]. Because of the small size of the available tumor tissues explanted from patients, it was not always possible to perform all the transcriptomic and proteomic analyses on the same samples.

2.3. Raman Spectroscopy

Raman spectra were collected using a Renishaw InVia Micro-Raman spectrometer, equipped with a solid-state diode laser source at 532 nm (nominal output power of 100 mW) and a confocal microscope (Leica 50× Long Working Distance objective), used both to focalize the incident laser beam and to collect the back-scattered light. Elastically scattered light is rejected by using a holographic edge filter, while inelastically scattered intensity is dispersed by a diffraction grating (1800 grooves/mm) on a Peltier cooled 1024 × 256 pixel Charge-Coupled Device (CCD) detector. The final instrumental resolution is of the order of 1 cm^{-1}. To prevent photo damage, the laser power at the sample was controlled by neutral density filters. The spectra were collected in the 100–3600 cm^{-1} range. On each sample, 5 measurement points were selected and 5 scans per point were accumulated for a total integration time of 50 s. The experimental conditions and data acquisition were controlled by Wire (Renishaw) software, which also allowed the preliminary data reduction (e.g., background and fluorescence subtraction). The MATLAB and ORIGIN 9.0 software were used to analyze the data.

2.4. Extraction of Carotenoids from the Thyroid Tissue

The carotenoids extraction was performed according to the procedure proposed by Peng and colleagues [17]. Briefly, healthy and tumoral thyroids were weighted and incubated in 560 µL of phosphate buffered saline solution containing 1 mg of butylated hydroxytoluene (BHT, Merck KGaA, Darmstadt, Germany) and 70 µL of a collagenase (50 mg/mL; Carlo Erba, Milano, Italy) at 37 °C for 1 h using a shaker (500 rpm). After a manual homogenization using glass potters, 70 µL proteinase K (20 mg/mL; Merck KGaA) was added to the samples, which were further incubated at 37 °C for 30 min using a shaker. After vortexing for 1 min, the lysate was mixed with 700 µL of an ethanolic buffer containing 1% SDS and 0.1% BHT. The solution was vortexed with n-hexane (Merck KGaA) and then centrifugated at 13,000 rpm for 30 s. The extraction with n-hexane was repeated twice, and the supernatants were then collected and dried under a nitrogen atmosphere. The extracts were solubilized in 120 µL methanol:chloroform 1:1, 80 µL of which were immediately injected into the High Performance Liquid Chromatography (HPLC). All the procedures were performed in the dark, under red light.

2.5. High Performance Liquid Chromatography (HPLC)

Carotenoids were separated following the method from [18] with minor modifications, using a YMC C30 column (250 mm × 4.6 mm i.d., 5 µm particle size) (YMC, Kyoto, Japan) kept at 35 °C and a flow of 1 mL/min. The mobile phase consisted of: (A) methanol, and (B) tert-butyl methyl ether. The following gradient was used (35 min): 90% A for 2 min, followed by a linear gradient to 80% A for 8 min and then to 30% A for 10 min, which was then maintained for further 15 min. The photodiode array was set at 450 and 472 nm. α-carotene, β-carotene, lutein, and lycopene were recognized and quantified by comparison with analytical standards (Merck KGaA). Unidentified peaks were quantified using the calibration curve of β-carotene. To verify the quantification reliability of using a few tens of mg of tissue, the method was verified by measuring the recovery of carotenoids from tissues spiked with known concentrations of β-carotene (the recovery rate was 96 ± 10%) and by performing up to four serial repetitions of n-hexane extraction (the extractions beyond the second did not yield detectable β-carotene).

2.6. Total RNA Extraction

Total RNA was extracted from the healthy and pathological lobes of each patient. RNA was obtained from two formalin-fixed paraffin-embedded (FFPE) tissue sections of 5 µm thickness each, using the RNAeasy® FFPE kit (Qiagen, Hilden, Germany) and following the manufacturer's instructions. Briefly, sections were deparaffinized using heptane and methanol. After centrifugation, pellets were air-dried and then resuspended into PKD buffer (Qiagen) containing 10 µL proteinase K (>600 mAU/mL; provided with the kit). After a first incubation at 56 °C for 15 min and a second incubation at 80 °C for 15 min, samples were put on ice for 3 min and finally centrifuged at 13,000 rpm for 15 min. The DNA was removed by incubation of the collected supernatant with DNAse for 15 min. After the addition of RBC buffer, samples were loaded on a RNeasy MinElute spin column and centrifuged. Finally, RNA was eluted using 30 µL RNase-free water, and quantified with Nanodrop 2000 (Thermo Fisher Scientific, Waltham, MA, USA).

2.7. Transcriptomic Analysis Using the PCR Array "Human Retinoic Acid Pathway"

Three micrograms of total RNA were reverse transcribed using the RT2 First Strand Kit (Qiagen), following the manufacturer's instructions. The expression of 96 genes was analyzed using the RT2 Profiler "Human Retinoic Acid Pathway" PCR Array 96-well format (cat no. 330231 PAHS-180Z; Qiagen) in combination with the RT2 SYBR Green PCR Master Mix (Qiagen), according to the manufacturer's protocol. Of these 96 genes, 84 genes belong to the human RA-related pathways, and 12 genes are either housekeeping genes (i.e., ß-actin (ACTB), ß-2-microglobulin (B2M), glyceraldehyde-3-phosphate dehydrogenase (GAPDH), hypoxanthine phosphoribosyltransferase 1 (HPRT1), and ribosomal protein lateral stalk subunit P0 (RPLP0)) or internal controls (i.e., human genomic DNA contamination, reverse transcription control, positive PCR control). The AriaMx Real-time PCR system (Applied Biosystems, Waltham, MA, USA) was used and the three steps of the cycling program were: 95 °C for 10 min for 1 cycle, followed by 40 cycles of 95 °C for 15 s and 60 °C for 1 min. The expression levels of each gene were quantified and reported as relative quantity (RQ) with respect to the calibrators represented by the housekeeping genes, according to the $2^{-\Delta\Delta Ct}$ method. Data analyses were performed using the Agilent AriaMx 1.71 software (Thermo Fisher Scientific) and the GeneGlobe software (Qiagen). The analysis of the differentially expressed genes (DEGs) between the healthy and PTC thyroids was performed using Venn diagrams (Bioinformatics & Evolutionary Genomics, http://bioinformatics.psb.ugent.be/webtools/Venn/; accessed on 13 January 2022). Visualization, interpretation, and analysis of the biomolecular pathways were performed using Reactome (https://reactome.org; accessed on 13 January 2022), a free, open-source, open-data, curated, and peer-reviewed knowledgebase [19].

2.8. Quantitative Real-Time Reverse Transcription–Polymerase Chain Reaction (qRT-PCR)

Total RNA was reverse transcribed and amplified with a GoTaq 2-step RT-qPCR system (Promega, Madison, WI, USA), following the manufacturer's instructions. The cDNA obtained was then amplified for the following transcripts: CRABP2 (forward: 5′-AGTGTCCAGTGCTCCAGCCTA-3′; reverse: 5′-CTGCAGCCACAGCAATCTTC-3′), CYP26B1 (forward: 5′-AGCTAGTGAGCACCGAGTGG-3′; reverse: 5′-GGGCAGGTAGCTCTCAAGTG-3′), DHRS3 (forward: 5′-CGTTGCTGGCAATCAGATCG-3′; reverse: 5′-CGCGGTTTCAAAGTGCAAGA-3′), RARγ (forward: 5′-CAGAGCAGCAGTTCTGAAGAGATA-3′; reverse: 5′-GACACGTGTACACCATGTTCTTCT-3′), RDH10 (forward: 5′-TGGGACATCAACACGCAAAGC-3′; reverse: 5′-TGCAAGTTACAGTGGGGCAGA-3′), RET (forward: 5′-CGCGACCTGCGCAAA-3′; reverse: 5′-CAAGTTCTTCCGAGGGAATTCC-3′), RXRβ (forward: 5′-AGCTCCCCCAGGATTCTC-3′; reverse: 5′-CAGGGAGTGACACTGTTGAGTTA-3′). The amplifications were performed using the SYBR Green system and the Rotor Gene RG-6000 Real-Time Thermocycler (Corbett Research, Qiagen GmbH), using the following thermal cycling conditions: 95 °C for 2 min followed by 40 cycles at 95 °C for 15 s and 60 °C for 1 min. The data were reported as RQ with respect to a calibrator sample (i.e., ACTB), according to the $2^{-\Delta\Delta Ct}$ method.

2.9. Biochemical Analysis

The frozen thyroids were weighted, homogenized, and sonicated in lysis buffer (20 mM Tris-HCl pH 8.0, 137 mM NaCl, 10 mM EDTA, 10% glycerol (*v/v*), 1% Triton™ X-100 (*v/v*), and protease inhibitors). Fifteen micrograms of protein extracts, previously quantified with the Bradford assay (Bio-Rad Laboratories, Hercules, CA, USA), were resolved by SDS-PAGE and transferred onto polyvinylidene fluoride (PVDF) membranes (Bio-Rad Laboratories). As reported elsewhere [20], after blocking with 3% BSA (*w/v*) dissolved in TBS buffer/0.5% Tween-20 (*v/v*), the membranes were probed overnight at 4 °C with the following antibodies purchased from Santa Cruz Biotechnology (Dallas, TX, USA): anti-catalase (CAT; sc-365738), anti-glutathione peroxidase-4 (GPx-4; sc-166570), anti-heme oxygenase-1 (HO-1; sc-136960), anti-NADPH oxidase 4 (NOX4; sc-518092), anti-NAD(P)H quinone dehydrogenase 1(NQO1; sc-32793), anti-oxoguanine glycosylase 1 (OGG1; sc-

376935), anti-superoxide dismutase-1 (SOD-1; sc-271014), anti-vinculin (sc-73614), anti-ß-actin (sc-47778), and anti-γH2AX (sc-517348). The membranes were then incubated for 1 h at room temperature with an anti-mouse HRP-conjugated secondary antibody (Bio-Rad Laboratories). All the experiments were technically repeated at least twice. Blots were acquired and processed using the ChemiDoc™ Imaging system (Bio-Rad Laboratories) [21]. Protein quantification was performed using the Image Lab software (version 6.0.1, Bio-Rad Laboratories).

2.10. Immunohistochemistry Analysis

At the time of surgery, the removed specimens were immediately submitted unfixed to the Pathology Unit at Fondazione Policlinico Universitario Campus Bio-Medico (Rome, Italy) in an appropriately labelled container. After gross evaluation, the specimens were formalin fixed, sampled, and processed for paraffin embedding. Four-micron sections from the paraffin block of the formalin-fixed specimens were used for immunohistochemical stain. Immunohistochemistry was performed with an automatized instrument (Dako Omnis, Agilent, Santa Clara, CA, USA) using an anti-SR-B1 rabbit monoclonal (clone EP1556Y, Abcam, Cambridge, UK) and revealed with peroxidase development. Negative controls were obtained by omitting the primary antibody; the positive control consisted of liver tissue. The thyroid lesions were subjected for SR-B1 immunohistochemistry. For each lesion, the section comprised both pathological tissue and normal parenchyma. Immuno-histochemical evaluation was conduct in a blind way by two pathologists. Images were acquired using a Nikon Eclipse Ni microscope equipped with a Nikon Digital Camera DS-Fi3. The original magnification was 20× (Nikon, Tokyo, Japan).

2.11. Gene Expression Analysis of External Database Repository

Gene expression normalized data (RNAseq) of the 474 PTC cases of the Cancer Genome Atlas (TGCA) project were downloaded from cBioPortal (https://www.cbioportal.org/datasets; accessed on 26 January 2022) [22]. The data were trimmed +1 to eliminate 0 values before statistical analyses. The gene expression data were median centered and log2 transformed. Hierarchical clustering and heatmap analyses were performed using Cluster 3.0 for Mac OS X (C Clustering Library 1.56; http://bonsai.hgc.jp/~mdehoon/software/cluster/software.htm; accessed on 26 January 2022) and Java Tree View (Version 1.1.6r4; http://jtreeview.sourceforge.net; accessed on 26 January 2022). The uncentered correlation and centroid linkage clustering method was used. Statistical analyses (including contingency plot) were performed using JMP 16 (SAS Institute S.R.L., Milano, Italy).

2.12. Statistical Analysis

The data were analyzed using GraphPad Prism 6 (version 6.01, GraphPad Software Inc., San Diego, CA, USA). The data were expressed as mean values ±standard deviations (SD). The statistical significance of differences was tested using the Student's two-tailed *t*-test (df: degree factor). p values of less than 0.05 were considered statistically significant.

3. Results

As previously reported, the Raman spectra of PTC tissues differ from those collected from both healthy thyrocytes and cells affected by other types of cancer [4]. In Supplementary Figure S1 is reported an exemplificative Raman spectrum derived from the TIR48 PTC patient. In the fingerprint region, the Raman spectra of PTC tissues are dominated, compared to the healthy counterpart, by intense bands at 1003, 1155, and 1515 cm^{-1}. These bands are due to the resonance effect of free carotenoids when excited at 532 nm. Other bands at 2155, 2301, and 2652 cm^{-1}, also related to carotenoids, are visible, along with bands at 1444, 1655, and 2818 cm^{-1}. The latter three bands are ascribed to fatty acid vibrational modes, namely CH$_2$ wagging, CC asymmetric stretching, and CH$_2$ symmetric stretching, respectively. With the aim of dissecting the molecular effects of carotenoid accumulation in PTC patients, here we performed a combination of analyses (i.e., HPLC, transcriptomic, oxidative stress analysis, and metanalysis) to clarify whether carotenoids are required to provide increased retinoic acid (RA) synthesis and signaling and/or to sustain antioxidant functions.

3.1. Determination of Carotenoids Content in Healthy and PTC Thyroids by HPLC

The presence of carotenoids in PTC patients was corroborated using the HPLC analyses of thyroids (Figure 1A). α-carotene, β-carotene, lutein, and lycopene were detected in the thyroids of all the four patients analyzed, together with five unidentified peaks attributable to molecules with absorbances measured at 450 and 472 nm (Figure 1B and Supplementary Table S1). The obtained results indicated that β-carotene (139.5 ± 67.6 and 147.0 ± 50.6 ng/g in healthy lobes and PTCs, respectively) is the most prevalent carotenoid found in healthy and PTCs thyroid lobes of the four analyzed patients (Figure 1B). Overall, for all the identified carotenoids, no significant differences were found between healthy thyroid lobes and the PTC counterparts of the analyzed patients (Figure 1B). Notably, no significant differences were scored between the total amount of the identified carotenoids (i.e., α-carotene, β-carotene, lutein, and lycopene) (242.5 ± 100.2 and 253.4 ± 87.6 ng/g in healthy lobes and PTCs, respectively) and the unidentified carotenoids (i.e., unknown 1 to 7) (217.1 ± 86.3 and 217.7 ± 53.9 ng/g in healthy lobes and PTCs, respectively) (Figure 1B and Supplementary Table S1).

A

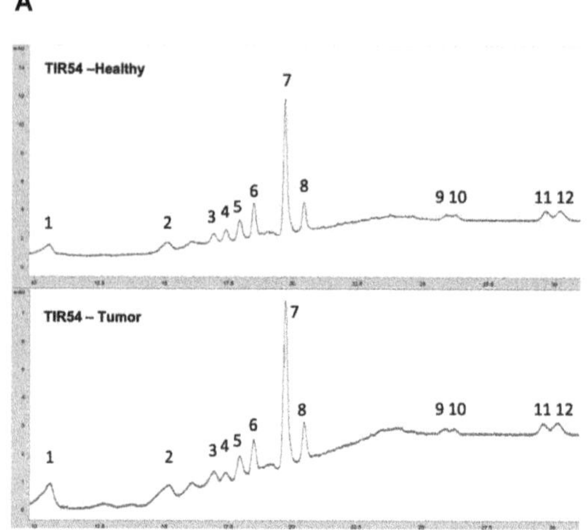

B

	Healthy *n = 4* (ng/g)	Tumor *n = 4* (ng/g)
α-carotene	36.5 ± 19.4	29.3 ± 16.7
β-carotene	139.5 ± 67.6	147.0 ± 50.6
Lutein	20.7 ± 1.08	32.4 ± 17.9
Lycopene	45.7 ± 25.2	44.6 ± 22.4
Total amount identified peaks	242.5 ± 100.2	253.4 ± 87.6
Unknown 1	31.0 ± 34.2	22.4 ± 19.4
Unknown 2	41.5 ± 32.8	55.0 ± 33.9
Unknown 3	29.4 ± 10.3	25.0 ± 10.4
Unknown 4	40.6 ± 5.3	39.4 ± 5.9
Unknown 5	40.0 ± 8.2	38.6 ± 5.3
Unknown 6	19.4 ± 5.0	19.2 ± 8.1
Unknown 7	15.2 ± 3.9	18.2 ± 5.7
Total amount unidentified peaks	217.1 ± 86.3	217.7 ± 53.9
Total amount carotenoids	**459.6 ± 93.3**	**471.2 ± 70.7**

Figure 1. HPLC profiles of carotenoids recorded in the healthy and tumoral thyroid lobes of PTC patients. (**A**) Representative carotenoids HPLC chromatogram: (1) lutein; (2) unknown 1; (3) unknown 2; (4) unknown 3; (5) unknown 4; (6) α-carotene; (7) β-carotene; (8) unknown 5; (9) unknown 6; (10) unknown 7; (11 and 12) lycopene. (**B**) Mean values of α-carotene, β-carotene, lutein, and lycopene concentrations (ng/g of thyroid tissue) quantified in the healthy and pathological lobes of the thyroids derived from four PTC patients (i.e., TIR22, TIR54, TIR77, and TIR94), together with seven unidentified peaks attributable to molecules with absorbances measured at 450 and 472 nm.

Of note, the carotenoids are Raman resonant only as free molecules and not when they are bound to other biomolecules, and HPLC measures the total amount of carotenoids irrespective of their binding status. Therefore, the discrepancy between the Raman and the HPLC analyses could be ascribed to the free/bound state of carotenoids in healthy versus PTC samples.

3.2. Evaluation of Carotenoids Metabolism in Healthy and PTC Thyroids

To further investigate the reason why the Raman spectra showed a marked increase in the band corresponding to free carotenoids and considering that β-carotene was the prevalent compound among the carotenoids identified by HPLC in the four analyzed patients, we tested the possibility that the different distributions of carotenoids between the healthy and the PTC counterparts could be due to a different metabolism of provitamin A carotenoids and more in the detail of the RA biosynthesis and related metabolism.

The gene expression profile related to the human RA pathway was performed in matched healthy and PTC counterparts by the RT^2 Profiler PCR Array "Human Retinoid Acid Pathway". The Supplementary Table S2 shows the complete list of the genes significantly modulated (fold change (FC) | >1.5 |) in each analyzed patient (i.e., TIR48, TIR50, TIR70, and TIR94). Supplementary Figures S2 and S3 show the heatmaps and the scatter plot, respectively, of the gene expression profile in each patient. Overall, compared to the control, 23 differentially expressed genes (DEGs) were found significantly modulated in all four patients and further 27 DEGs were found modulated in at least three patients (FC | >1.5 |) (Figure 2A). Of these, five genes were downregulated in all the four analyzed patients and 18 were downregulated in at least three out of the four patients (Figure 2B). Only one gene was commonly upregulated in three out of four PTC patients (Figure 2C).

To validate the RT^2 Profiler PCR Array results, RT-qPCR and immunoblot experiments were performed on selected DEGs. CRABP2, CYP26B1, DHRS3, RARγ, RDH10, RET, and RXRß were validated by RT-qPCR in the same patients tested by the RT^2 Profiler PCR Arrays (Figure 2D). In addition, the ALDH1A, CRABP2, DH10, and RARα protein levels were checked by immunoblots performed using the healthy and pathological lobes of the thyroids of three further PTC patients (i.e., TIR45, TIR46, TIR48) (Figure 2E). The results obtained confirmed the downregulation of selected DEGs in the PTC lobes compared to the matched healthy ones.

To perform an unsupervised discovery analysis of the pathways significantly modulated because of gene expression modulation in the analyzed patients, we used the Reactome software. The obtained results showed that both the overall significantly modulated DEGs as well as the significantly downregulated DEGs were enriched in several biological pathways, as reported in Supplementary Tables S3 and S4. In particular, "Signaling by RA", "RA biosynthesis pathway", "Signaling by nuclear receptors", "BMAL1:CLOCK, NPAS2 activates circadian gene expression", and two pathways related to HOX genes expression ("Activation of anterior HOX genes in hindbrain development during early embryogenesis" and "Activation of HOX genes during differentiation") were significantly downregulated and displayed low false discovery rate (FDR) values in PTCs compared to their relative counterparts (Supplementary Table S4). Protein–protein interaction networks involving the DEGs significantly modulated in the PTC patients were identified using the STRING database ([23]). A PPI network with 22 interaction pairs of the DEGs was identified (Figure 2F).

Mammalian scavenger receptors class B type 1 (SR-B1) expression is essential for the cellular uptake of carotenoids [24,25]. The immunohistochemistry analysis of the SR-B1 receptor showed that all the examined lesions, as well as the normal parenchyma, did not show any SR-B1 expression in thyrocytes (Supplementary Figure S4). This suggests that the increase in carotenoids levels as detected by RS [4,5] was not due to an increased uptake but possibly by a different carotenoid metabolism between the healthy and PTC lobes.

Overall, the obtained results indicate that despite HPLC analysis showed a comparable content of carotenoids in heathy and matched PTC thyroid lobes, the precursors of vitamin A in animals (i.e., α-carotene, β-carotene, and γ-cryptoxanthin) were converted to RA to a lower extent in PTC compared to healthy matched lobes [10]. This may explain the higher levels of free carotenoids detected by RS in PTC compared to the healthy counterparts [4,5], also considering that we did not observe an increased expression of the carotenoids receptor SR-B1.

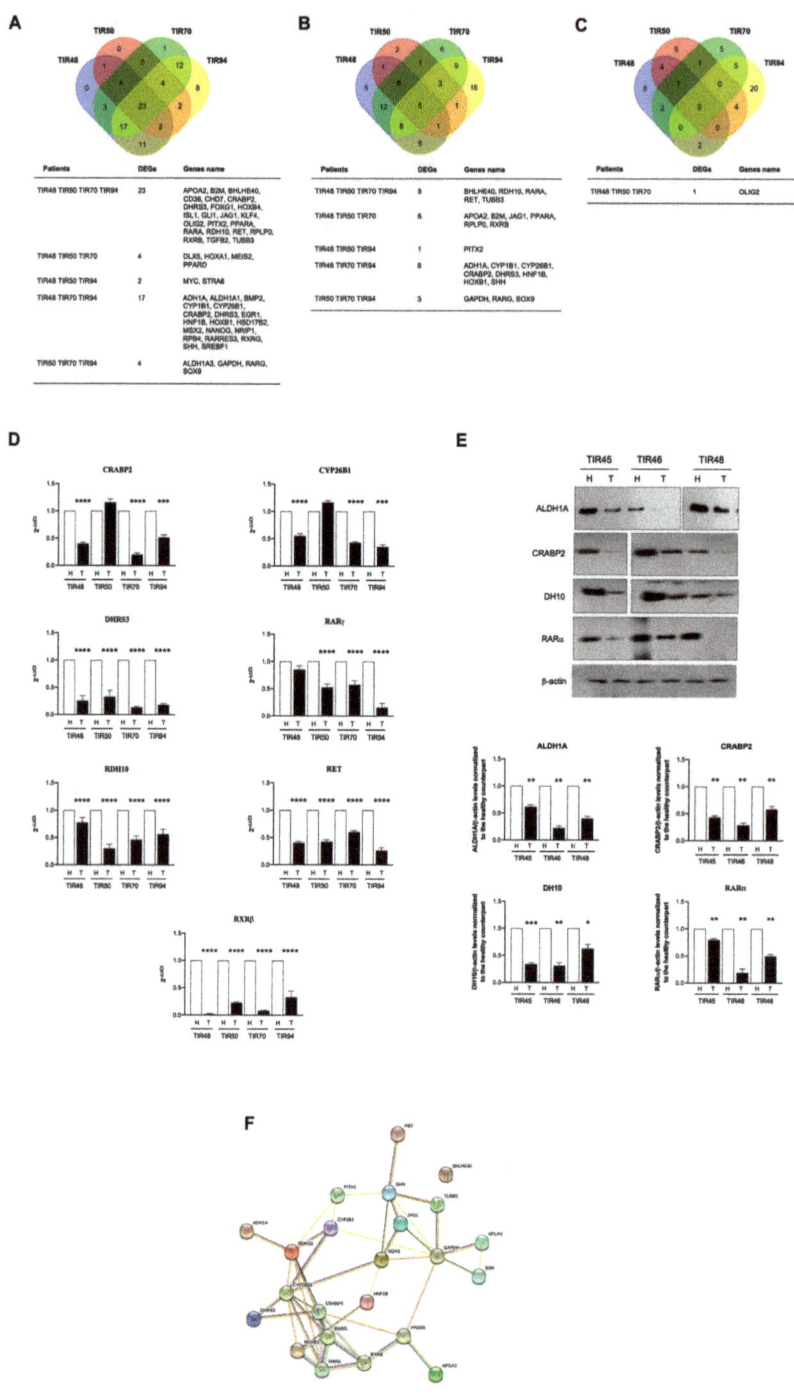

Figure 2. Transcriptome data analysis derived from PTC patients. Four PTC patients (i.e., TIR48, TIR50, TIR70, and TIR94) were analyzed by comparing the healthy and pathological lobes. (**A**) Venn

diagram of the differentially expressed genes (DEGs) that were found significantly modulated in the four PTC patients analyzed. Twenty-three DEGs were found significantly modulated in all four of the patients, and a further 27 DEGs were found modulated in at least three patients (foldchange (FC) | >1.5 |). (**B**) Venn diagram of the significantly down-regulated DEGs. Among the overall 23 transcripts significantly downregulated, five were found downregulated in all the three PTC analyzed patients and 18 were downregulated in at least three out of four PTC patients. (**C**) Venn diagram of the only significantly up-regulated DEG. Venn diagrams were calculated and drawn using the software available at the Bioinformatics & Evolutionary Genomics website (http://bioinformatics.psb.ugent.be/webtools/Venn/; accessed on 13 January 2022). (**D**) To validate DEGs, RT-qPCR experiments were performed using the RNA extracted from the healthy and PTC lobes of TIR48, TIR50, TIR70, and TIR94 patients. The expression levels of CRABP2, CYP26B1, DHRS3, RARγ, RDH10, RET, and RXRß genes have been reported as relative quantity in the PTC lobe with respect to the healthy one for each patient analyzed, according to the $2^{-\Delta\Delta Ct}$ method. Data are reported as mean ±SD of experiments repeated at least three times (Student's *t*-test, **** $p < 0.0001$, *** $p < 0.001$, with respect to the relative healthy lobe). (**E**) To validate DEGs, immunoblot experiments were performed using the protein lysates derived from the healthy and PTC lobes of TIR45, TIR46, and TIR48 patients. The expression levels of ALDH1A, CRABP2, DH10, and RARα proteins in the PTCs lobes have been normalized to the healthy lobe of each patient analyzed. Both representative images of the immunoblot and their quantification are reported. Graphs illustrate the mean ± SD of experiments repeated at least three times (Student's two-tailed *t*-test, * $p < 0.05$; ** $p < 0.01$; *** $p < 0.001$, with respect to the relative healthy lobe). (**F**) Protein–protein interactions network involving the DEGs significantly modulated in the PTC patients were identified using the STRING database. A PPI network with 22 interaction pairs of the DEGs was identified.

3.3. Impact of Retinoid Acid-Signature in a Cohort of Patients with Papillary Thyroid Cancer

Because the study may suffer due to the low number of available patients, to further investigate the biological and pathological significance of our identified "Human Retinoid Acid Pathway" signature we evaluated the expression of 84 RA-related genes derived from the 96-well PCR array in the framework of the Cancer Genome Atlas (TGCA) project [22] (Supplementary Table S5). An unsupervised hierarchical clustering analysis was performed to investigate the human RA pathway gene signature (n = 84) in the TCGA-PTC dataset (Figure 3). The obtained results revealed that within the 474 PTC available cases, the 84 gene RA pathway signature allowed the identification of two main clusters (i.e., Cluster 1 and 2) characterized by a heterogenous gene expression pattern (Figure 3A). Cluster 1 includes 225 PTC cases, whereas Cluster 2 includes 248 PTC cases. Besides PTC, the other two categories of TC with a larger incidence are the follicular variant of PTC (FV-PTC) and the follicular thyroid carcinoma (FTC). Intriguingly, FV-PTC (n = 99) failed predominantly in Cluster 1 (n = 86) compared to Cluster 2 (n = 13) (Figure 3B). Conversely, Cluster 2 was characterized mainly by the classical (n = 190) and tall cell (n = 29) PTC subtypes (Figure 3B). The analysis of the gene expression FC between Cluster 1 and Cluster 2 revealed that among the 84 genes belonging to the "Human Retinoid Acid Pathway", 58 were significantly modulated ($p < 0.05$). Among these, 42 genes (72.4%) were significantly downregulated in Cluster 1 compared to Cluster 2 ($p < 0.05$, Welch's *t*-test) (Supplementary Table S6).

Overall, these analyses allow the speculation that the subset of PTCs in which the RA-related pathways result in being significantly downregulated correlates more with FV-PTC, which was reported to have an intermediate clinical behavior between classical PTC (C-PTC) and FTC [26].

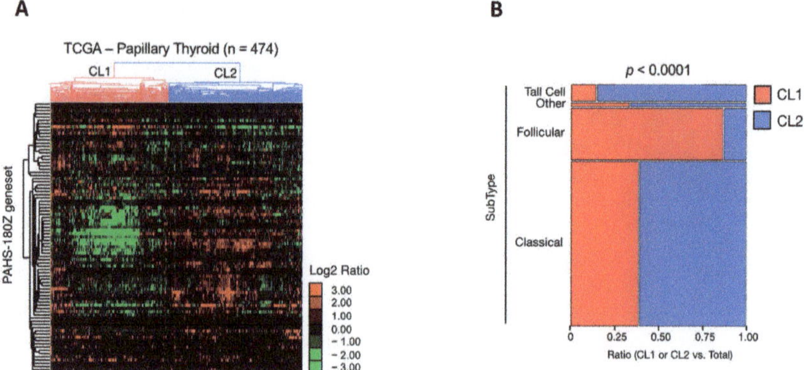

Figure 3. Gene expression metanalysis of external datasets using the RA-regulated genes. (**A**) Hierarchical clustering of "Human Retinoid Acid Pathway"-regulated genes in the framework of the TCGA-PTC dataset representing 474 PTC cases [22]. (**B**) Contingency analysis of PTC subtype distribution in CL1 or CL2 samples as in (**A**). Significance was calculated by the Likelihood Ratio test (JMP 16.0; SAS).

3.4. Antioxidant Response in Healthy and PTC Thyroids

Carotenoids are known to be potent scavengers of reactive oxygen species (ROS), thus providing protection against oxidative damage to photosynthetic and non-photosynthetic organisms at all levels of complexity [9,10,27–30]. However, newly formed carotenoids radical products can undergo further transformations, leading to a variety of secondary carotenoid derivatives that may no longer act as efficient antioxidants but rather as potentially harmful pro-oxidant agents. For instance, the carotenoid radical cation has strong oxidizing properties, has a relatively long lifetime (milliseconds) [31,32], and thus is able to interact with biological macromolecules [30].

To assess whether the possible fate of the carotenoids accumulated in the PTC is associated with the oxidative stress response, we analyzed the expression of oxidative stress markers and of enzymes involved in ROS scavenging in the thyroids of three patients (i.e., TIR45, TIR46, and TIR48). The immunoblot analysis of the H2AX histone phosphorylated at Ser19 (the phosphorylated form being named γH2AX) and of oxoguanine glycosylase 1 (OGG1) showed a significant increase of both these well-known markers of genomic oxidative stress in the tumors compared to the healthy counterparts of the three analyzed PTC patients (γH2AX: 4.2-mean fold increase; t = 10.321, df = 2, $p < 0.01$; OGG1: 1.3-mean fold increase; t = 5.564, df = 2, $p < 0.05$) (Figure 4A). The Supplementary Figure S5 reports the results obtained in each analyzed patient.

Next, we analyzed in the same patients the expression of the antioxidant enzymes. The obtained data indicate a significant reduction in the analyzed proteins in the tumor lobe compared to the relative healthy counterparts. In particular: (i) catalase (CAT) (0.2-mean fold decrease; t = 11.72, df = 2, $p < 0.01$); (ii) glutathione peroxidase-4 (Gpx-4) (0.4-mean fold decrease; t = 11.13, df = 2, $p < 0.01$); (iii) heme oxygenase-1 (HO-1) (0.4 mean fold-reduction decrease; t = 12.96, df = 2, $p < 0.01$); (iv) NADPH oxidase 4 (NOX-4) (0.2-mean fold decrease; t = 7.341, df = 2, $p < 0.05$); (v) NAD(*p*)H quinone dehydrogenase-1 (NQO-1) (0.3-mean fold decrease; t = 3.666, df = 2, not significant); (v) superoxide dismutase-1 (SOD-1) (0.3-mean fold decrease; t = 7.294, df = 2, $p < 0.05$) (Figure 4B). The Supplementary Figure S6 reports the results obtained in each patient analyzed.

Overall, these data indicate an increased oxidative stress status and a reduced antioxidant defense response in PTCs compared to healthy thyroids.

A

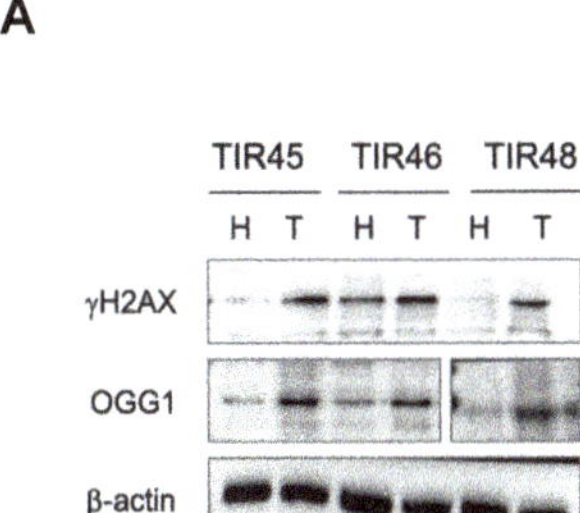

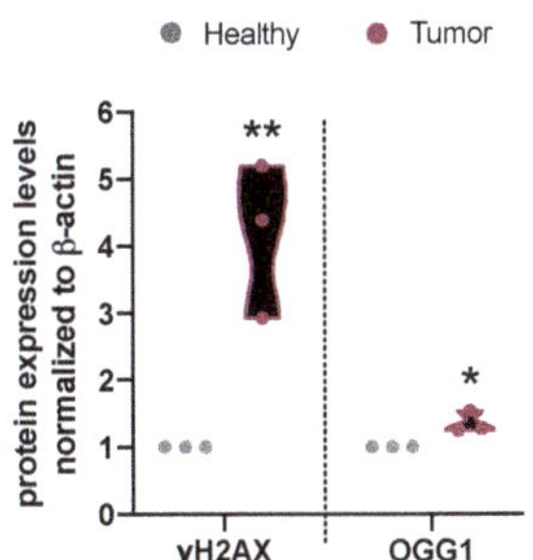

B

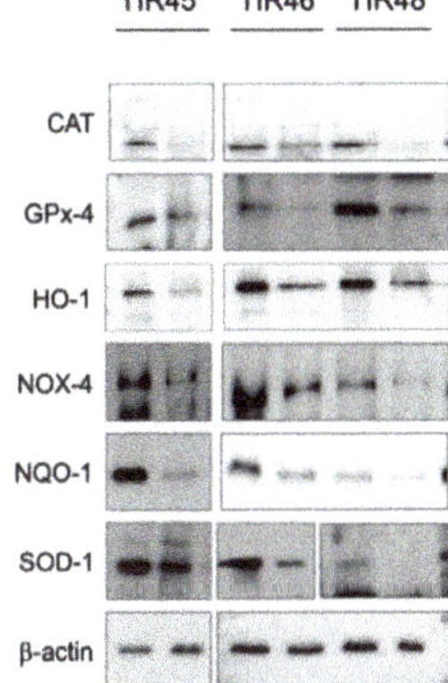

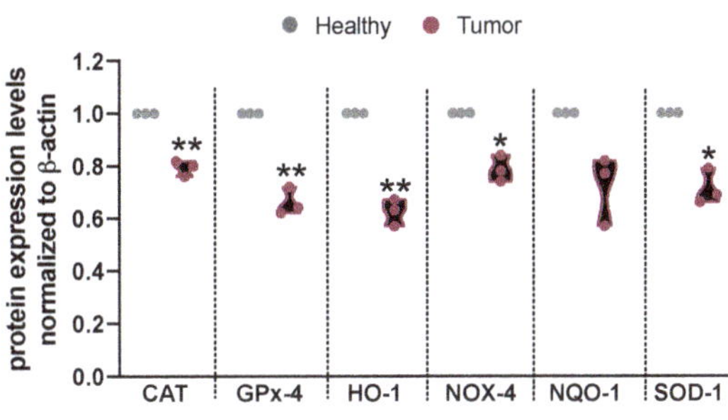

Figure 4. Oxidative stress and antioxidant response in PTCs. To evaluate the oxidative stress levels and the antioxidant response, protein lysates obtained from patients TIR45, TIR46, and TIR48 were analyzed by immunoblot. (**A**) Representative immunoblots are reported together with the relative quantification of pSer139-H2AX (γH2AX) and OGG1 proteins expression normalized to β-actin. Violin plots represent the mean value ±SD of the expression of each protein normalized to the healthy counterparts, in the three analyzed patients (Student's *t*-test, * $p < 0.05$; ** $p < 0.01$ with respect to healthy lobes). (**B**) Representative immunoblots are reported together with the relative quantification of CAT, GpX-4, HO-1, NOX-4, NQO-1, and SOD-1 proteins expression normalized to β-actin. Violin plots represent the mean value ± SD of expression of each protein normalized to the healthy counterparts, in the three analyzed patients (Student's *t*-test, * $p < 0.05$; ** $p < 0.01$ with respect to healthy lobes).

3.5. Dissecting the Metabolic Status of PTCs

With the aim of enriching the molecular characterization of PTCs and to evaluate whether PTC metabolism may be somehow differentiated from that of other TC subtypes considering the reported carotenoid accumulation [4,5], we inspected the DepMap datasets (https://depmap.org/portal/; accessed on 31 January 2022). This resource enables unbiased association analysis linking the cancer metabolome to genetic alterations, epigenetic features, and gene dependencies. In detail, we extrapolated from the DepMap portal the profile of 225 metabolites in 12 thyroid cancer cell lines (Figure 5). Three PTC (i.e., SW579, BCPAP, and BHT101), four FTC (i.e., CGTHW1, FTC238, ML1, TT2609C02), four anaplastic (i.e., 8305C, 8505C, CAL62; FTC133), one medullary (i.e., TT), and one thyroid sarcoma (i.e., S117) cell lines were profiled in the database. The data revealed that some of the metabolites involved in tryptophan metabolism were significantly higher in PTC cells compared to all the other thyroid cancer cell lines. In particular, the metabolites significantly modulated were: (i) anthranilic acid (3.8-fold increase; $p < 0.05$), a precursor to the amino acid tryptophan via the attachment of phosphoribosyl pyrophosphate to the amine group; (ii) NAD (3.2-fold induction; $p < 0.05$), an intermediate in the tryptophan-nicotinamide pathway [33]; (iii) 6-phosphogluconate (2.6-fold induction; $p < 0.05$), an intermediate in the pentose phosphate pathway that serves as an intermediate (e.g., erythrose 4-phosphate) and a cofactor in tryptophan biosynthesis [34]; and (iv) adenine (3.8-fold increase; $p < 0.05$), a nitrogen-containing base (Figure 5). Interestingly, the increased levels of anthranilic acid and NAD support a metabolism addressed toward glycolysis, as further reinforced by the statistically significant increase in phosphoenolpiruvate (PEP) observed in PTC cell lines compared to all the other thyroid cancer cells lines (2.4-fold increase; $p < 0.05$) (Figure 5). Notably, these results appear of interest because it has been reported that a high carotenoid content correlates with an increased biosynthesis of phenylalanine, tyrosine, and tryptophan ([35] (Supplementary Tables S7 and S8)). Therefore, it can be speculated that the reported accumulation of carotenoids in PTCs might possibly be linked with the increased tryptophan metabolism, which is increasingly being recognized as an important microenvironmental factor that suppresses antitumor immune response [36].

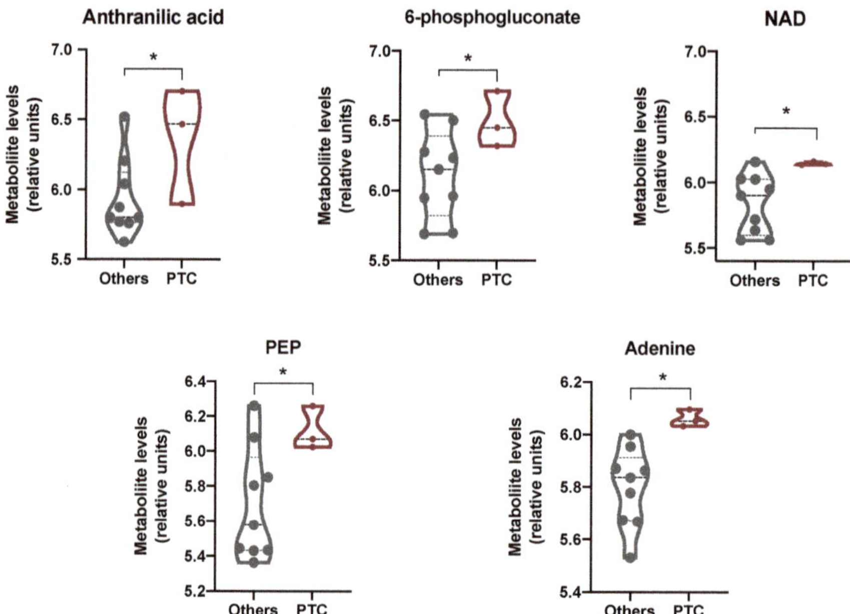

Figure 5. Dissection of the metabolic status of PTCs. We extrapolated from the DepMap portal (https://depmap.org/portal/; accessed on 31 January 2022) the profile of 225 metabolites in 12 thyroid cancer cell lines (i.e., three PTC (i.e., SW579, BCPAP, and BHT101), four FTC (i.e., CGTHW1, FTC238, ML1, TT2609C02), four anaplastic (i.e., 8305C, 8505C, CAL62; FTC133), one medullary (i.e., TT), and one thyroid sarcoma (i.e., S117)). Some of the metabolites involved in tryptophan metabolism (i.e., anthranilic acid, NAD, 6-phosphogluconate, adenine, and PEP) were found significantly higher in PTC cells compared to all the other thyroid cancer cell lines. Each dot in the plots represents the measured value of the indicated metabolite in one single thyroid cancer cell line (Student's *t*-test, * $p < 0.05$ with respect to healthy lobes). Crude and analyzed data are given in Supplementary Tables S7 and S8.

4. Discussion

Carcinogenesis is a multi-step process that results from an accumulation of injuries at several biological levels, which can include genetic and biochemical changes within cells. Our group has demonstrated that both PTC and FV-PTC are characterized by higher carotenoid levels compared to their healthy counterparts, as clearly detectable by RS [4,5]. The significant correlation between the subtype of TC and the carotenoids presence makes it possible to distinguish with the 90% of accuracy healthy tissues from and cancerous ones, as well as to discriminate between the three main high-incidence thyroid cancer variants (i.e., PTC, FTC, and FV-PTC) [4–6].

As Raman analysis does not allow us to distinguish which specific type of carotenoid accumulates in TC, here we performed HPLC analysis to dissect the profile of the carotenoids recorded in patients diagnosed with PTC. Surprisingly, the obtained results have shown that the carotenoids levels were comparable between the matched healthy and PTC lobes of all the analyzed patients. Although these findings are in contrast with the data obtained by RS, this difference can be reconciled considering the different experimental procedures of the two methods. Indeed, on one hand, RS is conducted on frozen tissue sections and allows the identification of free intracellular carotenoids. Due to their hydrophobic nature, the carotenoids released from the food matrix are dispersed into the gastrointestinal tract with the support of dietary lipids, bile-derived phospholipids, and bile salts. Thereafter, carotenoids are solubilized in the mixed micelles consisting of phospholipids, free fatty acids, monoacylglycerols, and bile salts. These solubilization steps represent a critical factor in carotenoids bioavailability and for the subsequent uptake by the intestinal cells [37]. Only once the carotenoids have been taken up by the cells does the RS analysis allow their detection in a carrier-free form into tissue sections. On the other hand, HPLC analysis requires a sample preparation in which the whole thyroid is lysed, and the proteins bound to carotenoids are degraded. Consequently, both the carotenoids localized into the vessels and those localized intracellularly can be analyzed by HPLC [38]. This implies that HPLC analysis allowed us to analyze a larger fraction of carotenoids compared to RS, indicating that their overall content in the extracellular and intracellular compartments does not differ, as expected, between the healthy and cancer thyroid lobes the same patients. On the contrary, the fraction of intracellular free carotenoids, as detected by RS, increases in PTC and FC-PTC patients [4–6].

As one of the major carotenoid normally found in human plasma and tissues is β-carotene [9,39], as also confirmed by the HPLC analyses of PTC patients performed here, we tested the possibility that the difference in carotenoids levels between the healthy and the PTC lobes could be due to a different metabolism of provitamin A [40]. The homeostasis of RA, which is the vitamin A derivative, is controlled by a complex metabolic pathway consisting of multiple reactions and enzymes. Briefly, dietary carotenoids such as β-carotene can be either converted to retinaldehyde (RAL) by β-carotene-15,15$'$-monooxygenase-1 within the enterocyte or absorbed unmodified by cells. Then, RAL can be either converted to RA by retinaldehyde dehydrogenases (RALDHs) or to retinol (ROL, also named vitamin A) by renital reductases [13,41]. Remarkably, all the above-mentioned reactions are reversible. As many tissues and organs, including the thyroid [42], express enzymes involved in the RA pathway, the β-carotene molecules delivered to these tissues can be converted in situ to retinoids [13]. Thyroid glandular cells specifically express the enzyme xanthine dehydrogenase (XDH) that, through a cellular retinoid-binding protein (CRBP)-dependent mechanism, directly oxidizes ROL to RA [42,43]. Importantly, some evidence supports a link between the thyroid gland and retinoids: both vitamin A deficiency and excess affect thyroid gland volume by affecting thyroid hormone synthesis in vivo [42,44,45]. Taking into consideration these complex pathways, if β-carotene is metabolized to RA its levels should be lower than if this metabolic pathway is inactive. Interestingly, the transcriptomic analysis performed here revealed that in all the analyzed PTC patients the overall pathways related to RA metabolism are significantly downregulated in PTC lobes compared to their healthy counterparts. This result suggests that the higher levels of carotenoids, and specifically that of β-carotene, found by RS and HPLC analyses in PTC patients, are not due to the increased uptake of carotenoids but rather to a different intracellular metabolic fate. This hypothesis is also sustained by the lack of expression of the SR-B1 receptor in PTC patients.

The beneficial and deleterious roles of the dietary carotenoids on human health may be due to several factors, but antioxidant and pro-oxidant processes seem to be particularly important. Under normal cellular conditions, carotenoids are powerful antioxidants [46,47] that act as efficient physical and chemical quenchers of singlet oxygen and as potent scavengers of other ROS [9,10,27–30]. However, the high levels of ROS that accumulate in cancer cells determine the possibility that carotenoids may act as prooxidants, causing ROS-mediated apoptosis. Indeed, the carotenoids radical products formed during the antioxidant processes can undergo further transformations, leading to a variety of secondary carotenoid derivatives that may no longer act as efficient antioxidants but rather as harmful pro-oxidant agents [30,31,48,49]. Accordingly, the well-known protective function of β-carotene and lycopene has been challenged by clinical trials where β-carotene supplementation in male smokers resulted in a significantly increased incidence of lung cancer [50]. Animal studies showed that diet influences the carcinogenic response to β-carotene and that β-carotene, per se, is pro-carcinogenic in UV carcinogenesis [51]. Interestingly, β-carotene has been shown to exhibit either limited antioxidant protection or to behave as a pro-oxidant under oxidative stress conditions [52]. Based on these considerations, we evaluated whether carotenoid accumulation in PTC patients could affect the oxidative status and the antioxidant response. Oxidative DNA damage caused by intracellular ROS is important in the pathology of a range of human diseases, including cancer [53–55]. The evaluation of the DNA damage levels performed here revealed the presence of significantly higher levels of DNA double-strand breaks (marked by γH2AX) in PTCs compared to the healthy counterparts. A frequently occurring mutagenic base lesion produced by ROS is 8-oxo-2′-deoxyguanosine (8-oxo-dG), which is repaired by OGG1. We found that OGG1 levels were also markedly increased in the PTC lobes of all the patients analyzed compared to the relative healthy lobes. These results agree with the observation that oxidant levels are significantly increased in patients with TC compared to the controls [56]. This high oxidative stress status agrees with data reporting increased levels of malondialdehyde (MDA) in PTCs, one of the final products of polyunsaturated fatty acid peroxidation in cells [57]. To evaluate the antioxidant response in PTC patients, the expression of six antioxidant molecules (i.e., CAT, GPx-4, HO-1, NOX-4, NQO-1, and SOD-1) was analyzed. The results obtained showed that all these antioxidant proteins and enzymes were downregulated in the PTC lobes compared to the healthy counterparts. Overall, these results agree with the oxidative stress status [58–60] and the decreased expression of glutathione peroxidase [60] and catalase [58,59] observed in TC, possibly contributing to the increase in the imbalance of the oxidant/antioxidant system in PTCs [61]. Interestingly, the unsupervised hierarchical clustering analysis of the gene expression profiles of the 474 PTC cases published in the framework of the TGCA project [22] indicate that the RA-related pathways resulted in being significantly downregulated in PTCs and in being associated with a worse prognosis. As we reported an increased oxidative stress status and a reduced antioxidant defense response in PTCs compared to matched healthy thyroids, it is possible to speculate that the downregulation of RA-related pathways causes carotenoid accumulation and, in turn, prooxidant effects in an oxidant microenvironment such as that present in cancer cells [30,31,48,49]. This may explain the worse prognosis of PTC patients in which the RA-related pathways result in being significantly downregulated.

To further dissect the metabolic status of PTCs that possibly correlates with the increased levels of carotenoids, we inspected the DepMap datasets to profile 225 metabolites in 12 thyroid cancer cell lines). With this aim, three PTC (i.e., SW579, BCPAP, and BHT101), four FTC (i.e., CGTHW1, FTC238, ML1, TT2609C02), four anaplastic (i.e., 8305C, 8505C, CAL62; FTC133), one medullary (i.e., TT), and one thyroid sarcoma (i.e., S117) cell lines were analyzed. Tryptophan catabolism in cancer is increasingly being recognized as an important microenvironmental factor that suppresses antitumor immune responses. Indeed, tryptophan catabolism creates an immunosuppressive milieu by inducing T-cell anergy and apoptosis through the accumulation of immunosuppressive tryptophan catabolites [62]. In PTCs, the increased levels of tumor-associated macrophages (TAM) correlate with lymph node metastasis [63], larger tumor size [64], and reduced survival [65]. Interestingly, the data reported here revealed that some of the metabolites involved in tryptophan metabolism (i.e., anthranilic acid, NAD, 6-phosphogluconate, and adenine) were significantly higher in PTC cells compared to all the other thyroid cancer cell lines analyzed. Of note, the high levels of anthranilic acid, NAD, and PEP in PTC cell lines compared to all the other thyroid cancer cell lines support the typical tumoral glycolytic metabolism. In addition, the high carotenoid content in PTCs correlates with the reported phenylalanine, tyrosine, and tryptophan biosynthesis [35].

5. Conclusions

Overall, the results obtained here suggest that in addition to the fact that carotenoids accumulation represents a hallmark of PTCs, the downregulation of RA-related pathways also appears implicated with PTC progression and severity, possibly representing a further diagnostic hallmark for this specific type of TC. Among the limitations of this study, the relatively low number of patients enrolled should be mentioned. Therefore, this was intended as a pilot study providing novel possible markers and possible therapeutic targets for PTC diagnosis and therapy. For the future, a larger study including a higher number of PTC patients will be necessary to further validate the molecular data reported here.

Supplementary Materials: The following supporting information can be downloaded at: https://www.mdpi.com/article/10.3390/antiox11081463/s1, Figure S1: Exemplificative Raman spectrum of a PTC patient, Figure S2: "Human Retinoic Acid Pathway" PCR array and heatmaps, Figure S3: Scatter plots derived from the PCR array, Figure S4: Representative immunohistochemistry image of the SR-B1 receptor expression in one PTC patient, Figure S5: Genomic oxidative stress in the three analyzed PTC patients, Figure S6: Antioxidant response in the three analyzed PTC patients, Table S1: Identification of α-carotene, β-carotene, lutein, and lycopene in the thyroids of the four patients analyzed, Table S2: List of the genes significantly modulated in the analyzed PTC patients, Table S3: Overall modulated DEGs sorted by *p*-value using the Reactome database, Table S4: Overall downregulated DEGs sorted by *p*-value using the Reactome database, Table S5: Unsupervised hierarchical clustering analysis of our dataset of "Human Retinoid Acid Pathway"-regulated genes with matched expression data in TCGA-PTC database, Table S6: Analysis of the gene expression fold change between Cluster 1 and Cluster 2 obtained from the analysis of 474 PTC cases, Table S7: Raw data extrapolated from the DepMap portal (https://depmap.org/portal/; accessed on 31 January 2022) referred to the profile of 225 metabolites in 12 thyroid cancer cell lines, Table S8: Metabolites whose levels significantly change ($p < 0.05$) between the PTC cell lines and the other thyroid cancer cell lines.

Author Contributions: R.L.S. and Y.C. performed the molecular biology and biochemical experiments; A.d.M. and P.A. supervised the molecular biology and biochemical experiments; G.P., B.G., and R.A. performed HPLC analysis and analyzed the data; M.D.G. performed the Raman spectra and analyzed the data; A.S. and M.A.R. coordinated the Raman spectra acquisition and data analysis; A.d.M. and F.A. analyzed the DepMap metabolic database and performed statistical analyses; F.B. analyzed the TCGA-PTC database and performed statistical analyses; M.V. prepared the frozen sections for Raman examination and the chilled specimens for biochemical analysis and performed routine staining and immunohistochemical reactions; P.C. and F.L. performed surgery; A.M.N., A.P., P.C. and F.L. enrolled the subjects; A.C. supervised the histological activities; A.C. and C.T. performed the histological diagnosis and evaluated the immunohistochemical staining; A.d.M. conceptualized and supervised the research, formally analyzed the data, and wrote, reviewed, and edited the manuscript. All authors have read and agreed to the published version of the manuscript.

Funding: The authors acknowledge financial support from Ministero della Salute, through the TIRAMA project (RF 2018 12366568). A.d.M., A.S., M.D.G., P.A. and M.A.R. gratefully acknowledge the Grant of Excellence Departments, MIUR (ARTICOLO 1, COMMI 314–337 LEGGE 232/2016).

Institutional Review Board Statement: The study was approved by the Ethical Committee of the University of Rome "Campus Biomedico" (UCBM) (prot. 33.15 TS ComEt CBM and prot. 31/19 PAR ComEt CBM) on the 26 July 2019. The informed consent was collected from patients before surgery. Enrolled patients were recorded in a codified file with an anonymous ID code, which was registered in the software database of the Pathology Unit of the UCBM. All experiments were performed in full accordance with the principle of Good Clinical Practice (GCP) and the ethical principles contained in the current version of the Declaration of Helsinki.

Informed Consent Statement: Informed consent was obtained from all subjects involved in the study.

Data Availability Statement: We will consider sharing de-identified, individual participant-level data that underlie the results reported in this Article on receipt of a request detailing the study hypothesis and statistical analysis plan. All requests should be sent to the corresponding author. The corresponding author and lead investigators of this study will discuss all requests and make decisions about whether data sharing is appropriate based on the scientific rigor of the proposal. All applicants will be asked to sign a data access agreement. The datasets used for the hierarchical clustering and heatmap analyses used to generate Figure 3 were obtained from the Cancer Genome Atlas (TGCA) project and downloaded from cBioPortal (https://www.cbioportal.org/datasets; accessed on 26 January 2022) [22]. All these datasets are available in Supplementary Tables S5 and S6. All the datasets used to generate Figure 5 were downloaded by the Broad Institute through the DepMap portal (https://depmap.org/portal; accessed on 31 January 2022) and are available in Supplementary Tables S7 and S8.

Conflicts of Interest: The authors declare no conflict of interest.

References

1. Kitahara, C.M.; Sosa, J.A. Understanding the ever-changing incidence of thyroid cancer. *Nat. Rev. Endocrinol.* **2020**, *16*, 617–618. [CrossRef]
2. Rossi, E.D.; Pantanowitz, L.; Hornick, J.L. A worldwide journey of thyroid cancer incidence centred on tumour histology. *Lancet Diabetes Endocrinol.* **2021**, *9*, 193–194. [CrossRef]
3. Fagin, J.A.; Wells, S.A., Jr. Biologic and Clinical Perspectives on Thyroid Cancer. *N. Engl. J. Med.* **2016**, *375*, 1054–1067. [CrossRef]
4. Sbroscia, M.; Di Gioacchino, M.; Ascenzi, P.; Crucitti, P.; Di Masi, A.; Giovannoni, I.; Longo, F.; Mariotti, D.; Naciu, A.M.; Palermo, A.; et al. Thyroid cancer diagnosis by Raman spectroscopy. *Sci. Rep.* **2020**, *10*, 13342. [CrossRef]
5. Sodo, A.; Verri, M.; Palermo, A.; Naciu, A.M.; Sponziello, M.; Durante, C.; Di Gioacchino, M.; Paolucci, A.; Di Masi, A.; Longo, F.; et al. Raman Spectroscopy Discloses Altered Molecular Profile in Thyroid Adenomas. *Diagnostics* **2021**, *11*, 43. [CrossRef]
6. Rau, J.V.; Graziani, V.; Fosca, M.; Taffon, C.; Rocchia, M.; Crucitti, P.; Pozzilli, P.; Muda, A.O.; Caricato, M.; Crescenzi, A. RAMAN spectroscopy imaging improves the diagnosis of papillary thyroid carcinoma. *Sci. Rep.* **2016**, *6*, 35117. [CrossRef]
7. Demmig-Adams, B.; Adams, W.W. Food and Photosynthesis: Antioxidants in photosynthesis and human nutrition. *Science* **2002**, *298*, 2149–2153. [CrossRef]
8. Eroglu, A.; Harrison, E.H. Carotenoid metabolism in mammals, including man: Formation, occurrence, and function of apocarotenoids. *J. Lipid Res.* **2013**, *54*, 1719–1730. [CrossRef] [PubMed]
9. Fiedor, J.; Burda, K. Potential Role of Carotenoids as Antioxidants in Human Health and Disease. *Nutrients* **2014**, *6*, 466–488. [CrossRef]
10. Toti, E.; Chen, C.-Y.O.; Palmery, M.; Villaño Valencia, D.; Peluso, I. Non-Provitamin A and Provitamin A Carotenoids as Immunomodulators: Recommended Dietary Allowance, Therapeutic Index, or Personalized Nutrition? *Oxid. Med. Cell. Longev.* **2018**, *2018*, 4637861. [CrossRef]
11. Giguère, V.; Ong, E.S.; Segui, P.; Evans, R.M. Identification of a receptor for the morphogen retinoic acid. *Nature* **1987**, *330*, 624–629. [CrossRef] [PubMed]
12. Petkovich, M.; Brand, N.J.; Krust, A.; Chambon, P. A human retinoic acid receptor which belongs to the family of nuclear receptors. *Nature* **1987**, *330*, 444–450. [CrossRef] [PubMed]
13. di Masi, A.; Leboffe, L.; De Marinis, E.; Pagano, F.; Cicconi, L.; Rochette-Egly, C.; Lo-Coco, F.; Ascenzi, P.; Nervi, C. Retinoic acid receptors: From molecular mechanisms to cancer therapy. *Mol. Asp. Med.* **2015**, *41*, 1–115. [CrossRef]
14. Tessler, F.N.; Middleton, W.D.; Grant, E.G.; Hoang, J.K.; Berland, L.L.; Teefey, S.A.; Cronan, J.J.; Beland, M.D.; Desser, T.S.; Frates, M.C.; et al. ACR Thyroid Imaging, Reporting and Data System (TI-RADS): White Paper of the ACR TI-RADS Committee. *J. Am. Coll. Radiol.* **2017**, *14*, 587–595. [CrossRef] [PubMed]
15. Gharib, H.; Papini, E.; Garber, J.R.; Duick, D.S.; Harrell, R.M.; Hegedüs, L.; Paschke, R.; Valcavi, R.; Vitti, P. American Association of Clinical Endocrinologists, American College of Endocrinology, and Associazione Medici Endocrinologi Medical Guidelines for Clinical Practice for the Diagnosis and Management of Thyroid Nodules—2016 Update Appendix. *Endocr. Pract.* **2016**, *22*, 1–60. [CrossRef]

16. Lloyd, R.V.; Osamura, R.Y.; Kloppel, G.; Rosai, J. Tumours of the thyroid gland. In *World Health Organization Classification of Tumours of Endocrine Organs*, 4th ed.; International Agency for Research on Cancer (IARC): Lyon, France, 2017; pp. 65–143.

17. Peng, Y.M.; Peng, Y.S.; Lin, Y. A nonsaponification method for the determination of carotenoids, retinoids, and tocopherols in solid human tissues. *Cancer Epidemiol. Biomark. Prev.* **1993**, *2*, 139–144.

18. Böhm, V. Use of column temperature to optimize carotenoid isomer separation by C30 high performance liquid chromatography. *J. Sep. Sci.* **2001**, *24*, 955–959. [CrossRef]

19. Fabregat, A.; Jupe, S.; Matthews, L.; Sidiropoulos, K.; Gillespie, M.; Garapati, P.; Haw, R.; Jassal, B.; Korninger, F.; May, B.; et al. The Reactome Pathway Knowledgebase. *Nucleic Acids Res.* **2018**, *46*, D649–D655. [CrossRef] [PubMed]

20. De Simone, G.; Di Masi, A.; Vita, G.M.; Polticelli, F.; Pesce, A.; Nardini, M.; Bolognesi, M.; Ciaccio, C.; Coletta, M.; Turilli, E.S.; et al. Mycobacterial and Human Nitrobindins: Structure and Function. *Antioxid. Redox Signal.* **2020**, *33*, 229–246. [CrossRef] [PubMed]

21. di Masi, A.; Leboffe, L.; Sodo, A.; Tabacco, G.; Cesareo, R.; Sbroscia, M.; Giovannoni, I.; Taffon, C.; Crucitti, P.; Longo, F.; et al. Metabolic profile of human parathyroid adenoma. *Endocrine* **2020**, *67*, 699–707. [CrossRef]

22. Cancer Genome Atlas Research Network. Integrated Genomic Characterization of Papillary Thyroid Carcinoma. *Cell* **2014**, *159*, 676–690. [CrossRef] [PubMed]

23. Szklarczyk, D.; Gable, A.L.; Nastou, K.C.; Lyon, D.; Kirsch, R.; Pyysalo, S.; Doncheva, N.T.; Legeay, M.; Fang, T.; Bork, P.; et al. The STRING database in 2021: Customizable protein-protein networks, and functional characterization of user-uploaded gene/measurement sets. *Nucleic Acids Res.* **2021**, *49*, D605–D612. [CrossRef]

24. Kiefer, C.; Sumser, E.; Wernet, M.F.; von Lintig, J. A class B scavenger receptor mediates the cellular uptake of carotenoids in *Drosophila*. *Proc. Natl. Acad. Sci. USA* **2002**, *99*, 10581–10586. [CrossRef] [PubMed]

25. Borel, P.; Lietz, G.; Goncalves, A.; de Edelenyi, F.S.; Lecompte, S.; Curtis, P.; Goumidi, L.; Caslake, M.J.; Miles, E.A.; Packard, C.; et al. CD36 and SR-BI Are Involved in Cellular Uptake of Provitamin A Carotenoids by Caco-2 and HEK Cells, and Some of Their Genetic Variants Are Associated with Plasma Concentrations of These Micronutrients in Humans. *J. Nutr.* **2013**, *143*, 448–456. [CrossRef] [PubMed]

26. Yu, X.-M.; Schneider, D.F.; Leverson, G.; Chen, H.; Sippel, R.S. Follicular Variant of Papillary Thyroid Carcinoma is a Unique Clinical Entity: A Population-Based Study of 10,740 Cases. *Thyroid* **2013**, *23*, 1263–1268. [CrossRef]

27. Mortensen, A.; Skibsted, L.H.; Sampson, J.; Rice-Evans, C.; Everett, S.A. Comparative mechanisms and rates of free radical scavenging by carotenoid antioxidants. *FEBS Lett.* **1997**, *418*, 91–97. [CrossRef]

28. Martin, H.D.; Ruck, C.; Schmidt, M.; Sell, S.; Beutner, S.; Mayer, B.; Walsh, R. Chemistry of carotenoid oxidation and free radical reactions. *Pure Appl. Chem.* **1999**, *71*, 2253–2262. [CrossRef]

29. Black, H.S.; Boehm, F.; Edge, R.; Truscott, T.G. The Benefits and Risks of Certain Dietary Carotenoids that Exhibit both Anti- and Pro-Oxidative Mechanisms—A Comprehensive Review. *Antioxidants* **2020**, *9*, 264. [CrossRef]

30. Shin, J.; Song, M.-H.; Oh, J.-W.; Keum, Y.-S.; Saini, R.K. Pro-oxidant Actions of Carotenoids in Triggering Apoptosis of Cancer Cells: A Review of Emerging Evidence. *Antioxidants* **2020**, *9*, 532. [CrossRef]

31. Kispert, L.D.; Konovalova, T.; Gao, Y. Carotenoid radical cations and dications: EPR, optical, and electrochemical studies. *Arch. Biochem. Biophys.* **2004**, *430*, 49–60. [CrossRef]

32. Chen, C.-H.; Han, R.-M.; Liang, R.; Fu, L.-M.; Wang, P.; Ai, X.-C.; Zhang, J.-P.; Skibsted, L.H. Direct Observation of the β-Carotene Reaction with Hydroxyl Radical. *J. Phys. Chem. B* **2011**, *115*, 2082–2089. [CrossRef]

33. Fukuwatari, T.; Shibata, K. Nutritional Aspect of Tryptophan Metabolism. *Int. J. Tryptophan Res.* **2013**, *6*, 3–8. [CrossRef]

34. Soderberg, T. Biosynthesis of ribose-5-phosphate and erythrose-4-phosphate in archaea: A phylogenetic analysis of archaeal genomes. *Archaea* **2005**, *1*, 347–352. [CrossRef]

35. Wei, X.; Chen, N.; Tang, B.; Luo, X.; You, W.; Ke, C. Untargeted metabolomic analysis of the carotenoid-based orange coloration in Haliotis gigantea using GC-TOF-MS. *Sci. Rep.* **2019**, *9*, 14545. [CrossRef]

36. Wei, Z.; Liu, X.; Cheng, C.; Yu, W.; Yi, P. Metabolism of Amino Acids in Cancer. *Front. Cell Dev. Biol.* **2021**, *8*, 603837. [CrossRef]

37. Nagao, A. Absorption and metabolism of dietary carotenoids. *BioFactors* **2011**, *37*, 83–87. [CrossRef] [PubMed]

38. Peng, Y.M.; Lin, Y.; Moon, T.; Baier, M. Micronutrient concentrations in paired skin and plasma of patients with actinic keratoses: Effect of prolonged retinol supplementation. *Cancer Epidemiol. Biomark. Prev.* **1993**, *2*, 145–150.

39. Perera, C.O.; Yen, G.M. Functional Properties of Carotenoids in Human Health. *Int. J. Food Prop.* **2007**, *10*, 201–230. [CrossRef]

40. Tang, X.H.; Gudas, L.J. Retinoids, retinoic acid receptors, and cancer. *Annu. Rev. Pathol.* **2011**, *6*, 345–364. [CrossRef] [PubMed]

41. Napoli, J.L. Physiological insights into all-trans-retinoic acid biosynthesis. *Biochim. Biophys. Acta-Mol. Cell Biol. Lipids* **2012**, *1821*, 152–167. [CrossRef]

42. Taibi, G.; Gueli, M.C.; Nicotra, C.M.A.; Cocciadiferro, L.; Carruba, G. Retinol oxidation to retinoic acid in human thyroid glandular cells. *J. Enzym. Inhib. Med. Chem.* **2014**, *29*, 796–803. [CrossRef]

43. Xu, P.; Huecksteadt, T.P.; Hoidal, J.R. Molecular Cloning and Characterization of the Human Xanthine Dehydrogenase Gene (XDH). *Genomics* **1996**, *34*, 173–180. [CrossRef]

44. Drill, V.A. Interrelations between thyroid function and vitamin metabolism. *Physiol. Rev.* **1943**, *23*, 355–379. [CrossRef]

45. Morley, J.E.; Melmed, S.; Reed, A.; Kasson, D.G.; Levin, S.R.; Pekary, A.E.; Hershman, J.M. Effect of vitamin A on the hypothalamo-pituitary-thyroid axis. *Am. J. Physiol. Endocrinol. Metab.* **1980**, *238*, E174–E179. [CrossRef]

46. Arathi, B.P.; Sowmya, P.R.-R.; Kuriakose, G.C.; Vijay, K.; Baskaran, V.; Jayabaskaran, C.; Lakshminarayana, R. Enhanced cytotoxic and apoptosis inducing activity of lycopene oxidation products in different cancer cell lines. *Food Chem. Toxicol.* **2016**, *97*, 265–276. [CrossRef]

47. Gansukh, E.; Mya, K.K.; Jung, M.; Keum, Y.-S.; Kim, D.H.; Saini, R.K. Lutein derived from marigold (*Tagetes erecta*) petals triggers ROS generation and activates Bax and caspase-3 mediated apoptosis of human cervical carcinoma (HeLa) cells. *Food Chem. Toxicol.* **2019**, *127*, 11–18. [CrossRef]

48. Burke, M.; Edge, R.; Land, E.J.; McGarvey, D.J.; Truscott, T. One-electron reduction potentials of dietary carotenoid radical cations in aqueous micellar environments. *FEBS Lett.* **2001**, *500*, 132–136. [CrossRef]

49. Wang, C.; Schlamadinger, D.E.; Desai, V.; Tauber, M.J. Triplet Excitons of Carotenoids Formed by Singlet Fission in a Membrane. *ChemPhysChem* **2011**, *12*, 2891–2894. [CrossRef]

50. Shekelle, R.B.; Liu, S.; Raynor, W.J.; Lepper, M.; Maliza, C.; Rossof, A.H.; Paul, O.; Shryock, A.M.; Stamler, J. Dietary vitamin a and risk of cancer in the western electric study. *Lancet* **1981**, *318*, 1185–1190. [CrossRef]

51. Black, H.S. Radical interception by carotenoids and effects on UV carcinogenesis. *Nutr. Cancer* **1998**, *31*, 212–217. [CrossRef]

52. Burton, G.W.; Ingold, K.U. β-Carotene: An Unusual Type of Lipid Antioxidant. *Science* **1984**, *224*, 569–573. [CrossRef]

53. Kubo, N.; Morita, M.; Nakashima, Y.; Kitao, H.; Egashira, A.; Saeki, H.; Oki, E.; Kakeji, Y.; Oda, Y.; Maehara, Y. Oxidative DNA damage in human esophageal cancer: Clinicopathological analysis of 8-hydroxydeoxyguanosine and its repair enzyme. *Dis. Esophagus* **2014**, *27*, 285–293. [CrossRef]

54. Bohn, T. Carotenoids and Markers of Oxidative Stress in Human Observational Studies and Intervention Trials: Implications for Chronic Diseases. *Antioxidants* **2019**, *8*, 179. [CrossRef]

55. Srinivas, U.S.; Tan, B.W.Q.; Vellayappan, B.A.; Jeyasekharan, A.D. ROS and the DNA damage response in cancer. *Redox Biol.* **2019**, *25*, 101084. [CrossRef]

56. Wang, D.; Feng, J.F.; Zeng, P.; Yang, Y.H.; Luo, J.; Yang, Y.W. Total oxidant/antioxidant status in sera of patients with thyroid cancers. *Endocr. Relat. Cancer* **2011**, *18*, 773–782. [CrossRef] [PubMed]

57. Lassoued, S.; Mseddi, M.; Mnif, F.; Abid, M.; Guermazi, F.; Masmoudi, H.; El Feki, A.; Attia, H. A comparative study of the oxidative profile in Graves' disease, Hashimoto's thyroiditis, and papillary thyroid cancer. *Biol. Trace Elem. Res.* **2010**, *138*, 107–115. [CrossRef]

58. Hasegawa, Y.; Takano, T.; Miyauchi, A.; Matsuzuka, F.; Yoshida, H.; Kuma, K.; Amino, N. Decreased expression of glutathione peroxidase mRNA in thyroid anaplastic carcinoma. *Cancer Lett.* **2002**, *182*, 69–74. [CrossRef]

59. Hasegawa, Y.; Takano, T.; Miyauchi, A.; Matsuzuka, F.; Yoshida, H.; Kuma, K.; Amino, N. Decreased expression of catalase mRNA in thyroid anaplastic carcinoma. *Jpn. J. Clin. Oncol.* **2003**, *33*, 6–9. [CrossRef]

60. Metere, A.; Frezzotti, F.; Graves, C.E.; Vergine, M.; De Luca, A.; Pietraforte, D.; Giacomelli, L. A possible role for selenoprotein glutathione peroxidase (GPx1) and thioredoxin reductases (TrxR1) in thyroid cancer: Our experience in thyroid surgery. *Cancer Cell Int.* **2018**, *18*, 7. [CrossRef]

61. El Hassani, R.A.; Buffet, C.; Leboulleux, S.; Dupuy, C. Oxidative stress in thyroid carcinomas: Biological and clinical significance. *Endocr.-Relat. Cancer* **2019**, *26*, R131–R143. [CrossRef]

62. Platten, M.; Wick, W.; Van den Eynde, B.J. Tryptophan Catabolism in Cancer: Beyond IDO and Tryptophan Depletion. *Cancer Res.* **2012**, *72*, 5435–5440. [CrossRef] [PubMed]

63. Fang, W.; Ye, L.; Shen, L.; Cai, J.; Huang, F.; Wei, Q.; Fei, X.; Chen, X.; Guan, H.; Wang, W.; et al. Tumor-associated macrophages promote the metastatic potential of thyroid papillary cancer by releasing CXCL8. *Carcinogenesis* **2014**, *35*, 1780–1787. [CrossRef]

64. Kim, S.; Cho, S.W.; Min, H.S.; Kim, K.M.; Yeom, G.J.; Kim, E.Y.; Lee, K.E.; Yun, Y.G.; Park, D.J.; Park, Y.J. The Expression of Tumor-Associated Macrophages in Papillary Thyroid Carcinoma. *Endocrinol. Metab.* **2013**, *28*, 192–198. [CrossRef] [PubMed]

65. Varricchi, G.; Loffredo, S.; Marone, G.; Modestino, L.; Fallahi, P.; Ferrari, S.M.; de Paulis, A.; Antonelli, A.; Galdiero, M.R. The Immune Landscape of Thyroid Cancer in the Context of Immune Checkpoint Inhibition. *Int. J. Mol. Sci.* **2019**, *20*, 3934. [CrossRef] [PubMed]

 antioxidants

Article

Biochemical Evaluation of the Effects of Hydroxyurea in Vitro on Red Blood Cells

Cristiane Oliveira Renó [1], Grazielle Aparecida Silva Maia [1], Leilismara Sousa Nogueira [1], Melina de Barros Pinheiro [2], Danyelle Romana Alves Rios [3], Vanessa Faria Cortes [1], Leandro Augusto de Oliveira Barbosa [1] and Hérica de Lima Santos [1,*]

[1] Laboratório de Bioquímica Celular, Campus Centro Oeste Dona Lindu, Universidade Federal de São João del-Rei, Divinópolis 35501-296, Minas Gerais, Brazil; cristianereno@outlook.pt (C.O.R.); graquimica@ufsj.edu.br (G.A.S.M.); leilismara@ufsj.edu.br (L.S.N.); cortesvf@ufsj.edu.br (V.F.C.); lbarbosa@ufsj.edu.br (L.A.d.O.B.)
[2] Laboratório de Análises Clínicas, Campus Centro Oeste Dona Lindu, Universidade Federal de São João del-Rei, Divinópolis 35501-296, Minas Gerais, Brazil; melinapinheiro@ufsj.edu.br
[3] Laboratório de Hematologia Clínica, Campus Centro Oeste Dona Lindu, Universidade Federal de São João del-Rei, Divinópolis 35501-296, Minas Gerais, Brazil; danyelleromana@ufsj.edu.br
* Correspondence: hlima@ufsj.edu.br; Tel.: +55-37-4690-4552 or +55-37-4690-4550

Citation: Renó, C.O.; Maia, G.A.S.; Nogueira, L.S.; de Barros Pinheiro, M.; Rios, D.R.A.; Cortes, V.F.; de Oliveira Barbosa, L.A.; de Lima Santos, H. Biochemical Evaluation of the Effects of Hydroxyurea in Vitro on Red Blood Cells. *Antioxidants* **2021**, *10*, 1599. https://doi.org/10.3390/antiox10101599

Academic Editors: Soliman Khatib, Dana Atrahimovich Blatt and Alessandra Napolitano

Received: 5 August 2021
Accepted: 28 September 2021
Published: 12 October 2021

Abstract: Hydroxyurea (HU) is a low-cost, low-toxicity drug that is often used in diseases, such as sickle cell anemia and different types of cancer. Its effects on the red blood cells (RBC) are still not fully understood. The in vitro effects of HU were evaluated on the biochemical parameters of the RBC from healthy individuals that were treated with 0.6 mM or 0.8 mM HU for 30 min and 1 h. After 30 min, there was a significant increase in almost all of the parameters analyzed in the two concentrations of HU, except for the pyruvate kinase (PK) activity. A treatment with 0.8 mM HU for 1 h resulted in a reduction of the levels of lipid peroxidation, Fe^{3+}, and in the activities of some of the enzymes, such as glutathione reductase (GR), glucose-6-phosphate dehydrogenase (G6PD), and PK. After the incubation for 1 h, the levels of H_2O_2, lipid peroxidation, reduced glutathione (GSH), enzymatic activity (hexokinase, G6PD, and superoxide dismutase (SOD) were reduced with the treatment of 0.8 mM HU when compared with 0.6 mM. The results have suggested that a treatment with HU at a concentration of 0.8 mM seemed to be more efficient in protecting against the free radicals, as well as in treating diseases, such as sickle cell anemia. HU appears to preferentially stimulate the pentose pathway over the glycolytic pathway. Although this study was carried out with the RBC from healthy individuals, the changes described in this study may help to elucidate the mechanisms of action of HU when administered for therapeutic purposes.

Keywords: hydroxyurea; red blood cells; energy metabolism; antioxidant system; reactive oxygen species

1. Introduction

Hydroxyurea (HU) is the main drug of choice for the treatment of sickle cell anemia (SCA) and it provides therapeutic benefits through its multiple mechanisms of action. It is easily synthesized by urea hydroxylation and acts by inhibiting an enzyme that is involved in the synthesis of ribonucleic acid. In addition to HU being used to treat hematological diseases, such as chronic myeloid leukemia, and its ability to generate nitric oxide, a potent vasodilator, also helps to relieve the patients' pain [1–3]. HU acts on the bone marrow and due to its cytotoxic effects, it decreases the production of blood cells. Furthermore, it stimulates the increase in the synthesis of HbF, which contributes to the reduction of painful crises and vaso-occlusive processes [2].

The drug is administered in different doses, which differ according to the patient and the disease to be treated. In SCA, the doses range from 15–35 mg/kg/day, and in resistant chronic myeloid leukemia (CML) from 1000–2000 mg/day. This drug is inexpensive. It is administered as a single agent and orally. Based on currently available data, the HU treatment fulfills the criteria established for SCA and it should be offered much more frequently [1,2,4]. The absorption of HU is fast and it occurs in the intestine, with wide body distribution [2,5,6] and maximum plasma doses, ranging between 0.2 mM and 0.8 mM, depending on the dose to which the patient is submitted [7].

Despite being widely used for many years, the effects of HU on human RBC have not been fully clarified. Some studies have linked the use of HU to the oxidation of hemoglobin, with increased lipid peroxidation in the membrane of these cells [8,9].

The main physiological role of the RBC is the transport of gases (O_2, CO_2) from the lungs to the tissues and vice versa, in addition to maintaining the systemic acid/base balance. They have important antioxidant systems, which contribute substantially to their functions and integrity. The damage of RBC integrity, as well as to their membranes, has been shown to contribute significantly to serious pathologies [10–12].

Owing to the importance of the RBC, particularly due to their great carrier potential for drug delivery, it is necessary to broaden the understanding of the HU actions in these cells under non-pathological conditions. This knowledge may be applied to diseases in which HU is included in pharmacotherapeutic protocols. In this context, the authors have aimed to describe the effects of HU in the RBC by evaluating some aspects that are related to the membrane profile, the energy metabolism, and the antioxidant system.

2. Materials and Methods

2.1. Ethics Statement

The present research project was approved by the Ethics Committee on Human Research of the Federal University of São João del Rei, Brazil (n° 2.977.566). All of the enrolled subjects signed an informative consent form.

2.2. Study Samples

This study included six volunteer and healthy adults aged between 24 and 35 years old, of both genders, who were selected from the community of the Federal University of São João del-Rei, Brazil. As the aim has been to evaluate the effect of hydroxyurea on healthy erythrocytes and not characterize its biological effect, the number of samples was small, but this is common practice in similar experiments [13–15]. All of the participants were healthy. This was confirmed by self-reported normal blood tests and an absence of hemoglobin S, in addition to no clinical symptoms. The exclusion criteria were a personal, or a family history of vascular disease, coagulopathies, anemia, the use of daily medications, or current or recent pregnancy. Thirty mL venous blood samples were collected after 8–12 h of fasting when using EDTA as an anticoagulant and were transported between 2 and 8 °C. These were equally distributed into three groups, including (1) control (with no HU), (2) addition of 0.6 M HU, and (3) addition of 0.8 M HU. All of the samples were analyzed at 30 and 60 min after adding HU. The concentrations chosen above were based on the therapeutic regimen of HU, which was adopted for patients with vaso-occlusive crises, that is, 15–35 mg/kg and a plasma peak of between 0.2–0.8 mM [7]. During the incubation period, the samples were kept at 4 °C and after this, the red blood cells concentrate was separated by centrifugation at $700 \times g$ and at 4 °C for 10 min when using a CT18000R (Cientec®) centrifuge (Belo Horizonte, MG, Brazil). The supernatant of each sample was removed, aliquoted, and stored in a freezer at −20 °C until required.

2.3. Preparation of Hemolysate

All of the RBC concentrates were subjected to three successive washes, then doubling the initial sample volume with 0.9% NaCl solution, and centrifuging for 10 min at $700 \times g$. After the last wash, each pellet of RBC was lysed, in the proportion of 1:20 when using a hemolysis solution containing 0.27 M EDTA and 0.007 mM β mercaptoethanol. This was then frozen in acetone in a freezer at -20 °C. The hemolysate was thawed at room temperature and kept on ice for further enzymatic analyses.

2.4. Determination of the Hemoglobin (Hb) Concentration

The determination of the Hb concentration in the hemolysates and plasma was performed by the Cyanomethemoglobin Method, which was adapted from Dacie and Lewis [16], accessed on 6 March 2019 when using Drabkin's solution (0.9 mM potassium ferricyanide and 1.5 mM potassium cyanide), as well as a hemoglobin standard of known concentration (10 g/dL). The reading was performed by scanning on a GENESYS™ 10 UV/Vis spectrophotometer (Thermo Scientific®, Waltham, MA, USA), at 540 nm for 5 min. The Hb concentrations of the samples were calculated by using the calibration factor that was obtained by the absorbance of the Hb standard.

2.5. Oxidative Stress Indicators

2.5.1. Lipid Peroxidation

To investigate lipid peroxidation, the plasma levels of the thiobarbituric acid reactive substances (TBARS) were determined. For this, 100 µL of plasma was added to a tube containing 1 mL of a solution that was made with 15% trichloroacetic acid, 0.38% thiobarbituric acid, and 0.25 N hydrochloric acid. The mixture was incubated at 100 °C for 30 min and then centrifuged at 3000 rpm for 10 min. The optical density (OD) was measured at 535 nm when using a GENESYS™ S10 UV/Vis spectrophotometer (Thermo Scientific®, Waltham, MA, USA). The malondialdehyde (MDA) content was determined by using a standard curve that was constructed in a concentration range of 1 to 100 nM [17,18].

2.5.2. Determination of Fe^{3+} in the Residual Plasma Samples

The Fe^{3+} quantification was performed as described by Adams [19], with slight modifications. In a test tube, 150 µL of plasma was mixed with 1.5 mL of 1 M KSCN, and the volume was adjusted to 3.0 mL with 0.9% NaCl. The samples were then homogenized using Vortex Multifuncional K40-1020 (KASVI®, São José dos Pinhais, PR, Brazil), and the optical density was read at 480 nm (OD480) by using GENESYS™ 10 UV/Vis spectrophotometer (Thermo Scientific®, Waltham, MA, USA). A standard curve was constructed when using a 1 M $FeCl_3$ solution as a standard. The OD was stable for 15 min.

2.5.3. Quantification of Hydrogen Peroxide (H_2O_2)

The quantification of the H_2O_2 production was performed by adding 20 µL of hemolysate to a reaction medium containing 250 µM of ammoniacal ferrous sulfate, 25 mM of H_2SO_4, and 100 µM of xylenol orange in a 1 mL Eppendorf® tube. After vortexing for 10 s, the tube was at rest and protected from the light for 45 min, and then the product formed was read on a GENESYS™ 10 UV/Vis scanning spectrophotometer (Thermo Scientific®, Waltham, MA, USA) at a wavelength of 580 nm [20]. The production of H_2O_2 was calculated from a standard curve by using hydrogen peroxide pure (H_2O_2) in concentrations from 0 to 100 nM.

2.5.4. Determination of the Superoxide Dismutase (SOD) Activity

The SOD activity was determined through a method that was adopted in the authors' laboratory when using epinephrine in an alkaline medium since it is converted into adrenochrome, producing $O^{\cdot -2}$, which is a SOD substrate. The SOD activity was defined by assessing its ability to inhibit epinephrine oxidation. The 100% oxidation was defined in a tube containing a medium, with glycine (50 mM, pH 10) and 25 µL of epinephrine (60 mM pH $-$ 2). The other tubes contained the same medium but with different volumes of hemolysate (2–6 µL). The oxidation reaction was measured and the OD of each condition was read at 480 nm on a GENESYS™ 10 UV/Vis scanning spectrophotometer (Thermo Scientific®, Waltham, MA, USA). The enzyme activity was expressed in units required to inhibit 50% of the rate of the adrenochrome formation [21].

2.5.5. Determination of the Catalase Activity

The catalase activity was determined in RBC hemolysates and was based on the method of Aebi [22], with some modifications. Immediately before determining the activity of this enzyme, the hemolysate was again centrifuged and the supernatant was used for this purpose. Twenty (20) µL of this hemolysate supernatant was added to a cuvette containing 2 mL of 50 mM phosphate buffer (pH 7.0). The reactions were initiated by the addition of 3.4 µL of freshly prepared 30% H_2O_2. The decomposition of H_2O_2 was followed spectrophotometrically at 240 nm, using a GENESYS™ 10 UV/Vis scanning spectrophotometer (Thermo Scientific®, Waltham, MA, USA). The activity was estimated from the slope and expressed as micromoles of H_2O_2 decomposed per minute.

2.5.6. Determination of the Enzymatic Activity of Glutathione Peroxidase (GPx)

The GPx activity was determined by following the oxidation of NADPH when using hydrogen peroxide at a wavelength of 340 nm. The Flohe and collaborators [23] and the Nakamura and Hosada [24] methods were adapted by using a reaction medium, with a 50 mM phosphate buffer, 4 mM EDTA, 1 mM GSH, 1.25 mM NaN_3, 0.16 mM NADPH, and 5 µL of hemolysate. The total volume was adjusted to 1 mL with water. The reaction was initiated by the addition of 4 mM H_2O_2. Modifications in the OD at 340 nm were observed within 2 min. The enzyme activity was measured at nmol NADPH·min^{-1}·L/gHb [23,24].

2.5.7. Determination of the Content of Reduced Glutathione (GSH)

The GSH concentration was evaluated by adding 5 µL of hemolysate to a reaction medium containing 0.1 M phosphate buffer (pH 8.0) and DNTB/EDTA 10 mM. After 15 min at room temperature, the solution was read by a GENESYS™ 10 UV/Vis scanning spectrophotometer (Thermo Scientific®, Waltham, MA, USA) at a wavelength of 412 nm. The GSH concentration was measured by performing a standard curve (correction factor) at nmolGSH·L/gHb [25,26].

2.5.8. Determination of the Glutathione Reductase (GR) Activity

The GR activity was determined by following the method as described by Racker and collaborators [27], with adaptations. The reaction medium that was used contained a phosphate buffer at pH 7.6, 50 mM EDTA, 0.4 mM, 0.1 mM NADPH, 50 µL of hemolysate, and 1 mM glutathione disulfide (GSSG). The volume was completed with distilled water to 1 mL, and the absorbance of the mixture was read by a GENESYS™ 10 UV/Vis scanning spectrophotometer (Thermo Scientific®, Waltham, MA, USA) at a wavelength of 340 nm for 2 min. The enzyme activity was measured at ρmol NADPH·min^{-1}·L/gHb [27].

2.6. Energy Metabolism of the RBC

2.6.1. Determination of the Glucose-6-Phosphate Dehydrogenase (G6PD) Activity

The G6PD activity was measured by incubating 20 µL of hemolysate at 37 °C for 10 min in a reaction medium containing a buffer (0.1 M Tris-HCl and 0.5 mM EDTA, at pH 8.0), 0.01 M $MgCl_2$, 0.2 mM NADP, and 0.6 mM glucose-6-phosphate. The absorbance was measured in a GENESYS™ 10 UV/Vis spectrophotometer (Thermo Scientific®, Waltham, MA, USA) at 340 nm, and its variation per minute was evaluated, with and without the substrate (glucose-6-phosphate). The enzyme activity was quantified by reducing $NADP^+$ to NADPH. The enzymatic activity was calculated at Ul·gHb^{-1}·min^{-1}, where U was equal to 1 µmol $NADP^+$·min^{-1}·mL^{-1} [28].

2.6.2. Determination of the Hexokinase Activity (HEX)

The HEX activity was determined by using 50 µL of hemolysate in a reaction medium containing a buffer (0.1 M Tris-HCl and 0.5 mM EDTA, at pH = 8.0), 0.01 M $MgCl_2$, 0.2 mM NADP, 2 mM glucose, 2 mM of ATP, and 0.1 of a unit of glucose-6-phosphate dehydrogenase enzyme. The absorbance was measured in a GENESYS™ 10 UV/Vis spectrophotometer (Thermo Scientific®, Waltham, MA, USA) at 340 nm, and its variation per minute was evaluated, with and without ATP. The enzyme activity was quantified by the reduction of $NADP^+$ to NADPH. The enzymatic activity was calculated at $Ul \cdot gHb^{-1} \cdot min^{-1}$, where U was equal to 1 μmol of $NADP^+ \cdot min^{-1} \cdot mL^{-1}$ [28].

2.6.3. Determination of the Pyruvate Kinase (PK) Activity

The PK activity was determined by incubating 20 μL of hemolysate at 37 °C for 10 min in a reaction medium containing a buffer (0.1 M Tris-HCl and 0.5 mM EDTA, at pH = 8.0), 0.01 M $MgCl_2$, 0.1 M KCl, 0.2 mM NADH, 1.5 mM ADP, 6 units of lactate dehydrogenase enzyme, and 5 mM phosphoenolpyruvate (PEP). The absorbance was measured in a GENESYS™ 10 UV/Vis spectrophotometer (Thermo Scientific®, Waltham, MA, USA) at 340 nm, and its variation per minute was evaluated, with and without PEP. The enzyme activity was quantified by the oxidation of NADPH to $NADP^+$. The enzymatic activity was calculated at $Ul \cdot gHb^{-1}.min^{-1}$, where U was equal to 1 μmol of $NADP^+ \cdot min^{-1} \cdot mL^{-1}$ [28].

2.7. Statistical Analysis

The results were analyzed when using GraphPad Prism 5 software. For the analysis of data normality, the Shapiro-Wilk test (for any sample size) was used. For the data with a normal distribution, the Analysis of Variance (ANOVA) was used, followed by Tukey's multiple comparison test. Values of $p < 0.05$ were considered significant.

3. Results

3.1. Determination of the Oxidative Stress Indicators

Biological markers, such as Fe^{3+}, MDA, and H_2O_2 are frequently used to predict oxidative stress in the RBC. During a 30-min treatment, it was possible to observe a profile increase of these indicators in the treated groups when compared to the control group, as described below. Figure 1 illustrates that the exposure of the RBC to concentrations of 0.6 and 0.8 mM of HU resulted in an increase in plasma Fe^{3+} of about 39.9% and 28.6%, respectively, compared to the control group (Figure 1A) ($p < 0.05$). Figure 1B depicts the evaluation of lipid peroxidation when using the TBARS method. The MDA content increased by 52% in the HU-treated group at a concentration of 0.8 mM ($p < 0.05$). H_2O_2 was also measured because of its ability to react and damage the RBC membrane. When evaluating the total H_2O_2 content in the hemolysate, a significant increase of 25% and 28% were found in the concentrations of 0.6 mM and 0.8 mM, respectively (Figure 1C).

The authors also examined whether the alterations in the oxidative stress indicators were time-dependent. After an hour of incubation with 0.8 mM HU, it was observed that the Fe^{3+} content was 20% lower ($p < 0.05$) when compared with the control groups (Figure 2A). A significant decrease of 47% ($p < 0.05$) in the MDA content in the plasma that was treated with 0.8 mM HU was also found when compared with the control groups. When comparing the groups that were treated with HU (60 min), a significant decrease (41%) ($p < 0.05$) in the MDA was observed with the concentration of 0.8 mM HU, in relation to 0.6 mM (Figure 2B).

The hydrogen peroxide content in the hemolysate, as shown in Figure 2C, increased significantly in the treated group at both HU concentrations when compared to control. The increase was 15%, ($p < 0.001$) and 7% at concentrations of 0.6 mM and 0.8 mM ($p < 0.05$), respectively. A significant decrease of 7% ($p < 0.05$) in the H_2O_2 was also observed in the sample of hemolysate that was treated with 0.8 mM when compared to that treated with 0.6 mM (Figure 2C).

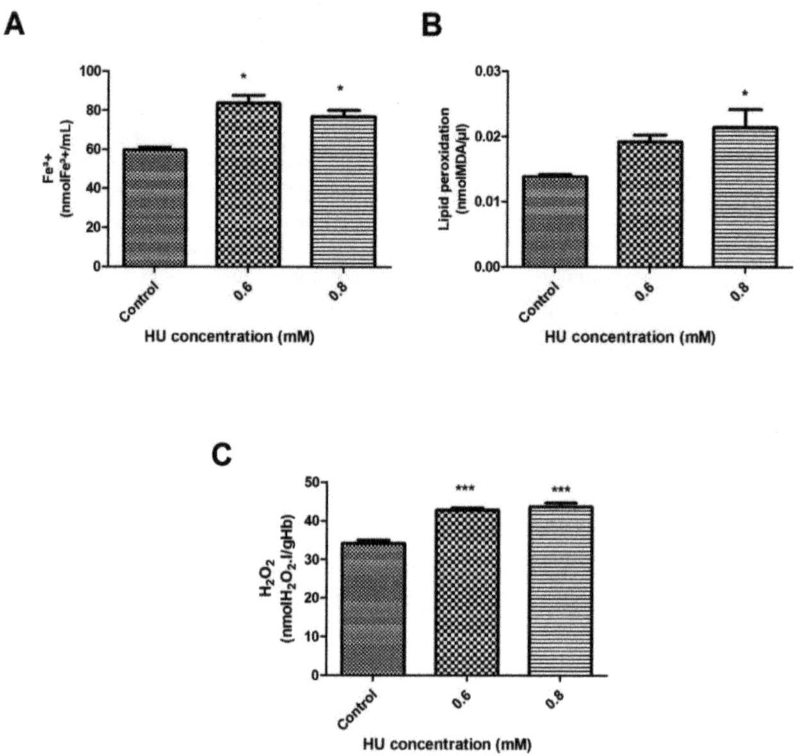

Figure 1. Oxidative stress indicators in the erythrocytes of the control samples that were treated with 0.6 and 0.8 mM HU ($n = 3$) for 30 min. (**A**) Fe^{3+} was measured in plasma; (**B**) Lipid peroxidation (MDA) was measured in plasma; (**C**) Hydrogen peroxide (H_2O_2) was measured in hemolysate. In all of the cases, * represents a comparison between the controls and the treated groups (* $p < 0.05$ and *** $p < 0.001$).

3.2. Antioxidant Defense of the RBC

As reported previously, oxidation is a possible mechanism of damage to the RBC, together with the enzymes of the antioxidant system, such as CAT, SOD, GPx, and GR, which act as protectors of these cells against attack by the reactive species. The possible oxidative effects of HU on the RBC were also measured by the antioxidant enzyme activities in the treated and control samples.

When evaluating the activity of the SOD enzyme, a significant increase of about 12% was observed after 30 min of incubation with HU, at both concentrations of HU when compared to the control group ($p < 0.001$) (Figure 3A). After 1 h of incubation, it was also possible to observe a significant increase ($p < 0.001$) of 13% and 8%, in the concentrations of 0.6 mM and 0.8 mM of HU, respectively, when compared to the control groups (Figure 4A). When comparing the two treated groups (30 and 60 min), a significant 5% increase occurred in the 0.6 mM HU (Figure 4A).

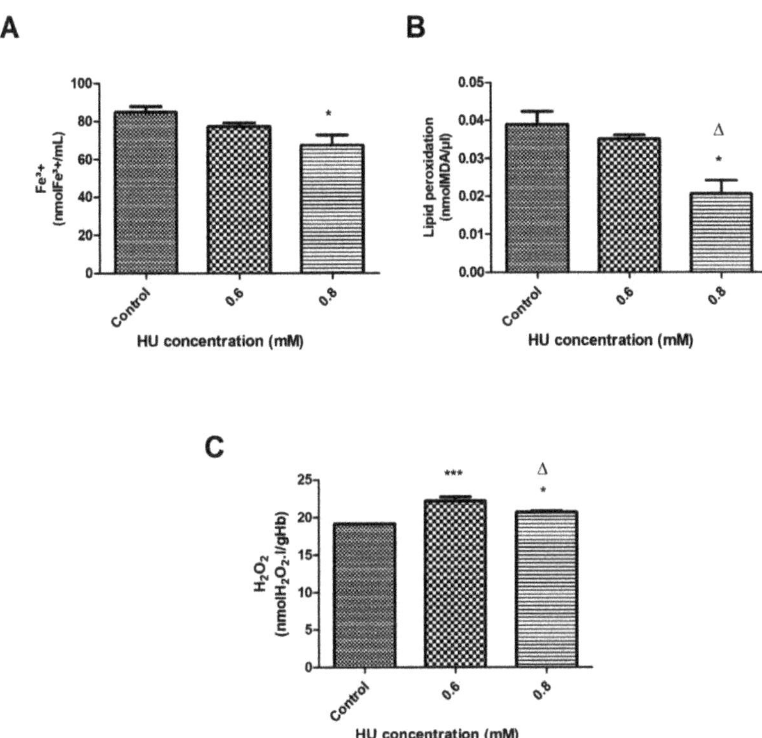

Figure 2. Oxidative stress indicators in the erythrocytes of the control samples that were treated with 0.6 and 0.8 mM HU ($n = 3$) for 1 h. (**A**) Fe^{3+} was measured in plasma; (**B**) Lipid peroxidation (MDA) was measured in plasma; (**C**) Hydrogen peroxide (H_2O_2) was measured in hemolysate. In all of the cases, * represents a comparison between the controls and the treated groups (* $p < 0.05$ and *** $p < 0.001$), while Δ represents a comparison between the groups treated with 0.6 and 0.8 mM HU ($p < 0.05$).

The catalase activity did not differ significantly among all of the examined groups (Figures 3B and 4B). The activity of the enzyme glutathione peroxidase showed a significant increase at both of the treatment times. The groups that were treated with HU showed an increase of 82% ($p < 0.05$) (30 min) and 15.5% ($p < 0.05$) (1 h) when compared with the control groups (Figures 3C and 4C).

The activity of glutathione reductase in the group that was treated with 0.6 mM HU (after 30 min) showed a significant increase ($p < 0.05$), which was 2 times greater when compared to the control groups (Figure 3D). The 1-h treatment presented a different profile, that is, the samples that were treated with 0.6 and 0.8 mM of HU obtained a significantly lower GR activity ($p < 0.05$) of about 26% and 17%, respectively, compared to the control groups (Figure 4D). The GSH content increased significantly ($p < 0.05$) in the two HU concentrations over 30 min at 62% and 52% in the samples that were treated with 0.6 and 0.8 mM, respectively (Figure 3E). After the 1-h treatment, it was possible to observe a significant increase ($p < 0.05$) of 6% but only at a concentration of 0.6 mM, while a decrease of 6% ($p < 0.05$) was observed in the treated group with 0.8 mM HU when compared to the control groups (Figure 4E). At a concentration of 0.6 mM HU, there was a significant increase of 12% ($p < 0.05$) in the GSH content when compared to the sample that was treated with 0.8 mM HU (Figure 4E).

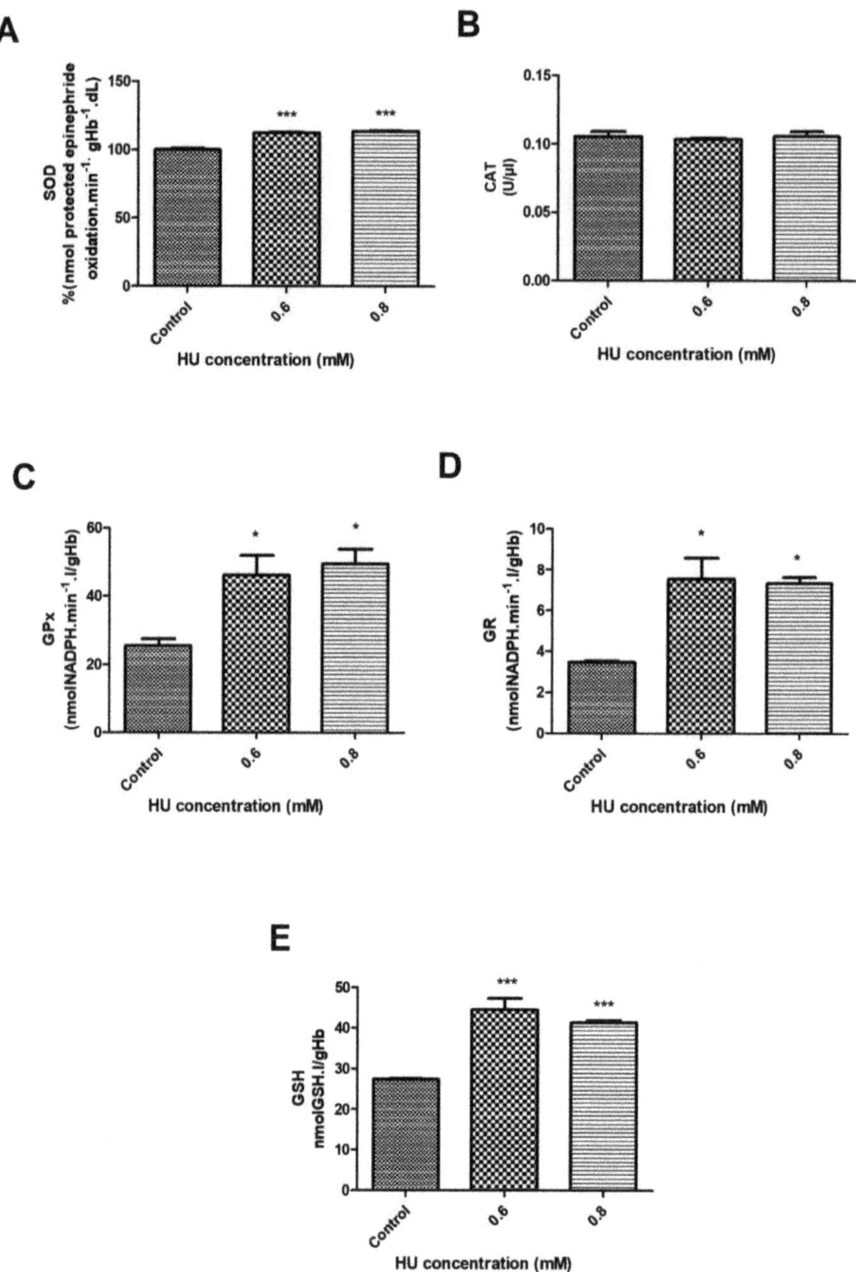

Figure 3. Enzymatic activity of the erythrocyte antioxidant system in the control samples that were treated with 0.6 and 0.8 mM HU ($n = 3$) for 30 min. Hemolysate was used to measure (**A**) Superoxide dismutase (SOD); (**B**) Catalase (CAT); (**C**) Glutathione peroxidase (GPx); (**D**) Glutathione reductase (GR); (**E**) Reduced glutathione. In all of the cases, * it represents a comparison between the control groups and the treated groups (* $p < 0.05$ and *** $p < 0.001$).

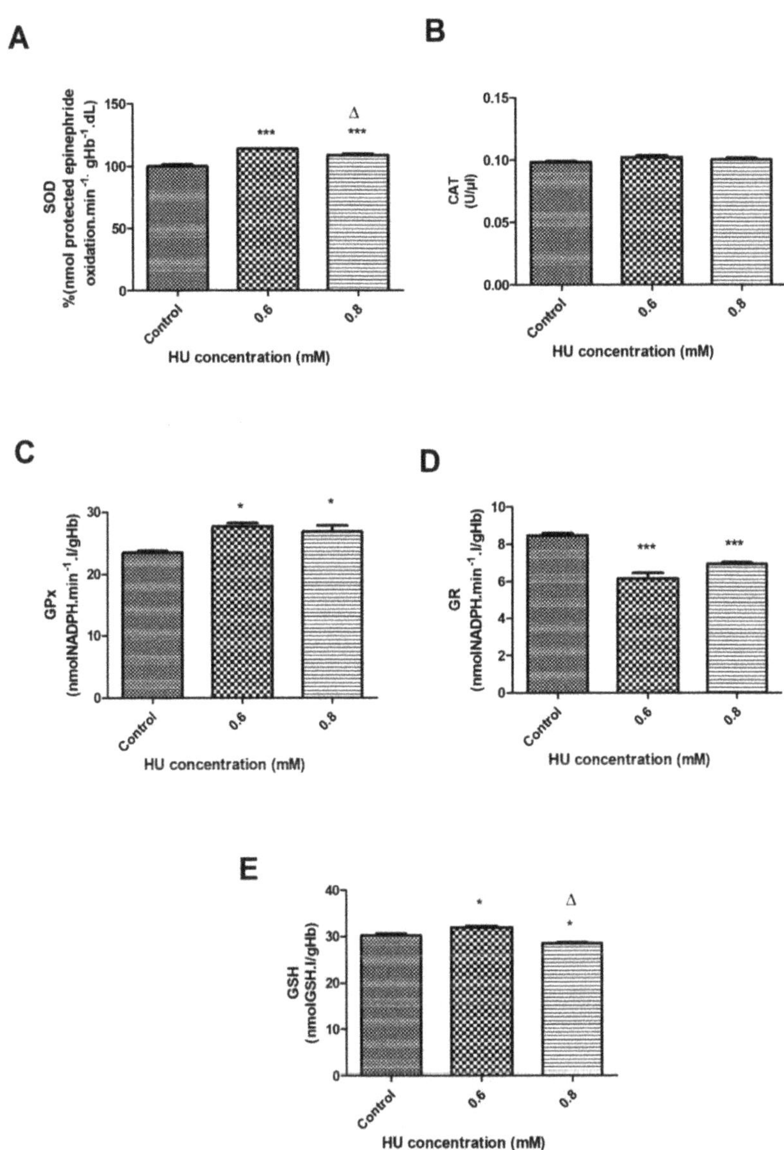

Figure 4. Enzymatic activity of the erythrocyte antioxidant system in the control samples that were treated with 0.6 and 0.8 mM HU (n = 3) for 1 h. Hemolysate was used to measure (**A**) Superoxide dismutase (SOD); (**B**) Catalase (CAT); (**C**) Glutathione peroxidase (GPx); (**D**) Glutathione reductase (GR); (**E**) Reduced glutathione. In all of the cases, * represents a comparison between the controls and the treated groups (* $p < 0.05$ and *** $p < 0.001$), while Δ represents a comparison between the groups that were treated with 0.6 and 0.8 mM HU ($p < 0.05$).

3.3. The Energy Metabolism of the RBC

The pathway (Glucose $\rightarrow$ Glucose 6-phosphate $\rightarrow$ 2 ATP + 2 lactate generation) uncouples the ATP generation from the oxygen consumption and it is the only source of metabolic energy for the mature human RBC. The primary outcome revealed that with 30 min of treatment with HU, the G6PD had a significant increase of 37.5% and 54.6% in its activity at concentrations of 0.6 mM and 0.8 mM, respectively ($p < 0.05$), compared to the control groups (Figure 5A).

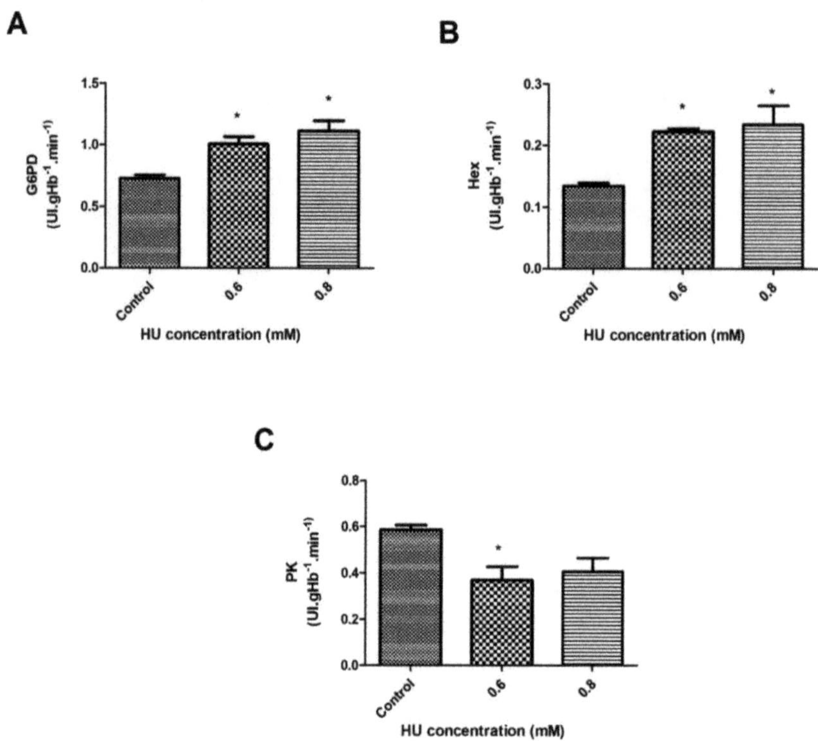

Figure 5. Enzymatic activity of the erythrocyte energy metabolism in the control samples that were treated with 0.6 and 0.8 mM HU ($n = 3$) for 30 min. Hemolysate was used to measure (**A**) Glycose-6-phosphate-dehydrogenase (G6PD); (**B**) Hexokinase (Hex); (**C**) Pyruvate kinase (PK). In all of the cases, * represents a comparison between the control groups and the groups that were treated with HU (* $p < 0.05$).

For the 1-h treatment, a significant increase in the G6PD activity of 30% ($p < 0.001$) was observed at a concentration of 0.6 mM HU, whereas at 0.8 mM of HU, the G6PD activity was lower by 23% ($p < 0.001$), compared to the control groups. When analyzing both of the HU concentrations, the G6PD activity was 42% lower when comparing 0.8 mM to 0.6 mM of HU (Figure 6A).

The hexokinase activity in the 30-min treatment had a significant increase of around 73% at both of the HU concentrations when compared to the control groups (Figure 5B). During the 1-h treatment, it was not possible to observe significant differences between the control groups and the treated groups. There was a significant decrease of about 77% ($p < 0.05$) in the hexokinase values of the sample that was treated with 0.8 mM, compared to that treated with 0.6 mM HU (Figure 6B).

Regarding the activity of pyruvate kinase (PK), it was possible to observe a significant reduction in the PK activity of about 23% ($p < 0.05$) after 30 min of incubation with 0.6 mM HU when compared to the control groups (Figure 5C). Within 1 h, a significant decrease of about 59% ($p < 0.05$) in the PK activity was also observed at a concentration of 0.8 mM, compared with the control groups (Figure 6C). To facilitate a data comparison, the results are summarized in Tables 1 and 2.

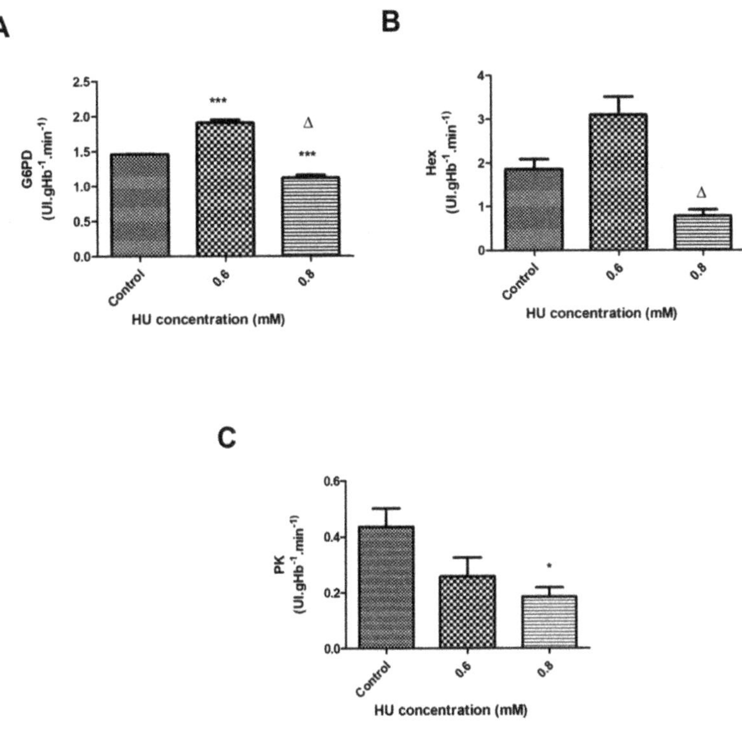

Figure 6. Enzymatic activity of the erythrocyte energy metabolism in the control samples that were treated with 0.6 and 0.8 mM HU ($n = 3$) for 1 h. Hemolysate was used to measure (**A**) Glucose-6-phosphate-dehydrogenase (G6PD); (**B**) Hexokinase (Hex); (**C**) Pyruvate kinase (PK). In all of the cases, * represents a comparison between the control groups and the groups that were treated with HU (* $p < 0.05$ and *** $p < 0.001$), while Δ represents a comparison between the groups that were treated with 0.6 and 0.8 mM HU ($p < 0.05$).

Table 1. Comparison of the effects of HU at different concentrations and treatment times on the erythrocytes of healthy individuals when in relation to the control groups.

Time of Treatment	30 min		1 h	
HU Concentration	0.6 mM	0.8 mM	0.6 mM	0.8 mM
Variables				
Fe^{3+}	↑	↑	-	↓
TBARS	-	↑	-	↓
H_2O_2	↑	↑	↑	↑
SOD	↑	↑	↑	↑
CAT	-	-	-	-
GPx	↑	↑	↑	↑
GR	↑	↑	↓	↓
GSH	↑	↑	↑	↓
G6PD	↑	↑	↑	↓
HEX	↑	↑	-	-
PK	↓	-	-	↓

TBARS = Thiobarbituric acid reactive substances, SOD = Superoxide dismutase; CAT = Catalase; GPx = Glutathione peroxidase; GR = Glutathione reductase; GSH = Reduced Glutathione; G6PD = Glucose-6-phosphate dehydrogenase; HEX = Hexokinase; PK = Pyruvate kinase; ↑ = a significant increase; ↓ = a significant decrease; - = no difference.

Table 2. Comparison of the data for the different markers when treating the erythrocytes with 0.8 mM versus 0.6 mM HU for 1 h.

Concentration	0.8 mM
Fe^{3+}	-
TBARS	↓
H_2O_2	↓
SOD	↓
CAT	-
GPx	-
GR	-
GSH	↓
G6PD	↓
HEX	↓
PK	-

TBARS = Thiobarbituric acid reactive substances, SOD = Superoxide dismutase; CAT = Catalase; GPx = Glutathione peroxidase; GR = Glutathione reductase; GSH = Reduced Glutathione; G6PD = Glucose-6-phosphate dehydrogenase; HEX = Hexokinase; PK = Pyruvate kinase; ↑ = a significant increase; ↓ = a significant decrease; - = no difference.

4. Discussion

Hydroxyurea is a drug that is used in the treatment of various diseases, such as cancer and sickle cell anemia, in addition to having a high potential to treat other diseases [29]. In the present in vitro study carried out with the RBC from healthy individuals, changes in these cells were observed after incubations of 0.6 mM and 0.8 mM HU for 30 and 60 min.

The results showed that after 30 min of treatment with HU, at both of the concentrations, there was an increase in the Fe^{3+}, MDA, and H_2O_2 levels (Figure 1A–C, respectively). According to Iyamu and collaborators [9], after treating the blood in vitro with different concentrations of HU, there was an increase in the oxidation of hemoglobin A and hemoglobin S, the latter of which is present in the patients with sickle cell anemia. The oxidation of hemoglobin leads to the formation of methemoglobin; consequently to the release of Fe^{3+}, contributing to its increase in the plasma [30]. Fe^{3+} can increase the generation of reactive oxygen species, as well as reacting with the cell membranes, resulting in an increase in lipid peroxidation in both of the cases [31].

MDA is a by-product of the lipid peroxidation of unsaturated phospholipid chains [32], and this increased significantly after 30 min of treatment with 0.8 mM HU, (Figure 1B). In a previous study that was carried out on Wistar rats, which were treated with different doses of HU, the authors reported an increase in the MDA levels that were caused by the drug, proportional to its concentration [29].

Concerning the H_2O_2 content, a significant increase in the treated groups was observed, as seen in Figures 1C and 2C. Consequently, this can cause the production of other free radicals, such as the hydroxyl radical (OH). OH , when reacting with Fe^{3+}, is one of the most potent oxidants known, and it can accentuate the oxidative stress event [30,33]. Huang and collaborators [34], when performing a treatment in yeasts with different concentrations of HU, observed that there was an increase in the levels of the reactive oxygen species (ROS), especially H_2O_2. One hypothesis to explain the increase in the concentration of H_2O_2 consists of an increase in the activity of the superoxide dismutase enzyme (SOD), as was observed in the present study (Figures 3A and 4). This enzyme, which is present in the erythrocyte antioxidant system, was responsible for generating H_2O_2 by dismuting the superoxide anion to H_2O_2.

It has already been described that HU also exhibits a greater oxidative potential of hemoglobin when compared to hydrogen peroxide, whilst under the same conditions of drug concentration and time of exposure, as in this current study [9]. This fact reinforces the hypothesis that HU not only induces an increase in H_2O_2 but it is also capable of inducing the formation of other radicals, contributing to a higher oxidation rate.

After 1 h of treatment with 0.8 mM HU, the levels of some oxidative stress markers, such as Fe^{3+} and MDA, decreased significantly when compared to the controls (Figure 2A,B, respectively).

Within this context, it is important to mention another study that also treated the RBC from healthy individuals with potential oxidizing agents, and later with HU. The authors observed that this would have an antioxidant potential since there was a decrease in lipid peroxidation and methemoglobin formation. The finding that HU could react preferably with the radicals formed by such oxidizing agents, would result in protection for the erythrocyte membrane [7]. This data is interesting because it would explain, at least in part, the beneficial effects resulting from the use of HU in patients with diseases, such as sickle cell anemia, and this drug could contribute to the reduction of free radicals. In 2012, a study by Torres et al. [35] reported a decrease in lipid peroxidation in the patients with sickle cell anemia who used HU when compared to the non-users. Another study that was developed by Brose et al. [36] reported that the treatment of rat neurons with neurotoxic agents, such as H_2O_2, and subsequently treated with HU, resulted in a reduction of the neurotoxic stress caused by these agents, as well as inducing an increase in the ATP content in these cells.

When relating the studies above to the current work, we realized that both of the concentrations of HU that were used for the treatment of the erythrocytes had, at first, an oxidizing effect. Over time, the erythrocyte protection mechanisms seem to be activated, thus reducing the damage to these cells. In diseases where there is natural oxidative stress, such as sickle cell anemia, the patients can be treated with HU, as the use of this drug can benefit those individuals affected by the disease because of its ability to reduce such stress. A study that was carried out with patients with sickle cell anemia, treated or not with HU, observed that the patients that were treated with the drug had a lower number of ROS and greater erythrocyte deformability when compared to the untreated patients [37]. The greater deformability of the RBC can be a favorable effect since such cells would present greater malleability and pass through the capillaries more easily. Despite the significant increase in hydrogen peroxide, the catalase levels did not show significant changes in any of the conditions that were evaluated in this study (Figures 3B and 4B).

The enzymes that are responsible for removing endogenous H_2O_2 are peroxiredoxin II, an enzyme that is activated by concentrations of up to 0.41 mM of H_2O_2, and glutathione peroxidase, which can reduce H_2O_2, as well as the hydroperoxides and peroxynitrites [38]. In this study, the activity of the enzyme GPx was increased at both times of the treatment with HU, which may be related to the increase in H_2O_2 that was identified. Corroborating with the findings of this study, Choo et al. [39] demonstrated that a treatment with HU that was induced in the patients with sickle cell anemia, not only increased the GPx activity but also increased its expression.

The authors also observed that the GR activity increased at both of the treatment concentrations after 30 min (Figure 3D). This finding can be explained by an increase in the GPx activity in all of the samples treated with HU, which generated a greater amount of the substrate (GSSG), culminating in an increase in the activity of glutathione reductase, an enzyme that is responsible for maintaining the GSH levels in the RBC [40]. An increase in the GSH content resulting from the treatment of the two HU concentrations was observed in the current study after 30 min of incubation with the drug (Figure 3E). This finding is in line with the data reported by Teixeira and colleagues [41], who previously described an increase in the GSH that was induced by the treatment with HU in those patients with sickle cell anemia.

The activity of the G6PD and HEX enzymes was increased in both of the HU concentrations after 30 min of incubation. This data has suggested that the energy metabolism of the RBC was also stimulated by the presence of HU during the treatment times (Figure 5A,B). The PK activity was reduced at a concentration of 0.6 mM (Figure 5C), indicating that the pentose phosphate pathway was preferably more active under these conditions.

The fact that the pentose pathway was preferably active, instead of the glycolytic pathway, could explain the increase in $Fe^{3}+$ that was found by the authors (Figure 1A) after the treatment with HU for 30 min. The NADH that was produced in this pathway was not available for use by the enzyme, cytochrome-b5 reductase, which is responsible for most of the interconnection of $Fe^{3}+$ to Fe^{2+} (methemoglobin to hemoglobin) in the RBC.

The increased activity of HEX produced glucose-6-phosphate, a substrate for the G6PD enzyme, which plays an important role in the antioxidant system of the RBC since it produces NADPH [42], which is a cofactor for the production of GSH, one of the most important non-enzymatic antioxidants in the RBC. GSH was responsible for helping to protect the membranes against the superoxides, as well as other types of reactive oxygen species [40,43].

The induction of an increase in the G6PD activity was an important effect of HU since patients with sickle cell anemia have a restricted flow of NADPH. At low oxygen tensions, HbS binds very strongly to the cytoplasmic domain of band 3, a protein that is responsible for regulating glycolysis. Thus, the enzymes of the glycolytic pathway that are linked to the cytoplasmic domain of band 3 (phosphofructokinase and glyceraldehyde-3-phosphate dehydrogenase), remain in the cytoplasm, creating competition between the glycolytic and pentose pathways for the substrate (glucose- 6-phosphate), restricting the formation of NADPH [44]. In sickle cell anemia, the greater the release of the ROS, the stronger is the link between the cytoplasmic domain of band 3, HbS, and the hemicromes that are formed by the denaturation of hemoglobin. Increased G6PD activities enhance the production of NADPH and this might help to lessen the clinical effects caused by sickle cell anemia.

After the exposure of the RBC to HU for 1 h, a decrease in the GR activity was observed (Figure 4D) at the 0.6 and 0.8 mM HU concentrations, while there was an increase in the GSH content with 0.6 mM, followed by a decrease of this enzyme when using 0.8 mM of the drug (Figure 4E). In a previous study by this current group, a reduction of about 41% in the GSH content and 17% in the GR activity was observed in the blood samples from the patients with sickle cell anemia when treated with HU, compared to the patients who did not use this medicine [45].

These results might be explained by the reduction in the GSH levels and the activity of the enzyme G6PD, which was responsible for the production of NADPH (cofactor for GR) in the concentration of 0.8 mM HU; consequently, this would decrease the activity of the GR. In addition, the enzyme saturation might also explain the decreased activity of this enzyme. The G6PD might have had a decreased activity because there was no increase in the production of its substrate (glucose-6-phosphate) by the enzyme hexokinase, which remained without significant changes (Figure 6B) between the controls and those treated with HU.

Raththagala et al. [46] reported a significant increase in the ATP release in the HU-treated rabbit RBC after 20 min, due to the induction of greater RBC deformability. In these circumstances, the greater release of ATP could indicate a greater functioning of the enzymes in the glycolytic pathway. In this study, the authors did not find significant differences in the activity of the hexokinase enzyme after 1 h of incubation with 0.8 mM HU, although there was a reduction in the activity of pyruvate kinase at this concentration (Figure 6C). This data corroborates the possible inactivation of the glycolytic pathway under these conditions and that it was also reduced after the treatment with HU for 30 min. This has allowed the authors to infer that the treatments with HU most likely preferentially stimulated the pentose pathway and not the glycolytic pathway.

The $Fe^{3}+$ content did not change significantly at the concentration of 0.6 mM HU when in relation to the controls, and it decreased when compared to the treatment with 0.8 mM HU. This result allows the authors to suggest that the increase in the activity of the enzyme, NADPH-methemoglobin reductase, helped in the process of converting $Fe^{3}+$ to Fe^{2+} (methemoglobin to hemoglobin). The decrease in the content of $Fe^{3}+$ might mean that the index of the reactive oxygen species decreased, as well as the oxidation of hemoglobin since a decrease in lipid peroxidation at a concentration of 0.8 mM was also identified.

When comparing the data that was obtained from the RBC that were treated with HU at concentrations of 0.6 and 0.8 mM after 1 h, it can be assumed that the latter concentration appeared to be more efficient in the fight against the free radicals. The experimental condition of the RBC that was incubated for 1 h with 0.8 mM HU was able to reduce the levels of the oxidative stress markers (hydrogen peroxide and lipid peroxidation—Figure 2B,C). It is also worth noting that 0.8 mM HU was the only dose capable of decreasing the Fe^{3+} content when compared with the control groups (Figure 2A).

Another interesting change when comparing both of the HU concentrations that were used in this study was the significant decrease in the SOD activity. This reduction, which was associated with the decrease in the concentration of $Fe^{3}+$, allows the authors to infer that the oxidation process of hemoglobin decreased at the concentration of 0.8 mM HU after 1 h of incubation since Fe^{3+} and the superoxide anion (substrate for SOD) were products of the Hb oxidation [47,48].

It is important to highlight the difference that was identified between the HU concentrations that were used in this study. The activities of the HEX and G6PD enzymes (Figure 6A,B) and the GSH content (Figure 4E) decreased when the RBC were incubated with 0.8 mM when compared to 0.6 mM HU. Two hypotheses are to be considered: (1) the enzymatic saturation during the process of protection against the ROS. GR, which produces GSH, did not have the necessary cofactor (NADPH) produced by G6PD, which in turn required the production of glucose-6-phosphate by hexokinase to function; (2) the amount of the reactive oxygen species decreased in the presence of 0.8 mM of HU so that the production of GSH and NADPH was not stimulated, a fact that was confirmed by the decrease in lipid peroxidation (Figure 6B). The differences between the results obtained by the two HU concentrations (0.6 and 0.8 mM) have led the authors to reflect on what would be the best dose to be used by the patients with diseases of which HU is a part of the therapeutic scheme. In diseases, such as sickle cell anemia, in which the existence of oxidative stress is one of the main causes related to the clinical symptoms, the dose of 0.8 mM might have a better effect on the antioxidant system and the production of ROS; consequently, reducing the oxidative stress and therefore, benefiting the patients.

Although this study was carried out with the RBC from healthy individuals, the changes described in this study might help to elucidate the mechanisms of action of HU when administered for therapeutic purposes for certain diseases. The evidence in the literature indicates that the study's results are similar to those reported in other studies that involve the treatment with HU, as in sickle cell anemia.

The effects of HU in those patients with sickle cell anemia go far beyond the increase in HbF and the decrease in cell production by the bone marrow. Further studies are still needed to understand other possible effects of using this drug on the RBC and whether these effects are harmful or not for the patient. The data presented in this paper raises the need for further investigations, to expand the effects of HU through other laboratory markers, opening perspectives for future studies. As an example, a more in-depth study of the HU effects on the composition and modulation of the RBC membranes can be suggested.

Author Contributions: Conceptualization, M.d.B.P., D.R.A.R., V.F.C., L.A.d.O.B., H.d.L.S. and C.O.R.; methodology, M.d.B.P., D.R.A.R., V.F.C., L.A.d.O.B., H.d.L.S. and C.O.R.; analysis, C.O.R.; resources, M.d.B.P., V.F.C., L.A.d.O.B. and H.d.L.S.; data curation, M.d.B.P., V.F.C., L.A.d.O.B. and H.d.L.S.; writing—original draft preparation, C.O.R., G.A.S.M. and L.S.N.; writing—review and editing, M.d.B.P., V.F.C., L.A.d.O.B., H.d.L.S., D.R.A.R., G.A.S.M., H.d.L.S. and C.O.R.; supervision, H.d.L.S.; funding acquisition, M.d.B.P., V.F.C., L.A.d.O.B., H.d.L.S., D.R.A.R. All authors have read and agreed to the published version of the manuscript.

Funding: This study was financed in part by the Coordenação de Aperfeiçoamento de Pessoal de Nível Superior-Brasil (CAPES)-Finance Code 001, FAPEMIG (Fundação de Amparo à Pesquisa do Estado de Minas Gerais) APQ-00855-19, and National Council for Scientific and Technological Development-CNPq 305173/2018-9.

Institutional Review Board Statement: The study was conducted according to the guidelines of the Declaration of Helsinki, and approved by the Ethics Committee on Human Research of the Federal University of São João del Rei, Brazil (n° 2.977.566/2018).

Informed Consent Statement: Informed consent was obtained from all of the subjects involved in the study.

Data Availability Statement: Data is contained within the article.

Acknowledgments: The authors are thankful to Maria das Graças Carvalho (UFSJ) for the helpful evaluation of the manuscript.

Conflicts of Interest: The authors declare no conflict of interest.

References

1. Kovacic, P. Hydroxyurea (therapeutics and mechanism): Metabolism, carbamoyl nitroso, nitroxyl, radicals, cell signaling, and clinical applications. *Med. Hypotheses* **2011**, *76*, 24–31. [CrossRef]
2. Agrawal, R.K.; Patel, R.K.; Shah, V.; Nainiwal, L.; Trivedi, B. Hydroxyurea in sickle cell disease: Drug review. *Indian J. Hematol. Blood Transfus.* **2014**, *30*, 91–96. [CrossRef] [PubMed]
3. Legrand, T.; Rakotoson, M.G.; Galacteros, F.; Bartolucci, P.; Hulin, A. Determination of hydroxyurea in human plasma by HPLC-UV using derivatization with xanthydrol. *J. Chromatogr. B Anal. Technol. Biomed. Life Sci.* **2017**, *1064*, 85–91. [CrossRef]
4. Navarra, P.; Preziosi, P. Hydroxyurea: New insights on an old drug. *Crit. Rev. Oncol. Hematol.* **1999**, *29*, 249–255. [CrossRef]
5. Tracewell, W.G.; Trump, D.L.; Vaughan, W.P.; Smith, D.C.; Gwilt, P.R. Population pharmacokinetics of hydroxyurea in cancer patients. *Cancer Chemother. Pharmacol.* **1995**, *35*, 417–422. [CrossRef]
6. Walker, A.L.; Franke, R.M.; Sparreboom, A.; Ware, R.E. Transcellular movement of hydroxyurea is mediated by specific solute carrier transporters. *Exp. Hematol.* **2011**, *39*, 446–456. [CrossRef] [PubMed]
7. Agil, A.; Sadrzadeh, S.M.H. Hydroxy-urea protects erythrocytes against oxidative damage. *Redox Rep.* **2000**, *5*, 29–34. [CrossRef]
8. Malec, J.; Przybyszewski, W.M.; Grabarczyk, M.; Sitarska, E. Hydroxyurea has the capacity to induce damage to human erythrocytes, which can be modified by radical scavengers. *Biochem. Biophys. Res. Commun.* **1984**, *120*, 566–573. [CrossRef]
9. Iyamu, E.W.; Fasold, H.; Roa, D.; del Pilar Aguinaga, M.; Asakura, T.; Turner, E.A. Hydroxyurea-induced oxidative damage of normal and sickle cell hemoglobins in vitro: Amelioration by radical scavengers. *J. Clin. Lab. Anal.* **2001**, *15*, 1–7. [CrossRef]
10. Lorenzi, T.F. *Manual of Propaedeutic and Clinical Hematology*, 4th ed.; Guanabara Koogan: Rio de Janeiro, Brazil, 2006.
11. Thom, C.S.; Dickson, C.F.; Gell, D.A.; Weiss, M.J. Hemoglobin variants: Biochemical properties and clinical correlates. *Cold Spring Harb. Perspect. Med.* **2013**, *3*, a011858. [CrossRef]
12. Luzzatto, L.; Nannelli, C.; Notaro, R. Glucose-6-Phosphate Dehydrogenase Deficiency. *Hematol. Oncol. Clin. N. Am.* **2016**, *30*, 373–393. [CrossRef]
13. Arbos, K.A.; Claro, L.M.; Borges, L.; Santos, C.A.; Weffort-Santos, A.M. Human erythrocytes as a system for evaluating the antioxidant capacity of vegetable extracts. *Nutr. Res.* **2008**, *28*, 457–463. [CrossRef] [PubMed]
14. Petit, K.; Suwalsky, M.; Colina, J.R.; Contreras, D.; Aguilar, L.F.; Jemiola-Rzeminska, M.; Strzalka, K. Toxic effects of the anticancer drug epirubicin in vitro assayed in human erythrocytes. *Toxicol. In Vitro* **2020**, *68*, 104964. [CrossRef]
15. Morabito, R.; Romano, O.; La Spada, G.; Marino, A. H_2O_2-Induced Oxidative Stress Affects $SO_4=$ Transport in Human Erythrocytes. *PLoS ONE* **2016**, *11*, e0146485. [CrossRef]
16. Dacie, J.V.; Lewis, S.M. *Pratical Hematology*, 8th ed.; Churchill Livingstone: Edinburgh, UK, 1996.
17. Atmaca, G. Antioxidant effects of sulfur-containing amino acids. *Yonsei Med. J.* **2004**, *45*, 776–788. [CrossRef] [PubMed]
18. Costa, C.M.; Santos, R.C.C.; Lima, E.S. A simple automated procedure for thiol measurement in human serum samples. *J. Bras. Patol. Med. Lab.* **2006**, *42*, 345–350. [CrossRef]
19. Adams, P.E. Determining Iron Content in Foods by Spectrophotometry. *J. Chem. Educ.* **1995**, *72*, 649–651. [CrossRef]
20. Jiang, Z.Y.; Woollard, A.C.; Wolff, S.P. Hydrogen peroxide production during experimental protein glycation. *FEBS Lett.* **1990**, *268*, 69–71. [CrossRef]
21. Misra, H.P.; Fridovich, I. The role of superoxide anion in the autoxidation of epinephrine and a simple assay for superoxide dismutase. *J. Biol. Chem.* **1972**, *247*, 3170–3175. [CrossRef]
22. Aebi, H. Catalase in vitro. *Methods Enzymol.* **1984**, *105*, 121–126. [CrossRef]
23. Flohe, L.; Gunzler, W.A.; Schock, H.H. Glutathione peroxidase: A selenoenzyme. *FEBS Lett.* **1973**, *32*, 132–134. [CrossRef]
24. Nakamura, W.; Hosoda, S.; KazukoHayashib, K. Purification, and the properties of rat liver glutathione peroxidase. *Biochim. Biophys. Acta* **1974**, *358*, 251–261. [CrossRef]
25. Beutler, E.; Duron, O.; Kelly, B.M. An improved method for the determination of blood glutathione. *J. Lab. Clin. Med.* **1963**, *61*, 882–888.
26. Jollow, D.J.; Mitchell, J.R.; Zampaglione, N.; Gillette, J.R. Bromobenzene-induced liver necrosis. Protective role of glutathione and evidence for 3,4-bromobenzene oxide as the hepatotoxic metabolite. *Pharmacology* **1974**, *11*, 151–169. [CrossRef] [PubMed]
27. Racker, E. Glutathione reductase from bakers' yeast and beef liver. *J. Biol. Chem.* **1955**, *217*, 855–865. [CrossRef]
28. Beutler, E. *Red Cell Metabolism: A Manual of Biochemical Methods*, 3rd ed.; Grune e Stratton: New York, NY, USA, 1984.

29. Ahmad, M.F.; Ansari, M.O.; Jameel, S.; Wani, A.L.; Parveen, N.; Siddique, H.R.; Shadab, G. Protective role of nimbolide against chemotherapeutic drug hydroxyurea induced genetic and oxidative damage in an animal model. *Environ. Toxicol. Pharmacol.* **2018**, *60*, 91–99. [CrossRef] [PubMed]

30. Tsantes, A.E.; Bonovas, S.; Travlou, A.; Sitaras, N.M. Redox imbalance, macrocytosis, and RBC homeostasis. *Antioxid. Redox Signal.* **2006**, *8*, 1205–1216. [CrossRef]

31. Belcher, J.D.; Beckman, J.D.; Balla, G.; Balla, J.; Vercellotti, G. Heme degradation and vascular injury. *Antioxid. Redox Signal.* **2010**, *12*, 233–248. [CrossRef]

32. Benzie, I.F. Lipid peroxidation: A review of causes, consequences, measurement, and dietary influences. *Int. J. Food Sci. Nutr.* **1996**, *47*, 233–261. [CrossRef]

33. Rocha, S.; Gomes, D.; Lima, M.; Bronze-da-Rocha, E.; Santos-Silva, A. Peroxiredoxin 2, glutathione peroxidase, and catalase in the cytosol and membrane of erythrocytes under H_2O_2-induced oxidative stress. *Free Radic. Res.* **2015**, *49*, 990–1003. [CrossRef]

34. Huang, M.E.; Facca, C.; Fatmi, Z.; Baille, D.; Benakli, S.; Vernis, L. DNA replication inhibitor hydroxyurea alters Fe-S centers by producing reactive oxygen species in vivo. *Sci. Rep.* **2016**, *6*, 29361. [CrossRef]

35. Torres, L.S.; Silva, D.G.H.; Belini Junior, E.; Almeida, E.A.; Lobo, C.L.C.; Cancado, R.D.; Ruiz, M.A.; Bonini-Domingos, C.R. The influence of hydroxyurea on oxidative stress in sickle cell anemia. *Rev. Bras. Hematol. Hemoter.* **2012**, *34*, 421–425. [CrossRef]

36. Brose, R.D.; Lehrmann, E.; Zhang, Y.; Reeves, R.H.; Smith, K.D.; Mattson, M.P. Hydroxyurea attenuates oxidative, metabolic, and excitotoxic stress in rat hippocampal neurons and improves spatial memory in a mouse model of Alzheimer's disease. *Neurobiol. Aging* **2018**, *72*, 121–133. [CrossRef] [PubMed]

37. Nader, E.; Grau, M.; Fort, R.; Collins, B.; Cannas, G.; Gauthier, A.; Walpurgis, K.; Martin, C.; Bloch, W.; Poutrel, S.; et al. Hydroxyurea therapy modulates sickle cell anemia red blood cell physiology: Impact on RBC deformability, oxidative stress, nitrite levels, and nitric oxide synthase signaling pathway. *Nitric Oxide* **2018**, *81*, 28–35. [CrossRef]

38. Rhee, S.G.; Chae, H.Z.; Kim, K. Peroxiredoxins: A historical overview and speculative preview of novel mechanisms and emerging concepts in cell signaling. *Free Radic. Biol. Med.* **2005**, *38*, 1543–1552. [CrossRef] [PubMed]

39. Cho, C.S.; Kato, G.J.; Yang, S.H.; Bae, S.W.; Lee, J.S.; Gladwin, M.T.; Rhee, S.G. Hydroxyurea-induced expression of glutathione peroxidase 1 in red blood cells of individuals with sickle cell anemia. *Antioxid. Redox Signal.* **2010**, *13*, 1–11. [CrossRef] [PubMed]

40. Kurata, M.; Suzuki, M.; Agar, N.S. Antioxidant systems and erythrocyte life span in mammals. *Comp. Biochem. Physiol. B* **1993**, *106*, 477–487. [CrossRef]

41. Teixeira Neto, P.F.; Gonçalves, R.P.; Elias, D.B.D.; Araújo, C.P.; Magalhães, H.I.F. Analysis of oxidative status and biochemical parameters in adult patients with sickle cell anemia treated with hydroxyurea, Ceará, Brazil. *Rev. Bras. Hematol. Hemoter.* **2011**, *33*, 207–210. [CrossRef]

42. Van Zwieten, R.; Verhoeven, A.J.; Roos, D. Inborn defects in the antioxidant systems of human red blood cells. *Free Radic. Biol. Med.* **2014**, *67*, 377–386. [CrossRef]

43. Kalpakcioglu, B.; Senel, K. The interrelation of glutathione reductase, catalase, glutathione peroxidase, superoxide dismutase, and glucose-6-phosphate in the pathogenesis of rheumatoid arthritis. *Clin. Rheumatol.* **2008**, *27*, 141–145. [CrossRef] [PubMed]

44. Rogers, S.C.; Ross, J.G.; d'Avignon, A.; Gibbons, L.B.; Gazit, V.; Hassan, M.N.; McLaughlin, D.; Griffin, S.; Neumayr, T.; Debaun, M.; et al. Sickle hemoglobin disturbs normal coupling among erythrocyte O_2 content, glycolysis, and antioxidant capacity. *Blood* **2013**, *121*, 1651–1662. [CrossRef] [PubMed]

45. Reno, C.O.; Barbosa, A.R.; de Carvalho, S.S.; Pinheiro, M.B.; Rios, D.R.; Cortes, V.F.; Barbosa, L.A.; Santos, H.L. Oxidative stress assessment in sickle cell anemia patients treated with hydroxyurea. *Ann. Hematol.* **2020**, *99*, 937–945. [CrossRef] [PubMed]

46. Raththagala, M.; Karunarathne, W.; Kryziniak, M.; McCracken, J.; Spence, D.M. Hydroxyurea stimulates the release of ATP from rabbit erythrocytes through an increase in calcium and nitric oxide production. *Eur. J. Pharmacol.* **2010**, *645*, 32–38. [CrossRef]

47. Naoum, P.C.; Radispiel, J.; Moraes, M.S. Methaemoglobin spectrometric dosage without chemical or enzymatic interferers. *Rev. Bras. Hematol. Hemoter.* **2004**, *26*, 9–22.

48. Cimen, M.Y. Free radical metabolism in human erythrocytes. *Clin. Chim. Acta* **2008**, *390*, 1–11. [CrossRef] [PubMed]

MDPI

St. Alban-Anlage 66

4052 Basel

Switzerland

www.mdpi.com

Antioxidants Editorial Office

E-mail: antioxidants@mdpi.com

www.mdpi.com/journal/antioxidants